ENTOMOLOGY

Ecology and Biodiversity

Dr. B.K. Tyagi

Director in-Charge
Centre for Research in Medical Entomology (ICMR)
Madurai - 625 002, TN, India

Dr. Vijay Veer

Scientist 'F' & Joint Director
Defence Research & Development Establishment (DRDO)
Jhansi Road
Gwalior - 474002, MP, India

With a Foreword by:
Prof. M.S. Swaminathan, The Father of Green Revolution in India

Published by:
Scientific Publishers (India)
5-A, New Pali Road, P.O. Box 91,
Jodhpur – 342 001 (India)

E-mail: info@scientificpub.com
Website: www.scientificpub.com

Print 2024

ISBN: 978-81-7233-727-8
eISBN: 978-93-86347-54-1

Printed in India

This festschrift is dedicated in the honour of the doyen of entomological sciences and the "Father of Indian Entomology"

Professor Dr T.N. Ananthakrishnan

(b. 15th December, 1925)

A teacher, scientist and academic statesman of global significance in the entomological realms

Dedicated to

Teachers who made us what we are today – a focused and hardworking scientist dedicated to constitute the nation's strong backbone to generate good health and economy.

Prof. M.S. Swaminathan
Member of Parliament (Rajya Sabha)
Chairman,
M S Swaminathan Research Foundation

21, Rathna Nagar
Teynampet
Chennai - 600 018

Tel : +91 44 2254 2790 (Off)
+91 44 2434 5312 (Res)
Fax : +91 44 2254 1319
E-mail: chairman@mssrf.res.in

FOREWORD

I am extremely happy a Festschrift titled "Entomology: Ecology and Biodiversity" is being brought out to commemorate the 85th birthday of Prof T N Ananthakrishnan. We are indebted to Prof B K Tyagi for getting this most timely and authoritative book compiled. This is a fitting tribute to the life and work of one of the greatest scientists of India, who is a living legend.

I have known Prof Ananthakrishnan for more than fifty years. From the very first day of my meeting him, I developed the highest respect for his scholarship, commitment to science and devotion to both education and research. He developed an outstanding school of Entomology in the Loyola college of Chennai. Normally colleges do not nurture outstanding research schools, but the Loyola College has been an exception, thanks to the efforts of Prof Ananthakrishnan. Prof Ananthakrishnan has devoted his life to research on Thysanoptera on the one hand, and to breeding outstanding young entomologists on the other.

Towards the later part of his life he planned and organized an outstanding scientific symposium on a topic of national and global importance towards the end of every year. In recent years, these annual meetings, which used to remind me of famous series

Contd...

Delhi Address: 503, Brahmaputra Apartments, Dr B D Marg, New Delhi - 110 001, Tel.: 011-2376 6897 like the Cold Spring Harbor Symposia, were very well organized with the participation of some of the leading scientists of our country. At the end of each such meeting, Prof Ananthakrishnan compiled and brought out an outstanding publication.

Ecology and Biodiversity are subjects which are very dear to Prof Ananthakrishnan's heart. His emphasis is always been on the mainstreaming of ecology in entomology research. The only other scientist who has done so much for the advancement of science in India after formal retirement, is the late Prof C V Raman. Prof Raman used to organize every year an outstanding symposium and also deliver a scholarly lecture containing the data of the work he had done during that year. He will never repeat what he said in an earlier year. Prof Ananthakrishnan belongs to that category of rare scientists whose life revolves around the pursuit of excellence in science.

I am confident that this book will inspire large numbers of young scholars and scientists and lead them to take to research in the field of biodiversity and ecology as related to entomology. The importance of biodiversity for human security was highlighted by the United Nations by declaring 2010 as the International Year of Biodiversity. Also, an International Conference on Biodiversity was held at Nagoya in Japan this year. This conference resulted in the setting up of an Intergovernmental Platform on Biodiversity and Ecosystem Services (IPBES). The present book will be an important contribution to the work of IPBES.

I pray for the long life and continued scientific leadership of Prof T N Ananthakrishnan. He is a scientists' scientist and is a role model for the younger generation. I also thank Prof B K Tyagi and all the contributors to this excellent book for choosing the most appropriate way of expressing our admiration and gratitude to Prof Ananthakrishnan for his life long contributions to science, scholarship and society.

(M.S. Swaminathan)

Delhi Address: 503, Brahmaputra Apartments, Dr B D Marg, New Delhi - 110 001, Tel.: 011-2376 6897

PREFACE

There are more kinds of insects than all other animals taken together, most of which are found in the tropical regions of the world. Estimates of the actual number of insects now living stay anywhere above 3 million. The dominance of insects amongst all living organisms on earth, coupled with infinite wealth of knowledge so important for our own existence, is a fundamental scientific fact which is yet to be widely acknowledged. This dominance means that in numbers of species beyond our comprehension these animals permeate diverse and essential natural processes in Earth's terrestrial, aerial and freshwater ecosystems, contributing to the function of the natural world as a self-sustaining biological system. Invariably insects, joined by other jointed-legs, are an integral and complex part of the terrestrial and freshwater ecosystems with which the future of humans is inextricably linked. Borne some 400 million years ago, in the Devonian period, insects have of course achieved, to the utter envy of all other arthropods, a formidable diversity on Earth so much so that they have virtually stolen the entire show and captivated human attention for centuries. Generally, insects are beneficial organisms, however, many of them are important pests and/or vectors to a large number of parasites and other microbial pathogens to both human being, his associates and the plants. The burden caused by their infestation or infection run up to several million-zillion rupees annually. This scenario makes insects the most formidable competitor to man for his successful survival. Therefore, knowledge about extreme biodiversity and ecological relationships of these animals is a practical necessity – in the man's own interest to sustain the species.

This book, comprising 25 articles written on a variety of entomological subjects (e.g., morphology, taxonomy and systematics of selected insect orders, families, genera and/or species, physiology, behavior and ecology, control of pests, application of GIS in mapping and modeling, bioinformatics etc.) by experts in their respective field, will hopefully provide students, research scholars, academicians and scientists alike with an advanced introduction to entomology by way of original and novel information about insects' diversity and ecology. Detailed description of certain chosen insect groups will provide ample opportunity to the reader to comprehend not only the harmful nature of certain species, but also the beneficial aspects of their life. The book truly establishes a foundation in basic entomology through varied and diverse treatment to entomology, while focusing on specialized topics in insect ecology and biodiversity, forensic and medical entomology. Additionally, the book offers an updated bio-bibliography of Prof. Ananthakrishnan which must inspire young researchers and academicians towards performance with excellence!

Professor Dr. T.N. Ananthakrishnan, Emeritus Scientist and the Ex-Director, Entomology Research Institute, Chennai as well as Zoological Survey of India, Kolkata is a globally respected teacher and researcher, besides being world's one of the most reputed entomologists. He will be turning 85 on 15th December, 2010 – a great day indeed necessitating celebration of his huge academic and scientific accomplishments in diversified fields of entomology, spanning over nearly 65 years! As a scientific community, with which Prof. T.N. Ananthakrishnan was ever so intimately linked throughout his life, it will be only befitting to dedicate in his honour this book, **"Entomology: *Ecology & Biodiversity*"** - a subject very dear to him since his formative days in science albeit his world class status of a complete zoologist!

Dr. B.K. Tyagi
&
Dr. Vijay Veer

15th December, 2010

ACKNOWLEDGEMENT

We foremost wish to express our deep sense of gratitude to Prof. M.S. Swaminathan for generously contributing the most significant "*Foreword*" to the festschrift in the honour of Prof. T.N. Ananthakrishnan.

Such a massive treatise on entomology would not have been possible without the unstinting support and encouragement of our most revered and worthy contributors of the scientific writings and, therefore, we are foremost indebted to all of them.

Prof. T.N. Ananthakrishnan's biobibliography is a very diverse subject and has always been a challenge to make it updated since it incessantly increases in its wealth of knowledge and inspirational anecdotes every day! We have made every endeavour to make it complete both in its bibliographic and biographic compositions, authenticated further by Prof. Ananthakrishnan himself to whom our heartfelt thanks are always due.

We had sown the thoughtful seed of writing this book as a festschrift for celebrating the 85th birthday of Prof. T.N. Ananthakrishnan during the spring 2006. Since then, during last four years of this young sapling's full development to a blooming tree to finally bear the fruits, we have been working relentlessly to connect with a large number of selected scientists and academicians. The task undertaken was immense and needed a most conducive environment, besides plentiful time to devote to the project. To this effect our wives, Mrs. Ajita Tyagi and Mrs. Dr. Neena Vijay Veer, generously provided a most educative ambience and kept us tick to our commitment despite many constraints, and for which we are very grateful to them.

Lastly, but not the least, we are extremely thankful to Shri Pawan Sharma and Shri Tanay Sharma, Directors, Scientific Publishers (India), Jodhpur, for their unflinching trust in our capability and commitment, on one hand, and providing all help in timely bringing out this highly educative book for all strata of our learned society – the undergraduate and post-graduate students, the research scholars, the scientists and the academicians alike, on the other.

Dr. B.K. Tyagi
&
Dr. Vijay Veer

LIST OF CONTRIBUTORS

DR. A. DUTTA-GUPTA
Department of Animal Sciences,
School of Life Sciences,
University of Hyderabad,
Hyderabad - 500 046, A.P., India
E-mail: *apdgsl@uohyd.ernet.in*

Dr. A. VIJAY ANANAD
Department of Biochemistry,
M.I.E.T. Arts and Science College,
Tiruchirappalli, (T.N.), India

DR. AMIT KUMAR BANERJEE
Bioinformatics Group,
Biology Division, Indian Institute
of Chemical Technology,
Hyderabad, A.P., India

Dr. ARTI PRASAD
Insect Microbial and Herbal
Control Laboratory,
Department of Zoology,
MLSU, Udaipur - 313 001, Raj.
E-mail: *artimlsu@yahoo.co.in*

DR. B.K. TYAGI
Centre for Research in Medical
Entomology (ICMR),
4-Sarojini Street, Chinna
Chokkikulam,
Madurai - 625 005, (T.N.), India.
E-mail: *bk_tyagi@sify.com*

DR. BHANVI WADHAWAN
Department of Zoology,
Punjabi University,
Patiala - 147 002, Punjab, India.

DR. C. RAGHUNATHAN
Zoological Survey of India
Andaman and Nicobar Regional
Centre, Port Blair - 744 102,
Andaman and Nicobar Islands,
India.

DR. C. SELVA KUMAR
Sri Paramakalyani Centre for
Environmental Sciences,
Manonmaniam Sundaranar
University, Alwarkurichi- 627 412,
Tirunelveli, (T.N.), India.

DR. C. SIVAPERUMAN
Zoological Survey of India,
Andaman and Nicobar Regional
Centre, Port Blair - 744 102,
Andaman and Nicobar Islands,
E-mail:
c_sivaperuman@yahoo.co.in

DR. DEBJANI DEY
Division of Entomology
Indian Agricultural Research
Institute, New Delhi - 110 012,
E-mail: *ddeyiari@hotmail.com*

DR. DEVINDER SINGH
Department of Zoology,
Punjabi University,
Patiala - 147 002, Punjab, India.
E-mail: *devinder.ss.61@gmail.com*

DR. E.R. PRASAD
Department of Plant Sciences,
School of Life Sciences,
University of Hyderabad,
Hyderabad - 500 046, India.
E-mail: *prasader@rediffmail.com*

DR. J.S. BHATTI
Delhi
E-mail: *bhatti.js@gmail.com*

DR. K. ILANGO
Zoological Survey of India,
Southern Regional Centre,
130 Santhome High Road
Chennai - 600 028, India.
E-mail: *kilangozsi@rediffmail.com*

DR. K. PADMASREE
Department of Plant Sciences,
School of Life Sciences,
University of Hyderabad,
Hyderabad - 500 046, India.
E-mail: *kpssl@uohyd.ernet.in*

DR. K.A. SUBRAMANIAN
Western Regional Centre,
Zoological Survey of India,
Sector-29, PCNT (PO),
Pune - 411 044, M.S., India
E-mail: *subbuka.zsi@gmail.com*

DR. K.G. SIVARAMAKRISHNAN
Flat No. 3, Rams Appartments,
New No. 07, Natesan Street,
T-Nagar, Chennai - 17, (T.N.)
&
Sri Paramakalyani Centre for
Environmental Sciences Manonmaniam Sundaranar University,
Alwarkurichi - 627 412, India
E-mail: *kgskrishnan@gmail.com*

DR. LAURENCE A. MOUND
CSIRO Entomology,
Canberra, Australia.
E-mail: *Laurence.Mound@csiro.au*

DR. M. ARUNACHALAM
Sri Paramakalyani Centre for
Environmental Sciences,
Manonmaniam Sundaranar
University, Alwarkurichi - 627 412
Tirunelveli, (T.N.), India.
Email: *kgskrishnan@gmail.com*

DR. M. GOVINDARAJU
Department of Environmental
Biotechnology, Bharathidasan
University, Tiruchirappalli (TN)
E-mail: *mgrasu@gmail.com; mgrasu@bdu.ac.in*

DR. MARIAMMA JACOB
Department of Zoology,
University of Kerala,
Kariavattom P.O.
Thiruvananthapuram - 695 581,
Kerala, India.
E-mail: *mjacob1953@yahoo.com*

DR. MATHEW M. OOMMEN
Department of Zoology,
University of Kerala,
Kariavattom, P.O.
Thiruvananthapuram - 695 581,
Kerala, India.
E-mail:
Mathew_M_Oommen@hotmail.com

DR. N. BAKTHAVATSALAM
National Bureau of Agriculturally
Important Insects,
Post Bag No. 2491, H.A.Farm Post,
Bangalore - 560 024, India.

DR. NEELIMA ARORA
Centre for Cellular and Molecular
Biology, Hyderabad, A.P., India
E-mail: *neelimaiict@gmail.com*

DR. P. RANI LALITHA
Department of Environmental
Biotechnology,
Bharathidasan University,
Tiruchirappalli - 620 024, India

DR. P.G. BRINTHA
Nachiyar Silk House,
1/4 NMK Colony, TVS Tollgate,
Tiruchirappalli - 620 020, India,

DR. R. CHANDRASEKAR
Department of Entomology,
Agricultural Science Centre
University of Kentucky,
Lexington - 40506, KY, USA
Email: *koreanchandru@yahoo.com*

DR. RAMAKRISHNA
Zoological Survey of India,
Prani Vigyan Bhawan, M-Block,
New Alipore, Kolkata - 700 053.

DR. REJI GOPALAKRISHNAN
Indian Agricultural Research
Institute, New Delhi - 110 012
&
Defence Research Laboratory,
Tezpur - 784 001, Assam, India.

DR. S. SUNDAR
Sri Paramakalyani Centre for Environmental Sciences, Manonmaniam Sundaranar University, Alwarkurichi - 627 412 Tirunelveli, (T.N.), India.

DR. S.N. SURENDRAN
Department of Zoology, Faculty of Science, University of Jaffna, Jaffna - 400 00, Sri Lanka
E-mail: *noble@jfn.ac.lk*

DR. S.R. PALLI
Department of Entomology, Agricultural Science Centre University of Kentucky, Lexington - 40506, KY, USA

DR. SAJAL BHATTACHARYA
Post Graduate Department of Zoology, Asutosh College (University of Calcutta), Kolkata - 700026, India
E-mail: *sajal_58@rediffmail.com*

DR. SUBHASH CHANDER
Indian Agricultural Research Institute, New Delhi - 110012, India

DR. SURESH KUMAR SHAH,
Zoological Survey of India Andaman and Nicobar Regional Centre, Port Blair - 744 102 Andaman and Nicobar Islands, India

DR. U.S.N. MURTY
Bioinformatics Group, Biology Division, Indian Institute of Chemical Technology, Hyderabad, A.P., India

DR. VIJAY VEER
Defense Research Development Establishment, Jhansi Road, Gwalior - 474 002 (MP), India.
E-mail: *vijayveer50@yahoo.com*

DR. VIJAYAPRASAD R. KOMERA
Department of Animal Sciences, School of Life Sciences, University of Hyderabad, Hyderabad - 500 046, AP, India

DR. VINOD KHANNA
Zoological Survey of India, Northern Regional Centre 218, Kaulagarh Road, Dehra Dun - 248 195, Uttrakhand. India.
E-mail: *drvkhanna51@gmail.com*

CONTENTS

Chapter - 1

PROFESSOR DR. T.N. ANANTHAKRISHNAN: ENTOMOLOGIST AND POLYMATH

Bio-bibliographical glimpses into his extraordinary and outstanding achievements in insect taxonomy, evolution, ecology, physiology, insect-plant interaction and pest control

B.K. Tyagi* and Vijay Veer**

*Centre for Research in Medical Entomology (ICMR), 4-Sarojini Street, Chinna Chokkikulam, Madurai - 625 005 (TN), India
**Defence Research Development Establishment, Jhansi Road, Gwalior -474 002 (MP), India

Professor T.N. Ananthakrishnan, seen with Prof. R.C. Mahajan, Dr. R. Reuben, Dr. J. Mahanta and Dr. B.K. Tyagi, during the release of the Abstracts booklet of the 8th International Symposium on Vectors & Vector-borne Diseases (2006; Madurai, TN, India)

PART - I
LIFE AND WORKS

"Who comprehends his trust, and to the same
Keeps faithful with a singleness of aim;
And therefore does not stoop, nor lie in wait
For wealth, or honour, or for worldly state;
Whom they must follow, on whose head must fall,
Like showers of manna, if they come at all."

The life of Professor Dr. T.N. Ananthakrishnan, an utterly indefatigable entomological genius *nonpareil*, truly reflects the spirit radiated from these famous lines by William Wordsworth. Prof. Anathakrishnan is one of those rare and great scientists who by their energetic character not only practically educate the budding researchers in habits of industry, but by the example of diligence and perseverance which they set before them, largely influence the scientific activity in all directions and contribute in a great degree to form the national character. The national progress, it is aptly said, is after all the sum of individual industry, energy and uprightness and Dr. Ananthakrishnan is a glorious example in our own lifetime. He is a total biologist, bestowed with exceptionally diverse and deep knowledge of nearly all the conceivable disciplines of life-sciences, though entomology has been his greatest forte.

Professor Ananthakrishnan has made innumerable outstanding contributions to the understanding of insect (or, in view of vast phyletic coverage, the arthropod) taxonomy, insect-plant interactions and the application of such knowledge for developing crop protection strategies. One of his many phenomenal contributions following four decade-long epoch making researches on thrips (Class: Insecta; Order Thysanoptera) has been to the basic understanding of 'insect-weed-crop interactions' which has been extended to the field for increasing agricultural productivity all over India. The diversity, the versatility and the magnificence of Prof. Ananthakrishnan's scientific work can be measured, if any indication, by his more than 400 research papers in a large number of national and international journals plus 30 books written by him which are significant landmarks in the evolution entomological sciences. He has described new species in many invertebrate groups of animals, but most of these (>200) come from his favourite, Thysanoptera or thrips - the tiny insects that cause severe damage to plant tissues often by developing galls or other deformities, but some have been also recorded incidentally biting humans! He has impressed the world authorities by the scope of his contributions and the depth with which he has analysed an almost infinite variety of the thrips. His steady investigative activities have resulted in remarkable eclectic publications that scan the biological kingdom and include taxonomy, systematics, morphology, bionomics, behaviour, distribution, reproductive biology, chemical cues and ecology *etc*. In the field of Thysanoptera, Prof. Ananthakrishnan has been one of two or three scientists in the world who is qualified to analyse in great depth and scope these important insects. It is no exaggeration, therefore, when the world famous entomologist Professor Douglas Whitman of the Illinois State University, USA referred to him as "*a brilliant biological treasure*." A great visionary he exhibited a comprehensive world-view of entomology even at that time when little was known about thrips' impact on agriculture entomology wherein he has made a conspicuous and indelible mark to be emulated by the thousands of young entomologists in India and elsewhere.

Prof. T.N. Ananthakrishnan was born on 15th December, 1925, at Mangalore. He did his high school classes at St Aloysius College High School from 1938 to 1941 and intermediate studies at Government Victoria College, Palghat, between 1941 and 1943. He then moved to Madras Christian College, Tambaram for his BSc (Hons.) degree course in Zoology during the tenure of Dr. Alexander John Boyd as principal. He won the Buckie Fellowship for pursuing his Honours course. It was at Madras Christian College that his passion for intensive and extensive field work began to take shape under encouragement of Dr. John P. Joshua who took him out periodically for field work, in addition to annual trips to the Gulf of Mannar. He passed his BSc (Hons.) in 1946 and Dr. Boyd offered him Demonstrator's post on a monthly salary of Rs. 47.50. He has by this time resolved to research, which ultimately remained with him as a mission and passion throughout his career, and even today. Only a few months ago after achieving that magic 80th year of age, he, with the support of his numerous colleagues, students and scholars, laid foundation of the country's first such entomological society, the *ENTOMOLOGY ACADEMY OF INDIA*, which is truly represented democratically by all the major disciplines of the day, including 'medical arthropodology' – all for the cause of creating a healthy and sustained entomological research activity in India.

In his early research days, Prof. Ananathakrishnan was acquainted to thrips by Dr. T.V. Ramakrishna Iyer, the great Indian entomologist and an international expert on thrips. Impressed by the ceaseless energy, perseverance and inquisitiveness of the ever agile youth in Dr. Ananthakrishnan, Dr. Ramakrishna Iyer gave him his lifelong collection of identified and invaluable material for reference. Dr. Ananthakrishnan became a temporary lecturer in zoology at the Madras Christian College in 1947-48, during which period Mr. Samuel Raj, his senior colleague, guided him in the field work. During weekends both of them used to walk from Tambaram to St. Thomas Mount and back, collecting insects, so much so that Dr. Ananthakrishnan became conversant with the more common insects. Together they spent late hours every day studying the collected material. They also used to visit the nearby Vandalur Hills which was known for its exquisite flora and fauna in those days. His interest in botany was kindled and augmented by two very brilliant teachers Prof. K. R. Venkatasubban and Dr. K. R. Rangaswamy Iyengar. In 1948, Dr. Boyd recommended the name of Dr Ananthakrishnan to Fr Jerome D'Souza who was then the principal of the Madras Loyola College. Thus at the age of 23 he began his career as lecturer in Zoology and this was the beginning of the Zoology section of the Natural Science Department in Loyola College with Fr Rapinat, a great botanist who interacted with Dr. Ananthakrishnan in many ways. In this phase of his career, Dr. Ananthakrishnan was greatly influenced by Fr Lourdu Yeddanapalli, Prof. Ekambaranathan, Dr. Rao Bahadur Y. Ramachandra Rao and Prof. M. S. Mani. He had the good fortune of assisting Prof. Ekambaranathan in bringing out two volumes of 'Manual of Zoology' which became the only classic Indian work in zoology for the BSc students for nearly two generations. Thanks to the British Council, he visited the British Museum of Natural History, London and several overseas universities, besides his memorable visit to the Mecca of Thysanopterists, LINZ in Austria where he had occasions to study thrips under Prof. H. Priesner, the greatest among Thysanopterists.

In 1957, at a young age of 32, Dr. Ananthakrishnan became a professor of Zoology at Loyola College. Growing under the careful sight of the great academician Dr. Lakshmanasamy Mudaliar, who was then the Vice Chancellor of Madras University, Dr. Ananthakrishnan was able to complete his M.Sc., Ph.D. and D.Sc. degrees from the University of Madras within a short span of 6 years! In 1960 he went to England for studying thrips at the British Museum of Natural History, London. This was the beginning of his association with experts in his field from different parts of the world. A versatile and dynamic orator, Dr. Ananthakrishnan was always the most sought after entomologist to offer lectures – in as many as 22 universities in the USA and in the UK, France, Germany *etc*.

In 1964 he relinquished his position as Professor and Head of the Department of Zoology to start the 'Entomology Research Centre', his brainchild and a future *alma mater* of a large number of entomological geniuses. From then on the Centre developed into the Entomology Research Institute in 1985 under his outstanding scientific leadership. By his tireless exertions Dr. Ananthakrishnan has developed this Institute into a research Institute of national and international significance. He has successfully guided 48 students to their Ph.D. degree in as diverse subjects as chemical ecology of insect-plant interactions, insect-weed-crop interactions, cecidiology, bioecology, insect taxonomy and systematics, and biodiversity. It is noteworthy that since the establishment of a full fledged Entomology Reseach Institute, taxonomic work was superceded by chemical ecology in Insect-plant interactions – a subject of great national interest to save crops from pestilent impacts, till he left the Institute in 1998. During his tenure at the ERI the *sanctum santorum* became an unparalleled rendezvous of great scientific colloqiua, and Prof. Ananthakrishnan brought in there scientists of international repute such as Bernard Crespi, a noted cecidologist from Michigan, Ann Arbor, USA so as to grow an ambience of high quality of research and academic excellence amongst scholars. With so much of academic and scientific responsibilities and commitments, Dr. Ananthakrishnan, being himself an avid cricket lover, never misses an opportunity to buck up the Team India against their opponent at the famous Chepauk stadium! His deep interest and knowledge in classical Indian music is also well known.

An internationally reputed entomologist and zoologist, Professor Ananthakrishnan has a highly chequered professional career, both as a scientist and a teacher, spanning over a period of nearly 60 years during which time he successfully had a host of coveted positions including, to mention a few just to illustrate, the Professor and Head of the Department of Zoology, Loyola College, and Founding Director of the Entomology Research Institute located at the Loyola College campus at Chennai, and the Director of the Zoological Survey of India at Kolkata. Of many significant advancements that Dr. Ananthakrishanan had brought about in the scientific area of various instituites and organizations he has been associated with, perhaps the most glittering example hails from his tenure as the Director of Zoological Survey of India where he introduced ecologically oriented programmes through initiating four new research stations, the Freshwater Research Station at Hyderabad, the Western Ghats Station at Calicut, the Estuarine Research Station at Berhampur and Sunderbans Station in West Bengal. Efforts were made during his period as Director to have an integrated approach through collaborative efforts with other national institutions. He is also a keen observer of Nature and, therefore, the world's highest nature conservation body, the International Union for Conservation of Nature & Natural Resources (IUCN)/Species Survival Commission (SSC) conferred on him the *Certificate of Appreciation* in recognition of his years of important work, in 2004. As a member of the Menon Committee of the Government of India on the Silent Valley, he visited various parts of the valley on a fact finding mission and was involved in several scientific discussions relating to the ecological and biosystematic aspects of the area, which was subsequently saved.

He is a Fellow of the Indian National Science Academy, Indian Academy of Sciences and the Indian Academy of Agricultural Sciences. He, along with Prof. M.S. Swaminathan, is also the patron of the recently founded Entomology Academy of India. He has won many awards, most notable of which are Kidwai Memorial Award of ICAR (1972), Jawaharlal Nehru Fellowship (1980), INSA Senior Scientists' Award (1987), Pitambar Pant Award of the Ministry of Environment and Forests (1986), Indian Ecology Society (1998), Plant Protection Association of India (1999), J.C. Bose Award of Indian Science Congress (2000) and Sundarlal Hora Medal of INSA (1999) and the life time achievement Award of Monsanto Research Centre, Bangalore (2002). For several decades he has been on the International Advisory Board of the Annual Review of Entomology, USA; Bulletin of Entomological Research, London, besides being on the editorial boards of leading biological journals in India including *Entomon* and *Oriental Insects.* In addition he has served on national research advisory committees/Task Force of ICMR, ICAR, CSIR, UGC, DST, DBT and Ministry of Environment and Forests. For many years during 1980s, Prof. Ananthakrishnan has been a Member of the Scientific Advisory Committee of the Centre for Research in Medical Entomology (ICMR). The Indian National Science Academy, New Delhi awarded him the golden Jubilee Commemorative Medal 2009, for Animal Sciences. At a special function held by the Chennai local Chapter of INSA, Prof. M.S. Swaminathan presented the medal, following which Prof. Ananthakrishnan presented the keynote lecture on "Implications of phenotypic plasticity in Insects." Prof. Ananthakrishnan's work on chemical Ecology with special reference to resistance mechanism in plants and more notably the pattern of polymorphism in mycophagous and cecidogenous insects, have revealed the intricacies of biological diversity in natural ecosystem. In this connection, his work on phenotypic plasticity in insect population, notably on thrips has increased relevance in biosystematic studies enabling better recognition of insect species in agriculture and forestry. In December 2009, Prof. Ananthakrishnan was honoured by the National Biodiversity Authority, Govt. of India, Chennai by requesting him to be the Chief Guest at a special function in which Dr. P.L. Gautam, Chairman, NBA, presided over the function.

Never knowing to stop, the retirement on superannuation from the ERI only added to his undousable fire to do more, to achieve more and teach more! He is credited for initiating a series of symposia popularly called, "Annual Discussion Meeting" from 2000 to 2009, under the chairmanship of Prof. M.S. Swaminathan. These Chennai Series lecture meetings became a hugely attended discourses by the entomologists in all major disciplines such as agriculture, horticulture, medical, pharmaceutical and so forth. All these Meetings were invariably presided by Prof. Swaminathan. In the late nineties to early twentieth century, he was involved as a consultant to Colgate-Palmolive Company, Piscataway, New Jersey in the field of medical entomology for a period of five years, in studies relating the efficacy of MNDA and DEET in mosquito control. A publication entitled "A comparison of mosquito repelling efficacy of methyl neodecanamide (MNDA) to that of DEET" was published in

Journal of Cosmetics Research. In 2003 Prof. T.N. Ananthakrishnan was amongst the scientists of the Colgate-Palmolive to present a paper at the Congress of Entomology in York (UK). A former chief of the British Museum of Natural History, London, Dr. Lawrence A. Mound, whom he had known for several years, writes "Ananthakrishnan will be recommended primarily for his success at communicating through his books, useful generalization about thrips to his students, to scientific contemporaries and to future coworkers for placing his taxanomic expertise with the wider context of understanding complexities of biological relationship between insects and plants" (*excerpt from*: International Scientific Publishers, New Delhi & Backhugs Publishers, Heiden, 1997). Honours in various colours and forms poured on him all along his work; In 2010, a special function was held to know him on the eve of his departure to the USA, in the august presence of Prof. M.S. Swaminathan who, in a later letter, said of him, "I thank you sincerely for the time spared to be with us and for the inspiring scholarly lectures. You are a scientists' scientist and a role model for the younger generation. You have been a pillar of strength of the MSSRF."

A widely travelled scientist across the world, he has by his visiting lecture assignments enriched entomologically the varied audiences with his vast knowledge and matchless articulation. In the early nineties he was on a visit to Nairobi to participate in the Entomology congress held at the Jumo Kenyatta Univrsity which provided opportunities to interact with African entomologists. Following this he visited Tokyo to meet with his thrips counterparts Kazhuo Haga and Iwao Kudo, long time thrips associates. On return, Prof. Swaminathan, then DG, International Rice Research Institute, Philippines invited Prof. Ananthakrishnan to visit the Institute and give a talk. He was visited at the ERI by Prof. Heinz Remboldt of the Max Planck Institute of Biochemistry, Martinsried, Munick followed by his visit to the former's Institute to give a lecture on Chemical Ecology in Insect-plant Interaction. During his regular visits to the USA, he had the pleasure of meeting Prof. Carl Schafer, the noted Hemipterist at Connectiant, Storrs, besides visiting to Prof. Maramorosch. At the Cornell University, Itaca, he then had occasion to give two lectures, besides visits to the Illinois Natural History Survey to meet his good friend and Thysanopterist, Dr. Lewis Stennard, who in later years visited India, including the ERI. Prof. Carl Maramorosch also visited the ERI to give a lecture. Professor Ananthakrishnan had associated with some of the distinguished cecidologists at the Cecidological Institute, Strasbourg, France, besides participating in the International Congress of Cecidology where the top brains among cecidologists were present. One of the distinguished cecidologists with whom he had occasion to meet was Dr. Josephy Shorthouse of the University of Sudburg, Canada who besides inviting him to Sudburg to present some lectures, also invited Prof. Ananthakrishnan for writing a chapter on "Thrips' galls" for his upcoming book on cecidology. Over the years Prof. Ananthakrishnan's former student Dr. Ananthanarayanan Raman, presently in Australia, has been closely associated, helping him with his publication. List and instances of his peregrinations across the globe in the quest and propagation of knowledge continued for ever, since even at the time of writing these lines he is in the USA - still thinking and writing, and sending emails on regular basis!

In spite of his international stature of a scientist, Prof. Ananthakrishnan is foremost a teacher. Generations of students and younger scientists hold him in veneration and admiration as a 'father figure' who has rare qualities of mind and heart. Many owe their career in the field of science to his personal example as a scientist and teacher. For the world, he is a living academic statesman of global significance in the field of entomology. For him aiming for achievements has been everything and never even once the thought about success could penetrate his mind.

Today, on 15th December, 2010, Prof. Ananthakrishnan has turned 85 - *an eighty five year old young man*, whose ceaseless enthusiasm and charm of entomological science has mellowed down the imprints of time and age. Entomologically speaking Prof. T.N. Ananthakrishnan is immortalized in the annals of insects' sciences by the scores of species in diverse animal groups that have been named in his honour by both his students and colleagues alike across the globe!

ARTHROPOD TAXA NAMED AFTER DR. ANANTHAKRISHNAN

A. Thrips (Order: Thysanoptera)

Genera

Ananthakrishnana Bhatti, 1967

Ananthakrishnaniella Stannard, 1970

Ananthakrishnanothrips Bournier, 1985

Species

Exothrips ananthakrishnani Bhatti 1975

Filipinothrips ananthakrishnani Tyagi, Kumar & Mound, 2009

Helionothrips ananthakrishnani Wilson, 1975

Holothrips ananthakrishnani Okajima, 1987

Hydatothrips ananthakrishnani Bhatti, 1973

Leeuwenia ananthakrishnani Varatharajan & Sen 2000

Liothrips (Liothrips) ananthakrishnani Sen, 1976

Plectrothrips ananthakrishnani Okajima, 1981

Psalidothrips ananthakrishnani (Johansen, 1986)

Terthrothrips ananthakrishnani Kudo, 1978

Xylaplothrips ananthakrishnani (Bournier, 1994)

B. Other than Thrips

Louse, Menoponidae (Phthiraptera)

Myrsidea ananthakrishnani Rai, 1978

Robber fly, Asilidae (Diptera)

Microstylum ananthakrishnani Joseph & Parui, 1984

Tree hopper, Membracidae (Homoptera)

Hypsauchenia ananthakrishnani Ananthasubramanian, 1996

Whitefly, Aleyrodidae (Hemiptera)

Africaleurodes ananthakrishnani Dubey & Sundararaj, 2006

Flower bug, Anthocoridae (Hemiptera)

Physopleurella anathakrishnani

Spider Oxyopidae (Arachinida)

Peucetia ananthakrishnani Murugesan *et al.*, 2006

Truly, his immortality in the zoological scientific history will be written in golden letters for he is a truly 'Complete Zoologist' of the modern era where naming species from diverse animal realms spring up naturally more as a matter of worship and devotion by myriads of his students, reminding the famous couplets of medieval Hindi poet Rahim, in veneration of the teacher;

Guru Gobind dou khare,
Kake lagoon paavn,
Balihari guru aapno,
Gobind diyo bataye.

One of his highly placed admirers of the IARI, Delhi, Dr. Vilayanoor Venkataraman Ramamurthy, during the 5th Annual Discussion Meeting, coinciding with his 80th birth anniversary celebrations, read out an ode which speaks volumes of reverence with which Prof. Ananthakrishnan is held in the hearts of thousands of his students, friends, colleagues and well wishers all over the world:

"Professor Tarakad Narayan Ananthakrishnan
today is eighty years to be.

Crusader for the cause of Entomology,
Pioneer in fostering ecological approach towards taxonomy,
Evolutionist of the interdisciplinary thought.

Promoter of the policy "To be one's own master is positive approach",
Advocate for the concept "Chance favours the prepared mind",
Practitioner of philosophy "Nothing more dangerous than
wanting to rest",
Pursuer of the thought "Persistence shall in spite of handicaps",
Champion for the cause "One shall aim achievement, forget success."

Enthusiast in showing that concept can be made a reality,
Campaigner for excellence through dedication and determination,
Warrior in fighting with courage during adversity,
Instiller of zest for science with curiosity about nature,
Idealist in oneness of commitment: Skill, intellectual, emotional.

An incredible individual,
Legacy he leaves is much more than awards he received.

All of us,
Who are fortunate to have known,
This extraordinary Professor,
Have gained a great deal.

Pray God for his long healthy life,
Plead to grace him to be immortal,
At least for the cause of Entomology."

To this, on his 85th birthday, we now most happily wish to add;
"You live for thousands of years,
Each year to have a thousand year!"

Through his many publications and students, ideas, energy, and work Prof. Ananthakrishnan will continue to reverberate and influence the course of world science for decades to come. We must always remember that we stand upon the shoulders of giants. For the young and budding entomologists in India and elsewhere in the world Prof. Ananthakrishnan is an epitome of hard work and application. His deep sense of sincerity for commitments is exemplary. Even at a mature age of 85, Prof. Ananthakrishnan's agility and clarity of mind dominates the most energetic in the science of entomology. He is incessantly contributing to the world of entomology through his writings and *Annual Discussion Meetings*, and knows no bounds to achieve his aims. And, we are reminded of those most famous and ever inspiring lines by Robert Frost:

The woods are lovely, dark and deep
But I have promises to keep,
And, miles to go before I sleep,
And, miles to go before I sleep.

PART - II

NEW GENERA AND SPECIES DESCRIBED BY PROF. T.N. ANANTHAKRISHNAN Ph.D., D.Sc., FNA, FASc, FNAAS

Phylum Arthropoda

Class Insecta

Order Thysanoptera

I. Suborder Tubulifera

New Genera

Ablemothrips Ananthakrishnan 1969
Aclystothrips Ananthakrishnan 1972
Aeglothrips Ananthakrishnan 1969
Aesthesiothrips Ananthakrishnan 1961
Agnostothrips (*Erythrinothrips*) Ananthakrishnan 1956
Alloiothrips Ananthakrishnan 1964
Araeothrips Ananthakrishnan 1976
Ayyarothrips Ananthakrishnan 1972
Azaleothrips Ananthakrishnan 1964
Bamboosiella Ananthakrishnan 1957
Bournieria Ananthakrishnan 1966
Bunothrips Ananthakrishnan 1969
Byctothrips Ananthakrishnan 1973
Calamothrips Ananthakrishnan 1967
Callothrips Ananthakrishnan 1969
Carissothrips Ananthakrishnan 1964
Chlarathrips Ananthakrishnan 1967
Corycidothrips Ananthakrishnan 1972
Cradothrips Ananthakrishnan 1973 subgenus of *Elaphrothrips*
Crotonothrips Ananthakrishnan 1968
Decothrips Ananthakrishnan 1969
Dixothrips Ananthakrishnan 1969
Ecacanthothrips *erythrinus* Ananthakrishnan 1956

Elaphridia Ananthakrishnan 1964

Eothrips *distinctus* Ananthakrishnan 1968

Erythrinothrips Ananthakrishnan 1956, subgenus of *Agnostothrips*

Gigantothrips *gardneri* Ananthakrishnan 1961

Glubothrips Ananthakrishnan 1969

Habrothrips Ananthakrishnan 1968

Kochummania Ananthakrishnan 1969

Loyolaia Ananthakrishnan 1964

Lygothrips Ananthakrishnan 1964

Mutothrips Ananthakrishnan & Swaminathan 1980

Neopodothrips Ananthakrishnan 1965, subgenus of *Podothrips*

Ocnothrips Ananthakrishnan 1969

Ocythrips Ananthakrishnan 1972

Opidnothrips Ananthakrishnan 1971

Paxillothrips Ananthakrishnan 1961

Percipiothrips Ananthakrishnan 1964

Percnothrips Ananthakrishnan 1967

Phaeothrips Ananthakrishnan 1969

Phenothrips Ananthakrishnan 1968

Phiarothrips Ananthakrishnan 1968

Phorinothrips Ananthakrishnan 1968

Priesneriana Ananthakrishnan 1956

Psenothrips Ananthakrishnan 1968

Pyknothrips Ananthakrishnan 1964

Ramachandraiella Ananthakrishnan 1964

Salothrips Ananthakrishnan 1976

Saucrothrips Ananthakrishnan 1968, subgenus of *Podothrips*

Schedothrips Ananthakrishnan 1969

Segnothrips Ananthakrishnan 1965

Socothrips Ananthakrishnan 1972

Sphingothrips Ananthakrishnan 1972

Stannardiana Ananthakrishnan 1964

Stannardothrips Ananthakrishnan 1965

Stigmothrips Ananthakrishnan 1964

Stinothrips Ananthakrishnan 1969

Synergothrips Ananthakrishnan 1972

Takahashia Ananthakrishnan 1970 (Taiwan) (junior homonym; see *Varshneyia* Ananthakrishnan 1973)

Telothrips Kudô & Ananthakrishnan 1974, subgenus of *Meiothrips*

Uredothrips Ananthakrishnan 1969

Varshneyia Ananthakrishnan 1973 (Taiwan) (syn. *Takahashia* An. 1970)

Xenothrips Ananthakrishnan 1965

New Species

Ablemothrips maxillatus Ananthakrishnan 1969

Aclystothrips aberrans Ananthakrishnan 1973

Aclystothrips priesneri Ananthakrishnan 1972

Adraneothrips disjunctus Ananthakrishnan 1972

Adraneothrips elegans Ananthakrishnan 1972

Adraneothrips madrasensis Ananthakrishnan 1968

Adraneothrips stannardi Ananthakrishnan 1969

Aeglothrips denticulus Ananthakrishnan 1969

Aesthesiothrips jatrophae Ananthakrishnan 1961

Agnostothrips (*Erythrinothrips*) *indicus* Ananthakrishnan 1956

Agrothrips insolitus Ananthakrishnan 1969

Alloiothrips nigrisetis Ananthakrishnan 1964

Allothrips bicolor Ananthakrishnan 1964

Allothrips indica Ananthakrishnan 1958

Allothrips montanus Ananthakrishnan 1968

Apterygothrips dempax Bhatti & Ananthakrishnan 1978

Apterygothrips pini Ananthakrishnan 1960

Araeothrips longisetis Ananthakrishnan 1976

Arrhenothrips acuminatus Ananthakrishnan 1969

Arrhenothrips brevis Ananthakrishnan & Swaminathan 1980

Ayyarothrips abstrusus Ananthakrishnan 1972

Azaleothrips amabilis Ananthakrishnan 1964

Bactrothrips luteus Ananthakrishnan 1973

Bamboosiella bicoloripes Ananthakrishnan 1957

Barythrips mathuri Ananthakrishnan 1961

Bournieria indica Ananthakrishnan 1966

Bunothrips cruralis Ananthakrishnan 1969

Byctothrips ayyari Ananthakrishnan 1973

Calamothrips fastigatus Ananthakrishnan 1967

Callothrips ascitus Ananthakrishnan 1969

Carathrips orientalis Ananthakrishnan 1969

Carissothrips nigrescens Ananthakrishnan 1964

Cercothrips seshadrii Ananthakrishnan 1964

Chlarathrips tersus Ananthakrishnan 1967

Corycidothrips inquilinus Ananthakrishnan 1972

Crotonothrips coorgensis Ananthakrishnan 1976

Crotonothrips davidi Ananthakrishnan 1972

Crotonothrips dissimilis Ananthakrishnan 1976

Crotonothrips gallarum Ananthakrishnan 1968

Crotonothrips memecylonicus Ananthakrishnan 1976

Crotonothrips parvus Ananthakrishnan 1976

Decothrips anacardii Ananthakrishnan 1969

Diaphorothrips spinosus Ananthakrishnan 1959

Diaphorothrips thevetii Ananthakrishnan 1957

Diceratothrips (*Diceratothrips*) *brevisetosus* Ananthakrishnan & Jagadish 1970

Diceratothrips (*Diceratothrips*) *longisetis* Ananthakrishnan & Jagadish 1970

Diceratothrips (*Dichaetothrips*) *uredinis* Ananthakrishnan & Jagadish 1970

Diceratothrips (*Dichaetothrips*) *usitatus* Ananthakrishnan & Jagadish 1970 (=*Dich. indicus* Ananthakrishnan 1961)

Dichaetothrips indicus Ananthakrishnan 1961 (junior honomym; see *Dicerato. usitatus* Aanthakrishnan & Jagadish 1970)

Dixothrips onerosus Ananthakrishnan 1969

Dolichothrips confusus Ananthakrishnan 1968

Dolichothrips flaviantennatus Seshadri & Ananthakrishnan 1954

Dolichothrips (*Dolicholepta*) *fulvus* Ananthakrishnan 1964

Dolichothrips (*Dolicholepta*) *inquilinus* Ananthakrishnan 1954

Dolichothrips (*Dolicholepta*) *malhavii* Ananthakrishnan 1961

Dolichothrips (Dolichothrips) montanus Ananthakrishnan 1964

Dolichothrips (Dolicholepta) rambhutanae Ananthakrishnan 1960

Ecacanthothrips ramakrishnai Ananthakrishnan 1952

Elaphrothrips (Cradothrips) Ananthakrishnan 1973

Elaphrothrips eranthemi Seshadri & Ananthakrishnan 1954

Elaphrothrips (Cradothrips) insignis Ananthakrishnan 1973

Elaphrothrips micidus Ananthakrishnan 1973

Elaphrothrips notabilis Ananthakrishnan 1973

Elaphrothrips secus Ananthakrishnan 1973

Eothrips sirumalaiensis Ananthakrishnan 1968

Fulgorothrips faurei Ananthakrishnan 1964

Glubothrips mucidus Ananthakrishnan 1969

Gynaikothrips bengalensis Ananthakrishnan 1973

Gynaikothrips cecidii Ananthakrishnan 1968

Gynaikothrips exilis Ananthakrishnan & Jagadish 1969

Gynaikothrips imitator Ananthakrishnan 1968

Gynaikothrips microchaetus Ananthakrishnan & Jagadish 1969

Gynaikothrips schefflericola Ananthakrishnan & Viswanathan 1975

Habrothrips curiosus Ananthakrishnan 1968

Haplothrips longisetosus Ananthakrishnan 1955

Haplothrips (Chiraplothrips) priesneri Ananthakrishnan 1961

Hoplandrothrips corticis Ananthakrishnan 1972

Hoplandrothrips graminis Ananthakrishnan 1964 (= *H. priesneri* An. 1956, *H. indicus* Ananthakrishnan 1959)

Hoplandrothrips indicus Ananthakrishnan 1959 (junior homonym; see *H. graminis* Ananthakrishnan 1964)

Hoplandrothrips priesneri Ananthakrishnan 1956 (junior homonym; see *H. graminis* Ananthakrishnan 1964)

Hoplothrips indicus Ananthakrishnan 1960

Hoplothrips nemorius Ananthakrishnan 1971

Horistothrips dentipes Seshadri & Ananthakrishnan 1954

Ischnothrips typicus Ananthakrishnan 1967

Ischyrothrips menoni Ananthakrishnan 1950

Karnyothrips fungulus Ananthakrishnan 1973

Kochummania excelsa Ananthakrishnan 1969

Leeuwenia ramakrishnae Ananthakrishnan 1970

Leeuwenia vorax Ananthakrishnan 1969

Liophlaeothrips ablusus Ananthakrishnan 1972

Liophlaeothrips amoenus Ananthakrishnan 1966

Liophlaeothrips cecidii Ananthakrishnan 1964

Liophlaeothrips curtus Ananthakrishnan & Jagadish 1969

Liophlaeothrips nitidus Ananthakrishnan 1968

Liophlaeothrips pavettae Ananthakrishnan & Jagadish 1969

Liophlaeothrips pictus Ananthakrishnan 1968

Liophlaeothrips reperticus Ananthakrishnan & Muraleedharan 1974

Liophlaeothrips segnis Ananthakrishnan & Jagadish 1969

Liophlaeothrips succinctus Ananthakrishnan & Jagadish 1969

Liophlaeothrips tenebrosus Ananthakrishnan & Jagadish 1969

Liothrips abstrusus Ananthakrishnan & Muraleedharan 1974

Liothrips aequilus Ananthakrishnan & Jagadish 1969

Liothrips associatus Ananthakrishnan & Jagadish 1969

Liothrips ater Ananthakrishnan & Jagadish 1969

Liothrips cecidii Ananthakrishnan & Jagadish 1969

Liothrips digressus Ananthakrishnan 1972

Liothrips emulatus Ananthakrishnan 1976

Liothrips epacrus Ananthakrishnan & Muraleedharan 1974

Liothrips eugeniae artabotrys Ananthakrishnan & Varadarasan 1978

Liothrips exiguus Ananthakrishnan 1972 (junior homonym; see *L. minys* Ananthakrishnan 1972)

Liothrips flavescens Ananthakrishnan & Jagadish 1969

Liothrips fragilis Ananthakrishnan 1976

Liothrips furvus Ananthakrishnan & Jagadish 1969

Liothrips himalayanus Ananthakrishnan & Jagadish 1970

Liothrips indicus Ananthakrishnan 1972

Liothrips inquilinus Ananthakrishnan & Jagadish 1967

Liothrips jogensis Ananthakrishnan & Jagadish 1967

Liothrips (Liothrips) kolliensis Ananthakrishnan 1972

Liothrips machilus Ananthakrishnan & Varadarasan 1978

Liothrips malabaricus Ananthakrishnan & Jagadish 1967

Liothrips minys Ananthakrishnan 1972 (syn. *L. tenuis* An. 1969, *L. exiguus* Ananthakrishnan 1972)

Liothrips morindae Ananthakrishnan & Muraleedharan 1974

Liothrips morulus Ananthakrishnan & Jagadish 1970

Liothrips mucronis Ananthakrishnan & Jagadish 1967

Liothrips (Zopyrothrips) nanus Ananthakrishnan 1972

Liothrips nubilis Ananthakrishnan & Jagadish 1969

Liothrips ramakrishnae Ananthakrishnan & Jagadish 1969

Liothrips retusus Ananthakrishnan 1976

Liothrips seshadrii Ananthakrishnan & Muraleedharan 1974

Liothrips (Zopyrothrips) sordidus Ananthakrishnan 1972

Liothrips (Zopyrothrips) spectabilis Ananthakrishnan 1972

Liothrips strigosus Ananthakrishnan & Jagadish 1969

Liothrips tenuis Ananthakrishnan 1969 (junior homonym; see *L. minys* Ananthakrishnan 1972)

Liothrips tersus Ananthakrishnan & Jagadish 1967

Liothrips variabilis Ananthakrishnan & Jagadish 1967

Loyolaia indica Ananthakrishnan 1964

Machatothrips corticosus Ananthakrishnan 1972

Machatothrips indicus Ananthakrishnan & Jagadish 1970

Machatothrips silvaticus Ananthakrishnan 1972

Macrophthalmothrips splendidus Ananthakrishnan 1968

Malacothrips lucidus Ananthakrishnan 1965

Malacothrips madrasensis Ananthakrishnan 1964

Meiothrips (Telothrips) Kudô & Ananthakrishnan 1974

Meiothrips menoni Ananthakrishnan 1964

Meiothrips (Telothrips) nepalensis Kudô & Ananthakrishnan 1974

Mesicothrips (Mesicothripoides) Ananthakrishnan 1967

Mesicothrips (Mesicothripoides) inquilinus Ananthakrishnan 1967

Mesothrips bianchii Ananthakrishnan 1976 (Australia)

Mesothrips cracens Ananthakrishnan 1968

Mesothrips elaeocarpi Ananthakrishnan 1976 (Indonesia)

Mesothrips extensivus Ananthakrishnan & Jagadish 1969

Mesothrips fulmeki Ananthakrishnan 1976 (Indonesia)

Mesothrips manii Ananthakrishnan 1972

Mesothrips memecylonicus Ananthakrishnan 1976 (Indonesia)

Mesothrips moundi Ananthakrishnan 1976 (Hong Kong)

Mesothrips orientalis Ananthakrishnan 1976 (Malaysia)

Mesothrips vitis Ananthakrishnan 1976 (Indonesia)

Mimothrips orientalis Ananthakrishnan 1966

Mutothrips validus Ananthakrishnan & Swaminathan 1980

Nanothrips typicus Ananthakrishnan 1964

Neoheegeria montana Ananthakrishnan & Jagadish 1970

Neosmerinthothrips inquilinus Ananthakrishnan 1960

Neothrips lepidus Ananthakrishnan 1971

Nesothrips alius Ananthakrishnan 1970

Nesothrips diversus Ananthakrishnan 1972

Nesothrips falcatus Ananthakrishnan 1967

Nesothrips indicus Ananthakrishnan 1968

Nesothrips robustus Ananthakrishnan 1964

Neurothrips indicus Ananthakrishnan 1964

Ocnothrips indicus Ananthakrishnan 1969

Ocythrips rarus Ananthakrishnan 1972

Odontoplothrips cecidii Ananthakrishnan 1965

Oedaleothrips ramamurthii Ananthakrishnan 1964

Oedaleothrips ramamurthii bicolor Ananthakrishnan 1966 (junior homonym; see *Oed. ra. indicus* Ananthakrishnan 1973)

Oedaleothrips ramamurthii indicus Ananthakrishnan 1973 (syn. *Oed. ra. Bicolor* Ananthakrishnan 1966)

Opidnothrips corticulus Ananthakrishnan 1971

Paxillothrips longicauda Ananthakrishnan 1961

Percnothrips turbinatus Ananthakrishnan 1967

Phaeothrips mimicus Ananthakrishnan 1969

Phenothrips decoratus Ananthakrishnan 1968

Phiarothrips reperticus Ananthakrishnan 1968

Philothrips indicus Ananthakrishnan 1968

Phlaeothrips nilgiricus Ananthakrishnan 1968

Phorinothrips levis Ananthakrishnan & Varadarasan 1978

Phorinothrips loranthi Ananthakrishnan 1968

Phorinothrips minusculus Ananthakrishnan 1972

Plectrothrips eximius Ananthakrishnan 1969
Podothrips (Neopodothrips) Ananthakrishnan 1965
Podothrips (Saucrothrips) Ananthakrishnan 1968
Podothrips bicolor Seshadri & Ananthakrishnan 1954
Podothrips (Neopodothrips) distinctus Ananthakrishnan 1965
Podothrips (Kentronothrips) moultoni Ananthakrishnan 1965
Podothrips (Kentronothrips) placitus Ananthakrishnan 1966
Podothrips (Saucrothrips) scitulus Ananthakrishnan 1968
Polyphemothrips cracens Ananthakrishnan 1968
Polyphemothrips enormis Ananthakrishnan 1969
Polyphemothrips fumidus Ananthakrishnan 1972
Polyphemothrips megacephalus Ananthakrishnan 1969
Polyphemothrips mirandus Ananthakrishnan 1969
Polyphemothrips ruidus Ananthakrishnan 1969
Polyphemothrips stannardi Ananthakrishnan 1972
Polyphemothrips subtilis Ananthakrishnan 1972
Praepodothrips cymbapogoni Ananthakrishnan 1956
Praepodothrips nigrocephalus Ananthakrishnan 1964
Praepodothrips priesneri Ananthakrishnan 1955
Pyknothrips reticulatus Ananthakrishnan 1964
Ramachandraiella minuta Ananthakrishnan 1964
Salothrips indicus Ananthakrishnan 1976
Schedothrips tumidus Ananthakrishnan 1969
Segnothrips trivandrensis Ananthakrishnan 1965
Socothrips verrucosus Ananthakrishnan 1972
Sophiothrips nigrus Ananthakrishnan 1971
Stannardiana variegata Ananthakrishnan 1964
Stannardothrips longirostris Ananthakrishnan 1965
Stephanothrips adnatus Ananthakrishnan 1972
Stigmothrips bambusae Ananthakrishnan 1964
Stigmothrips bhowalii Ananthakrishnan 1971
Stigmothrips consimilis Ananthakrishnan 1969
Stigmothrips infirmus Ananthakrishnan 1971
Stigmothrips limpidus Ananthakrishnan 1964

Stigmothrips nilgiriensis Ananthakrishnan 1971
Stigmothrips pteris Ananthakrishnan 1971
Strepterothrips orientalis Ananthakrishnan 1964
Symphyothrips aberrans Ananthakrishnan 1971
Symphyothrips associatus Ananthakrishnan 1968
Synergothrips prolatus Ananthakrishnan 1972
Teuchothrips eugeniae Seshadri & Ananthakrishnan 1954
Teuchothrips orientalis Ananthakrishnan 1968
Teuchothrips priesneri Ananthakrishnan 1964
Thilakothrips fimbriata Ananthakrishnan 1953
Thlibothrips inquilinus Ananthakrishnan & Varadarasan 1978
Thlibothrips malloti Ananthakrishnan 1961
Uredothrips indicus Ananthakrishnan 1969
Uredothrips tirumalaiensis Ananthakrishnan 1969
Xenothrips luteus Ananthakrishnan & Kudô 1974
Xenothrips luteus exastis Ananthakrishnan & Kudô 1974 (Thailand)
Xenothrips malabaricus Ananthakrishnan 1965
Xenothrips opacus Ananthakrishnan & Kudô 1974
Xylaplothrips bicolor Ananthakrishnan 1964
Xylaplothrips debilis Ananthakrishnan & Jagadish 1971
Xylaplothrips emineus Ananthakrishnan & Jagadish 1971
Xylaplothrips flavitibia Ananthakrishnan & Jagadish 1969
Xylaplothrips flavus Ananthakrishnan 1964
Xylaplothrips fungosus Ananthakrishnan & Jagadish 1969
Xylaplothrips graminellus Ananthakrishnan & Jagadish 1969
Xylaplothrips inquilinus Ananthakrishnan 1966
Xylaplothrips jogensis Ananthakrishnan & Jagadish 1969
Xylaplothrips ligs Ananthakrishnan & Jagadish 1971
Xylaplothrips longus Ananthakrishnan & Jagadish 1969
Xylaplothrips micans Ananthakrishnan & Jagadish 1971
Xylaplothrips mucidus Ananthakrishnan & Jagadish 1971
Xylaplothrips nayari Ananthakrishnan 1958
Xylaplothrips nefrens Ananthakrishnan 1972
Xylaplothrips orientalis Ananthakrishnan & Jagadish 1969

Xylaplothrips pellucidus Ananthakrishnan 1968

Xylaplothrips pusillus Ananthakrishnan & Jagadish 1969

Xylaplothrips rubiginosus Ananthakrishnan & Jagadish 1971

Xylaplothrips tener Ananthakrishnan & Jagadish 1969

Xylaplothrips varius Ananthakrishnan & Jagadish 1969

II. Suborder Terebrantia

New Genera

Aduncothrips Ananthakrishnan 1963, subgenus of *Erythrothrips*

Agriothrips Ananthakrishnan 1966

Amalothrips Ananthakrishnan 1967

Amphithrips Ananthakrishnan 1965

Aroidothrips Ananthakrishnan 1960

Cerothrips Ananthakrishnan 1961

Charientothrips Ananthakrishnan 1967

Euidothrips Ananthakrishnan 1968

Micothrips Ananthakrishnan 1965

Mictothrips Ananthakrishnan & Jagadish 1967, subgenus of *Chloethrips*

Ornothrips Ananthakrishnan 1965

Pellothrips Ananthakrishnan 1969

Pongamiothrips Ananthakrishnan 1962, subgenus of *Taeniothrips*

Priesneriola Ananthakrishnan 1964

Sporangiothrips Daniel 1985

Xynothrips Ananthakrishnan & Jagadish 1967, subgenus of *Chloethrips*

New Species

Aeolothrips nigricornis Ananthakrishnan 1968

Agriothrips brevisetosus Ananthakrishnan 1966

Amalothrips flaccidus Ananthakrishnan 1967

Amphithrips argutus Ananthakrishnan 1965

Anaphothrips (Neophysopus) sakimurai Ananthakrishnan 1961

Aroidothrips longistylus Ananthakrishnan 1960

Bathrips jasmini Ananthakrishnan 1968

Bolacidothrips graminis indicus Ananthakrishnan 1966

Bolacothrips bicolor Ananthakrishnan 1960

Cerothrips minutus Ananthakrishnan 1961

Charientothrips nilgiricus Ananthakrishnan 1967

Chirothrips loyolae Ananthakrishnan 1959

Chirothrips maximi Ananthakrishnan 1957

Chirothrips ramakrishnai Ananthakrishnan 1957

Chloethrips (*Mictothrips*) Ananthakrishnan & Jagadish 1967

Chloethrips (*Xynothrips*) Ananthakrishnan & Jagadish 1967

Chloethrips aureus Ananthakrishnan & Jagadish 1967

Chloethrips bicolor Ananthakrishnan & Jagadish 1967

Chloethrips dissidens Ananthakrishnan & Jagadish 1967

Chloethrips glandularis Ananthakrishnan & Jagadish 1967

Chloethrips graminis Ananthakrishnan & Jagadish 1967

Dendrothripiella stannardi Ananthakrishnan 1958

Dendrothrips cibarius Ananthakrishnan 1965

Dorcadothrips similis Ananthakrishnan 1968

Erythrothrips (*Aduncothrips*) Ananthakrishnan 1963

Euidothrips apsarus Ananthakrishnan 1968

Exothrips madrasensis Ananthakrishnan 1956

Exothrips shweta Bhatti & Ananthakrishnan 1978

Holarthrothrips indicus Bhatti & Ananthakrishnan 1978

Melanthrips affluens Ananthakrishnan 1966

Melanthrips baileyi Ananthakrishnan 1965

Merothrips indicus Bhatti & Ananthakrishnan 1975

Merothrips strasseni Ananthakrishnan 1970 (nomen nudum)

Micothrips fasciatus Ananthakrishnan 1965

Microcephalothrips brevipalpis var. *armatus* Ananthakrishnan 1956 (unavailable name)

Mycterothrips moultoni Seshadri & Ananthakrishnan 1954

Mycterothrips nimbus Ananthakrishnan 1965

Ornothrips fuscus Ananthakrishnan 1965

Oxythrips kochummani Ananthakrishnan 1969

Perissothrips aureus Ananthakrishnan 1954

Priesneriola oneillae Ananthakrishnan 1964

Projectothrips bhattii Ananthakrishnan 1973
(syn. *Proj. pandani* Ananthakrishnan 1962)

Projectothrips pruthii pandani Ananthakrishnan 1953
(junior homonym)

Projectothrips pandani Ananthakrishnan 1962
(not *pandani* An. 1953)

Ramaswamiahiella kallarensis Ananthakrishnan 1960

Rhopalandrothrips nilgiriensis Ananthakrishnan 1960

Rhopalandrothrips orchidii Ananthakrishnan 1961

Salpingothrips hoodi Ananthakrishnan 1969

Scirtothrips pteridicola Ananthakrishnan 1968

Scirtothrips titschacki Ananthakrishnan 1965

Sericothrips (Hydatothrips) boerhaaviae Seshadri & Ananthakrishnan 1954

Sericothrips graminis Ananthakrishnan 1956

Sporangiothrips acuminatus Daniel 1985

Taeniothrips (Pongamiothrips) Ananthakrishnan 1962

Taeniothrips ditissimus Ananthakrishnan & Jagadish 1966

Taeniothrips fulvus Ananthakrishnan & Jagadish 1969

Taeniothrips obscurus Ananthakrishnan & Jagadish 1966

Taeniothrips orchidi Ananthakrishnan 1968

Taeniothrips vinnulus Ananthakrishnan & Jagadish 1969

Thrips brunneus Ananthakrishnan & Jagadish 1968

Thrips emulatus Ananthakrishnan 1979

Thrips exhuberans Ananthakrishnan & Jagadish 1968

Thrips gracilis Ananthakrishnan & Jagadish 1968

Thrips (Athrips) heveae Ananthakrishnan & Jagadish 1968

Thrips hispidus Ananthakrishnan & Jagadish 1966

Thrips ignobilis Ananthakrishnan & Jagadish 1969

Thrips kodaikanalensis Ananthakrishnan & Jagadish 1966

Thrips montanus Ananthakrishnan & Jagadish 1968

Thrips tenebricus Ananthakrishnan & Jagadish 1968

Zonothrips luridus Ananthakrishnan 1968

Order Heteroptera

Family Anthocoridae

Buchananiella carayoni Muraleedharan & Ananthakrishnan, 1974

Orius shyamavarna Muraleedharan & Ananthakrishnan, 1974

Orius trivandrensis Muraleedharan & Ananthakrishnan, 1974

Scoloposcelis asiaticus Muraleedharan & Ananthakrishnan, 1974

Order Homoptera

Family Membracidae

Gargara albitarsis Ananthasubramanian & Ananthakrishnan, 1975

Gargara madrasensis Ananthasubramanian & Ananthakrishnan, 1975

Gargara malabarica Ananthasubramanian & Ananthakrishnan, 1975

Gargara rustica Ananthasubramanian & Ananthakrishnan, 1975

Lepotocentrus bauhiniae Ananthasubramanian & Ananthakrishnan, 1975

Lepotocentrus major Ananthasubramanian & Ananthakrishnan, 1975

Lepotocentrus mangiferae Ananthasubramanian & Ananthakrishnan, 1975

Lepotocentrus moringae Ananthasubramanian & Ananthakrishnan, 1975

Lepotocentrus nigra Ananthasubramanian & Ananthakrishnan, 1975

Lepotocentrus rhizophagus Ananthasubramanian & Ananthakrishnan, 1975

Lepotocentrus varicornis Ananthasubramanian & Ananthakrishnan, 1975

Otinotus obliquus Ananthasubramanian & Ananthakrishnan, 1975

Oxyrhachis brevicornutus Ananthasubramanian & Ananthakrishnan, 1975

Oxyrhachis krusadiensis Ananthasubramanian & Ananthakrishnan, 1975

Oxyrhachis minusculus Ananthasubramanian & Ananthakrishnan, 1975

Telingana nigroalata Ananthasubramanian & Ananthakrishnan, 1975

Tricentrus decornis Ananthasubramanian & Ananthakrishnan, 1975

Tricentrus pilosus Ananthasubramanian & Ananthakrishnan, 1975

Tricentrus purpureus Ananthasubramanian & Ananthakrishnan, 1975

Order Arachnida

Pseudoscorpionida

Sternophorus indicus Murthy and Ananthakrishnan, 1977 (synonymized with *Afrosternophorus ceylonicus* (Beier)

Sternophorus (Sternophorus) transiens Murthy and Ananthakrishnan 1977 (synonymized with *Afrosternophorus ceylonicus* (Beier)

PART - III

BIBLIOGRAPHY OF PROFESSOR DR. T.N. ANANTHAKRISHNAN

Ananthakrishnan, T.N. 1947a. An anomalous antenna in *Rhipiphorothrips cruentatus* Hood from Tambaram, South India. Journal of Bombay Natural History Society, 47 (2): 393–394, figs. A, B. Bombay.

Ananthakrishnan, T. N. 1947b. Observations on the habits of the conchostracan *Caenestheria* sp. recorded from Tambaram (S. India). Journal of Bombay Natural History Society, 47 (2): 396.

Ananthakrishnan, T. N. 1949. Lectures on Intermediate Human Physiology. 57 pp. Subrahmanyam Educational Institute, Madras. Published by S. Viswanathan.

Ananthakrishnan, T. N. 1950 (Dec.). Description of *Ischyrothrips menoni,* sp. nov. (Thysanoptera –Tubulifera) from South India. Indian Journal of Entomology (1949), 11 (1) : 9–10, fig. 1a–c. New Delhi.

Ananthakrishnan, T.N. 1952a. A new species of thrips – *Ecacanthothrips ramakrishnai* sp. nov. (Thysanoptera Tubulifera) from South India. Proceedings of the Indian Academy of Sciences, (Section B), 35 (1): 38–41, figs. 1–3. Bangalore.

Ananthakrishnan, T. N. 1952b. Variations in the mouth parts of some Thripidae (Thysanoptera–Terebrantia). Indian Journal of Entomology (1951), 12 (2) : 237–239. New Delhi.

Ananthakrishnan, T. N. 1953a (Jan.). Notes on some Thysanoptera from South India. Indian Journal of Entomology (1952), 14(3) : 197–201, figs. 1–5. New Delhi.

Ananthakrishnan, T.N. 1953b (Apr.). A new species of thrips, *Thilakothrips fimbriata* sp. n. (Thysanoptera–Tubulifera), from South India. Proceedings of the Royal Entomological Society of London, Series B, 22 (3–4) : 33–34, fig. 1A–C. London.

Ananthakrishnan, T. N. 1953c. A note on the male genital armature of some Terebrantia (Thysanoptera). Indian Journal of Entomology, 15 (3): 247–250, figs. 1–6. New Delhi.

Margabandhu, V. & Ananthakrishnan, T. N. 1953. Annotated list of Indian Thysanoptera (Supplement). Indian Journal of Entomology, 15 (3): 183–190. New Delhi.

Ananthakrishnan, T. N. 1954a. The bionomics of the *Mimusops* gall thrips *Arrhenothrips ramakrishnae* Hood. Agra University Journal of Research (Science), 3 (2): 463–473, figs. 1–14. Agra.

Seshadri, A. R. & Ananthakrishnan, T. N. 1954 (Sept.). Some new Indian Thysanoptera – I. Indian Journal of Entomology, 16(3): 210–226, figs. 1–5. New Delhi.

Ananthakrishnan, T.N. 1955a. On some internal organs of two tubuliferous Thysanoptera. Agra University Journal of Research (Science), 4 (1): 35–39, figs. 1–4. Agra.

Ananthakrishnan, T.N. 1955b. Host preferences in *Retithrips syriacus* (Mayet). Agra University Journal of Research (Science), 4(1):283–288, figs. 1–2. Agra.

Ananthakrishnan, T.N. 1955c (Apr.). Notes on *Thrips palmi* Karny, attacking *Sesamum indicum*. Journal of Bombay Natural History Society, 52 : 951–952, figs. 1–2. Bombay.

Ananthakrishnan, T.N. 1955d (Aug.). Studies on some Indian Thysanoptera – I. Annals and Magazine of Natural History, Series 12, 8 : 608–612, figs. 1–2. London.

Ananthakrishnan, T.N. 1955e. On the genus *Tiarothrips* Priesner (Thysanoptera Tubulifera). Zoologischer Anzeiger, 155 (3–4) : 83–87, fig. 1. Leipzig.

Ananthakrishnan, T.N. 1955f. Notes on *Pseudodendrothrips dwivarna* (R. & M.) from South India. Indian Journal of Entomology, 17(2) : 213–216, fig. 1A–F. New Delhi.

Ananthakrishnan, T.N. & Ananthasubramanian, K.S. 1955. Observations on temperature, humidity and light in relation to spinning in *Parembia* (Embioptera). Indian Journal of Entomology, 17(2) : 279–280. New Delhi.

Ananthakrishnan, T.N. 1956a. *Erythrinothrips,* nov. – a new subgenus of *Agnostothrips* Moulton (Tubulifera, Thysanoptera) from India. Indian Journal of Entomology (1955), 17(3) : 341–342, figs. 1–2. New Delhi.

Ananthakrishnan, T.N. 1956b (June). Studies on some Indian Thysanoptera – II. Proceedings of the Royal Entomological Society of London, Series B, 25 (5–6) : 75–78, figs. 1–3. London.

Ananthakrishnan, T.N. 1956c. Occurrence of apterous, brachypterous and macropterous forms in *Haplothrips apicalis* Bagnall. Current Science, 25(7) : 225–226, fig. 1, table 1. Bangalore.

Ananthakrishnan, T. N. 1956d. A Survey of our Present Knowledge of Indian Thysanoptera. Pp. 1–8. Amra Press: Madras.

Ananthakrishnan, T. N. 1956e. Thysanopteran fauna of some grasses. 1. *Andropogon pertusus*. Zoologischer Anzeiger, 156 (1–2) : 29–33. Leipzig.

Ananthakrishnan, T. N. 1956f. On the incidence of *Retithrips syriacus* (Mayet) on castor, in Madras (S. India). Zoologischer Anzeiger, 157 (1–2) : 33–35, 1 graph, 1 table. Leipzig.

Ananthakrishnan, T. N. 1956g. Studies on some Indian Thysanoptera III. Zoologischer Anzeiger, 157 (7–8) : 130–139, figs. 1–5. Leipzig.

Ananthakrishnan, T. N. & Ananthasubramanian, K. S. 1956. Some Embioptera from South India. Indian Journal of Entomology, 18 (2) : 165–168, figs. 1–2. New Delhi.

Ananthakrishnan, T. N. 1957a (Mar.). *Bamboosiella* novo gen. (Phlaeothripidae, Tubulifera) from India. Entomological News, 68 (3) : 65–68, fig. 1.

Ananthakrishnan, T. N. 1957b (Sept.). Studies on Indian Thysanoptera IV. Zoologischer Anzeiger, 159 (5–6) : 92–102, figs. 1–6. Leipzig.

Ananthakrishnan, T. N. 1957c. A study on the forms of *Haplothrips* (*Trybomiella*) *apicalis* Priesner. Proceedings of the Royal Entomological Society of London, Series B, 26 (9–10): 143–148, figs. 1–2, tables 1–6. London.

Ananthakrishnan, T. N. 1957d. On the taxonomic status of the Indian species of *Fulmekiola* Karny (Thysanoptera–Terebrantia). Indian Journal of Entomology (1956), 18 (4) : 336–339, figs. 1–2. New Delhi.

Ananthakrishnan, T. N. 1958a (Dec.). Two new species of tubuliferous Thysanoptera from India (Thysanoptera, Phlaeothripidae). Proceedings of the Entomological Society of Washington, 60 (6) : 277–280, figs. 1–6. Washington.

Ananthakrishnan, T.N. 1958b (Dec.). *Dendrothripiella stannardi* sp. nov. (Thysanoptera Terebrantia) from Kodaikanal Hills (S. India). Journal of the Zoological Society of India (1957), 9 (2) : 216–221, figs. 1–5. Calcutta.

Ananthakrishnan, T.N. 1959. Studies on some Indian Thysanoptera V. Zoologischer Anzeiger, 162 (9–10) : 313–324, figs. 1–7. Leipzig.

Ananthakrishnan, T.N. 1960a (Jan.). A remarkable instance of sexual dimorphism in a new species, *Rhopalandrothrips nilgiriensis* (Thysanoptera: Terebrantia). Pan-Pacific Entomologist, 36 (1) : 37–40, figs. 1–2.

Ananthakrishnan, T.N. 1960b. *Neosmerinthothrips inquilinus* sp. nov., an inquiline Tubuliferan (Thysanoptera) from India. Bulletin of Entomology, No. 1 : 32–33. Loyola College: Madras.

Ananthakrishnan, T.N. 1960c (Dec.). Thysanoptera from the Nilgiri and Kodaikanal Hills (South India). Journal of Bombay Natural History Society, 57 (3) : 557–578, figs. 1–6. Bombay.

Ananthasubramanian, K.S. & Ananthakrishnan, T.N. 1960. Biology of *Oligotoma minuscule* Enderlein (Oligotomidae, Embioptera). Bulletin of Entomology, No. 1 : 18–21, tables I–II. Loyola College : Madras

Ananthakrishnan, T.N. 1961a (Mar.). New Tubuliferan Thysanoptera from India. Journal of Zoological Society of India (1960), 12(2): 250–258, figs. 1–5.

Ananthakrishnan, T. N. 1961b (Aug.). A review of some grass-infesting thrips from India with a description of a new species. Journal of Bombay Natural History Society, 58 (2) : 420–428, figs. 1–2. Bombay.

Ananthakrishnan, T. N. 1961c. Studies on some Indian Thysanoptera – VI. Zoologischer Anzeiger, 167 (7–8) : 259–271, figs. 1–5. Leipzig.

Ananthakrishnan, T. N. 1961d. Allometry and speciation in *Ecacanthothrips* Bagnall. Proceedings of the Biological Society of Washington, 74 : 275–280, figs. 1–2, table 1. Washington.

Ananthakrishnan, T. N. 1962a (Aug.). Some little known Indian Terebrantia (Thysanoptera). Proceedings of the Royal Entomological Society of London, Series B, 31 (7–8) : 87–91, figs. 1–4. London.

Ananthakrishnan, T. N. 1962b. On the dimorphic species *Pygmaeothtips columniceps* Karny (Ins., Tubulifera: Thysanoptera). Senckenbergiana Biologica, 43 (5) : 363–368, figs. 1–11. Frankfurt/Main.

Ananthakrishnan, T. N. 1962c. Alary polymorphism in two species of Thysanoptera. Proceedings of 1st All India Zoological Congress 1959, Jabalpur, p. 2.

Ananthakrishnan, T. N. 1962d. Alary polymorphism in two species of Thysanoptera. Proceedings of 1st All India Zoological Congress 1959, Jabalpur, pp. 473–475, figs. 1a–d, 2 tables. [see 1962c].

Sakimura, K. & Ananthakrishnan, T. N. 1962. A study of the Indian species of the genus *Chaetanaphothrips*. Bulletin of Entomology, No. 3 : 48–56, figs. 1–13. Loyola College: Madras.

Ananthakrishnan, T. N. 1963 (Dec.). The Terebrantian Thysanoptera of the Indo-Ceylonese region. Treubia, 26 (2) : 73–122. Bogor.

Ananthasubramanian, K.S. & Ananthakrishnan, T.N. 1963a. The microarthropods of the pasture soil with notes on the seasonal fluctuations of *Symphylella* sp. (Myriapoda). Bulletin of Entomology, No. 4 : 26–32, tables I–II, 3 figs. Loyola College : Madras.

Ananthasubramanian, K. S. & Ananthakrishnan, T. N. 1963b. Bionomics of the viviparous cockroach, *Trichoblatta sericea* (Saussure) from India (Orthoptera : Blattoidea). Bulletin of Entomology, No. 4 : 33–37, tables I–II. Loyola College : Madras.

Ananthakrishnan, T.N. 1964a (Mar.). Indian Terebrantia – I. Bulletin of Entomology, No. 5 : 49–55, figs. a–c, 1 table. Loyola College: Madras.

Ananthakrishnan, T.N. 1964b. Thysanopterologica Indica – I. Entomologisk Tidskrift, 85 (1–2) : 99–120, figs. 1–11. Uppsala (Sweden).

Ananthakrishnan, T.N. 1964c. Thysanopterologica Indica – II. Entomologisk Tidskrift, 85 (3–4) : 218–235, figs. 1–4. Uppsala (Sweden).

Ananthakrishnan, T.N. 1964d. A contribution to our knowledge of the Tubulifera (Thysanoptera) from India. Opuscula Entomologica, Supplement 25 : 1–120, figs. 1–19.

Ananthakrishnan, T.N. 1965a (Apr.). Indian Terebrantia – II (Thysanoptera: Insecta). Bulletin of Entomology, No.6: 15–29, figs. 1-8. Loyola College: Madras.

Ananthakrishnan, T. N. 1965b (Apr.). Polymorphism and the taxonomy of Thysanoptera. Bulletin of Entomology, No. 6 : 57–61, figs. 1–2, table 1. Loyola College : Madras.

Ananthakrishnan, T. N. 1965c. Thysanopterologica Indica – III. Entomologisk Tidskrift, 86 (1–2) : 49–63, figs. 1–5. Uppsala (Sweden).

Ananthakrishnan, T. N. 1966a (Jan.). Thysanopterologica Indica – IV. Bulletin of Entomology, No. 7 : 1–12, figs. 1–2. Loyola College : Madras.

Ananthakrishnan, T. N. 1966b (Jan.). Indian Terebrantia – III. Bulletin of Entomology, No. 7 : 26–33, figs. 1–3. Loyola College : Madras.

Ananthakrishnan, T. N. & Jagadish, A. 1966a (June). Coffee and tea infesting thrips from Anamalais (S. India) with description of two new species of *Taeniothrips* Serville. Indian Journal of Entomology, 28 (2) : 250–257, figs. 1–3. Delhi.

Ananthakrishnan, T. N. & Jagadish, A. 1966b. Studies on some species of the genus *Thrips* Linn. from India – I. Entomologisk Tidskrift, 87 (1–2) : 85–99, figs. 1–8, 1 table. Uppsala (Sweden).

Ananthakrishnan, T. N. 1967a. New Indian Thysanoptera. Annales de la Société Entomologique de France, Nouvelle Serie, 3 (1) : 227–237, figs. 1–6.

Ananthakrishnan, T. N. 1967b. Structural diversity and variation range in thrips populations. Bulletin of the National Institute of Sciences of India, No. 34 : 371–374. New Delhi.

Ananthakrishnan, T. N. 1967c. Allomorphic growth patterns in some Tubulifera (Thysanoptera). Proceedings 54th Indian Science Congress, Calcutta, Pt. III, Abstracts, pp. 459–460.

Ananthakrishnan, T. N. 1967d. Allomorphic growth patterns in some Tubulifera (Thysanoptera). Indian Journal of Entomology, 29 (1) : 61–64, figs. 1–2. New Delhi. [see 1967c].

Ananthakrishnan, T. N. 1967e (Nov.). On the multiple effects of oedymerism in the males of *Nesothrips falcatus* Ananthakrishnan. Current Science, 36 (22) : 610–611,fig. 1. Bangalore.

Ananthakrishnan, T. N. & Jagadish, A. 1967a. Studies on some *Taeniothrips* species from India I. Opuscula Entomologica, 32 (1–2) : 159–170, figs. 1–6, tables I–II. Lund (Sweden).

Ananthakrishnan, T. N. & Jagadish, A. 1967b. Studies on the genus *Chloethrips* Priesner from India. Zoologischer Anzeiger, 178 (5–6) : 374–388, figs. 1–4. Leipzig.

Ananthakrishnan, T. N. & Jagadish, A. 1967c. On some species of *Liothrips* Uzel from India. Bulletin of Entomology, 8 (2) : 130–142, figs. 1–6. Entomological Society of India : New Delhi. [The reprint bears page numbers 1–13].

Ananthakrishnan, T. N. 1968a (Jan.). Studies on new and little known Indian Thysanoptera. Oriental Insects (1967 Sept.), 1(1–2) : 113–138, figs. 1–8. Delhi.

Ananthakrishnan, T. N. 1968b (July). Thysanopterologica Indica – V. Oriental Insects, 2 (1) : 41–58, figs. 1–6. Delhi.

Ananthakrishnan, T. N. 1968c. Implications on the nature of polymorphism in Thysanoptera. Journal of Animal Morphology and Physiology (1967), 14 (1) : 140–141. Baroda. [Abstracts of papers presented at the Seminar on Neuroendocrinology, held at Trivandrum, Oct. 20–23, 1966].

Ananthakrishnan, T. N. 1968d. Allomorphic growth patterns in relation to sex-limited polymorphism in some mycophagous Tubulifera (Thysanoptera). Current Science, 37 (20) : 577–578, table 1. Bangalore.

Ananthakrishnan, T. N. 1968e. Patterns of structural diversity in the males of some phlaeophilous Tubulifera (Thysanoptera). Annales de la Société Entomologique de France, Nouvelle Serie, 4 (2) : 413–418, figs. 1–4. Paris.

Ananthakrishnan, T. N. 1968f. Two new species of *Nesothrips* with further remarks on *Nesothrips robustus* (Thysan. Tubulifera). Annales de la Société Entomologique de France, Nouvelle Serie [new series], 4 (4) : 967–973, figs. 1–3. Paris.

Ananthakrishnan, T. N. 1968g. Indian Terebrantia – IV (Thysanoptera). Zoologischer Anzeiger, 180 (3–4) : 258–268, figs. 1–5. Leipzig.

Ananthakrishnan, T N. & Jagadish, A. 1968a. On the seasonal fluctuations and biology of *Anaphothrips flavicinctus* (Karny) on *Panicum maximum* in Madras. Journal of Bombay Natural History Society, 65 (1) : 243–248, tables 1–2, graphs I–III. Bombay.

Ananthakrishnan, T. N. & Jagadish, A. 1968b (Oct.). Studies on the species of the genus *Thrips* from India – II. Deutsche Entomologische Zeitschrift, Neue Folge [= new series], 15 (4–5) : 359–365, figs. 1–4.

Ananthakrishnan, T. N. 1969a (Jan.). On two new species of terebrantian Thysanoptera with remarks on a new genus *Pellothrips.* Oriental Insects (1968), 2 (2) : 201–204, figs. 1–6. Delhi.

Ananthakrishnan, T. N. 1969b (Mar.). Mycophagous Thysanoptera – I. Indian Forester, 95 (3) : 173–185, figs. 1–8. Dehra Dun.

Ananthakrishnan, T. N. 1969c (15 Aug.). New gall thrips from India (Ins., Thysanoptera, Phlaeothripidae). Senckenbergiana Biologica, 50 (3–4) : 179–194, figs. 1–14. Frankfurt/Main.

Ananthakrishnan, T. N. 1969d (Nov.). Mycophagous Thysanoptera – II. Oriental Insects, 3 (3) : 289–299, figs. 1–16. Delhi.

Ananthakrishnan, T. N. 1969e (Nov.). The genus *Polyphemothrips* Schmutz from India. Oriental Insects, 3 (3) : 301–310, figs. 1–13. Delhi.

Ananthakrishnan, T. N. 1969f. Indian Thysanoptera. viii + 171 pp., plates 1–10, figs. 1–38. CSIR Zoological Monograph No. 1. CSIR: New Delhi.

Ananthakrishnan, T. N. & Jagadish, A. 1969a (Jan.). Gall-inhabiting *Liothrips* Uzel from India. Oriental Insects (1968), 2 (2) : 205–217, figs. 1–23. Delhi.

Ananthakrishnan, T. N. & Jagadish, A. 1969b (June). Studies on Indian gall thrips (Thysanoptera). Marcellia, 36 (1–2) : 7–103, plates 1–32, 7 tables. Strasbourg (France).

Ananthakrishnan, T. N. & Jagadish, A. 1969c. On some species of *Taeniothrips* Serville and *Thrips* Linn. with further notes on *Hyalopterothrips roonwali* Bhatti from India. Zoologischer Anzeiger, 182 (1–2) : 113–121, figs. 1–4, table 1. Leipzig.

Ananthakrishnan, TN. & Jagadish, A. 1969d. Studies on the species of *Xylaplothrips* Priesner from India. Zoologischer Anzeiger, 182 (1–2) : 121–133, figs. 1–2. Leipzig.

Ananthakrishnan, T. N. 1970a (Apr.). Studies on the genus *Leeuwenia* Karny. Oriental Insects, 4 (1) : 47–57, figs. 1–5. Delhi.

Ananthakrishnan, T. N. 1970b. On some species of *Nesothrips* Kirkaldy with notes on the influence of oedymerism and gynaecoidism in species of allied genera of Cryptothripini (Tubulifera: Thysanoptera). Journal of Zoological Society of India, 22 (1–2) : 51–62, figs. 1–5, table 1.

Ananthakrishnan, T. N. 1970c. Some Aspects of Host Specificity in Indian Thysanoptera (Insecta). Prof. Ganapati Shas. Comm. Volume, pp. 1–5, 1 Table.

Ananthakrishnan, T. N. 1970d. *Kleothrips gigans* Schmutz – a study in intraspecific diversity (Thysanoptera: Tubulifera). Bulletin of Entomology (1969), 10 (2) : 130–133, table 1, figs. 1–2. Entomological Society of India : New Delhi.

Ananthakrishnan, T. N. 1970e. The Life of Insects. 46 pp., figs. 1–21. National Council of Educational Research and Training (NCERT) : New Delhi.

Ananthakrishnan, T. N. & Jagadish, A. 1970a (Nov.). On three new species of tubuliferous Thysanoptera from Darjeeling. Oriental Insects, 4 (3) : 259–264, figs. 1–3. Delhi.

Ananthakrishnan, T. N. & Jagadish, A. 1970b (Nov.). The species of *Dicerato-thrips* Bagnall and allied genera from India (Thysanoptera : Megathripinae : Insecta). Oriental Insects, 4 (3) : 265–280, figs. 1–5. Delhi.

Muralidharan [sic !], M. C. & Ananthakrishnan, T. N. 1970. On the free amino-acids of the haemolymph and regions of the alimentary canal of *Eyprepo-cnemis alacris alacris* (Serville) (Orthoptera : Insecta). Current Science, 39 (22) : 507–508, table 1. Bangalore.

Muralirangan, M. C. & Ananthakrishnan, T. N. 1970a. On the secretory activity of the cells of the hepatic caeca in the instars and adults of some acridids (Orthoptera:Insecta). Current Science, 39(13): 297–298, figs. 1–4. Bangalore.

Ananthakrishnan, T.N. 1971a. Trends in intraspecific sex-limited variations in some mycophagous Tubulifera (Thysanoptera). Journal of Bombay Natural History Society, 67 (3) : 481–501, figs. 1–6, tables 1–2, graphs 1–3. Bombay.

Ananthakrishnan, T. N. 1971b (June). Mycophagous Thysanoptera – III. Oriental Insects, 5 (2) : 189–207, figs. 1–9. Delhi.

Ananthakrishnan, T. N. 1971c. New species of *Stigmothrips* Ananthakrishnan with keys to Indian species. Journal of the Zoological Society of India, 23 (2) : 175–185, figs. 1–6.

Ananthakrishnan, T.N. 1971d. Thrips (Thysanoptera) in agriculture, horticulture and forestry – Diagnosis, bionomics and control. Journal of Scientific and Industrial Research, 30 (3) : 113–146, plates I–X. New Delhi.

Ananthakrishnan, T.N. & Jagadish, A. 1971. Further studies on the mycophagous species of *Xylaplothrips* Priesner. Zoologischer Anzeiger, 186 (3–4) : 259–267, figs. 1–5. Leipzig.

Muraleedharan, N. & Ananthakrishnan, T. N. 1971. Bionomics of *Montandoniola moraguesi* (Puton) (Heteroptera : Anthocoridae), a predator on gall thrips. Bulletin of Entomology, 12 (1) : 4–10, figs. 1–3, tables 1–3. Entomological Society of India: New Delhi.

Ananthakrishnan, T. N. 1972a. Further studies on Indian gall thrips. Marcellia (1971), 37 (3) : 111–127, pls. I–VI.

Ananthakrishnan, T. N. 1972b. Further studies on Indian gall thrips – II. Marcellia, 37 (5): 3–20, pls. I–VI.

Ananthakrishnan, T. N. 1972c (Dec.). Mycophagous Thysanoptera – IV. Oriental Insects, 6 (4) : 425–437, figs. 1–7. Delhi.

Ananthakrishnan, T. N. 1972d (Dec.). Mycophagous Thysanoptera – V. Oriental Insects, 6 (4) : 439–447, figs. 1–7. Delhi.

Ananthakrishnan, T. N. & Muraleedharan, N. 1972. Free amino acids in relation to host plant preferences in the polyphagous heliothripines *Rhipiphorothrips cruentatus* Hood and *Retithrips syriacus* (Mayet). Current Science, 41 (23) : 846–847. Bangalore.

Jagadish, A & Ananthakrishnan, T. N. 1972a. Taxonomic significance of the second instar larvae of some Indian Terebrantia (Thysanoptera : Insecta). Occasional Publication of the Entomology Research Unit, No. 1 : 1–31, figs. 1–58. Loyola College, Madras.

Jagadish, A & Ananthakrishnan, T. N. 1972b. Studies on the II instar larvae of some mycophagous Megathripinae – I (Thysanoptera : Insecta). Zoologischer Anzeiger, 188 (5–6) : 429–441, figs. 1–20. Leipzig.

Muralirangan, M. C. & Ananthakrishnan, T. N. 1972. Histological changes in the alimentary canal of *Eyprepocnemis alacris alacris* (Serville) [Orthoptera : Acrididae] during postembryonic development. Journal of the Zoological Society of India, 24 (1) : 1–27, text–figs. 1–3, tables 1–5, pl. 1.

Ananthakrishnan, T. N. 1973a (Apr.). Two homonyms in Thysanoptera. Oriental Insects, 7 (1) : 48. Delhi.

Ananthakrishnan, T. N. 1973b (Apr.). Mycophagous Tubulifera of India (Thysanoptera: Insecta). Occasional Publication of the Entomology Research Unit, Loyola College, No. 2, pp. 1–144, 51 text–figures, pls. I–XV, figs. A–K. Madras.

Ananthakrishnan, T. N. 1973c (20 July). Studies on some Indian species of the genus *Elaphrothrips* Buffa (Megathripinae: Tubulifera: Thysanoptera). Pacific Insects, 15 (2): 271–284, figs. 1–6.

Ananthakrishnan, T. N. 1973d (Dec.). Further studies on Indian gall thrips – III. Oriental Insects, 7 (4) : 539–546, figs. 1–3. Delhi.

Ananthakrishnan, T. N. 1973e. Thrips: Biology and Control. ix + 120 pp., 8 text-figs., 27 plates. Macmillan India: Delhi.

David, B. V. & Ananthakrishnan, T. N. 1973. Studies on the second instar larvae of Indian gall thrips. Occasional Publication of Entomology Research Unit, No. 3 : 1–30, pls. 1–11. Loyola College, Madras.

Viswanathan, T. R. & Ananthakrishnan, T. N. 1973a. On partial ovoviviparity in *Tiarothrips subramanii* (Ramakrishna) (Thysanoptera : Insecta). Current Science, 42 (18) : 649–650, fig. 1. Bangalore.

Viswanathan, T. R. & Ananthakrishnan, T. N. 1973b. Observations on the biology, ecology and behaviour of *Ecacanthothrips sanguineus* Bagnall Tubulifera : Thysanoptera). Current Science, 42(20):727–728, fig. 1. Bangalore

Ananthakrishnan, T. N. 1974a. *Aleurodothrips fasciapennis* (Franklin) predatory on coccids and aleyrodids. Journal of Bombay Natural History Society, 71 (1) : 157–160, fig. 1A, a–d. Bombay.

Ananthakrishnan, T. N. 1974b. Addenda. Kontyû, 42 (4) : 387. In: Kudô I & Ananthakrishnan TN, Kontyû, 42 (4) : 385–387.

[This 10-line 'Addenda' has the following correction: *Fulgorothrips breviceps* Bagnall and *F. faurei* Anan. should be placed in *Fulgorothrips* instead of in *Phoxothrips* where these were placed by Ananthakrishnan 1973b : 122–123].

Ananthakrishnan, T. N. & Kudô, I. 1974 (June). The species of the genus *Xenothrips* Ananthakrishnan (Thysanoptera, Phlaeothripidae). Kontyû, 42 (2) : 117–121, figs. 1–2. Tokyo.

Ananthakrishnan, T. N. & Muraleedharan, N. 1974a. On the incidence and effects of infestation of *Selenothrips rubrocinctus* (Giard) (Thysanoptera :Heliothripinae) on the free aminoacids of some susceptible host plants. Current Science, 43 (7) : 216–218, table I. Bangalore.

Ananthakrishnan, T. N. & Muraleedharan, N. 1974b (May). Studies on the *Gynaikothrips–Liophlaeothrips–Liothrips* complex from India. Oriental Insects, Supplement, No. 4 : 1–85, figs. 1–63. Delhi.

Kudô, I. & Ananthakrishnan, T. N. 1974. A new subgenus and species of *Meiothrips* Priesner (Thysanoptera, Megathripinae) from Nepal. Kontyû, 42 (4) : 385–387, 5 figs. Tokyo.

Muraleedharan, N. & Ananthakrishnan, T. N. 1974a. The genus *Buchananiella* Reuter from India with the description of a new species (Heteroptera : Anthocoridae). Oriental Insects, 8 (1) : 33–35, fig. 1. Delhi.

Muraleedharan, N. & Ananthakrishnan, T. N. 1974b. New and little-known species of Orius Wolff from India. Oriental Insects, 8(1): 37-41.

Muraleedharan, N. & Ananthakrishnan, T. N. 1974c. A new species of the genus *Scoloposcelis* Fieber (Heteroptera : Anthocoridae) from South India. Journal of Natural History, 8 : 511–512, figs. 1–2. London.

Muralirangan, M. C. & Ananthakrishnan, T. N. 1974. Taxonomic significance of the foregut armature in some Indian Acridoidea (Orthoptera). Oriental Insects, 8 (2) : 119–145, figs. 1–32.

Viswanathan, T. R. & Ananthakrishnan, T. N. 1974. Population fluctuations of 3 species of anthophilous Thysanoptera in relation to the numerical response of their predator, *Orius minutus* L. (Anthocoridae: Hemiptera). Current Science, 43 (1) : 19–20, figs. 1–2. Bangalore.

Ananthakrishnan, T. N. & Viswanathan, T. R. 1975a. Alary polymorphism in the gall thrips *Byctothrips ayyari* Ananthakrishnan (Tubulifera : Insecta). Current Science, 44 (17) : 632-633, tables I–II, fig. 1. Bangalore.

Ananthakrishnan, T. N. & Viswanathan, T. R 1975b (Dec.). A new gall thrips *Gynaikothrips schefflericola* sp. n. from India. Cecidologia Indica, 10 (3) : 163–166, figs. 1–3. Allahabad.

Ananthasubramanian, K. S. & Ananthakrishnan, T. N. 1975a. Relative growth in some species of Membracidae (Homoptera : Insecta). Current Science, 44 (23) : 857–859, table I. Bangalore.

Ananthasubramanian, K. S. & Ananthakrishnan, T. N. 1975b, Taxonomic, biological and ecological studies on Indian Membracids (Insecta : Homoptera) – I, Records of the Zoological Survey of India, 68 : 161–272, figs. 1–39, tables 1–3, Calcutta.

Ananthasubramanian, K. S. & Ananthakrishanan, T. N. 1975c. Taxonomic, Biological & Ecological studies of some Indian Membracids (Insecta: Homoptera) Part-II. Records of the Zoological Survey of India, 68 : 305-340, figs. 1–8, tables 1–12, Calcutta.

Bhatti, J. S. & Ananthakrishnan, T. N. 1975. The genus *Merothrips* in India (Thysanoptera : Merothripidae). Oriental Insects, 9(1): 31–43, 26 figs. Delhi.

Bhatti, J. S. & Ananthakrishnan, T. N. 1978. Studies on some Indian Thysanoptera. Entomon, 3 (2) : 229–238, 11 figs. Trivandrum.

Krishnamurthy, K. V.; Raman, A; & Ananthakrishnan, T. N. 1975. On the morphology of the ceratoneon thrips galls of *Schefflera racemosa* Harms. Marcellia, 38 (2–3) : 179–184, figs. 1–11.

Ananthakrishnan, T. N. 1976a (May). Two new genera and species of tubuliferan thrips (Thysanoptera: Insecta) from India. Proceedings of the Indian Academy of Sciences, (Section B), 83 (5) : 202–206, figs. 1–2. Bangalore.

Ananthakrishnan, T. N. 1976b (30 June). Studies on *Mesothrips* (Thysanoptera: Tubulifera). Oriental Insects, 10 (2) : 185–214, figs. 1–28. Delhi.

Ananthakrishnan, T. N. 1976c (July). New species of *Liothrips* inhabiting *Schefflera* galls from India, with a key to known species of *Liothrips* from *Schefflera* species. Ceylon Journal of Science (Biological Sciences), 12 (1) : 23–28, figs. 1–3.

Ananthakrishnan, T. N. 1976d (Nov.). New gall thrips of the genus *Crotonothrips* (Thysanoptera). Oriental Insects, 10 (3) : 411–419, figs. 1–4. Delhi.

Ananthakrishnan, T. N. 1976e. The distribution and host range of predatory thrips. Indian Journal of Plant Protection, 4 (1) : 67–68.

Ananthakrishnan, T. N. 1976f. Report of Work Done at the Entomology Research Unit 1965–1975, 12 pp. Loyola College, Madras. [pp. 7–12 are not numbered].

Ananthakrishnan, T. N. 1976g. Problems on host specificity in insects. Current Science, 45 (8) : 289–290. Bangalore.

Ananthakrishnan, T. N. 1976h. Insect Endocrinology. Pp. i–iv. Prof. K. K. Nayar Memorial Endowment Lecture, Commemoration Symposium, University of Kerala, Trivandrum.

Ananthakrishnan, T. N. & Thangavelu, K. 1976. The cereal thrips *Haplothrips ganglbaueri* Schmutz with particular reference to the trends of infestation on *Oryza sativa* and the weed *Echinochloa crusgalli.* Proceedings of the Indian Academy of Sciences (Section B), 83 (5) : 196–201, fig. 1. Bangalore.

Ananthakrishnan, T. N. & Viswanathan, T. R. 1976, 1983. General Animal Ecology. 324 pp. (1st ed.), 356 pp. (2nd ed.). Macmillan India : New Delhi.

David, B. V. & Ananthakrishnan, T. N. 1976. Host correlated variation in *Trialeurodes rara* Singh and *Bemisia tabaci* (Gennadius) (Aleyrodidae : Homoptera : Insecta). Current Science, 45 (6) : 223–225, tables I–II. Bangalore.

Nayar, K. K.; Ananthakrishnan, T. N; & David, B. V. 1976. General and Applied Entomology. x + 589 pp. Tata McGraw-Hill : New Delhi.

Viswanathan, T. R. & Ananthakrishnan, T. N. 1976. Aspects of host preference and succession in thrips infesting *Ruellia tuberosa.* Entomon, 1 (1) : 71–77, tables 1a, b, figs. 1–3. Trivandrum.

Ananthakrishnan, T. N. (editor). 1977a. Insects and Host Specificity. ix + 127 pp. Macmillan India : Delhi.

Ananthakrishnan, T. N. 1977b. Introduction. Pp. 1–4. In : Ananthakrishnan TN (ed.). Insects and Host Specificity, 127 pp. Macmillan India : Delhi

Ananthakrishnan, T. N. & Kandasamy, C. 1977. On the trends of infestation of two species of *Baliothrips* Uzel on paddy, maize and their weed hosts. Current Science, 46 (10) : 344–345, figs. 1–2. Bangalore.

Ananthakrishnan, T. N. & Raman, A. 1977. Aspects of host specificity in some cecidogenous insects. Pp. 51–56. In : Ananthakrishnan TN (ed.), Insects and Host Specificity. 127 pp. Macmillan India: Delhi.

Ananthakrishnan, T. N. & Swaminathan, S. 1977. Host–parasite and host–predator interactions in the gall thrips *Schedothrips orientalis* Anan. (Insecta : Thysanoptera). Entomon, 2 (2) : 247–251, figs. 1–2. Trivandrum.

Ananthakrishnan, T. N. & Thirumalai, G. 1977. The grass seed infesting thrips *Chirothrips mexicanus* Crawford on *Pennisetum typhoides* and its principal alternate host *Chloris barbata.* Current Science, 46 (6) : 193–194, fig. 1. Bangalore.

Ananthakrishnan, T. N.; Thirumalai, G.; & Kandasamy, C. 1977. Role of weeds in thrips infestation in *Oryza sativa.* p. 23. In : Second Oriental Entomology Symposium, March 23–27, 1977, Entomology Research Unit, Loyola College, Madras. [Abstracts, 110 pp.].

Ananthakrishnan, T. N. & Varadarasan, S. 1977a. *Androthrips flavipes* Schmutz (Insecta : Thysanoptera), a predatory inquiline in thrips galls. Entomon, 2 (1): 105–107, tables 1–2. Trivandrum.

Ananthakrishnan, T. N. & Varadarasan, S. 1977b. Reproductive efficiency in relation to the occurrence of polymorphism in *Arrhenothrips ramakrishnae* Hood (Thysanoptera : Insecta). Second Oriental Entomology Symposium, March 23–27, 1977, Entomology Research Unit, Loyola College, Madras. p. 102.

Krishnamurthy, K. V.; Raman, A; & Ananthakrishnan, T. N. 1977. Studies on plant galls from India. 2. Leaf galls of *Cordia obliqua* Willd. (*Cordia myxa* Linn.) (Boraginaceae). Ceylon Journal of Science (Biological Sciences), 12 (2): 73–84, pl. 1, fig. 1, figs. 2–10. Colombo.

Muralirangan, M. C. & Ananthakrishnan, T. N. 1977a. Biological studies on *Eyprepocnemis alacris alacris* (Serville) (Orthoptera : Acrididae): Growth in terms of weight, length and eye-stripe number. Entomon, 2 (2): 193–200, figs. 1-5.

Muralirangan, M. C. & Ananthakrishnan, T. N. 1977b. Morphological correlates in preferential feeding of *Eyprepocnemis alacris alacris* Serv. (Orthoptera : Acrididae). Pp. 35–39. In: Ananthakrishnan TN (ed.), Insects and Host Specificity. Macmillan India: New Delhi.

Murthy, V. A. & Ananthakrishnan, T. N. 1977. Studies on Indian Chelonethi (Pseudoscorpionida). Oriental Insects Monograph, 4: 1–210. Delhi.

Thirumalai, G. & Ananthakrishnan, T. N. 1977. A new pentatomid host for a pteromalid egg parasite and the nature of interspecific competition with other hymenopterous egg parasites. Current Science, 46 (24): 864–865, fig. 1. Bangalore.

Ananthakrishnan, T. N. 1978a. Exotic thrips pests and plant quarantine. Zoologiana, No. 1 : 43–48. Zoological Survey of India, Calcutta.

Ananthakrishnan, T. N. 1978b. Thrips galls and gall thrips. Zoological Survey of India, Technical Monograph, No. 1 : 1–69, figs. 1–11, pls. 1–26. Controller of Publications, Delhi. [pls. 1–26 are on 26 unnumbered pages after p. 69].

Ananthakrishnan, T. N. 1978c. Impact of deforestation and introduction of plantation crops on the thrips fauna of Western Ghats. Memoirs of School of Entomology, St. John's College, Agra, No. 6: 123–124.

Ananthakrishnan, T. N. & Thirumalai, G. 1978. Population fluctuations of three species of anthophilous Thysanoptera with notes on the biology of the seed feeding species *Chirothrips mexicanus* Crawford. Bulletin of Zoological Survey of India, 1 (2) : 197–201, figs. 1–3, pl. 6. Calcutta.

Ananthakrishnan, T. N. and Varadarasan, S. 1978 (Nov.). On some new gall inhabiting Thysanoptera. Oriental Insects, 12 (3) : 391–402, figs. 1–4, pls. 1–3. Delhi.

Muraleedharan, N. & Ananthakrishnan, T. N. 1978. Bioecology of four species of Anthocoridae (Hemiptera : Insecta) predaceous on thrips, with key to genera of anthocorids from India. Records of the Zoological Survey of India, Miscellaneous Publication, Occasional Paper, No. 11 : 1–32, figs. 1–18. Calcutta.

Raman, A., Ananthakrishnan, T. N., & Swaminathan, S. 1978. On the simple leaf galls of *Casearia tomentosa* Roxb. (Samydaceae) induced by *Gynaikothrips flaviantennatus* Moulton (Thysanoptera : Phlaeothripidae). Proceedings of the Indian Academy of Sciences, Section B (Plant Sciences –3), 87B (8) : 231–242, figs. 1–17, table 1. Bangalore.

Muralirangan, M. C. & Ananthakrishnan, T. N. 1978. On the growth rate of the ovary of *Eyprepocnemis alacris alacris* (Serville) during post embryonic development (Orthoptera : Acrididae). Proceedings of the Indian Academy of Sciences, Section B, (Animal Sciences 4), 87 (12) : 371–375, tables 1–2, fig. 1. Bangalore.

Ananthakrishnan, T. N. 1979a. Biosystematics of Thysanoptera. Annual Review of Entomology, 24: 159–183, fig. 1. Annual Reviews Inc.: Palo Alto (California).

Ananthakrishnan, T. N. 1979b. Diversity indices in relation to intrapopulation variation in two species of mycophagous tubuliferan Thysanoptera. Proceedings, Symposium Zoological Survey of India, 1 : 19–26, figs. 1–6, tables 1–2.

Ananthakrishnan, T. N. 1979c. Addenda. Kontyû, 47 (4) : 492. In : Kudô 1979, Kontyû, 47 (4) : 487–492.

[*Thrips emulatus* nom. nov., proposed for *T. brunneus* Ananthakrishnan & Jagadish 1968 (preoccupied by *T. brunneus* Ishida 1936)].

Ananthakrishnan, T. N. 1979d. Microarthropods and soil ecosystems. Journal of Bombay Natural History Society (1978), 75 (3) : 625–631, table 1. Bombay.

Ananthakrishnan, T. N. 1979e. Insects and microclimate. Zoologiana (1978), No. 2 : 29–32. Zoological Survey of India, Calcutta.

Ananthakrishnan, T. N. 1979f. Insect taxonomy at the Zoological Survey of India. Pp. 21–22. Workshop on Advances in Insect Taxonomy in the Orient, Manali (H. P.), October 9–12, 1979, 108 pp. (Published Oct. 9).

Ananthakrishnan, T. N. & Swaminathan, S. 1979. On the population trends of *Aneurothrips priesneri* Bhatti (Insecta : Thysanoptera) from the leaf galls of *Cordia obliqua*. Bulletin of Zoological Survey of India, 2 (1) : 91–94, figs. 1–2, table 1. Calcutta.

Raman, A. & Ananthakrishnan, T. N. 1979. On the developmental morphology of leaf fold galls of *Maytenus senegalensis* (Lam.) Excell. (Celastraceae), induced by *Alocothrips hadrocerus* (Karny) (Thysanoptera : Insecta). Proceedings of the Indian Academy of Sciences, Section B (Plant Sciences), 88 (2) : 103–107, figs. 1–5. Bangalore.

Ananthakrishnan, T. N. 1980a. Thrips. Pp. 149–164, figs. 1–2, tables I–II. In : Harris KF & Maramorosch K (eds.), Vectors of Plant Pathogens, 467 pp. Academic Press: New York.

Ananthakrishnan, T. N. 1980b. On some aspects of thrips galls. Bulletin de la Société Botanique de France, Actualités Botaniques, 127 (1) : 31–34, figs. 1–13.

Ananthakrishnan, T. N. 1980c. Thrips and agroecosystems. Proceedings of the Symposium on Environmental Biology, Trivandrum, pp. 82–85, pl. 1, fig. 1.

Ananthakrishnan, T. N. 1980d. Thrips, weeds and crops. Bulletin of Entomology (1978), 19 (1–2) : 14–17. Entomological Society of India: New Delhi.

Ananthakrishnan, T. N. & Sen, S. 1980. Taxonomy of Indian Thysanoptera. Zoological Survey of India, Handbook Series, No. 1, vii + 234 pp., text-figs. 1–16 and pls. I–XIX. Calcutta. [Text-figs. 12–16 and the plates appear on 25 unnumbered pages after p. 234].

Ananthakrishnan, T. N. & Swaminathan, S. 1980. A new genus and two new species of thrips inhabiting *Eugenia* galls (Phlaeothripidae : Thysanoptera). Bulletin of Zoological Survey of India, 2 (2–3) : 207–211, figs. 1–2, pl. III. Calcutta.

Haridass, E. T. & Ananthakrishnan, T. N. 1980a. Models for the predatory behaviour of some reduviids from Southern India (Insecta–Heteroptera–Reduviidae). Proceedings of the Indian Academy of Sciences (Animal Sciences), 89 (4) : 387–402, figs. 1–6. Bangalore.

Haridass, E. T. & Ananthakrishnan, T. N. 1980b. Functional morphology of the fossula spongiosa in some reduviids (Insecta–Heteroptera–Reduviidae). Proceedings of the Indian Academy of Sciences (Animal Sciences), 89 (5) : 457–466, figs. 1–5, table 1. Bangalore.

Haridass, E. T. & Ananthakrishnan, T. N. 1980c. Bionomics and behaviour of the reduviid bug, *Triatoma rubrofasciata* (De Geer), the vector of *Trypanosoma* (*Megatrypana*) *conorhini* (Donovan), in India (Insecta : Heteroptera). Proceedings of the Indian National Science Academy, (Biological Sciences, Part B), 46 (6) : 884–891, figs. 1–3, pl. 1. New Delhi.

Ananthakrishnan, T. N. (editor). 1981a. Recent Advances in Entomology in India. 161 pp. Entomology Research Institute, Loyola College, & S. Viswanathan : Madras.

Ananthakrishnan, T. N. 1981b. Glimpses of recent trends in entomological research. Pp. 1–4. In : Ananthakrishnan TN (ed.), Recent Advances in Entomology in India, 161 pp. Loyola College & S. Viswanathan : Madras.

Ananthakrishnan, T. N. 1981c. Impact of changing agroforestry practices on the environment of insects. Pp. 148–151. In : Sharma AK & Sharma A (eds.), Impact of the Development of Science and Technology on Environment. Indian Science Congress Association: Calcutta.

Ananthakrishnan, T. N. 1981d. Indian faunal wealth. Pp. 131–134. In : Majumdar EA (ed), Indian Heritage. Indian Council for Social Welfare : Bombay.

Ananthakrishnan, T. N. 1981e. Thrips–plant gall association with special reference to patterns of gall diversity in relation to varying thrips populations. Proceedings of the Indian National Science Academy, (Biological Sciences Part B), 47 (1) : 41–46, figs. 1–7, table 1. New Delhi.

Ananthakrishnan, T. N. & Daniel, A. M. 1981. Behavioural components in feeding, reproduction and dispersal of the grass seed-feeding thrips *Chirot-*

hrips mexicanus Crawford. Current Science, 50 (16) : 733–735, figs. 1–2, pl. 1. Bangalore.

Ananthakrishnan, T. N.; Varatharajan, R; & Gopinathan, K. 1981a. Pollination in *Wedelia chinensis* (Osbeck) Merr. and *Tridax procumbens* L. (Compositae) by thrips (Thysanoptera : Insecta). Proceedings of the Indian National Science Academy, (Biological Sciences Part B), 47 (2) : 159–165, table 1, figs. 1–3. New Delhi.

Ananthakrishnan, T. N.; Varatharajan, R; & Gopinathan, K. 1981b. Seasonal periodicity of thrips infesting some Compositae in relation to pollination. Proceedings of the Indian National Science Academy, (Biological Sciences Part B), 47 (6) : 811–815, figs. 1–2. New Delhi.

Gopinathan, K; Varatharajan, R.; & Ananthakrishnan, T. N. 1981. Incidence of *Microcephalothrips abdominalis* (Crawford) in relation to the pollination biology of the weed *Ageratum conyzoides* Linn. (Compositae). Proceedings of the Indian National Science Academy, (Biological Sciences Part B), 47 (4) : 505–509, fig. lA–D. New Delhi.

Haridass, E. T. & Ananthakrishnan, T. N. 1981a. Functional morphology of the salivary system in some Reduviidae (Insecta–Heteroptera). Proceedings of the Indian Academy of Sciences (Animal Sciences), 90 (2) : 145–160, figs. 1–5, table 1. Bangalore.

Haridass, E. T. & Ananthakrishnan, T. N. 1981b. Functional morphology of the pylorus and rectal glands in Reduviidae (Insecta–Heteroptera). Proceedings of the Indian Academy of Sciences (Animal Sciences), 90 (4) : 483–493, figs. 1–5. Bangalore.

Mukhopadhyay, A.; Ananthakrishnan, T.N.; & Thangavelu, K. 1981. Record of new host plants of *Oxycarenus laetus* Kirby (Lygaeidae : Heteroptera) with notes on its ecology from eastern India. Bulletin of Zoological Survey of India, 4 (1) : 117–118. Calcutta.

Muralirangan, M. C. & Ananthakrishnan, T. N. 1981a. *Eyprepocnemis alacris alacris* (Serville) – biology and functional morphology of the foregut in relation to feeding (Orthoptera : Acrididae). Proceedings of the Indian Academy of Sciences, (Animal Sciences), 90 (1) : 99–112, figs. 1–4. Bangalore.

Muralirangan, M. C. & Ananthakrishnan, T. N. 1981b. Aspects of population ecology and ecological energetics in Indian acridids. Pp. 61–65. In : Ananthakrishnan TN (ed.), Recent Advances in Entomology in India, 161 pp. Entomology Research Institute, Loyola College: Madras.

Raman, A. & Ananthakrishnan, T. N. 1981a. Recent advances in cecidological research in India. Pp. 67–77. In : Ananthakrishnan TN (ed.), Recent Advances in Entomology. Entomology Research Institute, 161 pp. Loyola College, Madras.

Swaminathan, S. & Ananthakrishnan, T. N. 1981b. Oviposition behaviour in two species of dryinid parasites (Dryinidae, Hymenoptera). Proceedings of the Indian Academy of Sciences (Animal Sciences), 90 (1) : 113–116, fig. 1. Bangalore.

Swaminathan, S.; Ananthakrishnan, T. N.; & Babujee, C. R. 1981. Epidermal morphology of five leaf galls caused by thrips (Thysanoptera : Insecta). Proceedings of the Indian National Science Academy, (Biological Sciences Part B), 47 (5) : 607–613, figs. 1–4, table 1. New Delhi.

Varadarasan, S. & Ananthakrishnan, T. N. 1981. Population dynamics and prey–predator/parasite relationships of gall-forming thrips. Proceedings of the Indian National Science Academy, (Biological Sciences Part B), 47 (3) : 321–340, figs. 1–6, tables 1–6. New Delhi.

Ananthakrishnan, T. N. 1982a. Species dynamics in mycophagous thrips of the Oriental region (Thysanoptera). Entomologia Generalis, 8 (1) : 35–41, figs. 1–11. Stuttgart and New York.

Ananthakrishnan, T. N. 1982b. Thrips and pollination biology. Current Science, 51 (4) : 168172, fig. 1, table 1. Bangalore.

Ananthakrishnan, T. N. 1982d, 1986e, 1989f. Bioresources Ecology. viii + 159 pp., 172 pp. (revised ed.), xii + 226 pp. (3rd ed.). Oxford & IBH: New Delhi

Ananthakrishnan, T. N., Daniel, A. M., & Suresh Kumar, N. 1982. Spatial and seasonal distribution patterns of some phytophagous thrips (Thysanoptera : Insecta) infesting *Ricinus communis* Linn. (Euphorbiaceae) and *Achyranthes aspera* Linn. (Amarantaceae). Proceedings of the Indian National Science Academy, (Biological Sciences Part B), 48 (2) : 183–189, figs. 1–4. New Delhi.

Swaminathan, S. & Ananthakrishnan, T. N. 1982. New natural enemy complex of some fulgoroids (Insecta : Homoptera) with biological studies of three hymenopterous parasites (Insecta: Hymenoptera). Proceedings of the Indian Academy of Sciences (Animal Sciences), 91 (2) : 177–187, figs. 1–4, tables 1–2. Bangalore.

Varadarasan, S. & Ananthakrishnan, T. N. 1982. Biological studies on some gall-thrips. Proceedings of the Indian National Science Academy, (Biological Sciences Part B), 48 (1) : 35–43, figs. 1–6, tables 1–2. New Delhi.

Varatharajan, R.; Gopinathan, K.; & Ananthakrishnan, T. N. 1982. Comparative efficiency of thrips in relation to other foraging insects in the pollination of *Cosmos bipinnatus* Cav. (Compositae). Proceedings of the Indian National Science Academy, (Biological Sciences Part B), 48 (6) : 735–739, fig. A–E, Table 1, New Delhi.

Ananthakrishnan, T. N. 1983. Adaptive strategies in cecidogenous Thysanoptera. Current Science, 52 (2) : 77–79, figs. 1–4. Bangalore.

Ananthakrishnan, T. N., Dhileepan, K., & Padmanabhan, B. 1983. Reproductive strategies and behavioural attributes in some sporophagous Idolothripinae (Tubulifera : Thysanoptera). Proceedings of the Indian Academy of Sciences (Animal Sciences), 92 (2) : 95–107, figs. 1–4, table 1. Bangalore.

Ananthakrishnan, T. N. & James, P. W. 1983. Feeding preferences in some sporophagous Idolothripinae (Tubulifera : Thysanoptera). Proceedings of the Indian National Science Academy, (Biological Sciences Part B), 49 (2) : 86–92, table 1, figs. 1–11. New Delhi.

Ananthakrishnan, T. N.; Padmanabhan, B.; & Dhileepan, K. 1983. Gut spore composition and influence of fungal host on the rate of mortality and post-embryonic development in *Tiarothrips subramanii* (Ramk.). Proceedings of the Indian Academy of Sciences (Animal Sciences), 92 (1) : 11–17, figs. 1–3, tables 1–2. Bangalore.

Ananthakrishnan, T. N.; Sanjayan, K. P.; & Raman, K. 1983. Impact of host plant diversity on the growth and fecundity of *Spilostethus hospes* (Fabricius) (Hemiptera : Lygaeidae). Annals of Entomology, 1 (1) : 1–6, figs. 1–2, tables 1–3. Dehra Dun.

Ananthakrishnan, T. N. & Suresh, G. 1983. Patterns of fungal resource utilization and feeding range in some mycophagous Tubulifera (Insecta : Thysanoptera). Proceedings of the Indian Academy of Sciences (Animal Sciences), 92 (4) : 285–291, figs. 1–2, table 1. Bangalore.

Raman, A. & Ananthakrishnan, T. N. 1983a. Studies on some thrips (Thysanoptera : Insecta) induced galls. 1. Developmental morphology. Proceedings of the Indian National Science Academy, (Biological Sciences Part B), 49 (5) : 313–358, pls. 1–2, figs. 1–22, table 1. New Delhi.

Raman, A. & Ananthakrishnan, T. N. 1983b. Studies on some thrips (Thysanoptera : Insecta) induced galls. 2. Fine structure of the nutritive zone. Proceedings of the Indian National Science Academy, (Series B), 49 (6) : 525–561, pls. 1–24. New Delhi.

Raman, A. & Ananthakrishnan, T. N. 1983c. On the developmental morphology of the rosette galls of *Acacia leucophloea* Willd. (Mimosaceae) induced by *Thilakothrips babuli* Ramk. (Thysanoptera : Insecta). Proceedings of the Indian Academy of Sciences (Animal Sciences), 92 (5) : 343–350, figs. 1–27. Bangalore.

Suresh, G. & Ananthakrishnan, T. N. 1983. Association of some sporophagous thrips in forest communities with plant pathogenic fungi, with some observations on their natural control agents. Indian Journal of Forestry, 6 (3) : 196–201, table 1, pl. 1 (figs. 1–6).

Ananthakrishnan, T. N. 1984a. Bioecology of Thrips. iv + 233 pp., 39 figs. Indira Publishing House: Michigan (U.S.A.).

Ananthakrishnan, T. N. (editor). 1984b. Biology of Gall Insects. xi + 362 pp. Oxford and IBH : New Delhi. [Thysanoptera : pp. 107–127, Raman A & Ananthakrishnan 1984; in Section 2].

Ananthakrishnan, T. N. 1984c. Adaptive strategies in cecidogenous insects. Pp. 1–9, figs. 1–3. In:Ananthakrishnan TN (ed.), Biology of Gall Insects, 362 pp. Oxford & IBH: New Delhi.

[Ananthakrishnan, T. N. 1984d]. (Anonymous) 1984. Entomology Research Institute, Loyola College, Madras. 17 pp.

This report has no author, but Ananthakrishnan as Director of ERI is obviously the author and publisher. The history of Entomology Research Institute is outlined, pp. 1–3. Publications from the Institute, 1975–1984, are listed under various heads, pp. 4–11, i.e. Insect–host plant interactions, Host-parasite–predator interactions,

Ecological Interactions and reproductive polymorphism in mycophagous insects, Cecidology, Pollination biology, Insect vector potential.

Ananthakrishnan, T. N. & Dhileepan, K. 1984. Thrips–fungus association with special reference to the sporophagous *Bactrothrips idolomorphus* (Karny) (Tubulifera : Thysanoptera). Proceedings of the Indian Academy of Sciences (Animal Sciences), 93 (3) : 243–249, fig. 1A–F. Bangalore.

Ananthakrishnan, T. N.; Padmanabhan, B.; Dhileepan, K.; & Suresh, G. 1984. Ecological interactions, species dynamics and reproductive biology of some mycophagous Thysanoptera. Entomology Research Institute, Occasional Publication No. 5 : 37 pp. + 18 unnumbered pages, 4 figs., 18 pls. Madras.

Ananthakrishnan, T. N. & Suresh, G. 1984. Thrips–fungus interactions with special reference to their vector potential. Proceedings of the Indian Academy of Sciences (Animal Sciences), 93 (4) : 315–322, figs. 1–4, table 1. Bangalore.

Parameswaran, R.; Ananthakrishnan, T. N.; & Ananthasubramanian, K. S. 1984. Outlines of Animal Physiology. 404 pp. Revised ed. S. Viswanathan: Madras.

Raman, A. & Ananthakrishnan, T. N. 1984a. Biology of gall thrips (Thysanoptera: Insecta). Pp. 107–127, figs. 1–2, tables 1–3. In: Ananthakrishnan TN (ed.), Biology of Gall Insects, 362 pp. Oxford and IBH: New Delhi.

Raman, A. & Ananthakrishnan T. N. 1984b. Adaptive strategies in insect gall systems. Pp. 111–112. In : Arya HC (ed.), Advancing Frontiers in Plant Sciences. Scientific Publishers: Jodhpur.

Suresh, G. & Ananthakrishnan, T. N. 1984. Some entomogenous fungal infections of a sporophagous tubuliferan *Elaphrothrips denticollis* (Insecta : Thysanoptera). Proceedings of the Indian Academy of Sciences (Animal Sciences), 93 (5) : 407–410, fig. 1A–F. Bangalore.

Suresh Kumar, N. & Ananthakrishnan, T. N. 1984. Predator–thrips interactions with reference to *Orius maxidentex* Ghauri and *Carayonocoris indicus* Muraleedharan (Anthocoridae : Heteroptera). Proceedings of the Indian National Science Academy, (Biological Sciences Part B), 50 (2) : 139–145, tables 1–3, figs. 1–2. New Delhi.

Swaminathan, S. & Ananthakrishnan, T. N. 1984. Population trends of some monophagous and polyphagous fulgoroids in relation to biotic and abiotic factors (Insecta: Homoptera). Proceedings of the Indian Academy of Sciences (Animal Sciences), 93 (1) : 1–8, figs. 1–3, tables 1–3. Bangalore.

Ananthakrishnan, T. N. 1985a. Resource partitioning in some phytophagous communities of thrips in terms of spatial and seasonal distribution. Pp. 33–37. In: Srivastava CBL & Goel SC (eds.), Recent Advances in Zoology. Rastogi & Co.: Meerut.

Ananthakrishnan, T. N. 1985b. Aspects of chemical ecology in insect host–parasitoid interactions. Proceedings of the National Seminar on Entomophagous Insects, Calicut, pp. 7–9. [Keynote address].

Ananthakrishnan, T. N., Dhileepan, K., & Padmanabhan, B. 1985. Behavioural responses in terms of feeding and reproduction in some grasshoppers (Orthoptera: Insecta). Proceedings of the Indian Academy of Sciences (Animal Sciences), 94 (5) : 443–461, figs. 1–8, tables 1–4. Bangalore

Ananthakrishnan, T. N. & Ramamurthi, B. N. 1965. Gall-inhabiting tubuliferous Thysanoptera – I. Journal of Bombay Natural History Society, 62 (2) : 266–278, figs. 1–16. Bombay.

Ananthakrishnan, T. N. & Sureshkumar, N. 1985. Anthocorids (Anthocoridae: Heteroptera) as efficient biocontrol agents of thrips (Thysanoptera : Insecta). Current Science, 54 (19) : 987–990, fig. 1A–D, table 1. Bangalore.

Gopinathan, K & Ananthakrishnan, T.N. 1985. Morphogenesis and histochemistry of some thrips (Thysanoptera : Insecta) –induced galls. Proceedings of the Indian National Science Academy, (Biological Sciences Part B), 51 (4) : 413–456, tables 1–5, figs. 1–2, pls. 1–20. New Delhi.

Raman, A. & Ananthakrishnan, T. N. 1985. Cecidogenous *Crotonothrips* (Thysanoptera) –*Memecylon* interactions : Host relations, nutritive tissue, tissue dynamics and cecidogenetic patterns. Proceedings of the Indian Academy of Sciences (Plant Sciences), 95(2) : 103–120, figs. 1–50. Bangalore.

Suresh Kumar, N. & Ananthakrishnan, T. N. 1985. *Geocoris ochropterus* Fabr. as a predator of some thrips. Proceedings of the Indian National Science Academy, (Biological Sciences Part B), 51 (2) : 185–193, tables 1–3, figs. 1–2. New Delhi.

Swaminathan, S.; Randhawa, N. S.; & Ananthakrishnan, T. N. (editors). 1995. Second Agricultural Science Congress, Proceedings. viii + 271 pp. National Academy of Agricultural Sciences : IARI, New Delhi.

Ananthakrishnan, T. N. (editor) 1986a. Dynamics of Insect–Plant Interactions. 170 pp. Loyola College: Madras. 7 chapters (pp. 1–117) have Ananthakrishnan as author/coauthor; references, pp. 140–170. [Thysanoptera : pp. 52–65, Ananthakrishnan, Sanjayan, & Daniel 1986; pp. 85–101, Ananthakrishnan & Gopinathan 1986a (in Section 2); pp. 102–117, Raman A & Ananthakrishnan 1986 (Section 4)].

Ananthakrishnan, T. N. 1986b. Nature of insect–plant interactions. Pp. 1–15. In : Ananthakrishnan TN (ed.), Dynamics of Insect–Plant Interactions, 170 pp. Loyola College: Madras.

[Ananthakrishnan 1986c (editor)] (Anonymous) 1986. Entomology Research Institute, Research Resumé, 38 pp. Loyola College: Madras.

[Although this report is anonymous, Ananthakrishnan is obviously the editor and publisher, as in the case of the subsequent reports, i.e. Ananthakrishnan 1988d, 1992g, 1996j]. [Insect–weed–crop interactions, pp. 9–13, 3 photos, 1 fig. (Thysanoptera, pp. 9–11). Insect–fungal interactions, pp. 13–17, 8 photos, 1 fig. (Thysanoptera). Behavioral ecology of predators and parasites, pp. 17–23, 7 photos (Thysanoptera, pp. 17, 18, 20. Pollination biology, pp. 23–28, 4 photos, 1 fig. (Thysanoptera). Cecidological studies, pp. 28–32, 6 photos (Thysanoptera).

Ananthakrishnan, T. N. 1986d. Bioresources Ecology. viii + 159 pp., 172 pp. (revised ed.), Oxford & IBH: New Delhi

Ananthakrishnan, T. N. 1986e. Adaptive specialization in relation to niche diversity in phytophagous and mycophagous thrips. Journal of Bombay Natural History Society, 83 Supplement: 72–77, text-fig. 1, pls. 1–3. Bombay.

(Anonymous) 1986 [Ananthakrishnan, T. N. 1986c (ed.)]. Entomology Research Institute. Research Resumé, 38 pp. Loyola College: Madras

Ananthakrishnan, T. N. & Balu, A. 1986. An unusual instance of a trophocyte and cytoplasmic nutritive link in the developing oocytes of *Dolicholepta gracilipes* Ramk and Marg (Thysanoptera : Phlaeothripidae). Current Science, 55 (22) : 1152–1154, fig. 1A–F. Bangalore.

Ananthakrishnan, T.N. & Gopinathan, K. 1986a. Insect pollinator–plant interactions. Pp. 85–101, fig. 8. In: Ananthakrishnan TN (ed.) Dynamics of Insect–Plant Interactions, 170 pp. Entomology Research Institute, Loyola College: Madras.

Ananthakrishnan, T.N. & Gopinathan, K. 1986b. Population-correlated morphism in thripsinduced galls of *Calycopteris floribundus* Lamk. Current Science, 55 (20) : 1049–1050, fig. 1A–C, table 1. Bangalore.

Ananthakrishnan, T.N. & Raman, K. 1986. Receptor systems and sensory perception in phytophagous insects. Pp. 37–51, figs. 2–5. In : Ananthakrishnan TN (ed.), Dynamics of Insect–Plant Interactions, 170 pp. Loyola College: Madras.

Ananthakrishnan, T. N.; Sanjayan, K. P.; & Daniel, A. M. 1986. Insect–weed–crop interactions. Pp. 52–65, figs. 6–7. In: Ananthakrishnan TN (ed.), Dynamics of Insect Plant Interactions, 170 pp. Loyola College: Madras. [Thysanoptera, pp. 56, 57–60, figs. 6–7].

Ananthakrishnan, T. N.; Sanjayan, K. P.; & Suresh Kumar, N. 1986. Host plant preferences in *Cyrtacanthacris ranacea* Stoll in some malvaceous hosts in terms of food utilization. Proceedings of the Indian National Science Academy, (Biological Sciences, Part B), 52(3): 351–357, figs. 1–2, tables 1–3. New Delhi.

Ananthakrishnan, T. N.; Suresh Kumar, N.; & Sanjayan, K. P. 1986. Sensillar diversity, density and distribution during post-embryonic development of *Cyrtacanthacris ranacea* Stoll and their role in feeding. Proceedings of the Indian Academy of Sciences (Animal Sciences), 95 (2) : 117–124, figs. 1–3. Bangalore.

Dhileepan, K. & Ananthakrishnan, T. N. 1986. The pleuropodium in the embryos of two species of viviparous sporophagous species of tubuliferan thrips (Thysanoptera : Insecta). Current Science, 55 (21) : 1105–1108, figs. 1–2. Bangalore.

Raman, A. & Ananthakrishnan, T. N. 1986a. Insect–plant gall interactions. Pp. 102–117, figs. 9–11. In : Ananthakrishnan TN (ed.), Dynamics of Insect–Plant Interactions, 170 pp. Loyola College: Madras.[Thysanoptera : pp. 102–104, 108, 110–111, 112, 114–115].

Raman, K. & Ananthakrishnan, T. N. 1986b. Mechanisms of host plant selection in phytophagous insects. Pp. 16–36, fig. 1. In : Ananthakrishnan TN (ed.), Dynamics of Insect–Plant Interactions, 170 pp. Loyola College: Madras.

Raman, K. & Ananthakrishnan, T. N. 1986c. Insect–plant interactions with respect to resistance of host plants to insect attack. Pp. 66–84. In: Ananthakrishnan TN (ed.), Dynamics of Insect – Plant Interactions, 170 pp. Loyola College: Madras.

Ananthakrishnan, T. N. 1987a. Allelochemics in insect–plant interactions. Pp. 77–81. In: Alternatives to Synthetic Insecticides in Integrated Pest Management Systems. (Proceedings of Symposium held at Madurai, March 30–31, 1987). [Guest Lecture].

Ananthakrishnan, T.N. 1987b. Biosystematic studies in India – progress, prospects and constraints. Insect Science and its Application, 8(4–6) : 423–424. ICIPE Science Press: Nairobi (Kenya). [Thysanoptera : p. 424].

Ananthakrishnan, T.N. 1987c. Implications of polymorphism in Thysanoptera. Insect Science and its Application, 8 (4–6) : 435–439, tables 1–2. ICIPE Science Press: Nairobi (Kenya).

Chandraguru, T.; Ananthakrishnan, T.N.; & Gopinathan, K. 1987. Age-correlated biochemical changes in two thrips-induced galls. Current Science, 56 (6) : 283–287, tables 1–2, figs. 1–2. Bangalore.

Dhileepan, K. & Ananthakrishnan, T. N. 1987a. Ovarian polymorphism in relation to reproductive diversity and associated histological and histochemical attributes in some sporophagous tubuliferan Thysanoptera. Proceedings of the Indian Academy of Sciences (Animal Science), 96 (1) : 1–13, figs. 1–4. Bangalore.

Dhileepan, K. & Ananthakrishnan, T. N. 1987b. Impact of sex-limited and alary polymorphism on spermathecal diversity and reproductive behaviour in some mycophagous Tubulifera (Insecta : Thysanoptera). Proceedings of the Indian National Science Academy, (Biological Sciences Part B), 53 (4) : 329–336, pls. 1–2, figs. 1–3, tables 1–4. New Delhi.

Gopinathan, K & Ananthakrishnan, T. N. 1987. Population density correlated morphism and growth in some thrips galls. Proceedings of the Indian National Science Academy, (Biological Sciences Part B), 53 (1) : 11–20, figs. 1–5, tables 1–5. New Delhi.

Sanjayan, K. P. & Ananthakrishnan, T. N. 1987a. Analysis of lygaeid–host seed interaction in terms of quantitative food utilisation, feeding behaviour and some chemical parameters involved in feeding preferences. Phytophaga, 1 : 35–55, tables 1–7, figs. 1–2, pls. 1–3. Oxford & IBH : New Delhi.

Sanjayan, K. P. & Ananthakrishnan, T. N. 1987b. Host preference of some acridids (Insecta : Orthoptera) in relation to some biochemical parameters. Proceedings of the Indian Academy of Sciences (Animal Sciences), 96 (1) : 15–21, table 1, fig. 1. Bangalore.

Suresh, G. & Ananthakrishnan, T.N. 1987. Biochemical correlates in *Rhipiphorothrips cruentatus – Terminalia catappa* interactions with special reference to leaf infestation patterns. Proceedings of the Indian Academy of Sciences (Animal Sciences), 96 (2) : 117–127, figs. 1–4, tables 1–2. Bangalore.

Sureshkumar, N. & Ananthakrishnan, T. N. 1987. Biotic interactions in relation to prey–predator relationship with special reference to some thrips species (Thysanoptera : Insecta). Journal of Entomological Research, New Delhi, 11 (2) : 192–202, tables 1–9.

Ananthakrishnan, T. N. 1988a. Oocyte–follicle cell dynamics in *Arrhenothrips ramakrishnae* Hood (Insecta: Thysanoptera) – I. Proceedings of the Indian Academy of Sciences (Animal Sciences), 97 (1) : 1–12, figs. 1–10. Bangalore.

Ananthakrishnan, T.N. 1988b. Oocyte–follicle epithelial cell dynamics in *Arrhenothrips ramakrishnae* Hood (Insecta: Thysanoptera) – II. Observations on the preterminal oocytes. Proceedings of the Indian Academy of Sciences (Animal Sciences), 97 (1) : 13–27, figs. 1–11. Bangalore.

Ananthakrishnan, T. N. 1988c. Electron microscopic studies on the follicle epithelial celloocyte cortex interface in *Arrhenothrips ramakrishnae* Hood (Insecta: Thysanoptera). Proceedings of the Indian Academy of Sciences (Animal Sciences), 97 (4) : 289–299, figs. 1–7. Bangalore.

Ananthakrishnan, T. N. 1988d. Micropinocytosis in the oocyte of *Arrhenothrips ramakrishnae* Hood (Insecta: Thysanoptera). Current Science, 57 (22) : 1243–1246, figs. 1–3. Bangalore.

Ananthakrishnan, T. N. 1988e. Insect–plant interactions – Problems and perspectives. Pp. 1–11. In: Ananthakrishnan TN & Raman A (eds.). Dynamics of Insect–Plant Interaction: Recent Advances and Future Trends, 223 pp. Oxford & IBH: New Delhi.

Ananthakrishnan, T. N. 1988f. Insect–plant interactions – Future outlook. Pp. 219–223. In: Ananthakrishnan TN & Raman A (eds.). Dynamics of Insect–Plant Interaction : Recent Advances and Future Trends, 223 pp. Oxford & IBH: New Delhi.

Ananthakrishnan, T. N. 1988g. Dynamics of the litter ecosystem in natural and interfered forests. International Journal of Ecology and Environmental Sciences, 14 (1) : 61–66, figs. 1–2. New Delhi.

[Mycophagous thrips in litter: p. 63, fig. 1 on p. 62, fig. 2 on p. 64].

Ananthakrishnan, T.N. 1988d (editor). Research Highlights 1963–1988, Entomology Research Institute, Loyola College, Madras. iii + 55 pp. Madras. (Edited and published by the Director, ERI, Madras, October 1988).

[Names of contributing authors are not stated]. [Thysanoptera : pp. 4–8 (thrips–crop–weed interaction), 16–20 (thrips–parasite relationships), 22–32 (pollination biology), 36 (trophic diversity of litter insects), 43–55.

ChE. Suresh, G. & Ananthakrishnan, T. N. 1988. Leaf-age correlated changes in oxidative enzymes in *Retithrips syriacus* (Mayet) infested *Ricinus communis*. Current Science, 57 (13) : 744–746, figs. 1–2. Bangalore.

Ananthakrishnan, T. N. 1989a. Perspectives in Insect Mycophagy. Dr. C. P. Alexander First Memorial Lecture. 17 pp. Department of Zoology, University of Delhi: Delhi. [Thrips mycophagy: pp. 10–13].

Ananthakrishnan, T. N. 1989b. Bioresources Ecology. xii + 226 pp. (3rd ed.). Oxford & IBH: New Delhi.

Ananthakrishnan, T. N & Raman, A. 1989. Thrips and Gall Dynamics. iv+ 120 pp. Oxford & IBH : New Delhi. (Simultaneously distributed by E. J. Brill, Leiden, Netherlands).

Ananthakrishnan, T. N. 1990a. Reproductive Biology of Thrips. 158 pp. Indira Publishing House: Michigan (U.S.A.).

Ananthakrishnan, T. N. 1990b. Integrated entomology: fact or fiction? Current Science, 59 (3) : 138–142, 1 fig. Bangalore.

Ananthakrishnan, T. N. 1990c. Facets of chemical ecology in insect–plant interactions: An overview. Proceedings of the Indian Academy of Sciences (Animal Sciences), 99 (3) : 177–183. Bangalore.

Ananthakrishnan, T. N. 1990d. Workshop Manual on Insect–Plant Interactions. 122 pp. Entomology Research Institute, Loyola College: Madras. [Interspersed, blank, are 53 of these pages].

Ananthakrishnan, T. N. 1990e. Entomology research in India: Thrust areas for the future. Current Science, 59 (7) : 342–343. Bangalore. Report on seminar held at the Entomology Research Institute, Madras, on 5 March 1990.

Ananthakrishnan, T. N. 1990f. Chemical ecology in biological control. Current Science, 59 (24) : 13191322, 1 fig. Bangalore.

Ananthakrishnan, T. N.; Annadurai, R. S.; Senrayan, R.; & Murugesan, S. 1990. Differential impact of tannic acid and pyrogallol on nutrition and reproduction of two co-existing pests, *Heliothis armigera* (Hubner) and *Spodoptera litura* (Fabricius) (Lepidoptera : Noctuidae). Phytophaga, 3 (1) : 55–70, tables 1–8, figs. 1–2. S. Viswanathan: Madras.

Ananthakrishnan, T. N. & Balu, A. 1990. Aspects of spermiogenesis in *Arrhenothrips ramakrishnae* Hood (Insecta : Thysanoptera). Proceedings of the Indian Academy of Sciences (Animal Sciences), 99 (2) : 91–102, figs. 1–12. Bangalore.

Ananthakrishnan, T. N.; Senrayan, R.; Annadurai, R. S.; & Murugesan, S. 1990. Antibiotic effects of resorcinol, gallic acid and phloroglucinol on *Heliothis armigera* Hubner (Insecta : Noctuidae). Proceedings of the Indian Academy of Sciences (Animal Sciences), 99 (1) : 39–52, figs. 1–2, tables 1–7. Bangalore.

Ananthakrishnan, T. N. (editor). 1991a. National Symposium on Emerging Trends in the Biological Control of Phytophagous Insects, January 19–21, 1991, Madras, India. Synopses. 115 pp. Entomology Research Institute, Loyola College: Madras.

Ananthakrishnan, T. N. 1991b. Emerging trends in the biological control of phytophagous insects: Some highlights. Pp. 1–4. In: Ananthakrishnan TN (ed.), National Symposium on Emerging Trends in the Biological Control of Phytophagous Insects, January 19–21, 1991, Madras, 115 pp.

Ananthakrishnan, T. N. 1991c. Chemical ecology and behavioural inputs in insect control. Pp. 109–114, fig. 1, pls. 1–2. In: Sen C & Dutta S (eds.), Biotechnology in Crop Protection (1989), xx + 292 pp. Bidhan Chandra Krishi Viswvidyalaya: Kalyani (West Bengal).

Ananthakrishnan, T. N. & Jacob, J. P. 1991. Bioenergetics in some acridids. Annals of Entomology, 9 (2) : 95–99, 1 table. Dehra Dun.

Ananthakrishnan, T. N.; Senrayan, R.; Murugesan, S. & Annadurai, R. S. 1991. Kairomones of *Heliothis armigera* and *Corcyra cephalonica* and their influence on the parasitic potential of *Trichogramma chilonis* (Trichogrammatidae : Hymenoptera). Journal of Biosciences, 16 (3) : 111–119, figs. 1–5, tables 1–3. Bangalore.

Dhileepan, K. & Ananthakrishnan, T. N. 1991. Patterns of life cycles in relation to reproductive diversity in some polymorphic mycophagous thrips (Tubulifera: Thysanoptera). Proceedings of the Zoological Society, 44 (1) : 33–38, figs. 1–3, tables 1–2. Calcutta.

Ananthakrishnan, T. N. 1992a. Dimensions of Insect–Plant Interactions. 184 pp. Oxford & IBH: New Delhi.

[Thysanoptera : pp. 83–86 (weed–crop interactions); 96–103 (thrips as pollinators); 109–118, 121–122 (galls); 146–184 (references)].

Ananthakrishnan, T. N. (editor). 1992b. Emerging Trends in Biological Control of Phytophagous Insects. xii + 255 pp. Oxford & IBH: New Delhi.

Ananthakrishnan, T. N. 1992c. Introduction. Pp. ix–xii. In: Ananthakrishnan TN (ed.), Emerging Trends in Biological Control of Phytophagous Insects, 255 pp. Oxford & IBH: New Delhi.

Ananthakrishnan, T. N. 1992d. Chemical ecology in biological control. Pp. 59–67, fig. 1. In: Ananthakrishnan TN (ed.), Emerging Trends in Biological Control of Phytophagous Insects, 255 pp. Oxford & IBH: New Delhi.

Ananthakrishnan, T.N. 1992e. Chemodynamics of insect–plant interactions. Palaeobotanist, 41: 144–148. Birbal Sahni Institute of Palaeobotany: Lucknow.

Ananthakrishnan, T. N. 1992f. Biotechnology in insect control. Current Science, 62 (12) : 766–768, figs. 1–2. Bangalore.

Ananthakrishnan, T. N. 1992g (editor). Research Highlights, Decennium 1982–1992, Entomology Research Institute, Loyola College, Madras. 40 pp.

[Thysanoptera : *Retithrips syriacus* (p. 3, role of secondary metabolites of host plants in regulating life cycle patterns of the thrips populations; p. 5, gas chromatographic analysis of host leaves and anal exudates of the thrips on different hosts); pp. 18–20, *Arrhenothrips ramakrishnae*, structural and behavioural changes of developing oocytes; spermiogenesis); p. 22, thrips galls, *Austrothrips cochinchinensis* on Calycopteris floribundus, *Liothrips viticola* on Vitis lanceolaria, *Phorinothrips loranthi* on Loranthus elasticus; pp. 27–40, publications 1982–1992.].

Ananthakrishnan, T. N. 1992h. Unique aspects in the biology of thrips-induced galls. Pp. 185–195, figs. 1–10, tables 1–2. In: Shorthouse JD & Rohfritsch O (eds.), Biology of Insect-Induced Galls, 285 pp. Oxford University Press: New York.

Ananthakrishnan, T. N.; Gopichandran, R. & Gurusubramanian, G. 1992. Influence of chemical profiles of host plants on the infestation diversity of *Retithrips syriacus.* Journal of Biosciences, 17 (4) : 483–489, tables 1–3. Bangalore.

Ananthakrishnan, T. N.; Gopichandran, R. & Subramaniam, V. R. 1992. Chemical ecology of litter breakdown in *Tectona grandis* and its impact on the dynamics of natural insect communities at varying altitudes. International Journal of Ecology and Environmental Sciences, 18 : 63–74, figs. 1–3, tables 1–4. New Delhi.

Ananthakrishnan, T. N.; Raman, K. & Sanjayan, K. P. 1982. Comparative growth rate, fecundity and behavioural diversity of the dusky cotton bug, *Oxycarenus hyalinipennis* Costa (Hemiptera : Lygaeidae) on certain malvaceous host plants. Proceedings of the Indian National Science Academy, (Biological Sciences, Part B), 48 (5) : 577–584, figs. 1–2, tables 1–3. New Delhi.

Ananthakrishnan, T. N. & Senrayan, R. 1992. Phytochemical-indúced responses, a vital factor governing insect host –parasite/predator interactions. Phytophaga, 4 (2) : 87–94. S. Viswanathan: Madras.

Annadurai, R. S.; Murugesan, S.; Senrayan, R.; Gurusubramanian, G. & Ananthakrishnan, T. N. 1992. Tritrophic interactions in *Heliothrips armigera* Hubner and its natural enemy systems: A chemical ecological approach (Noctuidae : Lepidoptera). Pp. 83–101, figs. 1–5, tables 1–10. In : Ananthakrishnan TN (ed.), Emerging Trends in Biological Control of Phytophagous Insects, 255 pp. Oxford & IBH: New Delhi.

Gopichandran, R.; Gurusubramanian, G. & Ananthakrishnan, T. N. 1992. Influence of chemical and physical factors on the varietal preference of *Retithrips syriacus* (Mayet) (Panchaetothripinae : Thysanoptera) on different cultivars of cotton. Phytophaga, 4(1): 1–9, tables 1–6. S. Viswanathan: Madras.

Senrayan, R.; Murugesan, S.; Annadurai, R. S. & Ananthakrishnan, T. N. 1992. Host nutrition and fitness of *Trichogramma chilonis* Ishii (Trichogrammatidae: Hymenoptera). Pp. 69–82, tables 1–8, figs. 1–4. In : Ananthakrishnan TN (ed.), Emerging Trends in Biological Control of Phytophagous Insects, 255 pp. Oxford & IBH : New Delhi.

Sivamani, E.; Rajendran, N.; Senrayan, R.; Ananthakrishnan, T.N. & Jayaraman, K. 1992. Influence of some plant phenolics on the activity of □-endotoxin of *Bacillus thuringiensis* var. *galleriae* on *Heliothis armigera.* Entomologia Experimentalis et Applicata, 63(3): 243–248. Kluwer Academic: Belgium.

Ananthakrishnan, T. N. 1993a. Bionomics of thrips. Annual Review of Entomology, 38 : 71–92. Annual Reviews Inc.: Palo Alto (California).

Ananthakrishnan, T. N. 1993b. The role of thrips in pollination. Current Science, 65 (3) : 262–264, fig. 1. Bangalore.

Ananthakrishnan, T. N. 1993c. Biological diversity – concepts and conservation strategies. Hexapoda, 5 (2) : 5–8. Madras.

Ananthakrishnan, T. N. 1993d. Insects in Social Forest Stands. 24 pp. (unnumbered), figs. 1–16 (colour photos). Entomology Research Institute, Loyola College: Madras.

Ananthakrishnan, T. N. 1993e. Background. Pp. 1–4. In : Chemical Ecology of Phytophagous Insects – The Changing Scenario, 30th Anniversary Commemorative Symposium, 17–19 Sept. 1993, Entomology Research Institute, Loyola College, Madras. (Abstracts of Presentations, 50 pp.).

Ananthakrishnan, T. N. 1993f. Changing dimensions in the chemical ecology of phytophagous insects: Role of infochemicals on their behavioural diversity. Pp. 5–6. In : Chemical Ecology of Phytophagous Insects – The Changing Scenario, 30th Anniversary Commemorative Symposium, 17–19 Sept. 1993, Entomology Research Institute, Loyola College, Madras. (Abstracts of Presentations, 50 pp.).

Ananthakrishnan, T. N. 1993g. Changing dimensions in the chemical ecology of phytophagous insects: Role of infochemicals on the behavioural diversity. Pp. 1–20, tables 1–3, fig. 1. In: Ananthakrishnan TN & Raman A (eds.), Chemical Ecology of Phytophagous Insects, 332 pp. Oxford & IBH: New Delhi

Ananthakrishnan, T. N. & Gopichandran, R. 1993. Chemical Ecology in Thrips–Host Plant Interactions. vi + 125 pp. Oxford & IBH: New Delhi.

Ananthakrishnan, T. N.; Gopichandran, R.; & Subramaniam, V.R. 1993. Dynamics of micro arthropod communities in relation to the chemical environment of decomposing *Eucalyptus globulus* litter. International Journal of Ecology and Environmental Sciences, 19 : 83–92, tables 1–6, figs. 1–2. New Delhi.

Ananthakrishnan, T. N. & Raman, A. (editors). 1993. Chemical Ecology of Phytophagous Insects. vi + 332 pp. Oxford & IBH : New Delhi.

Ananthakrishnan, T.N. & Thomas, S.K. 1993. Chemical ecology of two heterogeneous forest litter ecosystems and related trophic interactions of insect communities. International Journal of Ecology and Environmental Sciences, 19 : 143–160, tables 1–7, figs. 1–5. New Delhi.

Ananthakrishnan, T. N. & Wesley, S. D. 1993. Headspace volatiles of macerated parts of some cotton cultivars in relation to *Helicoverpa armigera* Hubner (Lepidoptera : Noctuidae). Phytophaga, 5(1): 63–65, table 1. S. Viswanathan: Madras.

Babu, A. & Ananthakrishnan, T. N. 1993. Predatory efficiency, reproductive potential, and bioenergetics of *Coccinella transversalis* F. and *Menochilus sexmaculatus* F. (Coleoptera : Coccinellidae) in relation to aphid (Homoptera: Aphididae) prey quality. Phytophaga, 5 (2) : 121–133, figs. 1–5, tables 1–3. S. Viswanathan: Madras.

Gopichandran, R. & Ananthakrishnan, T. N. 1993a. Chemical ecology in relation to forest litter dynamics. Pp. 12–13. Chemical Ecology of Phytophagous Insects – The Changing Scenario, 30th Anniversary Commemorative Symposium, 17–19 Sept. 1993, Entomology Research Institute, Loyola College, Madras. (Abstracts of Presentations, 50 pp.).

Gopichandran, R. & Ananthakrishnan, T. N. 1993b. Chemical ecology in relation to forest litter dynamics. Pp. 261–274, fig. 1, tables 1–2. In : Ananthakrishnan TN & Raman A (eds.), Chemical Ecology of Phytophagous Insects, 332 pp. Oxford & IBH: New Delhi.

Gurusubramanian, G. & Ananthakrishnan, T. N. 1993a. Biochemical implications of host preference in *Frankliniella schultzei* (Trybom), a polyphagous, anthophilous thrips. Phytophaga, 5 (2) : 77–95, tables 1–7, figs. 1–4. S. Viswanathan: Madras.

Gurusubramanian, G. and Ananthakrishnan, T.N. 1994b. The pollen vector potential of anthophilous thrips. Journal of Advanced Zoology, 15 (1) : 6–17.

Gurusubramanian, G.; Gopichandran, R. & Ananthakrishnan, T. N. 1993. Biochemical correlates of host specificity of *Caliothrips indicus* (Bagnall) relating to colonization patterns in a crop–weed ecosystem. Proceedings of the Indian National Science Academy, (Part B, Biological Sciences), 59 (5) : 471–476, figs. 1–2, table 1. New Delhi.

Padmanabhan, B. & Ananthakrishnan, T.N. 1993. Functional morphology and histology of the gut of some mycophagous Tubulifera and related adaptations. Hexapoda, 5 (1) : 1–16, pls. I–VIII. S. Viswanathan: Madras

Peter, A. J. & Ananthakrishnan, T. N. 1993. Effect of azadirachtin on the food utilization efficiency and digestive enzyme activity in *Cyrtacanthacris tatarica* (Orthoptera : Acrididae). Phytophaga, 5 (1) : 35–43, tables 1–4. S. Viswanathan : Madras.

Ananthakrishnan, T. N. 1994a. Biological Control Agents of Social Forest Insects. Pp. 1–11, figs. 1–18 (colour photos) on 18 unnumbered pages. Entomology Research Institute, Loyola College: Madras.

Ananthakrishnan, T. N. 1994b. Profiles of insect diversity. Current Science, 66 (4) : 271–281, figs. 1–28 (colour photos). Bangalore.

Ananthakrishnan, T. N. 1994c. Viewpoints on the chemical ecology of insects. Phytophaga, 6 (2) : 69–72. S. Viswanathan: Madras.

Ananthakrishnan, T. N. 1994d. An evaluation of MNDA repellency to household pests found in India. (Abstract No. 143, in the Session on Management of pests and beneficial insects). 5th European Congress of Entomology, University of York, UK, 29 Aug.–2 Sept. 1994, Abstracts. Hosted by the Royal Entomological Society.

Ananthakrishnan, T. N. (editor). 1994e. Functional Dynamics of Phytophagous Insects. viii + 304 pp. Oxford & IBH : New Delhi.

Ananthakrishnan, T.N. 1994f. Introduction. Pp. 1–5. In : Ananthakrishnan TN (ed.), Functional Dynamics of Phytophagous Insects, 304 pp. Oxford & IBH : New Delhi.

Ananthakrishnan, T.N.; Venkatesan, S. & Sridhar, S. 1994. Energetics and resource allocation in the lepidopteran complex infesting the cotton plant in terms of cultivar diversity and plant chemical quality. Pp. 95–113, figs. 1–6. In : Ananthakrishnan TN (ed.), Functional Dynamics of Phytophagous Insects, 304 pp. Oxford & IBH : New Delhi.

Ananthakrishnan, T.N.; Wesley, S.D.; Peter, A.J. & Marimuthu, S. 1994. Ecological interactions of *Helicoverpa armigera* (Hubner) and its natural enemies. International Journal of Ecology and Environmental Sciences, 20 (3) : 317–331, tables 1–6, figs. 1–5. New Delhi.

Jayaraj, S.; Ananthakrishnan, T.N. & Veeresh, G. K. (editors). 1994. Biological Pest Control in India : Progress and Perspectives. 101 pp. Rajiv Gandhi Institute for Contemporary Studies : New Delhi.

Ananthakrishnan, T. N. (editor). 1995a. Biological Control of Social Forest and Plantation Crops Insects. xii + 225 pp. Oxford & IBH : New Delhi.

Ananthakrishnan, T. N. 1995b. Introduction. Pp. 1–6. In : Ananthakrishnan TN (ed.), Biological Control of Social Forest and Plantation Crops Insects, 225 pp. Oxford & IBH : New Delhi.

Ananthakrishnan, T. N. 1995c. Polymorphism and behavioural diversity of insect vectors of plant diseases. Pp. 177–181. In : Swaminathan S, Randhawa NS, & Ananthakrishnan TN (eds.), Second Agricultural Science Congress, Proceedings, 271 pp. New Delhi.

Ananthakrishnan, T. N.; Wesley, Daniel S. & Peter, A. J. 1995. Kairomonal influence in *Acacia nilotica* as signals for efficient parasitization by *Sinophorus psycheae* (Hymenoptera: Ichneumonidae) on the bagworm *Eumeta cramerii* (Lepidoptera : Psychidae). Pp. 55–64, figs. 1–3, table 1. In : Ananthakrishnan TN (ed.), Biological Control of Social Forest and Plantation Crops Insects, 225 pp. Oxford & IBH: New Delhi.

Peter, A. J. & Ananthakrisnan, T. N. 1995. Impact of azadirachtin on the haemodynamics of *Cyrtacanthacris tatarica* L. (Acrididae: Orthoptera). Journal of Entomological Research, 19 (4) : 285–290, table 1, figs. 1–2.

Thomas, S. K.; Jayakumar, A. & Ananthakrishnan, T. N. 1995. Dynamics of insect communities at varying altitudes in the shola forests of Kodaikanal Hills, in relation to the chemical diversity of litter. International Journal of Ecology and Environmental Sciences, 21 (2) : 109129, tables 1–6, figs. 1–4. New Delhi.

Ananthakrishnan, T.N. (editor). 1996a. Biotechnological Perspectives in Chemical Ecology of Insects. vi + 299 pp. Oxford & IBH : New Delhi.

Ananthakrishnan, T.N. 1996b. Introduction. Pp. 1–3. In : Ananthakrishnan TN (ed.), Biotechnological Perspectives in Chemical Ecology of Insects, 299 pp. Oxford & IBH : New Delhi.

Ananthakrishnan, T.N. 1996c. Chemical signals, patterns of defence gene expression and molecular recognition in insect–plant interactions. Pp. 4–14, table 1. In : Ananthakrishnan TN (ed.), Biotechnological Perspectives in Chemical Ecology of Insects, 299 pp. Oxford & IBH : New Delhi.

Ananthakrishnan, T.N. 1996d. Forest Litter Insect Communities : Biology and Chemical Ecology. viii + 174 pp. Oxford & IBH : New Delhi. [Thysanoptera : pp. 38–41 (mycetophagous litter thrips), 48–56 (table 9 on p. 50, fig. 15 on p. 55), 65–66 (predators on thrips), 113 (trophic diversity), 120 (table 19), 121, 122 (table 20); references, pp. 150–165].

Ananthakrishnan, T.N. 1996e. Molecular messengers in insect biocommunication: Modality and relevance in biological control. Current Science, 70 (3) : 215–218, figs. 1–4. Bangalore.

Ananthakrishnan, T.N. 1996f. Challenges to efficient technology in the biological control of insects. Pp. 1–2. Symposium on Technological Advances in the Biocontrol of Insects, 11th Oct. 1996, Abstracts and Technical Programme. Entomology Research Institute, Loyola College, Madras.

Ananthakrishnan, T.N. (editor). 1996g. Proceedings, National Symposium on Biochemical Bases of Host Plant Resistance to Insects, iv + 246 pp. National Academy of Agricultural Sciences, IARI : New Delhi.

Ananthakrishnan, T.N. 1996h. Biochemical profiles of host plant resistance to insects. Pp. 1–13, tables 1–6. In : Ananthakrishnan TN (ed.), Proceedings, National Symposium on BiochemicalBases of Host Plant Resistance to Insects, 246 pp. New Delhi.

Ananthakrishnan, T.N. 1996i. Technological advances in the biological control of insects. Current Science, 71 (12) : 941–945, fig. 1. Bangalore.

Ananthakrishnan, T.N. (editor). 1996j. Research Highlights, Decennium 1985–1995, Entomology Research Institute, Loyola College, Madras. 46 pp. Edited and published by the Director, ERI, Madras [names of contributing authors are not stated].

[Thysanoptera: pp. 3–5 (*Retithrips syriacus,* hosts, gas chromatographic analysis of host leaves and anal exudates), 25–27 *(Arrhenothrips ramakrishnae,* oogenesis and spermiogenesis), 29–31 (thrips galls), 33–46 (publications 1985–1996)].

[Ananthakrishnan 1996k (editor)] (Anonymous) 1996. Insect – Natural Enemy Interactions : An Outline of ERI Contributions. Pp. 1–20. Entomology Research Institute, Loyola College, Madras.

Ananthakrishnan, T. N. 1996l. An evaluation of MNDA repellency to household pests found in India. XX International Congress of Entomology, Firenze, Italy, Aug. 25–31, 1996. Proceedings, p. 564.

[(Abstracts of Oral Communications, for the Session on 'Urban and Stored Products Entomology'; Abstract number 18–054). [Proceedings, pp. I–LVI and 1–820]. The title and the content of this abstract are the same as 1994d.]

Ananthakrishnan, T.N., 1996. Biochemical basis of Host plant Resistence in insects. Natural Academy of Agricultural Sciences.

Gopichandran, R. & Ananthakrishnan, T.N. 1996a. Biochemical correlates in varietal resistance of *Capsicum annuum* to infestation by *Scirtothrips dorsalis* Hood. Pp. 65–88, tables 1–5. In: Ananthakrishnan TN (ed.), Proceedings, National Symposium on Biochemical Bases of Host Plant Resistance to Insects, 246 pp. National Academy of Agricultural Sciences, IARI : New Delhi.

Gopichandran, R. & Ananthakrishnan, T. N. 1996b. Dynamics of age-correlated induced resistance of chilli plant cultivars to *Scirtothrips dorsalis* (Hood)

(Thysanoptera : Thripidae) infestations. Pp. 201–222, tables 1–6, figs. 1–4. In :

Ananthakrishnan TN (ed.), Biotechnological Perspectives in Chemical Ecology of Insects, 299 pp. Oxford & IBH: New Delhi.

Peter, A. J. & Ananthakrishnan, T. N. 1996. Neuroendocrine implications in azadirachtin-treated cotton grasshopper, *Cyrtacanthacris tatarica* Linn. (Acrididae : Orthoptera). Pp. 89–101, tables 1–3, figs. 1–4. In: Ananthakrishnan TN (ed.), Biotechnological Perspectives in Chemical Ecology of Insects, 299 pp. Oxford & IBH: New Delhi.

Ananthakrishnan, T. N. 1997a. Chemical interactions in biological control. Insect Environment, 2 (4) : 119–121.

Ananthakrishnan, T. N. 1997b. Allelochemical synergism and insect behavioural diversity. Current Science, 72 (9) : 628–630, fig. 1, table 1. Bangalore.

Ananthakrishnan, T. N. 1997c. Gallic and salicylic acids : Sentinels of plant defense against insects. Current Science, 73(7): 576–579, figs. 1–3. Bangalore.

Ananthakrishnan, T. N. 1997d. Insect biodiversity conservation with special reference to natural forest litter ecosystem. Pp. 307–316, table 1. In : Pushpangadan P, Ravi K, & Santosh V (eds.), Conservation and Economic Evaluation of Biodiversity, Vol. 1, 316 pp. Oxford & IBH : New Delhi.

Ananthakrishnan, T. N. 1998a. Basics of biocommunication in insect–plant interactions: Role of chemical signals. Pp. 1–6. In : Ananthakrishnan TN & Sen A (eds.), Biocommunication in Insects, 104 pp. Oxford & IBH : New Delhi.

Ananthakrishnan, T. N. 1998b. Modality and relevance of biocommunication in the biological control of insects. Pp. 38–44. In : Ananthakrishnan TN & Sen A (eds.), Biocommunication in Insects, 104 pp. Oxford & IBH : New Delhi.

Ananthakrishnan, T. N. (editor). 1998c. Technology in Biological Control. vi + 124 pp. Oxford & IBH : New Delhi.

Ananthakrishnan, T. N. 1998d. Introduction. Pp. 1–6. In : Ananthakrishnan TN (ed.), Technology in Biological Control, 124 pp. Oxford & IBH : New Delhi.

Ananthakrishnan, T. N. 1998e. Some technological inputs for increased efficacy in biological control. Pp. 7–14. In : Ananthakrishnan TN (ed.), Technology in Biological Control, 124 pp. Oxford & IBH : New Delhi.

Ananthakrishnan, T. N. 1998f. Insect herbivory, opportunistic interactions and strategic responses. Pest Management in Horticultural Ecosystems, 4 (1) : 1–6. Bangalore.

Ananthakrishnan, T. N. 1998g. Biotechnology and insect pest management. Pp. 209–220. In : Dhaliwal GS & Heinrichs EA (eds.), Critical Issues in Insect Pest Management, ix + 287 pp. Commonwealth Publishers : New Delhi.

Ananthakrishnan, T. N. 1998h. Insect gall systems: Patterns, processes and adaptive diversity. Current Science, 75 (7) : 672–676, figs. 1–5. Bangalore. [See also 2002b]. [Thysanoptera : p. 65; polymorphic and social behaviour].

Ananthakrishnan, T. N. 1998i. Changing trends in the biological control of insects. Pp. 1–6. Proceedings of the National Symposium on "Biological

Control of Insects in Agriculture, Forestry, Medicine and Veterinary Sciences". Bharathiar University, Coimbatore.

Ananthakrishnan, T.N., 1998. Biocommunication in Insects.Oxford & IBA, New Delhi

Ananthakrishnan, T. N.; Dhaliwal, G. S.; Arora, R.; Randhawa, N. S.; & Dhawan, A. K. 1998. Ecobehavioural diversity of biocontrol agents in the cotton ecosystem. Pp. 76–80. In: Ecological Agriculture and Sustainable Development. Vol. 2. [Proceedings of International Conference on Ecological Agriculture: Towards Sustainable Development, Chandigarh, India, 15–17 November, 1997.

Ananthakrishnan, T. N. & Sen, A. (editors). 1998. Biocommunication in Insects. vii + 104 pp. Oxford & IBH : New Delhi.

Ananthakrishnan, T.N., 1998. Insect Herbivory, Opportunistic interactions and strategic responses, in *Pest Management in Horticultural Ecosystems,* 4:1-6.

Ananthakrishnan, T.N., 1998. Biodiversity and insect pest management in critical issues in *Insect Pest Management*: 209-218

Ananthakrishnan, T. N. 1999a. Biotic interactions in changing environments: Insights into insect ecodynamics. Hexapoda, 11 (1–2) : 1–7. Madras.

Ananthakrishnan, T. N. 1999b. Induced responses, signal diversity and plant defence: Implication in insect phytophagy. Current Science, 76 (3) : 285–290. Bangalore.

Ananthakrishnan, T. N. 1999c. Behavioural dynamics in the biological control of insects: Role of infochemicals. Current Science, 77 (1) : 33–37. Bangalore.

Ananthakrishnan, T. N. 1999d. Multidimensional links in biodiversity research: An integrated exercise. Current Science, 77 (3) : 356–358. Bangalore.

Ananthakrishnan, T. N. 2000a. Changing landscapes, ecodiversity and insect faunal dynamics: The Tenmalai (Western Ghats) experience. Current Science, 78 (4) : 387–390, figs. 1–4. Bangalore.

Ananthakrishnan, T. N. 2000b. Phenotypic flexibility of plants and adaptive dynamics of Specialist generalist insects. Current Science, 78 (11) : 1303–1306. Bangalore.

Ananthakrishnan, T. N. 2000c. Expertise crisis in insect taxonomy: Need for introspection. Current Science, 79 (1) : 10. Bangalore.

Ananthakrishnan, T.N., 2000d. Biotic interactions in changing environments: Insights into insect ecodynamics, *Hexapoda*: 1-8.

Ananthakrishnan, T.N., 2000e. Phytochemicals as insect behaviour modifiers, *in Harwood Academic Publications*: 75-82.

Ananthakrishnan, T. N. 2001a. Frontiers of semiochemical research: Exploitation in crop protection strategies. 10 pp. FAO Endowment Lecture, Centre for Plant Protection Studies, Tamil Nadu Agricultural University, Coimbatore.

Ananthakrishnan, T. N. (editor). 2001b. Insects and Plant Defense Dynamics. xii + 253 pp. Oxford & IBH : New Delhi.

Ananthakrishnan, T. N. 2001c. Introduction. Pp. ix–xii. In: Ananthakrishnan TN (ed.), Insects and Plant Defense Dynamics, 253 pp. Oxford & IBH : New Delhi.

Ananthakrishnan, T. N. 2001d. Phytochemical defence profiles in insect–plant interactions. Pp. 1–22. In : Ananthakrishnan TN (ed.), Insects and Plant Defense Dynamics, 253 pp. Oxford & IBH : New Delhi.

Ananthakrishnan, T. N. 2001e. Molecular approaches to host plant defences against insects. Pp. 229–236. In : Ananthakrishnan TN (ed.), Insects and Plant Defense Dynamics, 253 pp. Oxford & IBH : New Delhi.

Ananthakrishnan, T. N. 2001f. Chemical ecology: the future. Pp. 237–242. In: Ananthakrishnan TN (ed.), Insects and Plant Defense Dynamics, 253 pp. Oxford & IBH : New Delhi.

Ananthakrishnan, T. N. 2001g. Molecular modalities in the manipulation of insect natural enemies : Role of biosemiotics. Current Science, 81 (5) : 442. Bangalore.

Ananthakrishnan, T. N. 2001h. Phytochemicals as insect behaviour modifiers. Pp. 73–82. In : Koul U & Dhaliwal GS (eds.), Phytochemical Biopesticides, x + 223 pp. Harwood Academic Publishers: Australia, Canada, etc.

Ananthakrishnan, T.N., 2001i. Signalling chemicals in insect behaviour: Role in insect-plant interactions *Bionotes*: 3: 24-28.

Ananthakrishnan, T.N., 2001j. Insects and plant defence Dynamics.Science Publishers Inc. Enfield, USA

Ananthakrishnan, T. N. 2002a. Future dimensions in entomology research. Current Science, 82 (7) : 783–785. Bangalore.

Ananthakrishnan, T. N. 2002b. Insects, Plants and Molecular Interactions. 120 pp. Madras Science Foundation, Chennai.

Ananthakrishnan TN. 2003a. Applied chemical ecology: induced and transgenic defences in insect–plant interactions. Current Science 84, 492–493. (A report on the third discussion meeting on Applied Chemical Ecology convened by T. N.Ananthakrishnan on 30 November 2002 at COSTED, Chennai).

Ananthakrishnan, T.N. 2003b. Insects and induced plant defense: Implications in biotechnology. Pp. 141–152. In : D'Souza L, Anuradha M, Hegde S, Nivas S, & Rajendra K (eds.), Biotechnology – Challenges and Prospects. Proceedings of the National Seminarcommemorating the Silver Jubilee of the Laboratory of Applied Biology, St. Aloysius College, Mangalore.

Polefka, T.H.; Liang, Li-Jung; & Ananthakrishnan, T.N. 2003. A comparison of the mosquitorepelling efficacy of methyl neodecanamide (MNDA) to that of Deet. Journal of Cosmetic Science, 54 : 283–288, tables I–II, fig. 1.

Ananthakrishnan, T.N. 2004a. Intraspecific diversity and species assessment: A biosystematic perspective. Perspectives on Biosystematics and Biodiversity, T.C.N. Commemorative Volume, pp. 3–9. Calicut University.

Ananthakrishnan, T.N. 2004b. Why chemical ecology : Emerging dimensions. Entomon, 29 (1) : 1–3. Trivandrum.

Ananthakrishnan, T.N. 2004c. Biodiveristy, biosignalling and biotechnology in insect-plant interactions, Current Science, 86 (3): 376-378. (A report on the 4th annual discussion meeting on the subject organized by T.N. Ananthakrishnan and held on 29 November 2003, at COSTED, Chennai).

Ananthakrishnan, T. N. 2004d. Changing dimensions of Entomological Research: Need for a holistic approach. Proceedings of the National Symposium on Frontier Areas of Entomological Research (eds. Subrahmanyam, B., Ramamurthy, V.V. & Singh, V.S.), Nov 5-7, 2003, IARI, New Delhi, 1-12.

David, B. V. & Ananthakrishnan, T. N. 2004. General and Applied Entomology. 2nd ed. xiv + 1184 pp. Tata McGraw-Hill : New Delhi. [See Nayar, Ananthakrishnan & David 1976].

Ananthakrishnan, T. N. 2005a. Bioengineering in biological control: Prospects and progress. Current Science, 881(1): 1717-1718.

Ananthakrishnan, T. N. 2005b. Book (4) reviews. (Scientist and Humanist: M. S. Swaminathan. R. D. Iyer. The Man Who Harvests Sunshine – The Modern Gandhi: M. S. Swaminathan. Andréi Erdélyi. Toward a Hunger-Free World: Life and Work of M. S. Swaminathan. Anwar Dil (ed). Science and Agriculture: M. S. Swaminathan and the Movement for Self-Reliance. S. Ramanujam *et al.* (eds.). Current Science, 89(2): 366-367.

Ananthakrishnan, T. N. 2005c. Book review. (Handbook of Industrial Crops. V. L. Chopra and K. V. Peter (eds). The Haworth Press Inc, New York). Current Science, 89(6): 1037-1038.

Ananthakrishnan, T.N. 2005d. Perspective and dimension of phenotypic plasticity in insects. pp. 1–23 in Ananthakrishnan, T.N. & Whitman, D. (eds.) Insect phenotypic plasticity diversity of responses. Enfield, New Hampshire, Science Publishers, Inc.

Ananthakrishnan, T.N. 2005e.Dimensions of molecular entomology. Current Science, 88(4): 541. (Based on the fifth Annual Discussion Meeting on Dimensions of Molecular Entomology held on 27 November 2004 in Chennai).

Ananthakrishnan, T. N. & Douglas Whitman (editors). 2005. Insect Phenotypic Plasticity – Diversity of Responses. Science Publishers, Inc., Enfield, USA and Plymouth, UK.

Ananthakrishnan, T.N. 2006a.Emerging technologies in resistance dynamics in insect–plant Interactions. Current Science, 90(4): 477-480. (A report on the VI Discussion Meet on Resistance Dynamics held on 2 December 2005 at COSTED, Chennai).

Ananthakrishnan, T.N. 2006b. Book review. (Heliothis/Helicoverpa Management: Emerging Trends and Strategy for Future Research. Hari C. Sharma (ed.). Oxford and IBH Publishing Company Pvt Ltd, New Delhi. 2005). Current Science, 90(4): 589-590.

Ananthakrishnan, T.N. 2006c. (review) Heliothis/Helicoverpa Management: Emerging Trends and Strategy for Future Research. Hari C. Sharma (ed.). Current Issue, 90 (4): 589-590.

Ananthakrishnan, T.N. & Sivaramakrishnan, K.G. 2006. Animal biodiversity: Patterns and processes. 264 pp., Scientific Publisher (India), Jodhpur.

Ananthakrishnan, T. N. 2007a. Semiochemicals in crop protection. Current Science, 92 (12): 882-885.(A report on the seventh Annual discussion meeting on semiochemicals in crop protection: ongoing technologies convened by T.N. Ananthakrishnan on 2nd Dec., 2006 at Centre for cooperation in Science on technology among developing societies, Chennai).

Ananthakrishnan, T. N. 2007b. Perespectives of molecular approaches to entomology, 1-9. In: Ananthakrishnan, T. N. (ed.). Dimensions of molecular entomology, 162 pp. University Press, Hyderabad.

Ananthakrishnan, T. N. (editor) 2007c. Dimensions of molecular entomology, 162 pp., University Press, Hyderabad.

Ananthakrishnan, T.N., 2007d. Ecological Entomology: Insect life in odd environment. Scientific Publishers (India). pp. 142.

Ananthakrishnan, T.N. and Alexander Jesudasan, R.W. 2007. Entomology Academy of India. Current Science, 93, 11: 1476-1477.

Ananthakrishnan, T.N. & Annadurai, R.S. 2007. Thrips-tospovirus interaction: biological and molecular implications. Cuurent Science, 92 (12): 1083-1086.

Ananthakrishnan, T.N. 2008. The social biology and adaptability in mycophagous tubuliferan thrips. Current Science, 94 (6): 708-709.

Ananthakrishnan, T.N. 2008. Ecological Entomology: Insect life in odd environment. Scientific Publishers (India). Pp. 142.

Ananthakrishnan, T.N. and Sivaramakrishnan, K.G. 2008. Ecological entomology insect life in odd environments. Scientific Publishers, Jodhpur (India). 142 pp.

TNA, 2009. Ecodynamics of Insect communities. Scientific Publishers (India). Part of Citation – INSA Award 2009.

Whitman, D.W. and Ananthakrishnan, T.N. (Eds) 2009. Phenotypic Plasticity of Insects: mechanism and consequences. Science Publishers, Enfield, NH, USA.

Ananthakrishnan T.N. 2009. Ecodynamics of insect communities. Scientific Publishers, Jodhpur (India).

Ananthakrishnan, T.N. and Alexander Jesudasan, R.W. 2009. Invasive insects in agriculture, medicine and forestry. Current Science, 97, 3: 300-301.

Ananthakrishnan, T.N. 2010a. Biodiversity and ecosystem level consequences in relation to thrips-host interactions. In: Insect Biodiversity: Functional dynamics and ecological perspectives, 179-198.

Ananathkrishnan, T.N. (Ed.) 2010b. Insect Biodiversity: Functional dynamics and ecological perspectives. Scientific Publishers (India)..pp. 228.

Ananathkrishnan, T.N. 2010c. Perspectives of changing biodiversity: Ecological and economic implications.(In: Insect Biodiversity: Functional dynamics and ecological perspectives. Ed. T.N. Ananathkrishnan), Scientific Publishers (India). pp. 1-12.

ACKNOWLEDGEMENT

Authors thank a large number of colleagues and friends, both in India and abroad, who have associated with Professor T.N. Ananthakrishnan in some way and provided the invaluable help of material related to his life and works. Most of the information has been inculcated from an isolated and authenticated publication on Prof. T.N. Ananthakrishnan published on the occasion of Dr. Tyagi's First Conference on Medical Arthropodology, held in Madurai, Tamil Nadu, India during October 15-16, 2007 (*cf.* Tyagi *et al.*, 2007).

Chapter - 2

THYSANOPTERA TAXONOMY - PURPOSE AND PRACTICE

Laurence A. Mound
CSIRO Entomology, Canberra, Australia
E-mail: *Laurence.Mound@csiro.au*

INTRODUCTION

Professor T.N. Ananthakrishnan will be remembered primarily through his many contributions to the taxonomy of the insect order Thysanoptera, the thrips. Between the years 1950 and 1980, according to cataloguing efforts by J.S. Bhatti (2004), TNA was responsible, either as sole or joint author, for erecting 76 new genus-group names and 320 species-group names. This is a monumental effort, involving great energy, determination, scholarship – and considerable financial investment. Modern biology involves many different study areas, from molecules to ecosystems, and the available pool of research workers and research funding is inevitably limited. There is thus competition between subjects for attention and financial support, and descriptive taxonomy is one area where funding has fallen over the past 50 years, as reflected in the number of experienced practitioners. This is not the place to re-examine why this should have happened. However, the opportunity is taken here to consider the objectives of insect taxonomists, and in particular how taxonomic studies on thrips generally have been approached. Unless we have a clear idea of what we as taxonomists want to achieve, of how we will achieve it, and for whose benefit, we can expect the rest of the biological community and its funding agencies to treat our studies as of limited priority.

Taxonomy is commonly perceived as being about naming and describing things, and many students seem to be taught this narrow view. But the purpose of taxonomy extends beyond describing and cataloguing. By examining the patterns of diversity, taxonomy is an essential tool in helping us to investigate how diversity has evolved, and how the processes involved in ecosystem diversification and maintenance might operate. Through this service function, taxonomy plays a major role in providing society with a factual basis for those many political and scientific decisions that are involved in conserving biological diversity. Vertebrate biologists may find it difficult to comprehend how little is known about the terrestrial diversity of small insects and mites. Thrips are prey for birds and lizards, pollinators of many plants, contributors to fungal decay and nutrient recycling, and pests and virus vectors on many crops; but despite their importance in these ecosystem roles, it is likely that scarcely 50% of the world's Thysanoptera species can be identified.

MINIMALIST AND COMPREHENSIVE APPROACHES

Consider the apparently simple process of "description". The objectives and methods deployed by taxonomists in preparing descriptions fall between two extremes. The minimalist approach involves publication of a few lines that meet the basic needs of the International Code of Zoological Nomenclature, the objective being validation of a new taxon name. The comprehensive approach involves a lengthy description with many illustrations, presumably with the objective of helping a reader to recognize this species when examining further specimens. Some authors with extensive experience of a group will craft a diagnosis that discriminates a new species in relatively few words, whereas some prefer to produce detailed lengthy descriptions. Unfortunately, lengthy descriptions by less experienced authors often repeat unnecessarily a range of characters that are present in all members of a genus and family.

Minimalist and comprehensive approaches exist also in the biological basis of species descriptions. Thus a new species may be based on a single wingless adult of unknown biology, or on a series of specimens of both sexes and immatures sampled from named host plants at different times and several locations. Adults within many thrips species vary greatly in size and colour, depending on developmental conditions; also, many species exhibit polymorphisms and sexual dimorphism, such that different morphs will not be recognized as congeneric if known only in isolation (Crespi *et al.,* 2004; Mound, 2005). Such intraspecific morphological plasticity has resulted in some species acquiring numerous synonyms, the synonymy rate for some authors exceeding 30% (Gaston & Mound, 1993). Unless a taxonomist has developed experience of intraspecific variation, both in space and time, description of a new species from one or a few specimens may well add more to taxonomic confusion than help to other biologists.

Mere length of a description is not an adequate measure of taxonomic utility. One of the longest descriptions of a thrips species ever published covered seven pages (the redescription of *Dendrothrips jeanneli* Bagnall, in Faure, 1964). But this failed to mention the one character state by which this species differs from almost all other Thripidae (the presence of 12, instead of six, sternal marginal setae). J.C. Faure was an important worker in the field of locust biology, and his taxonomic publications on thrips are exemplary in their detailed descriptions and superb illustrations – but he did not provide identification keys to facilitate studies by other biologists. Similarly J.D. Hood described over 1000 species of thrips (Hoebeke, 1994), but each was described in isolation with minimal discrimination from other species, no identification keys, and limited numbers of illustrations.

DO TAXONOMISTS HAVE A TARGET AUDIENCE?

The motivation of both Faure and Hood was probably "the service of science". This, however, is an unfocussed objective that does not identify any particular user community. Despite generating large amounts of new information, neither of them seemed interested in distilling that information into knowledge and a more general understanding. In current management terms, they did not identify a customer for their output. Some taxonomic authors find it expedient to issue their ideas and information in private or non-refereed publications. Such work is not necessarily poor science, but in rejecting the *mores* of the biological community their work is not targeted at any segment of that community. Taxonomy provides the essential reference base for the work of other biologists. Moreover, our work is funded, ultimately, through the tax-payer. We thus share a responsibility to make our descriptive taxonomy useful to others. This in no way limits the creativity of our science. Indeed, by targeting taxonomic work at the interests of other biologists, and particularly through collaborative studies with other individuals, the breadth and depth of creativity can be increased.

Collaborative studies, even manuscript refereeing, involve approaches to research that sometimes seem distrusted by taxonomists. But collaborators and referees are useful in stimulating each of us to consider our observations in new ways, and thus find further explanations for our data. However, collaborators and referees are useful only if they are encouraged to criticise constructively. A student or collaborator who always agrees with a lead scientist not only makes a limited contribution, but also fails to develop those critical faculties that are so essential to creative science.

OBJECTIVES AND PROCEDURES

Procedures in science usually involve the analysis of data, preferably from different sources, together with the erection and testing of hypotheses. In contrast, taxonomic science is unusual in that publication of primary data, taxon descriptions, is often accepted as a suitable objective. But taxonomic descriptions are only the starting point of our science, the real objective of which must be an understanding of diversity. The major museum collections at Frankfurt, London and Washington include large numbers of un-named and undescribed Thysanoptera taxa, often represented by few specimens and thus no biology. Describing new species from such specimens does little to increase our understanding of biological diversity. For example, providing a name for a few specimens of *Frankliniella* that seem to have rather longer setae than indicated in any available published description brings little benefit to anyone, especially considering that some of the most common species in this genus on crops in the neotropics cannot be securely identified. A more suitable scientific objective would be an analysis of the variation within and between samples, taken at different times of year and from a range of plant species. This could provide a more secure basis from which other biologists might be informed about the number of *Frankliniella* species that exist in an area, how they can be recognized, and how they partition the available plant resources.

It is not just the objectives of thrips taxonomy that are often weak; procedural weaknesses are commonly far more serious. These include the acquisition of primary data (field biology), the processing of data (statistical analysis), and the dissemination of information (publications). The lack of a biological basis for many thrips species is all too evident from the literature, with even recently published "host records" based on single specimens. Even when some consideration is given to variation, published discrimination between pairs of species commonly involves overlapping measurements of character states, with no indication of means or standard errors (Minaei & Mound, 2010a, 2010b). And although the most heavily used taxonomic publications are identification keys, it is easy to forget that such keys are limited, not just by the knowledge and experience of an author, but by the specimens available for study. Moreover, it is unusual for a taxonomist to have such keys tested before publication by students, collaborators and referees.

A CASE STUDY

For the purposes of this article it is useful to consider one important paper by the most outstanding of all thrips taxonomists, Herman Priesner. The genus *Liothrips* comprises over 290 described leaf-feeding species. In Asia, the basic reference work to the members of this genus (Priesner, 1953) treats more than 80 species, of which over 50 were described in that paper as new, none of them illustrated. It is relevant that Priesner never worked in the wet tropics. His extensive study was based on slide-mounted specimens that were given to him, and these specimens had been collected 20 or 30 years previously. Most of the species were described from few specimens; 40 of them were each based on no more than 5 specimens, including 10 based on single specimens. Many of these species are recorded from the same plant genera, such as *Schlefflera* or *Fagraea*, and more than a dozen species were recorded from

Piper. Despite this, there is no discussion of host associations, nor the significance of so many thrips species living on related, even identical, plant species, and there is no discussion of the structural variation among the thrips described.

Although monumental, and fundamental to Asian thrips work, Priesner's study suffers from all of the weaknesses considered above – lack of design in the acquisition of field data in order to test hypotheses; measurement differences quoted with no statistical support; and a key untested by any other worker prior to publication. Before we can begin to understand the radiation within the genus *Liothrips*, Priesner's basic work will require "ground-truthing" through comparison of the type specimens against variation amongst good series of specimens from known hosts. His study thus provides the basis for the best of modern taxonomic work to be carried out in the Asian area, investigating the plasticity of species through studies on variation within and between populations, on the same and different host plants (Mound & Kranz, 1997).

The traditional taxonomic approach, such as that on *Liothrips* discussed above, can usefully be contrasted with recent taxonomic work on thrips in Australia. These recent studies have targeted either the thrips diversity on particular plant genera, such as *Acacia* (Crespi *et al.*, 2004) or *Casuarina* (Mound *et al.*, 1998), or have targeted particular taxa for which an extensive body of information on host associations has been accumulated, such as *Anaphothrips* (Mound & Masumoto, 2009) and the Sericothripinae (Mound & Tree, 2009). Despite such studies, many Australian Thysanoptera remain undescribed – and will remain so until the essential field work is carried out to establish host associations and patterns of variation.

THE NEXT STEP

The emphasis in this article has remained firmly on descriptive taxonomy, together with the essential field studies on biology, behaviour and structural variation. Analytical methods, including cladistics and molecular work, have an important part to play in modern taxonomy, but such approaches are limited by the data available, particularly when applied to thrips. Such approaches will need far more extensive taxon sampling than is available at present, before they can make useful contributions to our understanding of the diversity and diversification of Thysanoptera.

Within descriptive taxonomy, morphologically based revisionary studies should never be viewed as the final word on a subject. They serve only to provide a basis for further and more comprehensive studies. The genus *Liothrips* mentioned above is typical of several thrips genera in Asia, each of which is ripe for extensive taxonomic study. But the objective must not simply be the description of specimens that look different. Good taxonomy is not about describing; it is about understanding. The account of the thrips fauna on *Acacia* trees in Australia (Crespi *et al.*, 2004) involved an attempt to understand why so many related species live on one plant genus, in what way their morphological differences are related to their biologies, and how this complex system evolved. The description of over 140 new species was just one of the mechanisms involved in that study, not the objective. Similar radiations of thrips species exist on several Asian plant genera, including *Bambusa*, *Fagraea*, *Ficus*, and *Schlefflera*. But whilst taxonomists limit themselves to describing, rather than understanding, we will remain ignorant of how these radiations have come about. There are many possible approaches to taxonomy, but the most influential are comprehensive and analytical studies that promote our understanding of how biological diversity has arisen, and how it might be maintained.

Professor Ananthakrishnan, through his studies on thrips morphological diversity, has laid a firm basis for good modern studies on these insects. This brief article is written in the hope that someone in Asia will take up the challenge that he has set. But in so doing, we should always remember the words of Isaac Newton "If I can see a little further it is because I stand on the shoulders of giants."

REFERENCES

Bhatti, J.S., 2004. T.N. Ananthakrishnan's taxa of Thysanoptera 1950 to 1980. *Thysanoptera* 2004: 85–181.

Crespi, B.J., Morris, D.C. & Mound, L.A., 2004 *Evolution of ecological and behavio-ural diversity: Australian* Acacia *thrips as model organisms.* Australian Biol-ogical Resources Study, Canberra & Australian National Insect Collection, Canberra. 328pp.

Faure, J.C., 1964. Thysanoptera of Africa – 9. *Revue de zoologie et de Botanique Africaine* 70: 53-91.

Gaston, K.J. & Mound, L.A. 1993. Taxonomy, hypothesis-testing and the biodiversity crisis. *Proceedings of the Royal Society, London* B251: 139-142.

Hoebeke, E.R. 1994. Joseph Douglas Hood (1889-1966): life and career of a renowned Thysanopterist. *Zoology (Journal of Pure and Applied Biology)* 4: 225-232.

Minaei, K. & Mound, L.A. 2010a. Grass-flower thrips of the genus *Chirothrips* (Thysanoptera: Thripidae), with a key to species from Iran. *Zootaxa* 2411: 33-43.

Minaei, K. & Mound, L.A. 2010b. Taxonomic problems in character state interpretation: variation in the wheat thrips *Haplothrips tritici* (Kurdjumov) (Thysanoptera: Phlaeothripidae) in Iran. *Deutsche Entomologische Zeitschrift* 57: 233–241.

Mound, L.A., 2005. Fighting, flight and fecundity: behavioural determinants of Thysanoptera structural diversity. Pp 81-105 in Ananthakrishnan TN & Whitman D. [eds] *Insects and phenotypic plasticity*. Science Publishers Inc. Enfield, NH, USA.

Mound, L.A., Crespi, B.J. & Tucker, A. 1998. Polymorphism and kleptoparasitism in thrips (Thysanoptera: Phlaeothripidae) from woody galls on *Casuarina* trees. *Australian Journal of Entomology* 37: 8-16.

Mound, L.A. & Kranz, B., 1997. optera and Plant Galls: Towards a Research Programme. Pp 11-24 in Raman, A. [ed.] *Ecology and Evolution of Plant-feeding Insects in Natural and Man-made Environments*. National Institute of Ecology, New Delhi; Backhuys Publishers, Leiden, Netherlands.

Mound, L.A. & Masumoto, M., 2009. Australian Thripinae of the *Anaphothrips* genus-group (Thysanoptera), with three new genera and thirty-three new species. *Zootaxa* 2042: 1-76. http://www.mapress.com/zootaxa/2009/f/zt02042p076.pdf

Mound, L.A. & Tree, D.J., 2009. Identification and host-plant associations of Australian Sericothripinae (Thysanoptera, Thripidae). *Zootaxa* 1983: 1-22.

Priesner, H., 1953. On the genera allied to *Liothrips* of the Oriental fauna, I. (Thysanoptera). *Treubia* 22: 357-380.

Chapter - 3

LATITUDINAL DISTRIBUTION OF STREAM INSECT COMMUNITIES OF THE WESTERN GHATS

K.A. Subramanian[1] and K.G. Sivaramakrishnan[2]

[1]Western Regional Centre, Zoological Survey of India, Sector-29, PCNT (PO), Pune-411 044, Maharashtra, India.
Email: subbuka.zsi@gmail.com
[2]Flat No. 3, Rams Appartments, New No. 07, Natesan Street, T-Nagar, Chennai-17, Tamil Nadu, India.
E-mail: kgskrishnan@gmail.com

ABSTRACT

Stream insect communities were sampled from 39 localities in the Western Ghats. A total of 15, 260 individuals belonging to 12 orders, 59 families and 94 genera were collected. The communities of northern latitude are characterized by widespread taxa. On the other hand, the taxa in southern latitude communities have restricted distribution. Stream insect communities in the Western Ghats show a clear latitudinal gradient. The diversity of stream insects decreases with increasing latitude. Average annual rainfall has no significant impact on the diversity of stream insects. However, the diversity of stream insects is significantly negatively correlated with number of dry months. Since the number of dry months is positively correlated with latitude in the Western Ghats, the observed latitudinal gradient of stream insects in the Western Ghats can be attributed to the number of dry months.

INTRODUCTION

The latitudinal gradient in species diversity is well known among ecologists though it was described after species-area curve (Rosenzweig, 1997). The well-known pattern is that the number of species of plants and animals encountered declines as you move away from the Equator, north or south (Rosenzweig, 1997). However, the pattern of distribution of stream insects across latitude is not very well understood in tropics. Results from various studies on the latitudinal gradient of benthic macroinvertebrate assemblages give equivocal results. The well known tropical-temperate comparison is that of Stout and Vandermeer (1975). This result, based on the extrapolation of species area is suspected (Vinson & Hawkins, 1998). Species richness in their study was only the projected numbers and not in their actual collections. This implies higher regional richness but not local richness (Vinson & Hawkins, 1998). Studies by Bishop (1973), Lake *et al.* (1994) and Pearson *et al.* (1986) suggest that tropical streams may be more diverse than the temperate ones and support findings of Stout and Vandermeer (1975). However, when studies by Illies (1969), Arthington (1990) and Lake *et al.* (1994) concluded that the temperate streams are diverse, Patrick (1961,1964) and Flowers (1991) argued that the richness between tropical and temperate streams are highly variable and that no strong latitudinal trend exists. A study on the stream insect communities of the Western Ghats found that, the generic diversity peaks in mid latitudes (Sivaramakrishnan, *et al.,* 2000). Results from these studies suggest that, the latitudinal gradient in aquatic insect diversity is not clearly established, especially in the tropics. With this background, the latitudinal gradient in aquatic insect diversity was explored in the streams of the Western Ghats. The Western Ghats provides an excellent opportunity to address this question because of its north-south orientation. The main objective of the present study was to explore the latitudinal gradient of aquatic insect communities in the Western Ghats.

METHODOLOGY

Study Sites

Aquatic insect communities were sampled from 39 localities of the Western Ghats (Map 1 and Table 1). Survey of India (SOI) topographic sheets (scale 1:50,000) and Global Positioning System (Garmin 12CX) were used to determine the latitude, longitude and altitude. Information on the stream order and river basin for the sampling sites were determined by using SOI maps. The average annual rainfall and number of dry months for the sites were extracted from bioclimatic maps of the Western Ghats (Pascal, 1982). The 39 study localities were categorized into 10 latitudinal zones. Sites were pooled in such a way that they fall within the same bioclimatic zone. Aquatic insect communities were compared within and between these zones.

Sampling Design

Data on stream insect communities were collected from August 1999 to February 2002. Previous studies have shown that stream insects are best sampled in the Western Ghats during post monsoon period (Sivaramakrishnan *et al.,* 2000). At each study site, the sampling spots were selected to represent stream orders, altitudinal range and riparian landuse. At each sampling spot, a stretch of approximately 100-150m was chosen for collection of samples. Target habitats, *viz.*, cascades, riffles and pools were sampled for stream insects within this stretch.

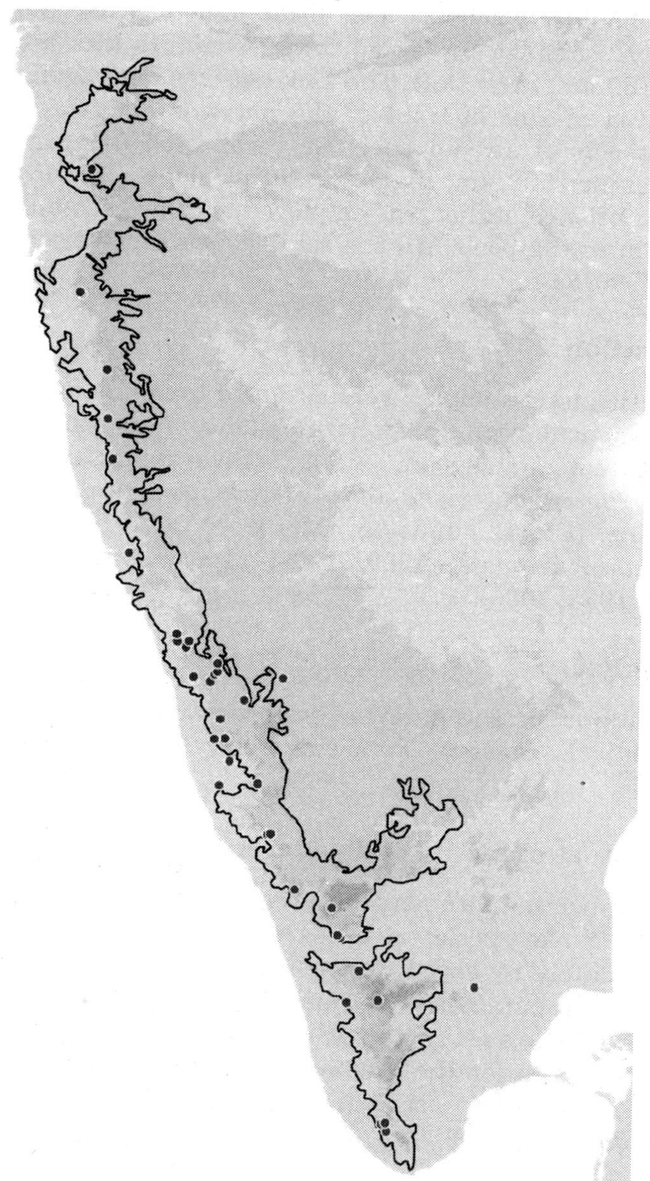

Map 1. Sampling sites in the Western Ghat region

Collection and Preservation

Stream insects were collected from three target habitats, *viz.* cascades, runs and pools. The methods employed for collecting stream insects from different habitats are outlined below. In cascades stream insects were sampled from 10 sqm area for one hour. Within the sampling area, stream insects were searched in all the possible substrata. Stream insects were searched visually and collected from substrata such as bedrocks, boulders, cobbles, leaf litter and dead wood. In runs stream insects were sampled by taking three, 1-minute kick-net samples (mesh opening: 180 μm; area 1m^2). The kick-net was held against water current and an area of 1m^2 in front of the net was disturbed for one minute. Stream insects of stream pools were collected using a nylon pond net (mesh opening: 500 μm; diameter: 30cm; depth: 15 cm). All out search method mentioned earlier was employed to collect stream insects from the substratum in the pools. All the samples were preserved in 70% ethanol for identification.

Identification

Aquatic insect samples were assigned to family and genus using keys for that particular group. Following keys were used for identification: *Ephemeroptera* (Dudgeon, 1999; Sivaramakrishan, unpublished); *Odonata, Plecoptera, Hemiptera, Megaloptera, Coleoptera, Diptera* and *Lepidoptera* (Fraser, 1933-36; Morse *et al.,* 1994; Dudgeon, 1999); *Hemiptera* (Thirumalai 1989, 1999; Morse *et al.,* 1994), *Trichoptera* (Wiggins, 1975, 1996).

Data Analysis

The input for the data analysis was relative abundance matrix of family and the genera. Following analysis was carried out for this objective.

(i) Estimators of family and generic richness

A number of parametric and non-parametric methods have been adopted to make species richness estimations and these methods have been adequately reviewed (Colwell and Coddington, 1994). Most commonly used non-parametric estimators are the Jackknife, the bootstrap method and Chao's estimators (S_1* and S_2*). Family and genetic richness of stream insects for the Western Ghats were calculated by Jackknife, bootstrap and Chao's estimators (S_1* and S_2*) using *Estimate S6b1a* software (Colwell, 2000).

(ii) Diversity indices

Alpha or point diversity and beta or differentiation diversity were used to measure the genus and family diversity within and between latitudinal zones. Alpha diversity was calculated by Shannon and Simpson's indices. Beta diversity was calculated by Jaccard's index.

(iii) Cluster analysis

Cluster analysis module of the software package STATISTICA ('99 edition) was used to examine similarity in community composition across latitudinal zones. 1-Jaccard's index was subjected to simple linkage Euclidean distances and results were plotted as dendrogram.

(iv) Rarefaction

It is not always possible to ensure equal sample sizes, especially while sampling insect communities. One way to overcome this is by rarefaction. Rarefaction is a method of working out the number of species that would be expected in samples of standard size (Magurran, 1988). The unbiased version of the rarefaction formula is used here for estimating rarefied family and generic richness (Magurran, 1988).

(v) Regression

Simple regression model of Visual Generalized Linear Model (Visual GLM) module of the software package STATISTICA was used to expolore richness gradients of taxa across latitude, average annual rainfall and number of dry months.

RESULTS

A total of 15, 260 individuals belonging to 12 orders, 59 families and 94 genera were collected in 293 sampling sessions from 39 localities in the Western Ghats. Of these 14,824 individuals were assigned to genus, 406 to families and remaining 30 to orders.

Estimation of family and generic richness

The figures 1 and 2, and Table 2 shows observed and estimated family and generic richness for the 39 sites. Table 3 & 4 gives observed and estimated family and generic richness for Ephemeroptera, Hemiptera, Trichoptera and other stream insects, called henceforth as Non-Ephemeroptera, Hemiptera and Trichoptera (Non-EHT).

Family and generic accumulation curves (observed and estimated) plotted against sites show that, the slope of the curve increases quite rapidly initially as new taxa are recorded with increasing number of sites. This steep rise gradually slows down by 20th site. The estimated species richness was slightly higher than the observed in all the taxa. This trend indicates that there might possibly be a few additional taxa, which were not collected and might have been missed either due to rarity or elusiveness. The gradual flattening of the curves, however, show that most of the existing taxa would have been sampled and hence can be used as a reasonable estimate of the total family and generic richness for the region sampled.

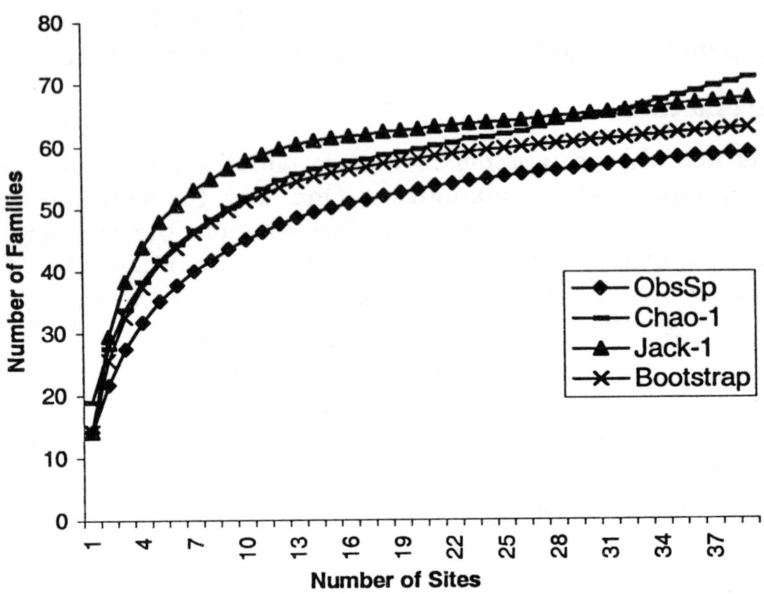

Fig 1. Observed and estimated family richness across 39 sites

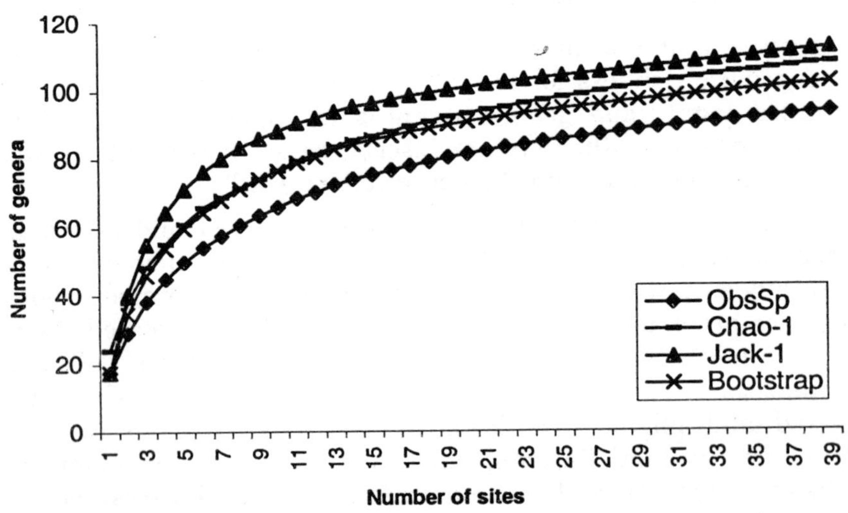

Fig 2. Observed and estimated generic richness across 39 sites

Table 1. Details of the Sampling Sites

Sl.No	Site Name	District	Altitude (m)	Latitude (°N)	Longitude (°E)	AARF (mm)	NDM
1	Neyyar	Trivandrum	635	8.58	77.22	2500	2
2	Aghastyamalai	Trivandrum	333	8.67	77.15	2500	2
3	Bonnacaurd	Trivandrum	892	8.68	77.20	2500	2
4	Thatekkad	Ernakulam	146	10.10	76.72	2500	3
5	Eravikulam	Idukki	2350	10.13	77.09	3594	2
6	Karandhamalai	Madurai	400	10.30	78.23	810	7
7	Topslip	Coimbatore	599	10.48	76.86	1795	4
8	Dhoni	Palakkad	350	10.88	76.63	2090	4
9	Meenvallom	Palakkad	400	10.90	76.60	2500	4
10	Upper Bhavani	Nilgiri	2300	11.22	76.52	2000	3
11	Muthapan Puzha	Calicut	530	11.43	76.08	3307	3
12	Makut	Coorg	205	12.08	75.75	3300	3
13	Aralarm	Kannur	100	12.09	75.79	3300	3
14	Panaje	Dakshinakannada	60	12.66	75.17	5500	4
15	Subrahmanya	Dakshinakannada	167	12.68	75.63	5500	4
16	Koyyur	Dakshinakannada	49	12.95	75.29	5500	4
17	Mala	Dakshinakannada	192	13.20	75.11	5500	4
18	Kudremukh	Chikamangalur	774	13.21	75.23	5500	4
19	Kigga	Chikamangalur	722	13.44	75.18	5500	4
20	Neralekoppa	Chikamangalur	650	13.67	75.45	2500	6
21	Hosanagara	Shimoga	513	13.88	75.05	2500	6
22	Nagodi	Shimoga	550	13.91	74.86	2500	6
23	Nittur	Shimoga	550	13.93	75.90	2500	6
24	Nellibedu	Shimoga	550	13.94	74.85	2500	6
25	Mavinahole	Uttara Kannada	550	13.97	75.10	3500	6
26	Nandihole	Shimoga	530	14.00	75.13	981	6
27	Haridravathi	Shimoga	525	14.10	75.13	981	6
28	Malemanne	Uttara Kannada	500	14.28	74.73	4000	6
29	Kathlekan	Uttara Kannada	525	14.28	74.75	4000	6
30	Kodigathe	Uttara Kannada	450	14.35	74.65	4000	6
31	Badal	Uttara Kannada	450	14.36	74.79	4000	6
32	Hulidevarakodlu	Uttara Kannada	150	14.44	74.64	4000	6
33	Bondla	South Goa	134	15.40	74.06	3000	6
34	Radha Nagari	Kholapur	550	16.50	73.85	3000	8

35	Amba	Raigad	550	16.98	73.78	3000	8
36	Koyana	Satara	630	17.57	73.76	5000	8
37	Tamhini	Pune	609	18.46	73.44	3500	8
38	Dongarvadi	Pune	600	18.48	73.42	3500	8
39	Triambak	Nasik	750	19.94	73.53	3000	8

AARF(mm): Average Annual Rainfall in mm; **NDM**: Number of Dry Months

Table 2. Observed and estimated order, family and generic richness

Sl. No	Taxa	No. Individuals	No. Observed	Chao-1	Jack-knife-1	Boot-strap
1	All Orders	15,260	12	12	12.97	12.39
2	All Families	15,260	59	71	67.77	62.96
3	All Genera	15,260	94	108.29	112.5	102.5

Table 3. Observed and estimated family richness across taxa

Sl. No	Taxa	No. Individuals	No. Observed	Chao-1	Jack-knife-1	Boot-strap
1	Ephemeroptera	3570	10	12	11.95	10.77
2	Hemiptera	2234	8	8.5	9.93	8.89
3	Trichoptera	6326	11	11.5	11.97	11.52
4	Non-EHT	3126	30	31.33	33.89	31.76
	Total	15,256*	59			

Table 4. Observed and estimated generic richness across taxa

Sl. No	Taxa	No. Individuals	No. Observed	Chao -1	Jack-knife-1	Boot-strap
1	Ephemeroptera	3570	18	21	21.89	19.66
2	Hemiptera	2234	22	26.5	28.75	24.97
3	Trichoptera	6326	17	17.5	19.92	18.25
4	Non-EHT	3126	37	38.8	41.87	39.53
	Total	15,256*	94			

*Total does not add up to 15,260, as 4 individuals could not be assigned to any locality.

Diversity

Diversity indices were calculated at the level of family and genera. Alpha diversity indices of genera were also computed for Ephemeroptera,

Hemiptera, Trichoptera and Non-EHT. The results are presented in Tables 5,6 & 7. The highest family and generic richness was in the 13° latitudinal zone, with 54 families and 81 genera. 19°N latitudinal zone had the lowest richness with 6 families and 6 genera. However, Shannon and Simpson's index was highest at 11°N and 13°N. The indices values were lowest for the latitudinal zones lying between 14°N-19°N and other zones had intermediate values. At the level of taxa, the generic diversity of Ephemeroptera, Hemiptera, Trichoptera and Non-EHT peaked at the latitudinal zones 13°-14°N. The latitudinal zones between 15°-19°N had the lowest and the zones between 9°-12°N had intermediate diversity.

Table 5. Alpha diversity indices for family and genera

Sl. No	Latitudinal Zone(°N)	Family			Genus		
		Observed	Simpsons	Shannon	Observed	Simpsons	Shannon
1	9	28	0.8814	2.5276	38	0.918	2.9166
2	10	37	0.9057	2.6932	51	0.9151	2.9188
3	11	32	0.9311	2.8922	42	0.9491	3.2125
4	12	22	0.861	2.3408	29	0.911	2.7315
5	13	54	0.9044	2.7425	81	0.9401	3.216
6	14	41	0.7844	2.2859	58	0.8552	2.7068
7	15	13	0.8565	2.0855	16	0.8603	2.1652
8	17	8	0.6374	1.4088	9	0.6835	1.5589
9	18	18	0.8982	2.4385	19	0.9012	2.4853
10	19	6	0.4899	0.9869	6	0.4899	0.9869

Table 6. Alpha diversity indices for Ephemeroptera and Hemiptera genera

Sl. No	Latitudinal Zone(°N)	Ephemerpotera			Hemiptera		
		Observed	Simpsons	Shannon	Observed	Simpsons	Shannon
1	9	7	0.6659	1.4222	6	0.5667	1.16
2	10	11	0.736	1.5887	8	0.68	1.4737
3	11	8	0.8066	1.7306	8	0.6472	1.3351
4	12	8	0.768	1.6185	3	0.7	0.9503
5	13	15	0.7848	1.767	16	0.7152	1.5157
6	14	13	0.8062	1.9451	8	0.8508	1.8775

7	15	5	0.7172	1.2994	3	0.7	0.9503
8	17	2	0.4667	0.6109	-	-	-
9	18	4	0.6791	1.1753	2	0.6	0.673
10	19	3	0.1565	0.3393	2	0.3429	0.5004

Table 7. Alpha diversity indices for Trichoptera and Non-EHT genera

Sl No	Latitudinal Zone(°N)	Trichoptera			Non-EHT		
		Observed	Simpsons	Shannon	Observed	Simpsons	Shannon
1	9	10	0.8146	1.8693	15	0.805	1.9487
2	10	12	0.8075	1.8636	20	0.7238	1.7088
3	11	11	0.8428	1.9687	15	0.8788	2.3069
4	12	5	0.4634	0.9443	13	0.8534	2.081
5	13	14	0.7267	1.6939	36	0.8757	2.5283
6	14	11	0.5667	1.1783	26	0.8867	2.528
7	15	1	0	0	7	0.694	1.3486
8	17	4	0.375	0.7456	3	0.7	0.9503
9	18	8	0.7586	1.6542	5	0.7409	1.4211
10	19	-	-	-	1	0	0

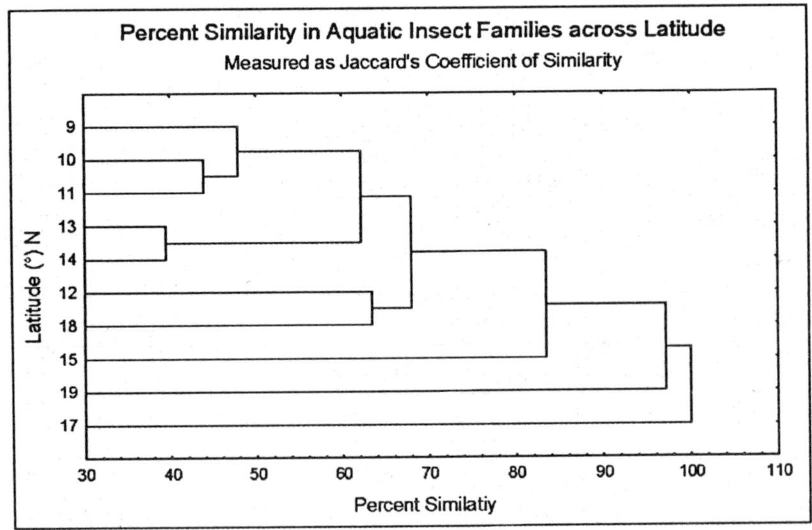

Figure 3. Percent similarity in family composition across latitude

Family and genus turn over or beta diversity across latitudinal zones was analyzed using Jaccard's index. The results of taxa turn over across latitudinal zones based on Jaccard's coefficient of similarity is depicted as dendrograms in Figures 3 & 4. Clustering with Jaccard's index shows three latitudinal groups at family and generic level. The zones 9°, 10°, 11°, 13° and 14°N formed the first group. The second group was formed by the zones 12° & 18°N. The third group, zones 15°, 17° and 19°N are distinct and does not form clear cluster with any latitudinal zone. It is interesting to note that the southern Western Ghats zone (12°N) groups with the northern Western Ghats zone (18°N). At the level of taxa, the stream insect community composition across latitude shows interesting results. The clustering patterns across latitude were different for Ephemeroptera, Hemiptera, Trichoptera and Non-EHT. While the Ephemeroptera formed two distinct clusters, north and south of 15°N, other taxa formed three (Hemiptera and Trichoptera) or four groups (Non-EHT). Community composition of Hemiptera genera at 9°N was very distinct from the rest of the zones and the genera of 12°N grouped with 18°&19°N. On the other hand, the Clustering of Trichoptera community was very similar to the overall grouping of the aquatic insect communities (family and genera) across latitude. Surprisingly, unlike other taxa, the Non-EHT genera at 12°N clustered with the southern Western Ghats zones (9°N-13°N). Community composition was similar between 15°N&18°N and the zones, 17°N &19°N were distinct.

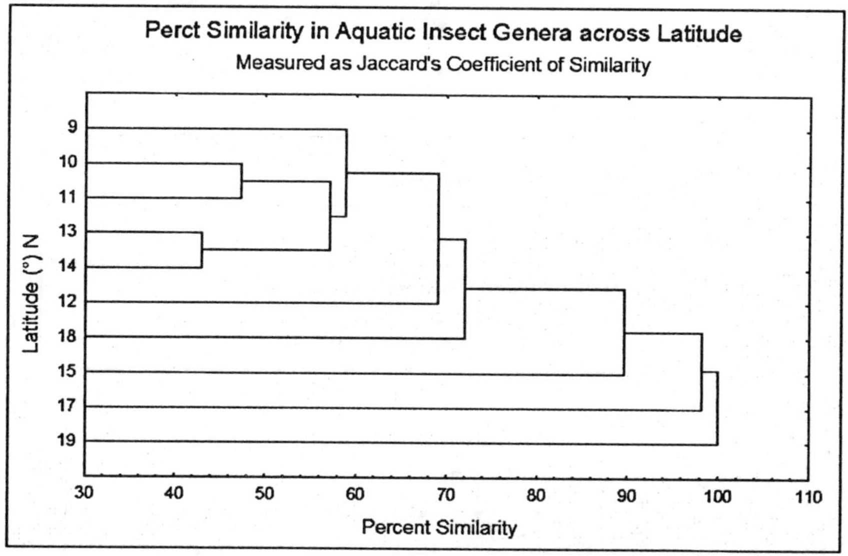

Figure 4. Percent similarity similarity in genera composition across latitude

Latitudinal gradients

Rarefied family and generic richness was used to normalize unequal sample sizes across the latitude. Latitudinal gradients across the Western Ghats were explored using maximum rarefied richness at the confidence interval of 0.01. Simple regression analysis of family, generic, Ephemeroptera, Hemiptera, Trichoptera and Non-EHT richness with latitude, average annual rainfall and number of dry months were carried out. Rarefied family and generic richness decreases significantly with increasing latitude (*R^2=0.21; F=8.81; $p<0.05$ & R^2=0.23; F=9.63; $p<0.05$*). In other words, southern latitude was richer than the northern latitude (Figures 5 & 6). However, Ephemeroptera, Hemiptera, Trichoptera and Non-EHT genera did not show any significant trend with latitude. The family and generic richness did not show any significant trend with the average annual rainfall. On the other hand, family and generic richness decreased significantly with the increasing number of dry months (*R^2=0.30; F=14.16; $p<0.05$ & R^2=0.31; F=14.95; $p<0.05$* (Figures 7 & 8).

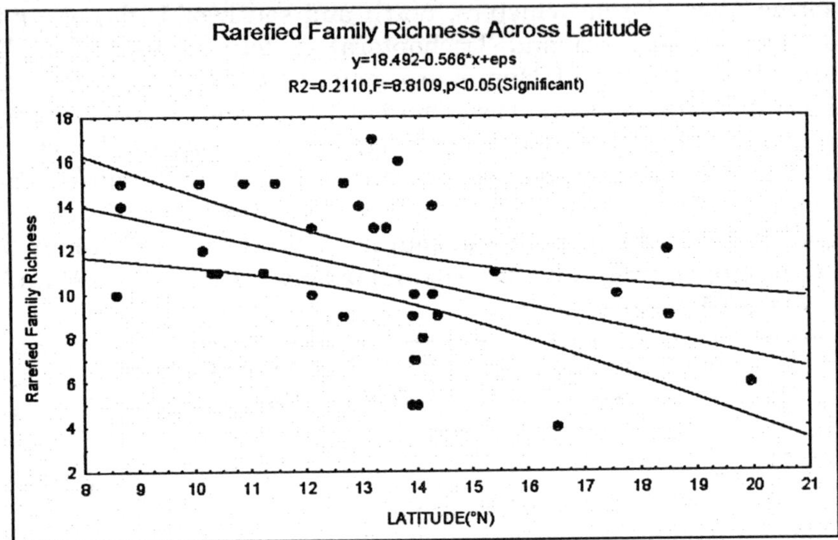

Figure 5. Rarefied family richness across latitude

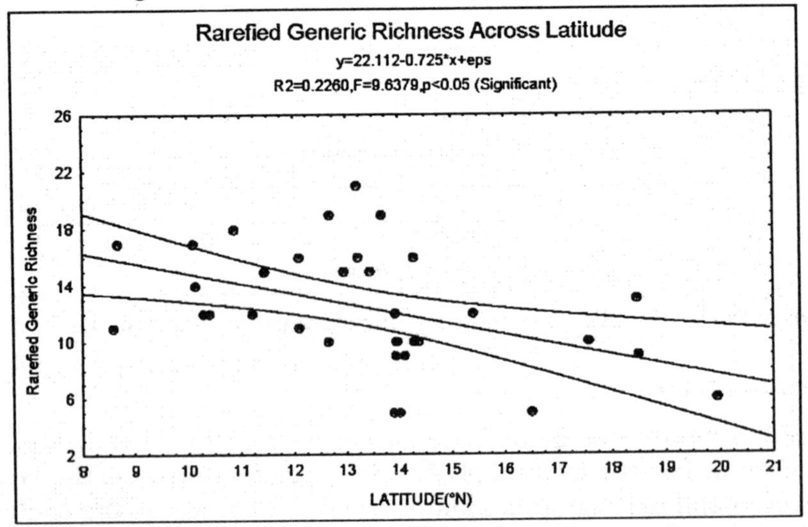

Figure 6. Rarefied generic richness across latitude

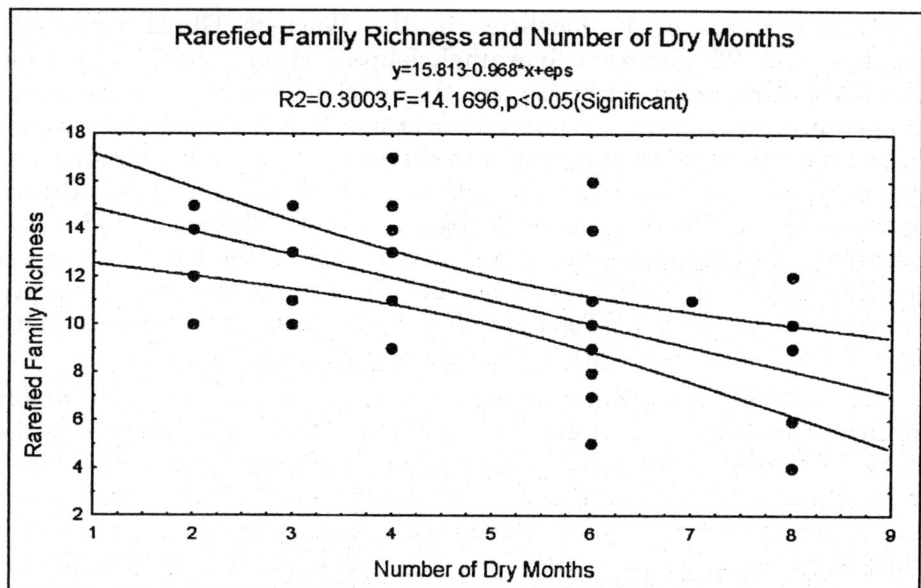

Figure 7. Rarefied family richness across number of dry months

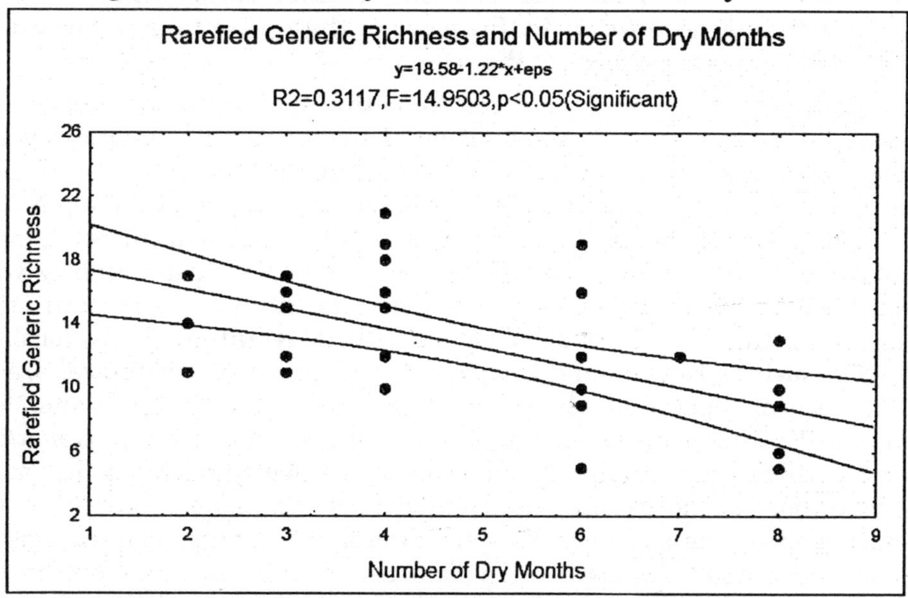

Figure 8. Rarefied generic richness across number of dry months

DISCUSSION

Diversity

The family and generic richness of stream insects reported for this study is higher than that of previously recorded. A study, restricted to the headwater riffles of 17 localities in the Western Ghats reported 45 families and 72 genera (Sivaramakrishnan *et al.,* 2000). This lower number of reported richness in the earlier study may be due to inadequate geographic and habitat coverage. The observed and estimated richness for the family and genera in the present study clearly shows that the sampling was adequate to address broad stream insect diversity patterns in the Western Ghats (Tables 2, 3 & 4). The estimated richness was slightly higher than the observed one both in the case of family and genera. Similar patterns were present for Ephemeroptera, Hemiptera, Trichoptera and Non-EHT genera. This higher estimated richness may be due to the absence of cryptic or rare taxa in the sample.

High diversity of family and genera for 13°N zone was also reported in the earlier study (Sivaramakrishnan *et al.,* 2000). This pattern remains more or less unchanged even when Ephemeroptera, Hemiptera, Trichoptera and Non-EHT were considered independently. This may be due to the influence of more number of sampling localities in the region. Moreover, the intensive study site (Kudremukh) focussing a drainage basin was located in this zone. This intensive study encountered extremely rare taxa such as *Neoephimeridae, Rheonanthus, Povilla, Micronecta and Hydaticus* etc. Presence of these rare taxa increased the family and generic richness of the region.

A clear pattern that emerges from the analysis of taxa turn over across latitudinal zones is the division of stream insect communities into northern and southern Western Ghats groups (Figures 3 & 4). Similar pattern is known from angiosperms, butterflies, and amphibians (Nair & Daniels, 1986; Gaonker, 1996; Daniels, 1992). In stream insect families and genera, the communities of southern latitudes cluster tightly when compared to the northern latitudes. One anomaly to this pattern is the study localities of 12°N, which clusters with 18°N. Interestingly, localities of 18°N such as Koyana and Dongarvadi still retain a substantial area in their natural vegetation when compared to sites at 15°N, 17°N and 19°N. The site Koyana is also known to harbour species such as *Bufo stomaticus* (toad), which have patchy distribution in the Western Ghats. In case of stream insects, the patchily distributed taxa which were present in this region and in the southern Western Ghats are the trichopteran genera *Glosossoma* and *Stenopsyche.* In the case of anurans, the presence of patchily distributed species in relict forest patches is attributed to past large-scale deforestation, which led to local extinction and discontinuous distribution (Daniels, 1992). Such factors could as well influence the distribution patterns of insects with less dispersal ability (Samways, 1996).

Genera turnover in Ephemeroptera, Hemiptera, Trichoptera and Non-EHT across latitude shows interesting results. A clear clustering of northern and southern Western Ghats communities is found only in

Ephemeroptera. Interestingly these divisions correspond to two major geological formations of the Ghats, the northern Deccan traps and the southern peninsular gneiss (Subrahmanya, 2001). Ephemeroptera are known for their importance in detecting biogeographic patterns owing to their ancient origin and poor dispersal abilities (Sivaramakrishnan, 1987). The communities of the northern latitude are represented by widely distributed genera such as *Baetis, Choroterpes, Epeorus* and *Teloganodes*. On the other hand, more restricted genera such as *Ecdyonurus, Caenis, Thalerosphyrus, Thraulus, Isonychia, Petersula* and *Edmundsulla* characterize the communities of the southern Western Ghats. It is interesting to note that, the genus *Petersula,* a member of *Kimminsula*-complex of Gondwanian origin is restricted to the southern Western Ghats. The *Kimminsula*-complex is restricted to India, Srilanka and Madagascar suggesting their geological kinship in origin (Sivaramakrishnan, 1984).

The auqatic Hemiptera of the Western Ghats clustered into three distinct communities. The community at 9°N was distinct from other latitudes. This is due to the presence of the genus *Perittopus*, in the streams of this latitude. The other two groups distinctly cluster into the northern and southern Western Ghats communities. The hemipteran fauna of the northern Western Ghats, as in Ephemeroptera were characterized by the presence of widespread genera such as *Metrocoris* and *Naucoris*. The only taxon, restricted to this region was an unidentified genus of *Corixidae*. The streams of southern Western Ghats were represented by genera such as *Amemboa, Rhagovelia, Timasius, Nychia* and *Enithares* with more restricted distribution.

The Trichoptera and non-EHT follow the general pattern of community organization of stream insects in the Western Ghats. The trend of presence of widespread taxa in the northern Western Ghats also seems to be true in the case of Trichoptera and non-EHT. Interestingly, Trichoptera were absent from the northern site (19°N). This could be either due to inadequate sampling or the Trichoptera is naturally absent from the ephemeral streams in this region. Further fieldwork is required to resolve this pattern. The widespread taxa present in the northern Western Ghats include: *Moselyana, Anisocentropus, Hydropsyche* and *Goerodes* (Trichoptera) and *Neoperla* (Plectoptera), *Dineutus* (Coleoptera), *Eubrianax* (Coleoptera), *Simulium*, and *Tipulidae* (Diptera). Taxa such as *Agapetus, Goera, Helicopsyche* and *Smicridea* (Trichoptera) and *Sandaracottus* (Coleoptera), *Philorus* (Diptera), *Euphea* (Odonata) and *Onychogomphus* (Odonata) were restricted to the southern Western Ghats. The trend of presence of more widespread taxa in northern Western Ghats and restricted taxa in the southern Western Ghats is also known for trees, odonates and anurans (Ghate *et al.,* 1998; Subramanian & Sivaramakrishnan, 2002; Daniels, 1992). The distribution patterns of stream insect communities across latitude are further discussed in detail below.

Latitudinal Gradients

A general pattern that emerges after correcting for the unequal sampling effort by rarefaction is that the southern latitude streams are richer than those of the northern latitudes. This pattern remains unchanged even at the level of families and genera (Figures 5 & 6). Though a few localities of the mid latitude streams are slightly more diverse, their influence on the trend is insignificant. The low diversity of northern Western Ghats is also known for angiosperms, odonates, butterflies, amphibians and birds (Nair & Daniels, 1986; Subramanian & Sivaramakrishnan, 2002, Gaonkar, 1996; Daniels, 1992; Ali & Ripley, 1983). The present study corroborates with the findings of earlier studies (Bishop, 1973; Stout & Vandermeer 1975; Pearson *et al.,* 1986; Lake *et al.,* 1994), which suggested that the lower latitude streams are more diverse than the high latitude streams.

The aquatic insect family and generic richness did not show any significant trend with average annual rainfall. Similar pattern have been observed for angiosperms, trees and amphibians (Nair & Daniels, 1986; Ghate *et al.,* 1998; Daniels, 1992) in the Western Ghats. However, the family and generic richness decreases significantly with increasing number of dry months as in the case of angiospersms, trees and amphibians (Figs. 7 & 8). A study on the odonates of the Western Ghats further clarify the significance of number of dry months in determining the diversity and distribution of stream insects in the Ghats. Subramanian and Sivaramakrishnan (2002) observed that the odonate communities of northern Western Ghats are characterized by species, which prefer to breed in temporary lentic water bodies. The community is dominated by libellulid (Libellulidae) species with wide geographic and habitat distribution. On the other hand, a large proportion of odonates from the southern and central Western Ghats breed in perennial lotic water bodies. More over, the endemicity of the central and southern Western Ghats odonates is very high (28 & 35% respectively) when compared to the northern Western Ghats (4%). Results from various taxa including a preliminary study on the stream insects (Sivaramakrishnan *et al.,* 2000) support the fact that the number of dry months indeed has a significant influence in determining the diversity and distribution organisms, including stream insects in the Western Ghats. Since the number of dry months is tightly positively correlated with latitude in the Western Ghats (Ghate *et al.,* 1998), the observed latitudinal pattern in stream insects could be a result of the difference in rainfall distribution between northern and southern latitudes. These results are further corroborated by a study on the diversity of Ephemeroptera, Plecoptera and Trichoptera genera at a global scale (Vinson & Hawkins, 2003). Vinson & Hawkins, (2003) shows that the coefficient of variation in monthly precipitation is important in determining local generic richness and better explain latitudinal diversity.

CONCLUSION

Stream insect communities in the Western Ghats show a clear latitudinal gradient. The diversity of stream insects decreases with increasing latitude. Average annual rainfall has no significant impact on the diversity of stream insects. However, the diversity of stream insects is significantly negatively correlated with number of dry months. Since the number of dry months is positively correlated with latitude, the observed latitudinal gradient of stream insects in the Western Ghats can be attributed to the number of dry months.

REFERENCES

Ali, S. & Ripley, S.D., 1983. Handbook of birds of India and Pakistan (compact edition). *Oxford University Press*. New Delhi.

Arthington, A.H., 1990. Latitudinal gradients in insect species richness of Australian lotic systems: a selective review. *Trop. Fresh. Biol.*: 2(2). 179-96.

Bishop, J.E., 1973. Limnology of a small Malayan River, Sungai Gombak. Monogr. Biol. 22, *Dr.W. Junk Publishers*, The Hague, 485 pp.

Colwell, R. K., 2000. Estimates: Statistical estimation of species richness and shared species from samples. Version 5.

Colwell, R.K. & Coddington, J.A., 1994. Estimating terrestrial biodiversity through extrapolation. *Phil. Trans. Royal Soc. Lond. B.* 345: 101-118.

Daniels, R.J.R., 1992. Geographical distribution patterns of amphibians in the Western Ghats, India. *J.Biogeography* 19: 521-529.

Dudgeon, D., 1999. Tropical Asian Streams-Zoobenthos, Ecology and Conservation. *Hongkong University Press*. Hongkong. 828pp.

Flowers, R.W., 1991. Diversity of stream living insects in northwestern Panama. *J. North Am. Benthol. Soc.* 10(3): 322-334.

Fraser, F.C., 1933-36. The fauna of British India, including Ceylon and Burma, Odonata, Vols.I-III. *Taylor & Francis Ltd.*, London.

Gaonkar, H., 1996. Butterflies of Western Ghats, India including Sri Lanka: a biodiversity assessment of a threatened mountain system. Unpublished Report submitted to Centre for Ecological Sciences, IISc, Bangalore. *Zoological Museum, Copenhagen, Denmark and Natural History Muesum, London*. 86 pp..

Ghate, U, Joshi, N.V. & Gadgil, M., 1998. On the patterns of tree diversity in the Western Ghats of India. *Curr. Sci.* 75(6): 594-603.

Illies, J., 1969. Biogeography and ecology of neotropical freshwater insects, especially those from running waters. In: Biogeography and Ecology of in

South America, (eds. EJ Fittkau, J Illies, H Sioli, GH Schwabe and H Klinge). 685-708. *Junk*, The Hague.

Lake, P.S., Schreiber, E.S.G., Milne, B.J. & Pearson R.G. 1994. Species richness in streams: patterns over time, with stream size and with latitude. *Verh. Int. Verein. Limnol.* 25:1822-1826.

Magurran, E. Anne., 1988. Ecological Diversity and its Measurement. *Croom Helm,* Australia. 179 pp.

Morse C.J, Yang Lianfang & Tian Lixin (ed.) 1994. Aquatic Insects of China useful for Monitoring Water Quality. *Hohai University Press*, Nanjiing People's Republic of China, 569 pp.

Nair, N. C. & Daniel, P., 1986. The floristic diversity of the Western Ghats and its conservation: a review. *Proc. Indian Acad. Sci.* (Anim. Sci./Plant Sci.) Suppl.: 127-163.

Pascal, J. P., 1988. Wet evergreen forests of the Western Ghats of India: ecology, structure, floristic composition and succession. Institute Francaise de Pondichery. Pondicherry, India. 345 pp.

Pearson, R,G., Benson, L.J. & Smith, R.E.W., 1986. Diversity and abundance of the fauna in Yuccabine creek, a tropical rainforest stream. In Limnology in Australia (Eds. P. de Deckker and W. D. Williams), pp. 329-42.

Rosenzweig, M. L., 1997. Species Diversity in space and time. *Cambridge University Press*. 436 pp.

Samways, M.J., 1996. Insects on the brink of a major discontinuity. *Biodiversity and Conservation* 5:1047-1058.

Sivaramakrishnan, K.G., 1984. A new genus and species of Leptophlebiidae: Atalophlebiinae from southern India (Ephemeroptera). *I.J. Entomol.* 26(3): 194-203.

Sivaramakrishnan, K.G., 1987. Continental drift and mayflies. *Biol. Education* 4: 7-10.

Sivaramakrishnan, K.G., Venkataraman, K., Moorthy, R.K., Subramanian, K.A. & Utkarsh, G., 2000. Aquatic Insect Diversity and Ubiquity of the Streams of the Western Ghats, India. *J. Indian Inst. Sci.* 80:537-552.

Stout, J., & Vandermeer J., 1975. Comparison of species richness for stream inhabiting insects in tropical and mid-latitude streams. *Am.Nat.* No: 109. 263-280.

Subrahmanya, K.R., 2001. Origin and evolution of the Western Ghats and the West Coast of India. *Memoir Geological Society of India* 47: 463-473.

Thirumalai, G., 1999. Aquatic and semi-aquatic Heteroptera of India. *Indian Association of Aquatic Biologists*. Publication No. 7. Hyderabad. pp 74.

Vinson, M.R. & Hawkins, C.P., 2003. Broad-scale geographical patterns in local stream insect genera richness. *Ecography* 26: 751-767

Vinson, M.R. & Hawkins, C. P., 1998. Biodiversity of Stream Insects: Variation at local, basin and regional scales. *Annu. Rev. Entomol.* 43: 271-293.

Wiggins, B., 1975. Larvae of the North American Caddisfly Genera (Trichoptera). *University of Toronto Press*, Toronto. 401 pp.

Wiggins, B., 1996. Larvae of the North American Caddisfly Genera (Trichoptera). 2nd edition. *University of Toronto Press*, Toronto. 457 pp.

Chapter - 4

HAPLOTHRIPS FUSCA MOULTON 1928 / *HAPLOTHRIPS GRAMINIS* HOOD 1912 IN HAWAII (INSECTA: TUBULIFERA: PHLAEOTHRIPIDAE)

J.S. Bhatti[1] and Dick M. Tsuda[2]

[1]New Delhi, Email: bhatti.js@gmail.com
[2]University of Hawaii/Manoa, CTAHR, Honolulu, HI, U.S.A.
Email: dmtsuda@hawaii.edu

ABSTRACT

The present study deals with the synonymy of *H. fusca* and *H. graminis* in Hawaii. The Hawaiian literature on the species is surveyed, and its geographical distribution and habitat in Hawaii are discussed.

The species name *Haplothrips fusca* Moulton 1928, a species described from Hawaii, was replaced in 1992 in Hawaiian literature by the species name *Haplothrips graminis* Hood 1912. An indication of a common identity of the two species was never published in printed literature.

The synonymy of *fusca* with *graminis* first appeared in 2002, on the internet. It was found during online search on *Haplothrips graminis* (leading to Hawaiian Arthropod Checklist query, in Nishida 2002— Online Database). Embedded in this database, *Haplothrips fusca* is listed as synonym under *H. graminis*.

Sakimura had, however, already recognized this synonymy around 1954. His detailed handwritten notes, which have now been seen, show that he studied and compared Hawaiian specimens with *H. graminis* collected by him in 1948 from Crystal City in Texas (a state from which Hood's type specimens originate). He concluded that both nominal species stand for the same biological species. This study was carried out presumably in 1954, when Sakimura started labeling Hawaiian specimens as *graminis* Hood.

The synonymy of *H. fusca* with *H. graminis* was never published by Sakimura. The Hawaiian Arthropod Checklist (Nishida 2002, Online Database) formalizes this synonymy.

Nishida, during preparation of the Hawaiian Arthropod Checklist, changed the species name *Haplothrips fusca* to *Haplothrips graminis* in his list, based on personal advice from Sakimura. Accordingly, the Checklist, when printed, had the nominal species *Haplothrips graminis*, but not *H. fusca*. This information was repeated in successive printed editions of the Checklist.

Sakimura's finding of the species from Lanai Island in 1957, not reported so far, is a new island record, based on specimens from the Senckenberg Research Institute, Frankfurt/Main, Germany.

INTRODUCTION

The nominal species *Haplothrips fusca* Moulton was described from Hawaii in 1928, and has been recorded in literature until now by the same name (Sakimura 1937, Sakimura & Krauss 1944, Krauss 1949, Swezey 1954, Zimmerman 1948, 1957; Beardsley & Tuthill 1959, CSIRO 2009, Mound 2010).

In 1992, when the Checklist of Hawaiian Terrestrial Arthropods was published (Nishida 1992: 230), the species name *Haplothrips fusca* Moulton 1928 was missing in the list, and the name *Haplothrips graminis* Hood 1912 appeared for the first time as a species found in Hawaii. The distribution for *H. graminis* in the Hawaiian islands, as given in the Checklist, was identical with the then known distribution of *fusca* in Hawaii. The data on *H. graminis* in the subsequent print editions of the Hawaiian Checklist (Nishida 1994, 1997, 2002) was identical to that published in 1992.

More recent taxonomic literature (CSIRO 2009, Mound 2010) lists *Haplothrips fusca* Moulton as a valid species. zur Strassen's working manuscript file on the species of Thysanoptera recorded from Hawaii (zur Strassen distribution lists: Hawaii, in CSIRO 2009, World Thysanoptera) includes the species *Haplothrips fuscus* Moulton (but not *graminis* Hood), since there was no published record of the synonymy of *fuscus* with *graminis*.

zur Strassen had already known, as far back as early 1975, that *graminis* occurs in Hawaii (and certainly also that *fusca* is a synonym), as evident from slides of *Haplothrips graminis* Hood, "K. Sakimura det.", present in zur Strassen collections in Frankfurt, mounted in Canada balsam on 18.II.1975 (as written on the slides), with the following data: "HAWAII-Archipel, Insel Lanai, Lanaihale, 100 m, 14.VIII.1957, von Gras *Paspalum conjugatum* Berg., K. Sakimura leg. (Balsam, 18.II.1975)" [= Hawaiian Islands: Lanai Island: Lanaihale, 100 m, 14.VIII.1957, from the grass *Paspalum conjugatum* Berg., K. Sakimura leg. (Balsam, 18.II.1975)].

Sakimura's finding of *graminis* on Lanai, reported below, is a new island record, the species has not been previously noted in literature from Lanai Island.

NISHIDA'S CHECKLIST OF HAWAIIAN TERRESTRIAL ARTHROPODS

The species name *Haplothrips graminis* first appeared in Hawaiian literature in 1992 (Nishida 1992: 230, Checklist of Hawaiian Terrestrial Arthropods), and then in the subsequent editions of the Checklist (Nishida 1994, 1997, 2002). The name *H. fusca* does not appear in the 1992 edition. In the index to species in the Checklist of 1994 (p. 244), 1997 (p. 219), and 2002 (p. 265), *fusca* is referred to *graminis*, i.e. "*fusca, Haplothrips* see *H. graminis*". There is no indication as to why *fusca* is referred to *graminis* (because of synonymy or due to misidentification).

The Introduction to the 1992 Checklist indicates that only published information on Hawaiian arthropods is included. Prior to putting together the first printed checklist, which was based on information compiled from the literature, Nishida had consulted with K. Sakimura regarding the Thysanoptera. Based on Sakimura's suggestion (= Sakimura, personal communication), Nishida changed, in the database, the name *fusca* to *graminis*, assuming that Sakimura would be publishing the synonymy. However, Sakimura passed away in 1989 before he could publish his finding that the names *fusca* and *graminis* stand for the same species.

The correction of *fusca* to *graminis* was carried over in subsequent versions of Nishida's checklist, still unsubstantiated by a publication.

After leaving Hawaii in 2001, Nishida finalized the print version of the Checklist 2002 (4th edition) in California. He had also left his citations database with the Bishop Museum, containing the files and records of the Arthropod Checklist prepared by him. The electronic version of Nishida's citations database was developed by the Museum and was published online, unknown to Nishida. This was put up by RLP as reviser ("revised 9 April 2002 by rlp" = Richard L. Pyle).

THE SYNONYMY OF *HAPLOTHRIPS FUSCA* MOULTON 1928 AND *HAPLOTHRIPS GRAMINIS* HOOD 1912

Moulton's species *H. fusca*, described from Oahu Island in Hawaii (Moulton 1928), was rediscovered on the Oahu Island by Sakimura during 1940 from the type host (Sakimura & Krauss 1944) and was also found on Molokai Island (Sakimura 1937, Sakimura & Krauss 1944), Niihau Island (Beardsley & Tuthill 1959), and Maui Island (Krauss 1949).

DMT (Dick M. Tsuda) looked through Sakimura's slides and research notes, and was able to gather the information given below. This information makes it clear that, by 1954, Sakimura had already concluded that the nominal species *H. fusca* is identical with *H. graminis*, and that he intended eventually to publish this synonymy. The conclusion on this synonymy was based on study of variation in Hawaiian specimens and in specimens of *graminis* collected by Sakimura in Crystal City, Texas in 1948 (see below).

(a) Six slides (collected 1933, 1935, 1940, 1940, 1944, 1944), originally labeled *Haplothrips fusca*, were corrected by Sakimura to *Haplothrips graminis*, probably in 1954, or later.

(b) 29 older slides bear the label *Haplothrips graminis*, det. K. Sakimura 1954.

(c) Specimens determined by Sakimura as *Haplothrips graminis* were confirmed from the U.S.N.M. in 1976:

Sakimura sent 3 slides, of specimens collected in Hawaii, to the U.S.N.M. in 1976 for confirmation of identification as *Haplothrips graminis*. These slides bear a noting on the side "confirmed May 1976 USNM":

— 1♀ (labelled *Haplothrips graminis* Hood), Molokai I., Maunaloa, 1380', Bermuda grass, July 14, 1964, beatings, K. Sakimura #4264.

— 4♀ (initially labelled *Haplothrips fusca* Moulton, but corrected to *Haplothrips graminis* Hood), Maui I., Kihei, on *Scirpus paludosus* var *digynus* inflorescence, Nov. 11, 1947, N.L.H. Krauss, #2346.

[This data matches with the record of *fusca* from Kihei (Krauss 1949: 321; where no details are given)].

— 1♀ (labelled *Haplothrips graminis* Hood), Oahu I., Puu Lanipo, 2500', grass sweeping, Sept. 16, 1945, Kay Sakimura, #1826.

(d) There is 1♀ paratype of *Haplothrips fusca* in the Sakimura collection, bearing the original data as noted below for Moulton's description of 1928.

[Hence, Sakimura's determination of *fusca* in Hawaii, i.e. Sakimura 1937, Sakimura & Krauss 1944, Krauss 1949, Beardsley & Tuthill 1959, would be based on comparison with this specimen obtained from Moulton].

(e) Sakimura's personal 1987 revised checklist had *Haplothrips graminis* Hood (= *Haplothrips fuscus* Moulton), but no reference is given. A copy of this checklist was also sent to Bianchi, as shown by a thank you letter for the checklist to Saki (as he was affectionately known to colleagues).

(f) Sakimura's handwritten notes contain two pages, copiously scribbled but clearly readable:

— one page with data on *H. graminis* specimens from Crystal City, Texas (U.S.A.) (from grass, sedge, and *Panicum obtusum*, Oct. 3, 1948, collected by Sakimura), and

— the second page with data on specimens of *H. fusca* from Hawaii (including a paratype ♀ received from Moulton).

9 slides, with 7♀ 3♂, of *Haplothrips graminis* from Crystal City, Texas, were located in Sakimura collections[1].
Noting on one of these slides: "det. K. Sakimura, July 1954".
The remaining slides bear the noting: "det. Aug. 1974".

Detailed characterization (and description of character variation in specimens from each of the two geographical areas, Crystal City and Hawaii) includes antennal colour, fore leg colour, various structural features used in the recognition of the species, and variation in the size of fore tarsal tooth in both sexes (including in gynaecoid and oedymerous males).
Based on study of variation, on the one hand in the Hawaiian specimens and on the other hand in the Texan specimens, Sakimura concluded that the specimens in each case represent the same species, irrespective of their geographical finding.
The notes dealing with the Hawaiian specimens have the following conclusion written in red ink on top of the page: "(*H. fuscus* Moulton 1928=) *H. graminis* Hood 1912, a new synonymy". These notes seem to originate from 1954 when he started labeling Hawaiian specimens as *graminis*.

(The data noted on the Texas specimens on these and other characters closely agree with those on Hawaiian specimens).

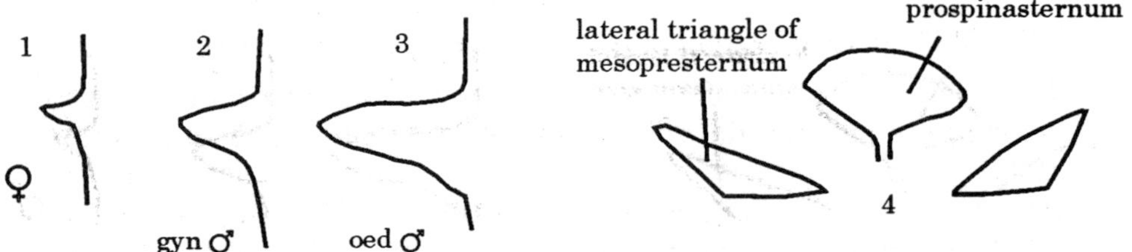

Figs. 1–4. *Haplothrips graminis*, specimens from Crystal City, Texas. (redrawn from Sakimura's pencilled sketches, found with the notes).
1–3, fore tarsal tooth of **1** – ♀, **2** – gynaecoid ♂, **3** – oedymerous ♂. Drawn to identical scale. — The tarsal tooth is tiny, 5 μ long in ♀ (slide no. 3496a-1), 9 μ in gynaecoid ♂ (3499a-1, in fully lateral view), 20 μ in oedymerous ♂ (3498a-1).
4, prospinasternum and mesopresternum.

---

[1] The Sakimura slide and reprint collections, including notes, are presently housed at the University of Hawaii/Manoa, CTAHR, Department of Plant & Environmental Protection Sciences, Honolulu, Hawaii. Contact person: Dick Tsuda <dmtsuda@hawaii.edu>.

LITERATURE OVERVIEW OF THE *H. GRAMINIS* (with particular reference to Hawaii)

***Haplothrips graminis* Hood 1912**, Proceedings of the Biological Society of Washington, 25: 69–70, fig. 3. USA (Texas).

Fig. 3 (on p. 69), head and prothorax, dorsal.

Haplothrips graminis was described from U.S.A. (Texas: Brownsville and Padre Island) and Mexico (Matamaros), based on 'a good series of both sexes'.

Hood found the species common in sweepings. He also noted that the type locality is Brownsville, Texas, although there is no indication of holotype designation.

[Lectotype ♀ (macropterous), USA: Texas, Brownsville (C. A. Hart), USNM Type 74687; subsequently designated by Pitkin (1978: 280). The date of collection of the lectotype is not stated].

***Haplothrips fusca* Moulton 1928**, Proceedings of Hawaiian Entomological Society, 7 (1): 124–125.

Described from holotype ♀, and paratypes 3♀ 2♂.

Hawaii: Oahu Island: Ft. Kamehameha, on *Batis maritima* flowers, May 8, 1927, F. X. Williams. (Moulton No. 2539).

[Synonymized by Nishida 2002, Online Database].

Haplothrips fusca Moulton. — **Sakimura 1937**, Proceedings of the Hawaiian Entomological Society, 9 (3): 420, 423.

p. 420. Oahu, Molokai. *Batis maritima*.

This is the first report from Molokai.

[The collection from Molokai was made in a wind trap on Molokai during 1933 but the host was not known there (see Sakimura & Krauss 1944: 119)].

p. 423. *Haplothrips fusca* is an incidental species and flower feeder.

Haplothrips fusca Moulton. — **Sakimura & Krauss 1944**, Proceedings of the Hawaiian Entomological Society, 12 (1): 119–120, 121.

Haplothrips fusca Moulton, p. 119–120 (collections from Molokai and Oahu).

Kaunakakai, Molokai: *Batis maritima*, flowers.

Near edge of Waikolu Valley, Molokai: *Cladium angustifolium* (a sedge), flowers.

Moderate infestations were observed at both places (in Molokai). *Batis* was growing in a shoreside swamp and *Cladium* was found along a roadside in the rain forest.

The type specimens were collected from *Batis maritima* on Oahu during 1927 (Moulton 1928), and the second collection [Sakimura 1937: 420] was made in a wind trap on Molokai during 1933 but the host was not known there.

The next specimens were caught, again on the type host, at Koko Head, Oahu during 1940. *Batis* and *Cladium* are the only known hosts, of which the former seems to be the preferred one. A colony mixed with *Haplothrips gowdeyi* was observed at Koko Head and also on an unknown host at Kaunakakai on these trips.

p. 121, found in Molokai (in list of thrips known from Maui and Molokai).

Haplothrips (Haplothrips) fusca Moulton. — **Zimmerman 1948,** Insects of Hawaii, Supplement, Vol. 2: 444.

Haplothrips (Haplothrips) fusca Moulton. Oahu (type locality: Fort Kamehameha), Molokai.
Immigrant, but not yet known elsewhere.
Host plants: *Batis maritima, Cladium angustifolium* (sedge).

Haplothrips fusca Moulton. — **Krauss 1949,** Proceedings of the Hawaiian Entomological Society, 13 (3): 321.

Haplothrips fusca Moulton is reported from Kihei on Maui Island. (No further data is given).
[The data on 4♀, Maui I., Kihei, on *Scirpus paludosus* var *digynus* inflorescence, Nov. 11, 1947, N.L.H. Krauss, #2346, now seen by Dick Tsuda, matches this record].
Krauss reports new island records of 14 species of thrips collected by him (identifications by Sakimura) on the islands of Lanai (October–November 1947), Molokai (November 3–5), and Maui (November 7–15).

Haplothrips fusca Moulton. — **Swezey 1954,** Bernice P. Bishop Museum, Special Publication, No. 44: 195.

p. 195. *Haplothrips fusca* Moulton: "Zimmerman (1948, vol. 2: 444) has recorded this thrips from *Cladium angustifolium*".

Haplothrips fusca Moulton. — **Zimmerman 1957,** Insects of Hawaii, Supplement, Vol. 6: 183.

p. 183. On p. 444 (of Zimmerman 1948), add Maui to distribution of *Haplothrips fusca.*

Haplothrips fusca Moulton. — **Beardsley & Tuthill 1959,** Proceedings of the Hawaiian Entomological Society, 17 (1): 56, 57.

p. 57. *Haplothrips fusca* Moulton (det. K. Sakimura, p. 56): sweeping *Scirpus* sp. and *Batis maritima*.
p. 56. Collected on the island of Niihau during August 1947. Determination by K. Sakimura.

Haplothrips graminis Hood. — **Nishida 1992,** Bishop Museum Technical Report, No. 1: 230. (Checklist of Hawaiian Terrestrial Arthropods).

p. 230. The species is adventive in Hawaii.
It has been found on the islands of Niihau, Oahu, Molokai, and Maui.
[There is no mention of *Haplothrips fusca*].

Haplothrips graminis Hood. — **Nishida 1994,** Bishop Museum Technical Report, No. 4: 217. (Checklist, Second edition).

p. 217. Contains identical information as in the 1992 Checklist.
p. 244. Entry for *Haplothrips fusca*, in the Index, reads: "*fusca, Haplothrips* see *H. graminis*".

Haplothrips graminis Hood. — **Nishida 1997,** Bishop Museum Technical Report, No. 12: 187. (Checklist, Third edition).

p. 187. Contains identical information as in the 1992 Checklist.

p. 219. Entry for *Haplothrips fusca*, in the Index, reads: "*fusca, Haplothrips* see *H. graminis*".

Haplothrips graminis Hood. — **Nishida 2002,** Bishop Museum Technical Report, No. 22: 226. (Checklist, Fourth edition).

p. 226. Contains identical information as in the 1992 Checklist.

p. 265. Entry for *fusca*, in the Index, reads: "*fusca, Haplothrips* see *H. graminis*".

Haplothrips graminis Hood. — **Nishida 2002,** (Online Database). Bishop Museum – Hawaiian arthropod checklist. [http://www2.bishopmuseum.org/HBS/checklist/query.asp?grp=Arthropod]

Haplothrips graminis Hood 1912; *Haplothrips fusca* is listed as synonym. Adventive. Found in Niihau, Oahu, Molokai, Maui.

The following "Literature citation list" is given in tabulated form, with 8 references, the species mentioned against each reference is "*Haplothrips graminis*":

Krauss 1949, Proc Haw Ent Soc, 13 (3): 321
Beardsley & Tuthill 1959, Proc Haw Ent Soc, 17(1): 57
Moulton 1928, Proc Haw Ent Soc, 7 (1): 124, 134
Sakimura 1937, Proc Haw Ent Soc, 9 (3): 420
Sakimura & Krauss 1944, Proc Haw Ent Soc, 12 (1): 119, 121
Swezey 1954, Spec Publ Bernice P Bishop Mus, 44: 195
Zimmerman 1948, Ins Haw, 2: 444
Zimmerman 1957, Ins Haw, 6: 183

However, the species name *Haplothrips fusca* alone appears in each of these publications; there is no mention of *Haplothrips graminis* in any of these.

ISLAND RECORDS

The Hawaiian Islands on which *Haplothrips graminis* has been recorded are listed below (from west to east).

Niihau Island. Beardsley & Tuthill (1959: 57).

Oahu Island. Ft. Kamehameha: Moulton (1928: 125).

Koko Head: Sakimura & Krauss (1944: 119).

Puu Lanipo: specimens in Sakimura collection (Dick Tsuda, see page 89 above).

Molokai Island. Sakimura (1937: 420).

Kaunakakai, and near edge of Waikolu Valley: Sakimura & Krauss (1944: 119).

Maunaloa: specimens in Sakimura collection (Dick Tsuda, see page 89 above).

Lanai Island. Lanaihale: new record. Specimens in zur Strassen collections, SMF, Frankfurt am Main, Germany (see page 86 above).

Maui Island. Kihei: Krauss (1949: 321).

RECORDED HABITAT OF *HAPLOTHRIPS GRAMINIS* IN HAWAII

Bataceae	
Batis maritima L., flowers	Moulton 1928: 125; Oahu Island (Ft. Kamehameha) Sakimura & Krauss 1944: 119; Molokai Island (Kaunakakai) Beardsley & Tuthill 1959: 57; Niihau Island, sweeping *B. maritima*
Cyperaceae	
Cladium angustifolium (1) (Gaudich.) Benth. & Hook.f. ex Drake), flowers	Sakimura & Krauss 1944: 119; Molokai Island (near edge of Waikolu Valley)
Scirpus paludosus var. *digynus* (2), inflorescence	Sakimura's collection of 1947, Maui Island (seen by Dick Tsuda, see page 89 above)
Scirpus (3) sp., sweeping	Beardsley & Tuthill 1959: 57; Niihau Island, sweeping
Poaceae	
Cynodon dactylon (L.) Pers., Bermuda grass	Sakimura's collection of 1964, Molokai Island (Maunaloa) (seen by Dick Tsuda, see page 89 above)
Paspalum conjugatum P. J. Bergius	Sakimura's collection of 1957; Lanai Island (specimens in zur Strassen coll., SMF, Frankfurt am Main, see page 89 above)
In wind trap (cf. Sakimura & Krauss 1944), host plant not known	Sakimura 1937: 420; Molokai Island

(1) *Machaerina angustifolia* (Gaudich.) T. Koyama is the currently accepted name of this plant. *Cladium angustifolium* is synonym.

(2) *Bolboschoenus maritimus* (L.) Palla is the currently accepted name of this plant. The combination *Scirpus paludosus* var *digynus* was not found in Hawaiian sources. The Manual (5) lists *Scirpus maritimus* var. *digynus* as synonym of *B. maritimus*. (4)

(3) This record could pertain to a species of *Bolboschoenus* or of *Schoenoplectus*. In the Manual of Flowering Plants of Hawai'i (Wagner et al. 1990), plants in the genus *Scirpus* in Hawaii were split into the genera *Bolboschoenus* or *Schoenoplectus* (otherwise recognized as subgenera of *Scirpus*). While *Scirpus* is still a recognized genus, with the split Hawaii has no taxa in the genus *Scirpus*. (4)

(4) Information from Clyde Imada, Research Specialist, Department of Natural Sciences/ Botany, Bishop Museum, Honolulu [— communicated by e-mail from Richard Pyle, Associate Zoologist, Database Coordinator, Hawaii Biological Survey, Bernice P. Bishop Museum, Honolulu, Hawaii].

(5) Wagner, W.L., Herbst, D.R. & Sohmer, S.H. 1990. Manual of Flowering Plants of Hawai'i. Vols. 1 & 2. 1853 pp. Bishop Museum Special Publications, Vol. 83. Honolulu (Hawaii). (cited from: hbs.bishopmuseum.org/pubs-online/bpbm-sp.html)

ACKNOWLEDGEMENT

Gordon Nishida, California, is thanked for information on his Hawaiian Checklist and for providing pdf copies of the relevant pages of the Checklist. The Hawaiian Arthropod Checklist has proved useful in tracing the *fusca–graminis* problem. It has also provided important leads to the Hawaiian literature on this species. Richard Pyle and Clyde Imada, Bishop Museum, Honolulu, are thanked for providing information on the currently recognized scientific names of some plants associated with *graminis* in Hawaii. Two slides of *graminis* from Lanai were received by JSB many years ago through the kindness of R. zur Strassen.

REFERENCES

Beardsley, J.W. & Tuthill, L.D. 1959. Additions to the known insect fauna of Niihau. Proceedings of the Hawaiian Entomological Society, 17 (1): 56–61. Honolulu (Hawaii).

Haplothrips fusca Moulton; sweeping *Scirpus* sp. and *Batis maritima*.

CSIRO 2009. zur Strassen distribution lists: Hawaii. World Thysanoptera – anic.ento.csiro.au (accessed 7 Sept 2010).

Haplothrips fuscus Moulton is listed, along with 8 other species of *Haplothrips*: *H. davisi* Bianchi, *H. doliicornis* (Bianchi), *H. fissus* Bianchi, *H. gowdeyi* (Franklin), *H. leucanthemi* (Schrank), *H. rosai* Bianchi, *H. sesuvii* Priesner, *H. (Trybomiella) robustus* Bagnall.

Hood, J.D. 1912. New genera and species of North American Thysanoptera from the South and West. Proceedings of the Biological Society of Washington, 25: 61–76. Washington, D.C.

Haplothrips graminis Hood is described as new species, p. 69–70. U.S.A. (Texas: Brownsville and Padre Island) and Mexico.

Krauss [N.L.H.] 1949. New thrips records. Proceedings of the Hawaiian Entomological Society, 13 (3): 321. Honolulu (Hawaii).

Krauss reports new island records of 14 species of thrips collected by him (identifications by Sakimura) on the islands of Lanai (October-November 1947), Molokai (November 3–5), and Maui (November 7–15).

Haplothrips fusca Moulton is reported from Kihei on Maui island. (Further data is not given, but see p. 87 above).

Moulton, D. 1928. Thysanoptera of the Hawaiian Islands. Proceedings of the Hawaiian Entomological Society, 7 (1): 105–134. Honolulu (Hawaii).

Haplothrips fusca Moulton is described as new species, p. 124–125. Oahu Island: Ft. Kamehameha, on *Batis maritima* flowers, May 8, 1927, F. X. Williams.

Mound, L.A. 2010. Thysanoptera (Thrips) of the World – a checklist. www.ento.csiro.au/thysanoptera/worldthrips.html (accessed 10 Aug. 2010).

Haplothrips (*Haplothrips*) *fuscus* Moulton 1928. USA: Hawaii.

Nishida, G.M. 1992. Checklist of Hawaiian Terrestrial Arthropods. Bishop Museum Technical Report, No. 1: viii + 262 p. Honolulu (Hawaii).

p. 230. *Haplothrips graminis* Hood, 1912. Adventive. Niihau, Oahu, Molokai, Maui.
[The above information is repeated in subsequent editions, 1994, 1997, 2002, of the Checklist].
p. 244 (entry in Index), refers to *H. graminis*.
[The name *H. fusca* does not appear in this Checklist].

(This reference and the references to subsequent editions of the Checklist listed below have been taken from literature citations available on the web, and have been confirmed by Nishida. However, the word 'Editor', appearing with Nishida's name in this and all subsequent editions, has been omitted, since, as it now becomes evident, Nishida is the sole author of the Checklist in all four editions).
[The data in the Checklist was compiled, in its entirety, by Nishida alone from published information.
In the 1st edition, 1992, Nishida is listed as 'Editor' on the title page and on the inside cover page, along with three 'Co-Editors'.
In a personal communication (e-mail, Aug 10, 2010) in response to my query, Nishida informed as follows: "In the first Checklist, in the preface, there were a number of people listed who reviewed certain groups, but they were not listed as authors of the sections. In 2nd through 4th editions ... I was the one responsible for the entire edition"].

Nishida, G.M. 1994. Checklist of Hawaiian Terrestrial Arthropods. Second edition. Bishop Museum Technical Report, No. 4: iv + 287 p.

p. 217. *Haplothrips graminis* Hood, 1912. Adventive. Niihau, Oahu, Molokai, Maui.
p. 244 (entry in Index), reads: *fusca*, *Haplothrips* see *H. graminis*.

Nishida, G.M. 1997. Checklist of Hawaiian Terrestrial Arthropods. Third edition. Bishop Museum Technical Report, No. 12: iv + 263 p.

p. 187. *Haplothrips graminis* Hood, 1912. Adventive. Niihau, Oahu, Molokai, Maui.

p. 219 (entry in Index), reads: *fusca, Haplothrips* see *H. graminis*.

Nishida, G.M. 2002. Hawaiian Terrestrial Arthropod Checklist. Fourth Edition. Bishop Museum Technical Report, No. 22: iv + 313 p.

p. 226. *Haplothrips graminis* Hood, 1912. Adventive. Niihau, Oahu, Molokai, Maui.

p. 265 (entry in Index), reads: *fusca, Haplothrips* see *H. graminis*.

Nishida, G.M. 2002. (Online Database). Bishop Museum – Hawaiian Arthropod Checklist. [http://www2.bishopmuseum.org/HBS/checklist/query.asp?grp = Arthropod] [Version: 9 April 2002].

Haplothrips graminis Hood 1912 (*Haplothrips fusca* is listed as synonym).
Adventive in Hawaii. Found in Niihau, Oahu, Molokai, Maui.
Literature citation list gives 8 references.

Pitkin, B.R. 1978. Lectotype designations of certain species of thrips described by J. D. Hood and notes on his collection (Thysanoptera). Proceedings of the Entomological Society of Washington, 80 (2): 264–295. Washington, D.C.

Haplothrips graminis Hood 1912. Lectotype is designated, on p. 280.

Sakimura, K. 1937. A survey of host ranges of thrips in and around Hawaiian pineapple fields. Proceedings of the Hawaiian Entomological Society, 9 (3): 419–427. Honolulu (Hawaii).

p. 420. *Haplothrips fusca* is listed from Oahu and Molokai islands. On *Batis maritima*.
This is the first report from Molokai.

Sakimura, K. & Krauss, N.L.H. 1944. Thysanoptera from Maui and Molokai. Proceedings of the Hawaiian Entomological Society, 12 (1): 113–122. Honolulu (Hawaii).

p. 119–120. *Haplothrips fusca* Moulton, collections from Molokai and Oahu.

p. 121. In list of thrips known from Maui and Molokai.

Swezey, O.H. 1954. Forest Entomology in Hawaii. An annotated Check-list of the Insect Faunas of the various components of the Hawaiian Forests. Bernice P. Bishop Museum Special Publication, No. 44. ix+266 pp. Honolulu (Hawaii).

p. 195. *Haplothrips fusca* Moulton: "Zimmerman (1948, vol. 2: 444) has recorded this thrips from *Cladium angustifolium*".

Zimmerman, E.C. 1948. Insects of Hawaii. Volume 2. Apterygota to Thysanoptera. 475 pp. University of Hawaii Press, Honolulu.

p. 444. *Haplothrips* (*Haplothrips*) *fusca* Moulton. Oahu (type locality: Fort Kamehameha), Molokai. Host plants: *Batis maritima, Cladium angustifolium* (sedge).

Zimmerman, E.C. 1957. Ephemeroptera–Neuroptera–Trichoptera and Supplement to Volumes 1 to 5. Insects of Hawaii, vol. 6: ix + 209 pp. University of Hawaii Press, Honolulu.

p. 183. On p. 444 (of Zimmerman 1948), Maui is to be added to distribution of *Haplothrips fusca*.

Chapter - 5

SPACIAL AND TEMPORAL DISTRIBUTION OF PHLEBOTOMINE SAND FLIES IN RELATION TO LEISHMANIASES IN INDIA

K. Ilango

Zoological Survey of India, Southern Regional Centre,
130 Santhome High Road, Chennai-600 028, India
E-mail: *kilangozsi@rediffmail.com*

ABSTRACT

The distribution of Phlebotomine sand flies (Diptera: Psychodidae) varies both spatially and temporally across the Indian Region. *Phlebotomus argentipes* is a proven vector of protozoan parasite *Leishmania donovani* causing visceral leishmaniasis (Kala-Azar) in north eastern parts of India, Nepal and Bangladesh. Alluvial soil type, high rain fall, temperature and humidity found in these areas are known to influence the biology of *P. argentipes* and thereby impacting the epidemics of Kala-azar. Temperate climate in Terai areas of Himalayan foot hills is a habitat for *Phlebotomus major* which is a well known vector of *Leishmania infantum* that causes infantile visceral leishmanisis. *Phlebotomus papatasi, P. salehi* and *P. sergenti* transmit 'Oriental sore or Delhi boils' in northwestern parts of India including Rajasthan and Punjab by *Leishmania major* and *L. tropica,* respectively. Both sand fly vectors and Leishmania parasites found in north western India are under the influence of arid/ semi arid atmospheric conditions with less rain fall, extremely cold weather in winter and high temperature in summer. Sand fly vector-borne diseases are far more susceptible to climatic changes than any other vector borne diseases as leishmaniae and their associated vectors present kaleidoscopic pattern of geographic distribution. This paper deals with the sand fly vectors influenced by the changes in environmental factors and climatic conditions that are directly associated with epidemics of Kala-azar recorded in W. Bengal and Bihar and the incidence of oriental sore in Punjab and Rajasthan.

INTRODUCTION

Leishmaniasis is transmitted by the protozoan parasite *Leishmania* (Kinetoplastidae: Mastigophora) through the bite of >30 species of sand flies of the subfamily Phlebotoimnae (Diptera: Psychodidae). The disease is among the vector borne diseases restricted to tropical and warm temperate climates found in northern Africa, the Middle East, parts of Europe, central South America and parts of Indian sub-continent. Leishmaniasis is mainly a zoonotic disease, affecting mostly rodents and dogs and the humans are incidental hosts. Both visceral leishmaniasis (popularly known as Kala-azar) and cutaneous leishmaniasis ('Oriental Sore' or Delhi boils') are endemic to India, Pakistan, Bangladesh, Nepal and Sri Lanka. Climate change such as El Nino influences shifts in the distribution and behavior of sand fly vectors that are causing resurgence of leishmanises by increasing the contact between the vectors and hosts (Cardenas *et al.*, 2006). This paper reports the review of leishmaniases burden, distribution of leishmania parasites, their associated sand fly vectors, influence of climate change on sand fly vectors and leishmaniases in India.

DISEASE BURDEN

Human population in 88 countries is affected by leishmaniases, with 12 million people infected worldwide and 2 million new cases recorded annually, of which 1.5 million for cutaneous leishmaniasis (CL), and 500, 000 for visceral leishmaniasis (VL). India is world's one of the most important foci for both visceral leishmaniasis and cutaneous leishmaniasis, accounting for 50% of the total burden of the VL disease.

1. In Indian sub-continent, about 200 million populations are at risk of VL.
2. Nearly 25,000–40,000 cases and 200–300 deaths are reported every year, but these are grossly underestimates.
3. Recent well-designed multicentric studies identified VL burden of 21 cases/10,000 among sampled population in Indian sub-continent (Bangladesh, India and Nepal).
4. This estimates 4,20,000 cases per 200 million risk population clearly indicating that the disease is highly under-reported.

DISTRIBUTION OF PHLEBOTOMINE SAND FLY VECTOR AND LEISHMANIA PARASITES

Phlebotomine sand flies occur in warm parts of the world and their origin was estimated to be in Permian. Of the 700 sand fly species recorded from the World, 70 species are vectors of 30 species of *Leishmania* that cause diseases to people. Sand flies are classified into 5 genera of which New World genera that include *Warileya, Brumptomyia* and *Lutzomyia* (causing mucocutaneous and visceral leishmaniases in the New World) and that old world comprises *Phlebotomus* (causing both cutaneous and visceral leishmaniases) and *Sergentomyia* (feeding on reptiles causing *Sauroleishmania*). Indian sand fly faunas represent 50 species, of which 6 species are leishmaniae vectors and 3 are species complexes. *Leishmania donovani* is transmitted by the bite of *Phlebotomus argentipes* in the endemic foci affecting young adults in north eastern and southern states of India, Bangladesh and Nepal). The disease is confined to man and no animal reservoirs are found. The peak density of *P. argentipes* reaches soon after the monsoon rains during August-October, while the disease season is from November to February and is often with epidemic cycle of 12-15 years of periodicity. *L. infantum* transmitted by the bite of *Phlebotomus major* affects young children in the foot hills of Himalayas and Jammu & Kashmir. Oriental sores or Delhi boils are cutaneous leishmaniasis casued by *Leishmania major* through the bite of *Plebotomus papatasi* and *Phlebotomus salehi* while *L. tropica* is caused by *Ph. sergenti* in (Rajasthan and Punjab). Some progress on the biology of *P. argentipes* in relation to VL outbreak has been attempted during the last 100 years but the knowledge of other sand fly vectors remains unknown in India.

RISK FACTORS OF LEISHMANIASES

The epidemics of leishmanisis is attributed to the risk factors such as environmental changes, deforestation, building of dams, new irrigation schemes, urbanization, temperature and climate changes and migration of non-immune people to endemic areas. The impact of climate variability including events such as El Nino Southern Oscillation (ENSO) associated with dry season leading to flood and drought in many parts of the world and La Nina with wet seasons has been reported to the increasing contact between sand fly vectors and hosts that led to epidemic proportion thus seriously hampering socioeconomic progress of the country. While increases in temperature are likely to spread and the development of both sand fly vectors and *Leishmania* parasites, the impact of climate change needs to be studied in endemic foci of cutaneous leishmanisis and visceral leishmaniasis in India.

INFLUENCE OF CLIMATIC FACTORS ON SAND FLIES AND LEISHMANIASES

During the last 100 years of weather pattern recorded in India the surface temperature has increased by 0.3 °C leading to change in spatial pattern of rain fall, cyclone events and the occurrence of more intense and frequent extreme temperature. Changes in climatic scenarios projected by 2050 are an increase in average temperature by 2-4 °C, decrease in number of rainy days by >15 days in western and central India and an increase by 5-10 days near foothills of Himalayas and north east India (Bhattacharya *et al.*, 2006). Such changes in climate could certainly influence spatial and temporal distribution of the sand fly vectors and leishmaniasis endemic foci in India.

Distribution of VL/CL in India is significantly less than the distribution of the sandfly vectors. The occurrence of disease transmission within the range of the vectors depends on vector abundance, vector survival, vector biting rate (i.e., gonotrophic cycle), the extrinsic incubation period, and the length of the transmission season. Each of these parameters is climate dependent, but the precise relationship with climate needs to be further studied and evaluated. Caution is also required in the interpretation of laboratory experiments as sand flies are able to evade extreme weather conditions in the field by their choice of resting site and time of activity.

Temperature and humidity are the two most important climatic factors for sand fly survival, development and activity. *Phlebotomus* sand flies can survive cold temperatures in diapause (overwintering), which is initiated by a combination of low temperature and reduced daylight and can last 4 to 8 months depending on location. The biting activity of sand flies is strongly seasonal and restricted to the summer months in most areas. Sand flies are very sensitive to temperature, and increases in temperature also may increase daily mortality rates. Temperature influences both blood-meal digestion and development of Leishmania in the sand fly. Engorged flies of *P. argentipes* maintained at 28° ± 1°C were only able to survive ovipostion, reefed and oviposit again (Short *et al.*, 1926). Thus the optimum temperature for the development of *L. donovani* in *P. argentipes* was expected that transmissions by bite were achieved only with flies incubated at 28°C. The best temperature for transmission of *L. tropica* in *P. papatasi* and *P. sergenti* was considered at 30°C. Thus Leishmania species show differences in infection at various temperatures with different sand fly vectors. It was also found that too high temperature was not only interfere with the establishment of parasite infection in sand fly midgut but also inhibit anterior migration of parasites to mouthparts of the sand fly vectors (Killick-Kendrick, 1979).

Phlebotominae are sensitive to sudden temperature changes and prefer regions with small differences between maximum and minimum temperatures. The mapping *P. orientalis* in Sudan found that the geographic distribution was best explained by mean annual maximum daily temperature and soil type. In European region, climate change is likely to push leishmaniasis distribution northwards in the future. The distribution of the vector sand fly is closely linked to climatic conditions because it is usually found south of latitude 45° N and less than 800 m above sea level, but recent findings have detected one species in Germany as high as 49 °N. Also, the incidence of infection has increased in France and Italy and new endemic areas have been detected in northern Croatia, northern Italy, Germany and Switzerland. In Italy climate change may expand the range of one vector (*P. perniciosus*) but decrease the range of another *(P. perfiliewi)*. A 3°C increase in temperature could increase the geographic and seasonal distribution of *P. papatasi* in southwest Asia, provided other ecological requirements are met. The effect of climate and other ecological factors on the distribution and prevalence of leishmanisis (VL & CL) in India were studied by Sivaramakrishnaiah & Ramanatham (1967) using published data of temperature, humidity and rainfall between 1943 and 1953. Based on the latter study, bioclimatic maps were produced classifying zones favorable for low and high incidence of both VL & CL in India. Satellite Remote sensing technology studied by Sudhakar *et al.* (2006) was used to compare the densities of *P. argentipes* with that of different land use cover types between VL endemic and non-endemic areas in Bihar state. The endemic areas had the features such as higher percentage of water bodies, higher proportion of marshes with succulent vegetation and distinct soil types, higher vector density during summer (March-June) than the non-endemic areas and that covered with predominantly thorny and hard-stemmed plants. However, biology of sand flies associated with climate change in both VL and CL endemic areas in the Indian sub-continent is fragmentary and far behind than the Neo-tropical leishmanisis foci and needs in depth study for disease control programs.

CONCLUSION

Climate variability and temperature change influence undoubtedly the biology of sand flies that in turn impact the disease transmission dynamics. Clearly further studies on sand fly biology and climate change at micro-level foci of both VL and CL in India are needed in order to timely intervention of disease outbreak.

REFERENCES

S. Battacharya, C. Sharma, R. C. Diman & A. P. Mitra, 2006. Climate change and malaria in India. *Current Science* 90 (3): 369-375.

Cardenas, R., C.M., Sandoval, A.J., Rodriquez-Morales & C.F. Paredes, 2006. Impact of climate variability in the occurrence of leishmaniasis in northeastern Colombia. *American Journal of Tropical Medicine and Hygiene* 75(2): 273-277.

R. Killick-Kendrick, 1979. Biology of *Leishmania* in phlebotomine sand flies, pp. 395-460. In: Biology of the Kinetoplastida, vol 2, (Eds, W. H. R. Lumsden and D. A. Evans). Academic Press: London/ New York.

Shortt, H.E., P. J. Barraud & A. C. Craighead, 1926. The life-history and morphology of *Herpetosoma donovani* in the sandfly *Phlebotomus argentipes*. *Indian Journal of Medical Research* 13: 947-959.

Sivaramakrishnaiah, K. & R. Ramanatham, 1967. Studies on the effect of climate on Leishmnaisis in India. *Indian Journal of Medical Research* 55(11): 1159-1172.

Sudhakar, S., T. Srinivas, A. Palit, S. K. Kar & S. K. Battacharya, 2006. Mapping of riskprone areas of kala-azar (*Visceral leishmaniasis*) in parts of Bihar state, India: an RS and GIS approach. *Journal of Vector-borne Disease* 43: 115-122.

Chapter - 6

A BRIEF CONSPECTUS AND RESEARCH PRIORTIES ON PLECOPTERA (STONEFLIES) OF INDIA

K.G. Sivaramakrishnan, M. Arunachalam, S. Sundar and C. Selva Kumar

Sri Paramakalyani Centre for Environmental Sciences, Manonmaniam Sundaranar University, Alwarkurichi - 627 412 Tirunelveli, Tamilnadu, India. E-mail: *kgskrishnan@gmail.com*

ABSTRACT

Plecoptera (or Stoneflies) is a small monophyletic order of hemimetabolous insects, with around 3,497 described species included under 286 genera and 16 families. Currently there are 117 described species under 26 genera and 8 families in India. They constitute a significant ecological component of the lotic ecosystems. Aspects of stonefly general biology, life cycle pattern, longitudinal diversity and distribution in India are briefly reviewed followed by an overview of systematics and patterns of phylogeny and biogeography. Future priorities in Plecoptera research are highlighted.

INTRODUCTION

Plecoptera (or Stoneflies) is a small monophyletic order of hemimetabolous insects, with around 3,497 described species included under 286 genara and 16 families. They occur on all continents except in Antarctica (Fochetti & Tierno de Figueroa, 2008). Currently, there are 117 described species under 26 genera and 8 families in India (Kailash Chandra & Sharma 2009). They constitute a significant ecological component of the lotic ecosystems. For Linnaeus, Plecoptera were simply a few individual species of his genus *Phryganea* (Linnaeus, 1758). Burmeister (1839) defined the order and gave it the name Plecoptera, when Pictet (1841) homologized wing veins across Plecoptera families. Species diversity of Plecoptera is highest in cool, fast flowing and rocky streams (Sivec & Yule, 2004). Their diversity declines rapidly from temperate Asian latitudes (nine families) to tropical latitudes (four or fewer families). Of the eight families of Plecoptera known from the mountainous north of India, viz., Capniidae, Leuctridae, Neumouridae, Taeniopterygidae, Chloroperlidae, Perlidae, Perlodidae and Peltoperlidae, only one Perlidae (possibly with one species of Amphinemuridae) is recorded so far from South India. Most species of Plecoptera are stenothermic, with water temperature playing an important role in their longitudinal distribution, egg hatching and adult emergence periods.

GENERAL BIOLOGY

Being a primitive hemimetabolous insect, stonefly has egg, nymphal (larval) and imaginal (adult) stages in its life history. Unfortunately, there is paucity of information on seasonality, life cycles, egg incubation strategies and general bionomics of tropical stonefly species in general and Indian species in particular. It is still an enigma whether the numerous morphospecies that are presently distinguished in the tropics are true biospecies or are there regional variation and morphoclines over vast tropical areas (Zwick, 2003)?

Adult stoneflies are rather inconspicuous insects, which usually hide in bushes or grass close to the water. They are soft bodied, with large head, 3-segmented tarsi, elongate filiform antennae, mouthparts with mandibles, a pair of compound eyes, 2 or 3 ocelli, 2 usually long anal cerci, 10-segmented abdomen with vestiges of the eleventh segment, and 2 pairs of membraneous large wings (although a few genera are apterous). At rest, they close their broad wings over the body, the front pair covering the fan-shaped hindpair, which is folded. Hence, the name of the insect order Plecoptera, which means "plaited or folded wings". Adults of stoneflies differ from those of mayflies not only in folding the wings but also in having much longer antennae and only two tails. Most are awkward fliers. Some adults do not feed. Others drink honeydew, the sugary excrement of aphids and some eat leaves, rotten wood or algae that encrust bark. Adults have a short lifespan from a few days to weeks. Mate searching behaviour by intersexual vibrational communication and ovipositional behavioural strategies are highly complex worth investigating.

The nymphs are similar to the adults in appearance, but do not have fully developed wings or genitalia. Wingpads develop gradually on the thorax with successive instars. As in the adults, tarsi are 3-segmented. Many stonefly nymphs co-occur in the same habitats with mayfly nymphs, but have much longer antennae, only 2 long tails, and their gills are clusters of filaments that are usually at the bases of the 3 pairs of legs on the thorax. Stonefly nymphs inhabit stony substrates in cool streams. They are often found on the surfaces of cobbles and boulders in riffles and in decaying leaf litter, submerged macrophytes as well as gravel and sandy substrates. Ecological investigations on stoneflies of tropical Asian streams are fragmented (Dudgeon, 1999). Perlid stonefly nymphs inhabiting south Indian streams are carnivores eating a wide variety of aquatic invertebrates, although small early instars may be detritivores, becoming increasingly predatory with increasing size. Study of predatorprey interactions involving stoneflies is a priority area of research in India, since the impact of predatory stoneflies on their prey could well be significant.

LIFE CYCLE PATTERN

In tropical Asia, stonefly life cycles appear to be aseasonal with adult emergence throughout the year while stoneflies in temperate regions typically have highly synchronized life cycles, with emergence of the entire population occurring within a few weeks. Many stoneflies remain in the vicinity of the stream or lake, mate on or near the ground under stones or in vegetation and therefore do not need the capability of flight. This applies to alpine Himalayan species where low air temperatures restrict flight activity. Sexual dimorphism occurs in several species whereby the males are short-winged while females have wings of normal length. This saves the males energy which can be utilized elsewhere, while normal-winged females once mated are able to function as the dispersal agency (Brittain, 1990). Egg structure is species specific. Ovipositional behaviour is complex with some species extruding eggs form the tip of the abdomen at the water surface, others using ovipositors to lay their eggs deeper in the substrate, yet others crawling under water and depositing their eggs under stones. Eggs hatch in a few weeks after they are laid. Species inhabiting intermittent streams may have egg diapause extending to 6 months or more (Sivec & Yule, 2004). Life cycle studies on Indian Stoneflies are virtually nil.

LATITUDINAL DIVERSITY AND DISTRIBUTIONS

Ideally factors that determine faunal diversity include the size of the area, habitat heterogeneity, adaptations to specific niches, ecological requirements, competition, predation, historical events, temperature, and latitude amongst others (Gaston, 2000). The spatial scale is a fundamental factor for determining latitudinal diversity (Boyero, 2003). Site diversity is usually assessed on local and regional scales. Republic of India spans from N 37° to S 08°. It has 10 Biogeographic zones viz., Trans-Himalayan, Himalayan, Desert, Semi-arid, Western Ghats, Deccan Peninsula, Gangetic plain, North-East India, Islands and Coasts (Alfred, 1998).

Based on data on stonefly distributions in India currently available, which is rather incomplete and with limitations of lack of comprehensive and uniform investigations in all latitudinal zones within the Republic of India, an attempt is made to analyse the latitudinal diversity profile of Plecoptera of India mainly at the generic level to focus the need for more intensive studies in future since the restrictions of many Plecoptera larvae to cool and clean streams and rivers make them potentially interesting as biological indicators. Stoneflies are absent in semi-arid desert and euryhaline coastal regions. They are speciose in stony, cool mountain streams particularly in the two hotspots of biodiversity, *viz.*, the Western Ghats and North–East, with maximum species richness in Trans- Himalayan and Alpine Himalayan lotic ecosystems. The two ecological factors that seem to be the most important in Plecoptera distributions are cold temperatures and water quality in addition to the basic patterns of colonization dictated by historical biogeographic events at the time of their origin and subsequent evolution and patterns of current distribution as a result of vicarience/dispersal. In a nutshell, diversity of Indian stoneflies increases progressively from southern peninsular tropics towards the northern subtropical, temperate and alpine upper Himalaya (Das, 1991, 1998).

Table 1. Distribution of Stoneflies in different biogeographic zones in India

Sl. No	Family	Trans-Himalayan (N 32° - 37°)	Himalayan (N 28° - 32°) (N 22° - 37°)	Desert (N24°- 28°)	Semi-arid (N 22° – 28°)	Western Ghats (N 08° – 22°)	Deccan Peninsula (N 12° – 18°)	Gangetic plain (N 24° – 28°)	North-East India (N 21° - 28°)	Islands (N07° - 14°)	Coasts(N08°- 22°)
1	Perlidae Genus: *Neoperla* *Phanoperla* *Chinoperla*			-	-		-	-	-	-	-
2	Capniidae			-	-	-	-	-	-	-	-
3	Leuctridae			-	-	-	-	-	-	-	-
4	Taeniopterygidae			-	-	-	-	-	-	-	-
5	Perlodidae			-	-	-	-	-	-	-	-
6	Chloroperlidae			-	-	-	-	-	-	-	-
7	Nemouridae			-	-	-	-	-	-	-	-
8	Peltoperlidae			-	-	-	-	-	-	-	-

(- Data not available)

SYSTEMATIC OVERVIEW

The order Plecoptera includes 16 families of which 4 families are included in the suborder, Antarctoperlaria present only in the southern

Hemisphere (hence, not represented in India) and the remaining 12 families are included in the suborder, Arctoperlaria distributed mostly in the Northern Hemisphere, 8 of which are recorded in Himalayan region. Systematic accounts of the stonefly families recorded from India are as follows:

- Family Perlidae: The taxonomic situation of this family was cleared by the significant publication entitled "Synopsis of the World Genera of Perlinae" by Sivec *et al.* (1988). The genera of Perlinae recorded from India are *Chinoperla* (1 species), *Claassenia* (1 species), *Kammimuria* (4 species), *Neoperla* (14 species), *Perla* (4 species), *Phanoperla* (11 species) and *Tyloperla* (1 species) (Kailash Chandra and Sharma, 2009). Perlids are the dominant stoneflies of India. Adult perlids are characterized by cerci longer than greatest width of pronotum, which is as wide as head. Branched gill remnants are located behind leg bases. Larval perlids are characterized by the presence of thoracic gills, a striking primitive feature and labial glossae which are rounded and globular, a specialized trait. They are mainly, predaceous, although smaller larvae may ingest algae and/or detritus (Hynes, 1976). Three genera of the tribe Perlini viz., *Neoperla, Phanoperla* and *Chinoperla* are recorded from the Himalayan region. The aforementioned genera have penetrated to the hill streams of southern tip of the peninsular India both in the Western Ghats and the Eastern Ghats.

Lack of the anterior ocellus is the most widely used diagnostic character of Neoperlini. They have distinctive male genitalia. The hemitergite x are transversely positioned and their medially-directed tips are covered with minute conical setae. Another diagnostic feature is the tubular sclerites connected with the attachment of the efferent ducts (Zwick, 1982a). Presently, only the strongly sclerotized, dorsally projecting sclerites at the sides of the basal penis opening suggest monophyly of the single large genus, *Neoperla*. All *Neoperla* have the primitive complete set of thoracic gills (including supracoxal ones), have an unmodified occiput, lack the retractor sclerite of the eversible penial sac, lack outgrowths on the anchor plate of the eggs, and males of most species have a modified tergite VII. Oriental Perlidae differ from other *Neoperla* in the angle at which the branches of Cu forked, in the size and placement of tentorial callouses on the head, and in the male genitalia.

Adult males of *Neoperla* are biocellate, hemitergites bifurcate, with anterior finger-like process extending from mesally directed basal piece, tergum 9 not sclerotized, with Y-shaped sclerite and Cu_2 of fore wings straight basally, arising from Cu_1, at an acute angle; tergum 7 typically with a lobe. Nymphs of *Neoperla* are biocellate, anterolateral submedian lobes without basal suture, with 2 pairs of PSC3 gills, posterior margin of mesosternum without setal fringe and lateral margins of pronotum with fringe incomplete; without strong bristles along wing pads and with anal gills. The eggs are striate, without outgrowths on anchor plate of eggs. *Neoperla* is found in North America and throughout the Oriental and Afrotropical regions. Several of the South Indian *Neoperla* are apparently endemic. Zwick (1981) has provided an excellent key to the males of south Indian Genera of Neoperlinae and species of *Neoperla*. Nine species of *Neoperla* are described from South India so far (Zwick, 1981; Stark & Sivec, 2008). More indepth survey in montane streams of the Western Ghats and the Northern Eastern Ghats and in North East India may widen the horizon of knowledge of neoperline stoneflies of India.

Phanoperla adults are of small to medium size wings, 6.1-18.6 mm long, yellowish to light brown, sometimes with darker marks. 2 (rarely 3) ocelli, pronotum very short, almost twice as long as wide. Cu_2 of front wing very short, its base strongly bent, forking out of Cu_1 almost at a right angle; tergum 7 without lobe. Nymphs of *Phanoperla* are usually bicellate, anterolateral submental lobes separated form prementum by transverse sutures; one pair PSC 3 gills; posterior margin of mesosternum with setal fringe. Abdominal sternites and tergites all fused to form rings. Cerci composed of very short segments densely covered with fine procumbent hairs, often void of erect setae; usually, cerci are strikingly short thick, and conical. Zwick (1982 b) notes that the posterior part of the head (anterior to the occipital ridge) of *Phanoperla* is pale and this is a distinctive feature of live specimens. *Phanoperla* eggs are variable, anchor plate always with finger or tentacle-shaped extensions carrying the usual granular, strong light-refracting bodies. *Phanoperla* is endemic to the Oriental Region (Zwick, 1982 b) occurring from Sri Lanka and India (including Nepal), through South-east Asia, Indonesia and the Philipp-ines to Hainan. The closely related genus *Chinoperla* ranges from south-ern China, Thailand and Peninsular Malaysia into Borneo and Sumatra and Westward into India. The nymphs of *Chinoperla* have yet to be described.

Chinoperla shares with *Phanoperla* the occipital mark (less distinct, margins not raised), the delicate retractor sclerite of the eversible penial sac, tentacular outgrowths on the anchor of the egg, and also lacks supracoxal gills. These details are all apomorphie. It also has brushes on abdominal sternites which are plesiomorphic and occur in most Perlinae and also in some *Neoperla*. *Chinoperla* differs from *Phanoperla* by a straight Cu_2 arising from Cu_1 at the usual oblique angle, by lack of a transverse vein between the anal veins in the hind wing, by presence of a forked anterior process on tergite IX, lack of minute conical setae on the same tergite and by structured details of the penis (Zwick, 1982a). *Chinoperla* plus *Phanoperla* definitely form a monophyletic group, and the monophyly of genus *Phanoperla* is supported by the shortened, curved

Cu_2. The genus *Chinoperla* ranges from southern China, Thailand, and Peninsular Malaysia into Borneo and Sumatra and Westward into India. *Chinoperla* species are united by the long, slender, cleft sclerite on T_8 and by the very short Sc vein in the forewing (Sivec *et al.,* 1988).

Among the perlid subfamily Acroneuriinae are included *Brahmana* (India and Nepal) *Gibosia* and *Kalidasia* (Peninsular Malaysia and India). Hopefully, this account will stimulate and facilitate continued collecting and research so badly needed on this family of stoneflies in India, many of which are undoubtedly endangered or extinct through habitat degradation (Sivec *et al.,* 1988).

- Family Capniidae: Six species of *Capnia* viz., *C. bifida, C. gibbera, C. hingstoni, C. manii, C. montana and C. pedestris* are recorded from India. Adult capniids are small and dark forms, walking on snow and ice near streams. No, long series of cross-veins are present in the wings. Wings are rarely absent. Tails are conspicuous. Nymphs are small, slender and cylindrical, often dark forms (3-6 mm). They occur in habitats ranging from tiny springs to moderately sized streams. Wing pads are occasionally absent. Hind wing pads are nearly parallel. Anal lobe of hind wing pads of some species extends over half the length of wing pad, thus giving the appearance of a thickened pad. Lateral margins of abdomen as viewed from above appear zigzagged. They are essentially detritivores.
- Family Leuctridae: Jewett (1958) recorded *Rhopalopsole magnicerca* from India. Wings of adults at rest conspicuously roll over the sides of the body. Tails are reduced and inconspicuous. Labial palps of nymphs sometimes extend beyond the body of the labium. Lateral margins of abdomen, as viewed from above, are relatively smooth and not conspicuously zigzagged. Nymphs occur in swift, rocky-bottomed streams.
- Family Taeniopterygidae: Subfamily Brachypterinae is represented in Himalayas in India by *Kyphopteryx* sp. and *Mesyatsia sp.* whereas subfamily Taenioptertginae is not recorded so far from India. Tarsal segments of adults 1 and 2 are about the same length. There is no apparent X in the fore wings, since the apical costal cross-vein is absent or distally located. Tails may be inconspicuous or conspicuous. Nymphs do not possess gills. Nymphs often occur in large rivers.
- Family Perlodidae: Subfamily Perlodinae is represented in Himalayas in India by *Filchneria amabilis, Filchneria shobhaae* and *Neofilchneria uncata*. Adults are similar to perlid stoneflies. They are often greenish or yellowish but sometimes darker. Vein A_2 of the fore wings forks at a point basal to the A_1-A_2 cross-vein.

Nymphs resemble perlids but lack the branched thoracic gills. Nymphs are often highly patterned forms and when mature, usually measure 8-16 mm excluding tails, although size is extremely variable.

- Family Chloroperlidae: This family is represented in Himalayas in India by *Xanthoperla kishanganga*. Adults are similar to perlodids, but vein A_2 in the fore wings either forks at a point beyond the placement of the A_1-A_2 cross-vein or does not fork at all. Anal lobe of the hind wings is sometime reduced or absent. Nymphs are often patternless, somewhat cylindrical forms. They usually measure 5-12 mm (excluding tails) when mature. Gills are generally lacking. Hind wing pads are almost always nearly parallel. Tails are almost always shorter than the abdomen.
- Family Nemouridae: Both the subfamilies Amphinemurinae and Nemourinae are represented in India. Eighteen species of *Amphinemura*, 14 species of *Indonemoura*, 9 species of *Mesonemoura* and one species each of *Protonemoura* and *Sphaeronemoura* are recorded from India (Kailash Chandra & Sharma, 2009). It is now the second most diverse family. X is usually located in the distal half of the fore wings, formed in part by the apical costal crossvein. Wings lie flat on the body and do not curve around the sides of the body. Tails are inconspicuous and reduced to one full segment. The nymphal body shape is stout. Hind legs, when stretched alongside abdomen do not reach abdominal apex. Wingpads of mature nymphs are divergent from body axis. Small rivers, streams and springs are the most common habitats of nemourid nymphs. Some prefer soft substrates, and some commonly are associated with leaf packs and other detritus. Mountain species often occur in cold, rocky-bottomed streams. They are primarily detritivores, devouring leaves and the resident microorganisms of those leaves. Nymphs of the subfamily Amphinemourinae possess highly branched cervical gills whereas those of the subfamily Nemourinae may lack cervical gills, have unbranched cervical gills, or more rarely possess cervical gills with less than 6 branches (McCafferty, 1981).
- Family Peltoperlidae: Although the Peltoperlidae extend into the Oriental Region, the distribution of the family as a whole is centered on Palaearctic East Asia and North America (Illies, 1965). Cerci of adults are not much longer than the greatest width of pronotum, which is much wider than the head. No remnants of nymphal gills are present. The nymphs are aberrant in their general appearance. They are 'Cockroach – like'. Thorax is much wider than either head or abdomen. 7 species of

Cryptoperla and one species of *Peltopteryx*, the genus established by Stark (1989) are recorded from India.

Perlids

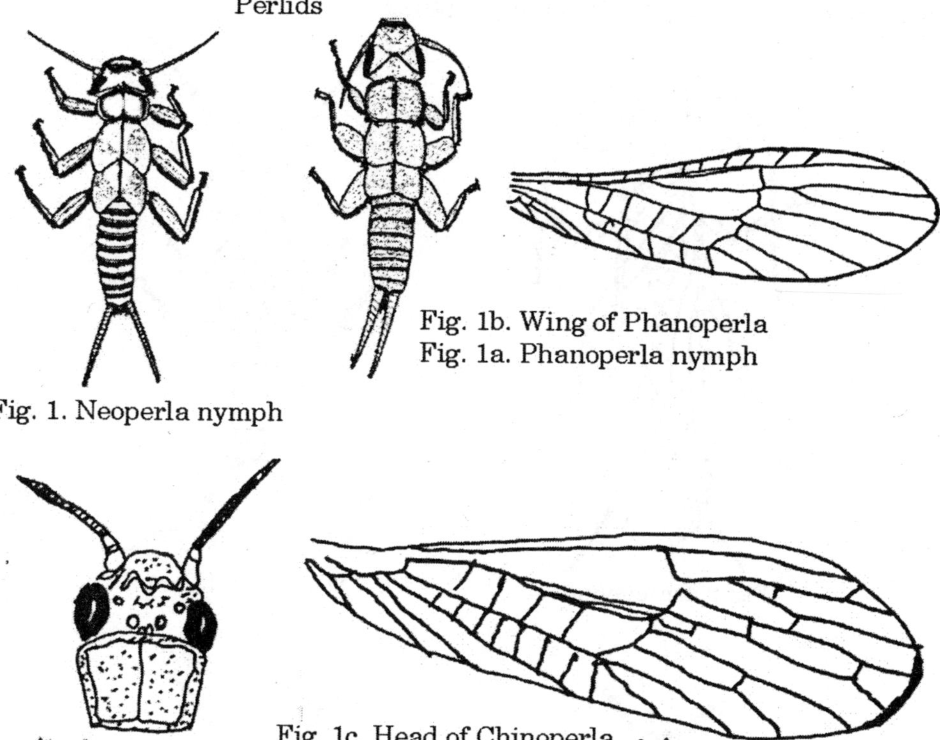

Fig. 1. Neoperla nymph

Fig. 1b. Wing of Phanoperla

Fig. 1a. Phanoperla nymph

Fig. 1c. Head of Chinoperla

Fig. 1d. Wing of Chinoperla

PHYLOGENY AND BIOGEOGRAPHY

Stoneflies are pterygote, neopteran, exopterygote insects. Although they are usually included among Polyneoptera (possibly they are the sister-group of the remaining Polyneoptera), the relationships of Plecoptera with other orders of this subdivision are poorly understood (Gullan & Cranston, 2005). The order includes 16 families. Zwick (2000) recognizes two large groups: Antarctoperlaria, present only in the Southern Hemisphere, and Arctoperlaria, distributed mostly in Northern Hemisphere. The first taxon includes 4 families; the latter includes 12 families belonging to two subgroups (Systellognatha and Euholognatha) with six families each.

The monophyly of Antarctoperlaria is supported by morphological characters, while monophyly of Arctoperlaria is supported by ethological and morphological characters related to mate finding: the intersexual vibrational communication (drumming, tremulation, rubbing or combinations of them) and/or the existence of structures associated to it in the male (lobes, hammers etc.) that are widely present in all families of this group except Scopuridae, and absent in Antarctoperlaria (Zwick, 2000). Among the existing families, only Notonemouridae seems to be a polyphyletic assemblage (McLellan, 2000).

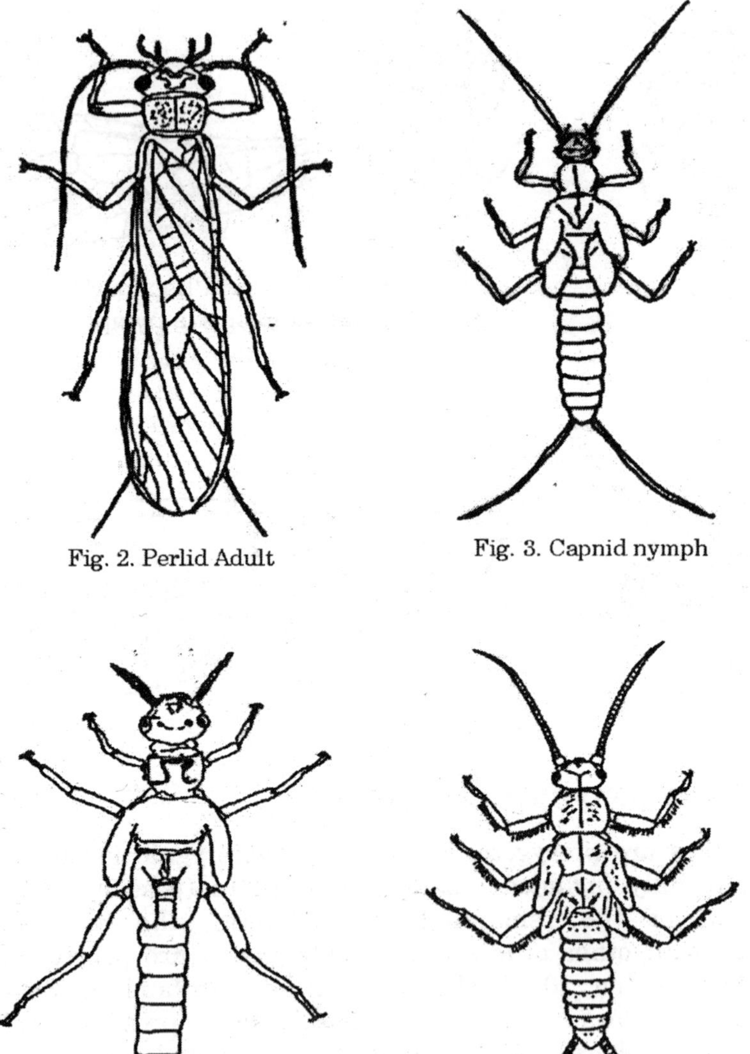

Fig. 2. Perlid Adult

Fig. 3. Capnid nymph

Fig. 4. Leutrid nymph

Fig. 5. Taeniopterygid nymph

According to Rohdendorf & Rasnitsyn (1980), recent stoneflies are derived from Plecoptera from the Permian Period. The oldest known Plecoptera fossils are from the early Permian Period, 258-263 million years ago (Shinitshenkova, 1997).

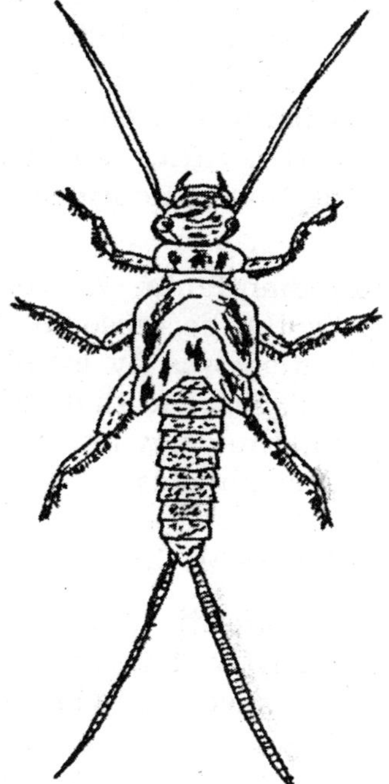

Fig. 6. Perlodid nymph

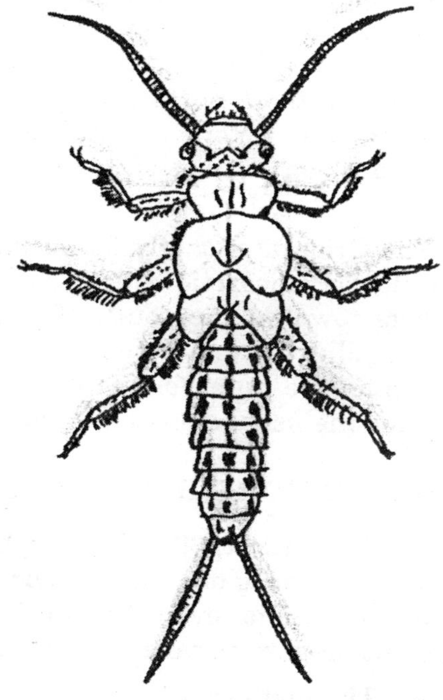

Fig. 7. Chloroperlid nymph

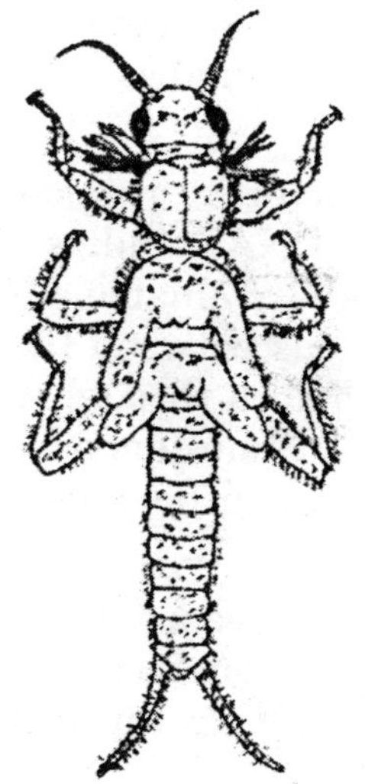

Fig. 8. Nemourid nymph

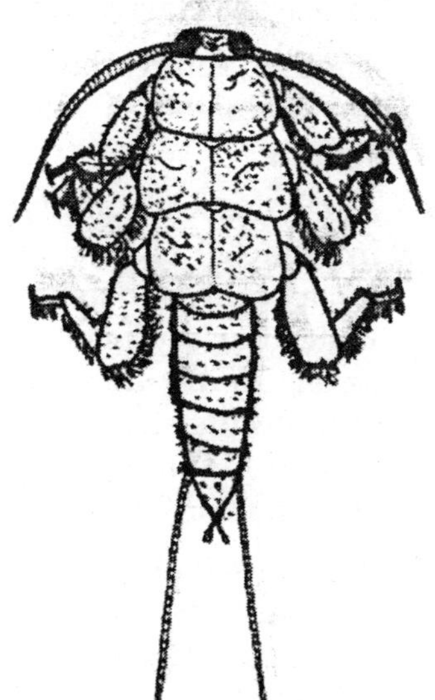

Fig. 9. Peltoperlid nymph

According to Illies (1965), Banarescu (1990) and Zwick (1990, 2000), Arctoperlaria and Antarctoperlaria originated as independent lines at the splitting of Pangea and the subsequent separation of Gondwanaland and Laurasia, at the end of Triassic Period. Antarctoperlaria possibly began their diversification before the continents separated, producing some sister-groups distributed in South America, Australia and New Zealand. The absence of Antarctoperlaria from South Africa and India may be interpreted as a later extinction event. These lands became warmer and drier during their northward journey after separation from Antarctica and Australia. Before the isolation of Malagasey, Africa was for sure colonized by one family of Antarctoperlaria, the Notonemouridae (Zwick, 1990).

The Oriental stonefly fauna inclusive of Indian fauna was colonized from close Palaearctic areas, as suggested by the decrease in species numbers towards the south. The first penetration was by members of three families – Nemouridae, Peltoperlidae and Perlidae that extended down into the rain forest of southeast Asia, thus having somewhat overcome the usual aversion of the order for warm climates.

FUTURE PRIORITIES IN PLECOPTERA RESEARCH

Peter Zwick (2003) has highlighted the priorities in Plecoptera research at the global level. They apply equally well to the study of stoneflies of India, even the alpha taxonomic knowledge of which is very poor. With regard to taxa distinctions,

- The methods of artificially everting male genitalia and using not only associated sclerites, but also the shape of the aedeagus may be used to clarify taxonomic dilemmas.
- Another refinement of taxonomic distinctions should be through the inclusion of behavioural evidence in studies of drumming signals.
- Allozyme work studying chemical character expressions to make inferences on genetics, as well as DNA studies analyzing nucleotide sequences are to be promoted. Invaluable advantage of molecular methods is to allow associations between life stages or sexes. Adult -larval association of Plecoptera through application of molecular methods should be adopted. Simultaneously, pairs in copulation or mating studies and drumming behaviour could be used for conformation.
- There is little information regarding the seasonality, life cycles and egg incubation strategies of Indian stoneflies. Such studies are urgently needed in the context of the biomonitoring potential of stoneflies.

With regard to taxa relationships, contributions to reconstruction of the cladistic relationships between taxa through application of molecular phylogenetics complementary to the traditional morphological phylogenetic relationships are urgently needed. In fact, no molecularly based phylogenetic system of Plecoptera has yet been proposed (Zwick, 2003).

ACKNOWLEDGEMENTS

K.G. Sivaramakrishnan and M. Arunachalam thank UGC for financial support through a major research project F.No. 32-635/2006 (SR) to Sri Paramakalyani Centre for Environmental Sciences, Manonmaniam Sundaranar University, Alwarkurichi, Tirunelveli, Tamilnadu, India.

REFERENCES

Alfred, J.R.B., 1998. Faunal Diversity in India: An Overview. Pp 1-9 In: Alfred, J.R.B. *Faunal Diversity in India.* ENVIS Centre Zoological Survey of India, Calcutta 497 pp.

Banarescu, P., 1990. Zoogeography of Fresh Water. *General Distribution and Dispersal of Freshwater Animal*, Vol. 1. Aula-Verlag, Wiesbaden.

Boyero, L., 2003. Multiscale patterns of spatial variation of stream macroinvertebrate communities. *Ecological Research* 18: 365-379.

Brittain, J.E., 1990. Life history strategies in Ephemeroptera and Plecoptera p. 1-12. In: Cambell (Ed.) *Mayflies and Stoneflies : Life Histories and Biology*, Kluwer Academic Publishers, Boston 366 pp.

Burmeister, H.C.C., 1839. *Handbuch der Entomologie. Berlin: Enslin*. pp. I-XII, 757-1050.

Das, B.C., 1991. Plecoptera. Pp. 259-260. In: Jairajpuri, M.S. Director, ZSI (ed.). *Animal Resources of India: Protozoa to Mammalia,* 694 pp.

Das, B.C., 1998. Plecoptera. Pp. 180-182. In: Alfred, J.R.B., Das, A.K and Sanyal, A.K. *Faunal Diversity in India.* ENVIS Centre Zoological Survey of India, Calcutta, 497 pp.

Dudgeon, D., 1999. Tropical Asian Streams, Zoobenthos, *Ecology and Conservation*, Hong kong University Press, Hong kong, 830 pp.

Fochetti, R. & J.M. Tierno de Figueroa, 2008. Global diversity of stoneflies (Plecoptera: Insecta) in freshwater. *Journ. Hydrobiologia* 595: 365-377.

Gaston, K.J., 2000. Global patterns in biodiversity. *Nature* 405: 220-227.

Gullan, P.J. & P.S. Cranston, 2005. *The Insects. An outline of Entomology.* 3rd ed. Blackwell Publishing Ltd., Oxford.

Hynes, H.B.N., 1976. The biology of Plecoptera. *Ann. Rev. Entomol.* 21: 135-153.

Illies, J., 1965. Phylogeny and zoogeography of the Plecoptera. *Ann. Rev. Entomol.* 10: 117-140.

Kailash Chandra & Sharma, R.M., 2009. An electronic checklist of Indian Stoneflies (Insecta: Plecoptera). *Zoological Survey of India.*

Linnaeus C., 1758. *Systema naturae.* Holmiae: Louv. *Salvii.* 10th ed. Vol. 1,2: 824 pp.

McCafferty, W.P., 1981. Aquatic Entomology. The Fishermem's and Ecologists' Illustrated Guide to Insects and their Relatives. *Science Books International* Boston, 448 pp.

McLellan, I.D., 2000. A revision of Cristaperla (Plecoptera: Notonemouridae) and some comments on Notonemouridae and its generic groups. *New Zealand Journal of Zoology* 27: 233-244.

Pictet F.J., 1841. Histoire naturelle generale et particulier des insectess nevropteres. *Famille des Perlides 1. Partie. Geneva (Kessman) and Paris (Bsilliere),* I-XIV, 1-423, 1-23 & 53 pls.

Rohdendorf, B.B. & A.P. Rasnitsyn, 1980. (Historical Development of Insect Class). *Trudy Paleontologischeskogo Instituta* 178: 1-267 (in Russian).

Shinitshenkova, N.D., 1997. Paleontology of stoneflies. In Landolt, P. & M. Sartori (eds), Ephemeroptera and Plecoptera, *Biology-Ecology-Systematics* MTL Fribourg, 561-565 pp.

Sivec, I., Stark, B.P. & Uchida, S., 1988. Synopsis of the world genera of Perlinae (Plecoptera: perlidae). *Scopolia* 16: 1-66.

Sivec, I. & Yule, C.M., 2004. Insecta: Plecoptera. Pp. 443-456. In: Yule, C.M and Sen, Y.H. (Eds.). *Freshwater Invertebrate of the Malaysian Region.* Monash University. Akademi Sains, Malaysia.

Stark, B.P. 1989. Oriental Peltoperlinae (Plecoptera): a generic review and description of a new genus and seven new species. *Entomologica Scandinavia* 19 (4): 503-525.

Zwick, P., 1981. The South Indian species of Neoperla (Plecoptera: Perlidae). *Oriental Insects* 15: 113-126.

Zwick, P., 1982a. Contribution to the knowledge of Chinoperla (Plecoptera: Perlidae: Neoperlini). *Aquatic Insects* 4: 167-170.

Zwick, P., 1982b. A revision of the Oriental stonefly genus Phanoperla (Plecoptera: Perlidae). *Systematic Entomology* 7: 87-126.

Zwick, P., 1990. Transantarctic relationships in the Plecoptera. p. 141-148. In: Campbell I.C. (ed.), *Mayflies and Stoneflies, Life History and Biology.* Kluwer Academic Publishers, London, 366 pp.

Zwick, P., 2000. Phylogenetic System and Zoogeography of the Plecoptera. *Annual Review of Entomology* 45: 709-746.

Zwick, P., 2003. Plecoptera research today: Questions to be asked in the new millennim. Pp. 245-251. In: Gaino, E (Ed.) *Research Updated on Ephemeroptera and Plecoptera*, University of Perugia, Italy.

Chapter - 7

DIVERSITY AND DISTRIBUTION OF ODONATES IN GREAT NICOBAR BIOSPHERE RESERVE (GNBR), ANDAMAN & NICOBAR ISLANDS

C. Sivaperuman, Suresh Kumar Shah, C. Raghunathan and Ramakrishna*

Zoological Survey of India, Andaman and Nicobar Regional Centre, Port Blair- 744 102, Andaman and Nicobar Islands, India
* Zoological Survey of India, Prani Vigyan Bhawan, M-Block, New Alipore, Kolkata - 700 053, India

ABSTRACT

The Great Nicobar Biosphere Reserve is the southernmost Island of Andaman and Nicobar Archipelago. This study was conducted during 2008 – 2010 and eleven different locations were surveyed to assess the diversity and distribution of Odonates. A total of twenty species belonging to 4 families were recorded during the study period. Highest number of species was observed from the family Libellulidae. The diversity index was varied in different locations. High similarity was between Gingen basti and East-west Road.

INTRODUCTION

The Order Odonata is divided into three suborders, the damselflies (Zygoptera), the dragonflies (Anisoptera) and the primitive dragonflies (Anisozygoptera). However, the odonatologists of the recent world commonly use the term "*dragonfly*" for the members of all the three suborders. The Odonata is one of the primitive and ancient insect orders. It is very diverse and is the second largest aquatic insect order. Dragonflies are predaceous, hemimetabolous and amphibiotic insects, which inhabit all kinds of freshwater habitats either permanent or temporary. Odonata experiences two totally different life styles. In almost all cases, the egg and larval stages are aquatic whereas the adults are terrestrial (Silsby, 2001). The prey of the adults consists mostly of the harmful insects of crops, orchards and forests and thus has a regulatory impact on the agro-forestry. Total of 470 species under 139 genera and 19 families were reported in India (Subramanian, 2009), of which 45 species under 31 genera and 11 families have been recorded in Andaman and Nicobar Islands by various workers (Fraser, 1933, 1934 & 1936; Chhotani *et al.*, 1983; Lahiri & Mitra, 1993). Recently, Mitra (2002) reported 32 species from Nicobar Group of Islands. The present study was made to describe the odonates of Great Nicobar Biosphere Reserve (GNBR).

STUDY AREA

The Great Nicobar Biosphere Reserve (GNBR) is the southernmost Island of Andaman and Nicobar archipelago. It lies between 6°45' and 7°15' N latitudes and 93°38' and 93°55' E longitudes in the Bay of Bengal (Fig. 1). The GNBR includes Campbell Bay and Galathea National Parks. It spreads over an area of 885 sq km. This reserve is known for its unique biodiversity and houses rich genetic germplasm resources.

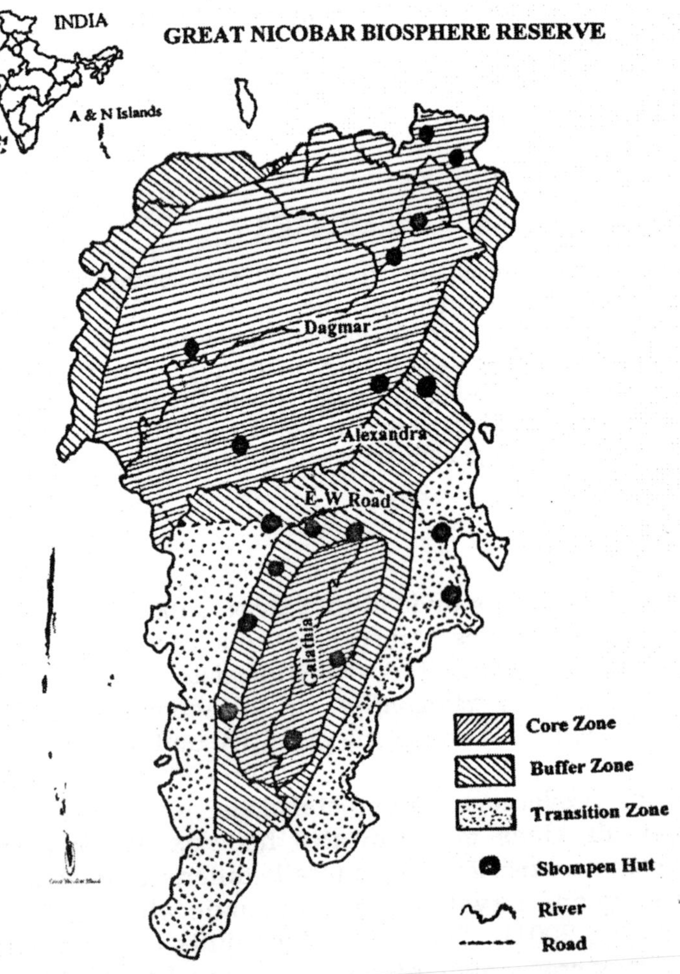

Fig. 1. Great Nicobar Biosphere Reserve

The Great Nicobar Biosphere Reserve represents the tropical rain forest in Andaman and Nicobar Islands of bio geographic region. The major forest area in this Biosphere Reserve is still in its virgin state and rich in species composition. The GNBR has rich heritage of faunal diversity and this area also inhabits one of the most endangered species *viz.*, Megapode (*Megapodius frycinet*) and Edible-nest swiftlet (*Collocalia fuciphaga*). Other important species include White Bellied Sea Eagle (*Haliaeetus leucogaster*), Nicobar Serpent Eagle (*Spilornis cheela*), Nicobar Pigeon (*Caloenas nicobarica*), Nicobar Parakeets (*Psittacula caniceps*).

METHODS

The study was conducted from 2008 through 2010 in eleven different locations at Great Nicobar Biosphere Reserve (GNBR) (Table 1). All specimens were captured during the day with a butterfly net. The samples were immediately stored in envelopes. Photographs were taken of almost all specimens in order to facilitate further identification. Odonates were identified based on standard systematic keys (Fraser, 1933, 1934, 1936; Subramaninan, 2004; Mitra, 2006).

Table 1. Coordinates of the different locations in the study area

Location	Coordinates	
	Latitude	Longitude
East West Road	06° 52.531' N	93° 52.213' E
B-quary	06° 59.708' N	93° 56.680' E
Mahar Nala	06° 59.584' N	93° 55.079' E
Amphibian Road	07° 00.406' N	93° 54.788' E
Kondul	07° 13.027' N	93° 42.936' E
Lakshman Beach	07° 01.249' N	93° 54.321' E
Kopenheat	06° 50.922' N	93° 47.983' E
Indira Point	06° 45.822' N	93° 49.607' E
Navidara	07° 07.564' N	93° 53.144' E
Gandhi Nagar	06° 50.196' N	93° 53.680' E
Mahar Nala	06° 59.518' N	93° 55.104' E

Species richness and abundance of odonates: Total number of odonates and number of individuals seen in each location were calculated using the census data and field observations.

Diversity indices: Diversity indices were calculated using the programme SPDIVERS.BAS developed by Ludwig & Reynolds (1988).

RESULTS

Species composition of odonates in GNBR

Twenty species of Odonates belong to 4 families were recorded from Great Nicobar Biosphere Reserve (Table 2) and highest number of species recorded from the family Libellulidae (15) followed by Coenagrionidae (3).

Table 2. List of species recorded in Great Nicobar Biosphere Reserve

Sl.No.	Common Name	Scientific Name
	Aeschnidae	
1.	Indian dukhawker	*Gycantha dravida* Lieftinck
	Libellulidae	
2.	Asian pintail	*Acisoma panorpoides paorpoides* Rambur
3.	Asian groundling	*Brachythemis contaminata* (Fabricius)
4.		*Brachythemis* sp. indet.
5.	Blue percher	*Diplocodes trivialis* (Rambur)
6.	Pruinosed blood tail	*Lathrecista asiatica asiatica* (Fabricius)
7.	Green skimmer	*Orthtrum sabina sabina* (Drury)
8.	Crimson- tailed Marsh hawk	*Orthetrum pruinosum neglectum* (Rambur)
9.	Wandering glider	*Pantala flavescens* (Fabricius)
10.	Variable flutterer	*Rhyothemis variegata variegata* (Linnaeus)
11.	Yellow striped flutterer	*Rhyothemis phyllis phyllis* (Sulzer)
12.	Dancing drop wing	*Trithemis pallidinervis* Kirby
13.	Indigo drop wing	*Trithemis festiva* (Rambur)
14.	Red grass hawk	*Neurothemis fluctuans* (Fabricius)
15.	Forest skimmer	*Neurothemis intermedia intermedia* (Rambur)
16.	Fulvous forest skimmer	*Neurothemis fulvia* (Drury)
	Coenagrionidae	
17.	Common blue tail	*Ischnura senegalensis* (Rambur)
18.	Orange tailed midget	*Agriocnemis femina oeyzae* Lieftinck
19.	Wandering wisp	*Agriocnemis pygmaea* (Rambur)
	Calopterygidae	
20.	Clear winged forest glory	*Vestalis gracilis* (Rambur)

Distribution of odonates

Species of odonates recorded in different locations during the period of study is given in Table 3. Out of twenty species, *Orthtrum sabina sabina, Lathercista asiatica asiatica, Agriocnemis pygmaea* and *Diplocodes trivialis* were recorded in all the sites.

Plate 1. Dragonflies of Great Nicobar Biosphere Reserve
(a) *Lathrecista asiatica asiatica* Fabricius, (b) *Orthetrum sabina sabina* (Drury), (c) *Rhythemis variegata variegata* Linnaeus, (d) *Neurothemis fluctuans* (Fabricius), (e) *Neurothemis intermedia intermedia* (Rambur), (f) *Trithermis pallidinervis* Kirby

Table 3. Distribution of the Odonates in different locations

Sl. No.	Species name	1	2	3	4	5	6	7	8	9	10	11
1	*Gycantha dravida* Lieftinck	√			√	√					√	
2	*Orthtrum sabina sabina* (Drury)	√	√	√	√	√	√	√	√	√	√	√
3	*Rhyothemis variegata variegata* (Linnaeus)	√		√	√	√	√		√			
4	*Brachythemis* sp.	√		√		√				√		√
5	*Lathrecista asiatica asiatica* (Fabricius)	√	√	√	√	√	√	√	√	√	√	√
6	*Trithemis pallidinervis* Kirby	√	√		√				√	√		
7	*Pantala flavescens* (Fabricius)			√	√					√		
8	*Ischnura senegalensis* (Rambur)	√		√			√	√				√
9	*Neurothemis fluctuans* (Fabricius)	√		√	√	√				√		
10	*Neurothemis intermedia intermedia* (Rambur)			√	√	√			√		√	
11	*Neurothemis fulvia* (Drury)	√				√				√		
12	*Vestalis gracilis* (Rambur)			√		√						
13	*Agriocnemis femina oeyzae* Lieftinck	√	√		√			√				√
14	*Diplocodes trivialis* (Rambur)	√	√	√	√	√	√	√	√	√	√	√
15	*Brachythemis contaminata* (Fabricius)	√		√		√						
16	*Orthetrum pruinosum neglectum* (Rambur)	√		√	√	√			√			
17	*Trithemis festiva* (Rambur)	√		√	√	√						
18	*Rhyothemis phyllis phyllis* (Sulzer)	√		√		√						
19	*Agriocnemis pygmaea* (Rambur)	√	√	√	√	√	√	√	√	√	√	√
20	*Acisoma panorpoides panorpoides* Rambur			√	√		√					

1- Amphibian Road; 2- B Query; 3- Gingen Basti; 4- Gandhi Nagar; 5- East West Road; 6- Indira Point; 7- Lakshman Beach; 8- Mahar Nala; 9- Navidera; 10- Kopen Heat; 11- Kondul

Diversity indices in different locations

Species diversity index (H') was highest in Gingen Basti (2.66) and lowest at B-quarry (1.67). Simpson Index (λ) was highest at Indira Point and Lakshman Beach (0.18) followed by Kopen heat (0.17), and B-quarry (0.16) (Table 4).

Table 4. Diversity indices of odonates in GNBR

Location	Richness indices		Diversity indices		Hill's Number		Evenness indices	
	R1	R2	λ	H'	N1	N2	E1	E2
Amphibian road	3.83	2.26	0.08	2.58	13.16	12.62	0.93	0.82
B-Quarry	1.62	1.28	0.16	1.67	5.31	6.36	0.93	0.88
Gingen basti	3.90	2.33	0.05	2.66	14.29	18.40	0.96	0.89
Gandhi Nagar	3.34	2.00	0.07	2.53	12.49	15.00	0.96	0.89
East-west Road	3.53	2.06	0.05	2.65	14.17	18.23	0.98	0.94
Indira Point	1.66	1.15	0.18	1.75	5.77	5.65	0.90	0.82
Lakshman Beach	1.73	1.41	0.18	1.59	4.88	5.57	0.89	0.81
Mahar Nala	1.92	1.30	0.15	1.92	6.85	6.81	0.93	0.86
Navidera	2.23	1.50	0.10	2.10	8.19	9.66	0.96	0.91
Kopen heat	1.53	1.18	0.17	1.68	5.35	5.82	0.94	0.89
Kondul	1.75	1.26	0.13	1.88	6.55	7.74	0.97	0.94

Similarity indices between the different locations

Similarity indices between the different locations were presented in (Table 5). The similarity index which is influenced by the most abundance species depicted the closet similarity of community between Gingen basti and East-west road, followed by Mahar nala and Gandhi nagar, the least similarity was observed between East-west road and B-quarry.

Table 5. Similarity indices of odonates in different locations in GNBR

	Amphibian road	B-Qiarry	Gingen basti	Gandhi Nagar	East-west Road	Indira Point	Lakshman beach	Mahar Nala	Navider	Kopen heat	Kondul
Amphibian road	0.00	47.22	51.55	52.53	56.31	39.08	38.24	52.27	48.84	34.21	46.91
B-Quarry		0.00	23.19	50.70	16.00	27.12	45.00	46.67	51.72	45.83	49.06
Gingen basti			0.00	58.33	62.00	35.71	27.69	35.29	43.37	27.40	43.59
Gandhi Nagar				0.00	45.10	37.21	23.88	59.77	51.76	50.67	32.50
East-west Road					0.00	31.11	22.54	39.56	40.45	37.97	35.71
Indira Point						0.00	47.27	37.33	21.92	34.92	47.06
Lakshman Beach							0.00	25.00	22.22	36.36	61.22
Mahar Nala								0.00	43.24	62.50	34.78
Navidera									0.00	35.48	50.75
Kopen heat										0.00	49.12
Kondul											0.00

DISCUSSION

A total of 20 species of Odonata under four families and 16 genera were recorded during the period of study. Out of these *Lathrecista asiatica asiatica, Neurothemis fluctuans, Orthetrum sabina sabina* were most common and abundance species in Great Nicobar Biosphere Reserve. The family Aeschnidae and Calopterygidae were represent only one species *Gycantha dravida* and *Vestalis gracilis* respectively. The species richness of odonates in GNBR is similar to other ecosystems in India (Table 6). The overall diversity index (H') was 2.86, which is comparable with other ecosystems of Javalis and Chiteri Sandal forest (Sharma *et al.,* 2007), Moichur forest of Rajaji Naitonal Park (Vashishth *et al.,* 2002). The high similarity index value indicated that two locations of GNBR are quite similar in odonate community structure. The croplands in Gandhi Nagar supported more number of species in terms of richness and abundance. The GNBR support a more diverse odonate community in the Andaman and Nicobar Islands.

Table 6. The species richness of GNBR's odonates in other ecosystems in India

Sl. No.	Name of the area/ecosystem	Number of species	Reference
1.	Rice field, Coimbatore	16	Gunathilagaraj *et al.,* 1999
2.	Courtallam, Tamil Nadu	14	Palot and Soniya, 2000
3.	Great Himalayan National Park	6	Uniyal *et al.,* 2000
4.	Annamalai University, Tamil Nadu	14	Asaithambi and Manickavasaram, 2002
5.	Irrigated rice field, Madurai, Tamil Nadu	12	Kandibane *et al.,* 2003
6.	Maryoor, Kerala	20	Sharma *et al.,* 2007
7.	Bengaluru, Karnataka	17	Sharma *et al.,* 2007
8.	Thangali, Karnataka	12	Sharma *et al.,* 2007
9.	Mandagadde, Karnataka	11	Sharma *et al.,* 2007
10.	Javadis, Tamil Nadu	16	Sharma *et al.,* 2007
11.	Chitteri, Tamil Nadu	8	Sharma *et al.,* 2007

ACKNOWLEDGEMENTS

We are thankful to the Ministry of Environment and Forests, Government of India. The authors are grateful to Dr. Dinesh Kannan, Divisional Forest Officer, Nicobar Forest Division, Campbell Bay for his support during the field survey.

REFERENCES

Asaithambi, M. & S. Manikavasaram, 2000. Odonata of Annamalai Unviersity, Annamalai Nagar, Tamil Nadu. *Zoos' Print Journal* 17(2): 704-706.

Chhotani, C., A.R. Lahiri & T.R. Mitra, 1983. Contributions to the odonate fauna (Insecta) of Andaman and Nicobar Islands with descriptions of two new species. *Rec zool.Surv India* 80: 467-494.

Fraser, F.C., 1933. Fauna of British India, including Ceylon and Burma. Odonata Vol. I, Taylor & Francis Ltd. London, 423pp.

Fraser, F.C. 1934. Fauna of British India, including Ceylon and Burma. Odonata Vol. II, Taylor & Francis Ltd. London, 398pp.

Fraser, F.C. 1936. Fauna of British India including Ceylon and Burma. Odonata Vol. III, Taylor & Francis Ltd. London, 461pp.

Gunathilagaraj, K., R.P. Soundrarajan, N. Chitra & M. Swamiappan, 1999. Odonata in the rice fields of Coimbatore. *Zoos' Print Journal* 14(6): 43-44.

Kandibane, M., N.R. Mahadevan & K. Gunathilagaraj, 2003. Odonata in irrigated rice field of Madurai, Tamil Nadu. *Zoos' Print Journal* 18(7): 1151-1156.

Lahiri,A.R. & B. Mitra, 1993. New records of dragonflies (Insecta) Odonata from Bay Islands. *J. Andaman Sci. Assoc.* 9: 96-99.

Ludwig, J. A. & J.F. Reynolds, 1988. Statistical Ecology, A Premier on Methods and Computing. A Wiley-Interscince publication, 337 pp.

Mitra, T.R., 2002. Note on zoogeography of Odonata (Insecta) of Nicobar Islands, Indian Ocean. *Rec zool.Surv India* 100(3-4): 183-188.

Mitra, T.R., 2006. Handbook on Common Indian Dragonflies (Insecta: Odonata): Zoological Survey of India. pp. iii+136,

Palot, M.J. & V.P. Soniya, 2000. Odonata from courtallam, Tamil Nadu, Southern India. *Zoos' Print Journal* 15(7): 3001-3003.

Sharma, G., R. Sundarraj & L.R. Karibasvaraja, 2007. Species diversity of Odonata in the selected provenances of Sandal in Southern India. *Zoos' Print Journal* 22(7): 2765-2767.

Silsby, J., 2001. Dragonflies of the world. Natural History Museum in association with CSIRO Publishing, UK and Europe.

Subramanian, K.A., 2009. *A Checklist of Odonata of India.* Zoological Survey of India.

Uniya, V.P., A. Mitra & P.K. Mathur, 2000. Dragonfly fauna (Insecta: Odonata) in Great Himalayan National Park, Western Himalaya. *Ann. For.* 8(1): 116-119.

Vashishth, N., P.C. Joshi & A. Singh, 2002. Odonata community dynamics in Rajaji National Park, India. *Fraseria* 7: 21-25.

Chapter - 8

REVISITING TYAGI'S THEORIES OF 'THE DESERT MALARIA (1995)' AND 'THE *AEDES ALBOPICTUS* - TRANSMITTED DENGUE SANS *AEDES AEGYPTI* (2006)', WITH REFERENCE TO EMERGENCE OF (1) *PLASMODIUM FALCIPARUM*-DOMINATED MALARIA IN THE THAR DESERT, AND (2) DENGUE FEVER IN KERALA STATE

B.K. Tyagi

Centre for Research in Medical Entomology (ICMR),
4-Sarojini Street, Chinna Chokkikulam, Madurai - 625002, TN, India
E-mail: *bk_tyagi@sify.com*

ABSTRACT

In mid-1990s and 2000s I, then serving as a scientist first in the Desert Medicine Research Centre (1988-2001) at Jodhpur in the Thar Desert and subsequently in the Centre for Research in Medical Entomology (2001~) at Madurai in southern India bracing Western Ghat region of Kerala State, propounded two theories (i) "The Desert Malaria" for the Thar Desert, and (ii) "The *Aedes albopictus*-transmitted dengue virus infection" sans *A. aegypti,* the primary vector, in the Western Ghat's mountainous and sylvatic environments. Both the Thar Desert and the Western Ghats present contrasting ecosystems, and can well serve commensurate models for comprehending pathways of evolution of the two major vector-borne diseases in the *terra incognito* for these infections. While the Thar Desert in the northwestern Rajasthan State of India, a unique physiography, had generally been typically characterized malriologically as a region with unstable malaria and potential for sporadic outbreaks dominated by *Plasmodium vivax.* The Western Ghat-bound region of the Kerala State in southern India was free from dengue until mid-1990s. Both these ecosystems have been subjected to human interventions and have drastically metamorphosed during past few decades, in a way that has resulted into inveigling new vectors of certain vectorborne diseases or accentuated likelihood of their exacerbation. Emergence of malaria, particularly *P. falciparum*-dominated malaria in the Thar Desert and dengue in the Kerala State present exceptional evidence-based models to advance our knowledge about the dynamics of these diseases in the wake of anthropogenic interventions as well as climatic vicissitudes.

(1) THE DESERT MALARIA

The Thar Desert ecosystem

The Great Indian Thar Desert, with a history of existence of only 10,000 years is understood to be one of the youngest deserts in the world. The geographic limits of the Thar region are generally understood to be between the irrigated lands of the Indus and Sutlej and the western edge of the Aravalli Hills (i.e., approximately between 24°-30° N and 68°-78°E), although in Rajasthan State where the desert in the country is maximally present, it is spread over nearly 75,000 km² (62% of the total land of the State) and harbouring a population of 12.8 million people (39% of the population of the State). The seasonal Luni river divides the desert into two parts; in the north a vast stretch of sand dunes or levees and in the south an expansive basin rimmed by flat- topped hills of sandstone and limestone. The climate of the Thar Desert is, generally speaking, of the tropical desert type with arid hot environment. Monsoon had been highly deficient and erratic with humidity ephemeral; the average rainfall being around 200 mm. The relative humidity is generally very low during the hot and dry periods of the year; during summer it dips to as low as 15-34%, making the summer very dry and hot.

The Thar Desert is fed by the nearly 10,000 km network of three major canal systems used mainly for agriculture and other vital needs including drinking.

(i) **The Gang canal project** which became fully operative in 1928, with its 1251 km distribution system in Sri Ganganagar district only. Along with the Bikaner canal feeder, it provides water to about 65% of the 300,000 ha of cultivable land.

(ii) **The Bhakra-Sirhind canal project,** with its 1219 km long distribution system, was started in 1955 to utilize a further 300,000 ha to cultivable land.

(iii) **The Indira Gandhi Nahar Pariyojana project (IGNP)**: The most important of all the canal systems is the IGNP which is one of the world's largest projects of its kind in the desert. It started in 1958 and irrigation of part of the area began in 1961. The project, when completed, will include a 649-km long main canal (of which 445 km is in the Thar Desert) and over 9000 km of distributaries.

Although only two of the three stages envisaged are completed by now, the main canal including feeder and 5500 km of distributaries is already irrigating 950,000 ha of land (approximately 50% of the irrigated area that is planned. This Herculean task, bringing Himalayan waters to the Thar Desert, aims at transforming vast stretches of western desert terrain into verdure. As a result, the annual agriculture production from the newly irrigated land now exceeds Rs. 6500 million. While several irrigation-intensive crops such as sugarcane, mustard, groundnut, etc., have been established in some of the canalized desert districts, the production of rice in Sri Ganganagar district, the most irrigated among all desert districts, has topped the list in the whole State (Fig. 1).

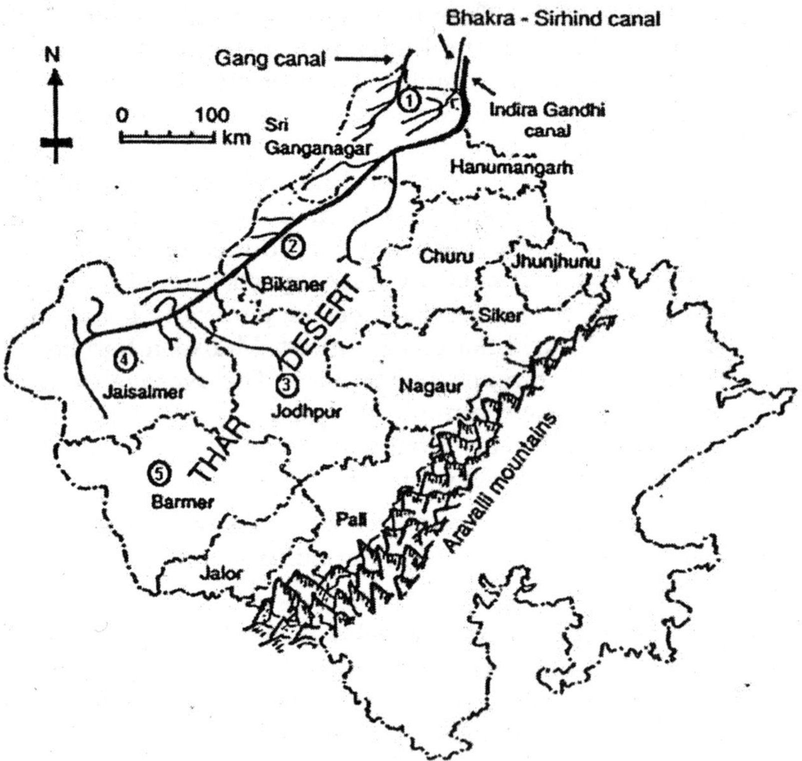

Figure 1. Twelve desert districts in the Thar Desert. Those districts depicted by encircled numerals have experienced repeated *P. falciparum* dominated malaria dpidemics; all falling under canal based irrigation, particularly IGNP.

Extensive long term canalization of the desert has resulted into marshy and/or boggish areas along much areas of the main IGNP canal. The problem of seepage water and water-logging have created serious mosquitogenic conditions. Besides the surplus irrigation water that accumulates on irrigated land, vast areas along the course of the Indira Gandhi main canal have been engulfed by seepage water from the irrigation canals on a perennial basis. An estimated 8600 ha in the Nalibed area, Sri Ganganagar/Hanumangarh district, is permanently inundated under the IGNP Stage I by the Ghaggar river floodwater while

1000 ha is converted into marshy land due to water- logging and excessive seepage from the canal in Stage II. Supply of canal water to the desert region over several decades has caused a rise of the water table and water-holding potential of the soil, and a plentiful growth of vegetation, particularly hydrophytic weeds. Such extensive breeding grounds have proved to be favourite sites for breeding by several different vector mosquitoes (Fig. 2), many of which were even not found in the interior of the Thar Desert.

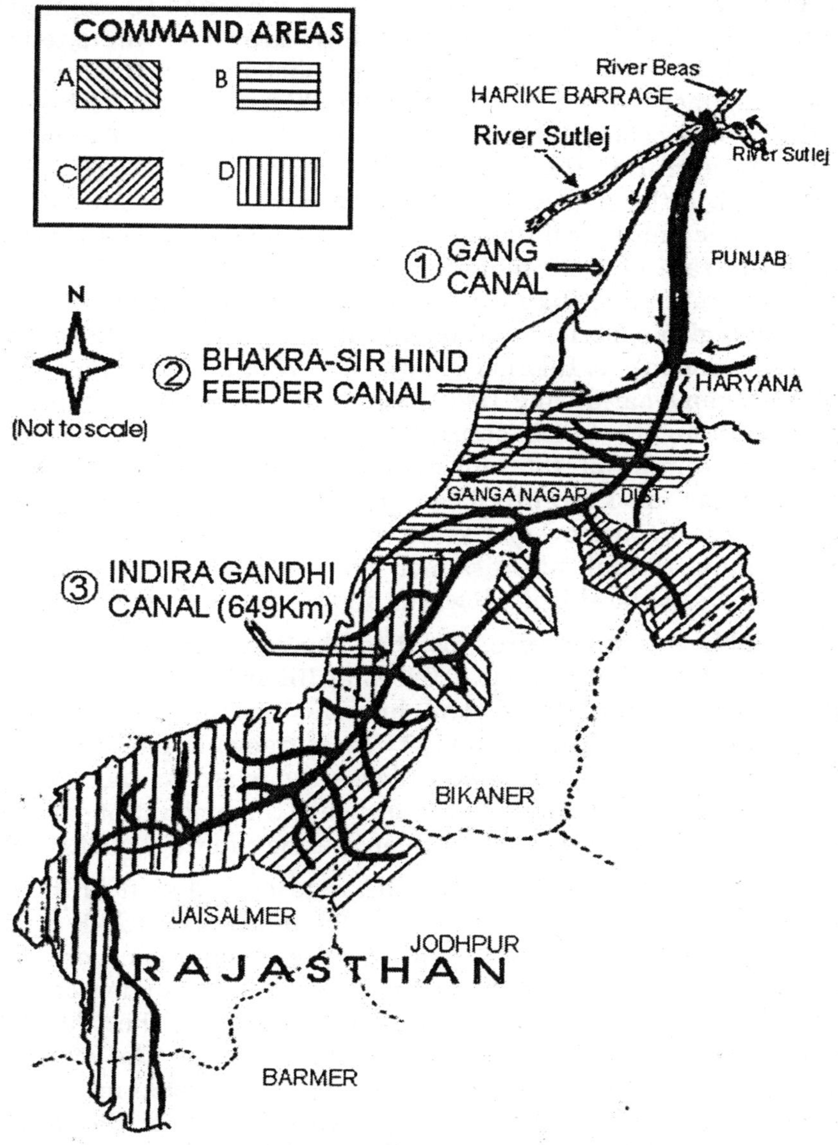

Fig. 2. Index plan of the Indira Gandhi Nahar Pariyojana (showing also the Gang canal and the Bhakra-Sir Hind Feeder canal).

Climate dynamics during past fifty years

A recent compilation of percentage departure of monsoon season rainfall (June to September) for the period 1901–2000 over west Rajasthan by Sikka & Kulshrestha (2001) and Dr. S.M. Kulshrestha (*pers. comm.*) depicts how significantly the drift has been over past 100 years. Rao (1996) have found an increase in the mean decadal rainfall in the Ganganagar district, which had been subjected to canal-based irrigation for a considerable period. The Gang canal and IG canal projects aimed to utilize 7.59 MAF of water from Ravi-Beas duo waters to irrigate an 1.143 mill. ha area. In the Ganganagar district, the Bikaner district and the Jaisalmer district, 708 ha, 775 ha, 37,022 ha and 84 ha of land, respectively, has been put under irrigation. The impact of irrigation in Rajasthan is seen from the vegetational and micro-climatic changes that have occurred in the region. Sri Ganganagar receives an annual rainfall of 243 mm. The annual rainfall of Sri Ganganagar showed an increasing trend, at the rate of 1029 mm/year, during the period 1926-1993. The mean annual rainfall of Bikaner is 286 mm with a coefficient of variation of 48%. The region experienced droughts once every 3 years (38% chance). The rainfall (annual) at Bikaner showed an overall decreasing trend at a rate of 0.147 mm/year during 1926-1993, which is forcibly attributable to consecutive droughts which occurred after 1961. Jaisalmer receives a mean annual rainfall of 185 mm with a coefficient of variation of 64%. The rainfall has until recently been erratic and poorly distributed with droughts occurring once every 4 years (24% chance). However, the overall trend in annual rainfall at Jaisalmer which remained the same during 1901-1993 seems to be shifting towards an escalation in recent years.

The ambient temperature, too, has shown decreasing trend being 0.039°C/year at Sri Ganganagar, 0.023°C at Bikaner, and 0.009°C/year at Jaisalmer. With regard to relative humidity in the Thar desert, Rao (1996) recorded 5-10% higher humidity measurements in cropped surfaces at 1-m height in the irrigated villages compared to the unirrigated cropped surfaces. This highlights the fact that the amount of canal water available in the desert region can influence atmospheric humidity.

These observations sufficiently indicate towards a shift in the abiotic and biotic characteristics of the Thar Desert under the IGNP command area as compared to that in the typical desertic environs. For example, the desert insect fauna, including mosquitoes, have undergone a sea-change. As is expected from a inhospitable and hostile desertic environment, but also the monotonous insect faunal life, of the Thar Desert, in fact very few scientific explorations were made in this unique ecosystem. Most of these researches revolved around the migratory locusts which used to enter the Thar Desert from Pakistan and cause extensive damage to the crops in the Indian part of the Thar Desert in northwestern Rajasthan. Such was the havoc created by this locust species to agriculture that a specific Locust Research Centre had to be stationed at Jodhpur for developing ways and means to cease the orientation of this menace and its control by forecasting to farmers. Another insect which has inhabited the Thar Desert for centuries is the dung beetle (genetrra *Scarabaeus* and *Gymnoplerus*) which play an important role in pasture ecosystem where they annually breakdown tones of animal dung and much of it incorporated into the soil to increase fertility and also to clean the environment. This vacuum of knowledge largely attributed to the monotonous life forms of insects in the Thar Desert where they were largely innocuous, although ectoparasites on camels and sheeps could have been aplenty and must be causing health concerns to these desert milching animals. Incidentally, only *An. stephensi* was the only anopheline mosquito that occurred in the Thar Desert until late 1980s. This mosquito is genetically adapted to desertic conditions, the cradle of which existed in the Middle-East from where it seemed to have advanced its distribution more eastwards into India. It was ultimately the IGNP scheme, coupled with the Gang Canal and the Bhakra-Hind Canal systems, which not only brought metamorphosis to the desert physiography but also the dynamics of both the malaria vectors and the disease in the Thar Desert.

The IGNP Command area and the physiographic vicissitudes in the Thar Desert topography and ecology

The Indira Gandhi Nahar Pariyojana, conceived in 1948, began to be constructed in 1958, and the first waters were flown in 1962. Extensive canalization with improper canal linings caused at many places huge areas flooded with stagnant waters. The IGNP has substantially changed the ground water hydrology in the command area. Before the project, in 1952, the ground water table in stage-1 was observed between 40-51 mbgl. Compared to this, in 1973, the water table was observed between 9.00 and 32.00 mbgl, which considering soil types available in the Thar Desert, resulted to large scale waterlogging due to rise in water table. The perennial presence of water sheet and the resultant greenery decreased ambient temperature (-0.52°C for 100 years), and increased rainfall. Ironically duststorms, once so frequent, became occasional as even events of droughts and famine decreased. Years of sustained canal water supply to the Thar Desert region changed the crop pattern as well. Till the turn of the first half of the 20th century Sorghum or Jwar (*Sorghum vulgare*), bajra or pearl millet (*Pennisetum* typhoides) moth (*P. aconifolius*), moong (*P. radiatus*) and gwar (*Cyamopsis psaraliodes*) etc. were the main desert crops completely dependent on the monsoon-fed rains. However, with the advent of extensive canal irrigation in the sixties, some irrigation crops got ushered into the Thar desert, such as cotton, wheat, mustard, groundnut, sugarcane and, in particular, paddy have also thrived making Sri Ganganagar in the Thar Desert the number one district in the

Rajasthan State for rice production. With the groundwater quality changing, the soil, properties which earlier was coarse-textured with low fertility, has also altered making the land more irrigable and productive (*cf.* Tyagi, 2004 and Ijumba & Lindsay, 2001).

MALARIA IN THE THAR DESERT DURING PRE & POST-1980S

Rajasthan State has been experiencing scourges of malaria for decades, but the disease intensity varied a great deal and mostly confined to north-east and south-western parts of Rajasthan. In fact there existed patches of hyper-endemicity which for known reasons had been experiencing incessant episodes of focal and/or widespread epidemics. Yet, the overall impression of the State has been that of a non-problematic area with few local outbreaks here and there. Based on malaria endemicity malaria in Rajasthan has been originally classified as given hereunder;

(i) Zone or Stratum II, signifying moderately refractory areas with high epidemic potential (*i.e.*, eastern plains and southern forested and mountainous districts of Rajasthan State), and

(ii) Zone III, signifying non-refractory areas with moderate to high epidemic potential (*i.e.*, western and northern districts of Rajasthan State, including whole of the Thar Desert). Obviously, malaria in this desertic zone was transmitted through *Anopheles stephensi*, the oldest amongst all anophelines ever existed in the Thar Desert, albeit later joined by the more virulent *An. culicifacies*, with the advent of the canalized irrigation waters.

It is interesting to note that in last nearly two decades the physiography of the Thar Desert, in particular, has undergone such a major change that the zoning of the region needs a re-characterization of its epidemiological potential (Tyagi, 1995). Under prevailing grim circumstances, the malaria transmisson dynamics in Rajasthan State particularly the Thar Desert region requires a serious thinking to device appropriate control methodologies (Tyagi & Chaudhary, 1997). In general, the malariometric data (1961-94) show that the overall number of positive cases in the State had increased 27 times, from 8494 cases in 1961 to 229,772 in 1994, while the targeted population increased only 2.5 times during the same period. It is also quite evident that while the slide positivity rate (SPR) had significantly increased from 1.43 in 1961 to 5.07 in 1994 (3.5–fold increase), the slide *P. falciparum* rate (SFR) had also increased as many folds only, from 0.55 to 1.95. However, the increase in either the SPR or the SFR was not steady and, therefore, no definite trend can be noticed. The maximum SPR was recorded in 1976, with several epidemics in non-desert parts of the State. From the mid-eighties onwards, however, the trend had become clearer and both the SPR and SFR had steadily increased. In fact, the highest ever SFR had been recorded in 1994 – the year when the State as well as the Thar Desert region experienced the maximum deaths due to malaria in their entire history. A cursory review of the malariometric data reveals it all (Table 1).

Anopheline fauna of the Thar Desert

Until about two decades from now the Thar Desert was hardly considered an important habitat for the malarial disease or its vectors. This was due essentially to very low prevalence of malaria and without any epidemic in the region. In the wake of such knowledge about malaria, no significant anopheline survey was ever made in the interior parts of the Thar Desert, and the fewer studies then directed on the subject have been from around the townships. Also, medical entomologists got more attracted to the forested and mountainous districts of the Udaipur zone in southern Rajasthan, being highly malarious. It is not surprising, therefore, that in the early days nearly all the major faunal surveys in the State were organized in the Udaipur zone districts only. Although, the first ever account of Indian anopheline was presented on a 'wall map' in 1912 by the Central Malaria Bureau, at Kasauli (subsequently transformed first into malaria Institute of India and, for a long time the National Institute of Communicable Diseases, Delhi, and finally the National Centre for Diseases Control, Delhi), it became possible only in the late twenties when for the first time four anopheline species from the then Rajaputana State were put on record (Covel, 1927). Later in his *magnum opus* on 'Family Culicidae: Tribe Anophelini' in the classic 'Fauna of British India' series, now largely out of date, Christophers (1933) described in detail the taxonomy and bionomics of these four species, *viz., An. stephensi, An. culicifacies, An. annularis* and *An. subpictus*, which were stated to be occurring in at least a few desert districts as well. Tyagi (2002) has postulated the pathway of orientation of *An. stephensi*, the oldest of all the anophelines in the Thar Desrt, having made it eastwards from the Middle-East xeric ecosystem, possibly through commercial reasons! Interestingly, 'tanka', the underground water reservoir found aplenty in the Thar Desert and originally supporting the breeding of *An. stephensi,* was got introduced in the 16th Century. This shows as to how *An. stephensi*, found it convenient to survive in the inhospitable climate of the desert. The total number of taxa in the State was first raised to seven and subsequently to 16 (Puri, 1955), without any change in the strength of the desert fauna. According to Puri (1960), most of the 16 anopheline species were restricted to the two southern forested districts, Sirohi and Udaipur, contributing 56% of the total fauna from the State, while the Thar Desert had a meager share of seven (44%) species from five of the then eleven desert districts. Interestingly, though Rao (1984) extensively dealt with the Indian anophelines, but no mention was made to either the Rajasthan State or the Thar Desert while discussing the regional distribution of the taxa.

Table 1. Malariometric data for Rajasthan State (1961-2000)

Year	Population (x000)	Blood slides collected	Blood slides examined	Positive cases	Pf cases	%Pf	ABER	API	SPR	SFR	Deaths
1961	17562	669202	595176	8494	3266	38.45	3.39	0.48	1.43	0.55	0
1962	18785	943159	1082815	3210	508	15.85	5.76	0.17	0.30	0.05	0
1963	19048	1779180	1743916	3813	1210	31.73	9.16	0.20	0.22	0.77	0
1964	20917	2200707	2184182	3164	826	26.10	10.44	0.15	0.14	0.04	0
1965	22001	2212563	2183592	2872	348	12.11	9.92	0.13	0.13	0.02	0
1966	22454	1863401	1857961	9680	532	5.29	8.27	0.43	0.52	0.03	0
1967	23040	1772313	1763870	23898	1221	5.10	7.66	1.04	1.35	0.07	0
1968	23850	1661789	1659286	14999	923	6.15	6.96	0.63	0.90	0.06	0
1969	24670	1485220	1471587	15487	1158	7.47	5.97	0.63	1.05	0.08	0
1970	25320	1741741	1736028	79788	4985	6.24	6.86	3.15	4.60	0.29	0
1971	25850	2184134	2174281	109773	8064	7.34	8.41	4.25	5.05	0.37	0
1972	26510	5037069	2215322	82517	8470	10.26	8.36	3.11	3.72	0.38	0
1973	26715	2233100	2162576	118012	16451	13.94	8.09	4.42	5.46	0.76	0
1974	27560	2694698	2688624	177596	13663	7.69	9.76	6.44	6.61	0.51	0
1975	27560	3186044	3062207	354567	31304	8.82	11.11	12.87	11.58	1.02	0
1976	28031	3578216	3572281	412776	24163	5.85	12.74	14.73	11.55	0.68	0
1977	29104	3318120	3318120	231862	12445	5.36	11.40	7.97	6.99	0.38	0
1978	29670	3260421	3226710	154549	8612	5.57	10.88	5.21	4.79	0.27	0
1979	31092	3515605	3515605	83394	4670	5.59	11.31	2.68	2.37	0.13	0
1980	31594	4208295	4208295	96118	15871	16.51	13.32	3.04	2.28	0.38	0
1981	32491	3992445	3991672	100694	14752	14.65	12.29	3.10	2.52	0.37	0
1982	33113	3284810	3284810	75320	12296	16.32	9.92	2.27	2.29	0.37	0
1983	33583	3161396	3161396	115177	35462	30.78	9.41	3.43	3.64	1.12	0
1984	34124	2901731	2901731	101955	20443	20.05	8.50	2.99	3.51	0.70	0
1985	34655	3037182	3037182	67040	12643	18.86	8.76	1.93	2.21	0.42	4
1986	34897	2941659	2941659	54618	13890	25.43	8.43	1.57	1.86	0.47	2
1987	35378	3219363	3219363	65523	13942	21.28	9.10	1.85	2.04	0.43	0
1988	35457	3493559	3493559	104109	29189	28.04	9.85	2.94	2.98	0.84	2
1989	35683	3074207	3074207	112316	24072	21.43	8.62	3.15	3.65	0.78	1
1990	35854	3567539	3567539	114689	32500	28.34	9.95	3.20	3.21	0.91	65
1991	43881	3179925	3179925	77573	16097	20.75	7.25	1.77	2.44	0.51	10
1992	43880	3833880	3833880	121499	41513	34.17	8.74	2.77	3.17	1.08	55
1993	44005	3644944	3644944	107797	26387	24.48	8.73	2.44	2.96	0.72	19
1994	44005	4855841	4855841	241255	94020	38.97	11.04	5.48	4.97	1.94	452
1995	44005	5181432	5181432	350780	45212	12.58	11.77	7.97	6.76	0.87	45
1996	44005	6591873	6591873	300547	72780	24.21	14.97	6.82	4.55	1.0	0

1997	44005	6000745	6000745	272670	19742	7.24	13.63	6.19	4.54	0.32	0
1998	44005	4977977	4977977	76438	10030	13.12	11.31	1.73	1.53	0.20	0
1999	45159	5223301	5223301	53154	5857	11.02	11.57	1.18	1.02	0.11	0
2000	44006	4969259	4969259	35973	3368	9.36	11.29	0.82	0.72	0.07	10

Two more species, *An. nigerrimus* (new to Sri Ganganagar) and *An. splendidus* (new to Rajasthan State), collected out of a total of seven species in Sri Ganganagar district, were added to the list of anopheles from the Thar Desert raising the total to nine (52.9%), with only two additions to the faunal strength of the state, including *An. d'thali,* new to Rajasthan, and elevating the fauna from the Thar to 10 (55.6%), and in the State to 18 (34.6% of the total anophelines recorded from India). Subsequently, *An. barbirostris* from Bikaner and *An. fluviatilis* from Jaisalmer have also been recoded making the latest inventory shooting to 12 (66.6% of the total taxa enlisted from Rajasthan. It is comprehensible from the distribution of taxa in each district that there stands a good possibility of discovering more new records from the Thar Desert, particularly the unexplored and/or less explored north-eastern and southern desert districts, provided long term surveys are undertaken, though at the moment the Thar Desert remains under represented from districts' point of view. Although debatable, with the introduction of *An. fluviatilis,* the total number of anopheline taxa stands to 13 species in the Thar Desert (Nagpal & Sharma, 1995; Tyagi, 2004). Based on the association of the type of malaria with the specific *Anopheles* species and the distribution of *Anopheles* vector species across the Thar Desert, Tyagi (2002) had propounded a new malariological classification which holds good even after a decade in the Thar Desert.

Emergence of *Plasmodium falciparum*-dominated malaria in the Thar Desert

The correlation of environmental changes brought about in the Thar Desert by extensive canalization and irrigational activities with malaria prevalence of varying intensities in different physiographic situations is so clearly discernible and demonstrable that the Thar Desert practically offers a inafalliable model to understand the causes underlying resurgence of malaria in a desert ecosystem and control of vector mosquitoes with proper water management.

Decades ago, particularly before the canalized irrigation had been ushered in the Thar Desert, most of the positive cases were reported from the hilly and forested terrain of Udaipur in the southern part of Rajasthan. Thus, while in 1986 the Udaipur zone had contributed 44.1% positive cases and 67.7% *P. falciparum* cases as 24.4% positives and 11.6% P. *falciparum* cases in the desert region, in 1994 only 14.1% positives (*P. falciparum* 26.6%) hailed from Udaipur zone compared to

53.3% positives from the desert districts with a whopping 62.5% *P. falciparum* cases. Amongst the then 11 desert districts, Barmer was the most malarious in 1994 (and even now in 2003!), contributing 31.3% positive cases and 35.8% *P. falciparum* cases (Tyagi *et al.*, 1995).

As in most districts of various States in the country during 1975, Sri Ganganagar district boasted of a relatively very high Slide Positivity Rate (SPR) (11.36%) and the Annual Parasite Incidence (API) (13.83), with an all-time low *Plasmodium falciparum* proportion (1.2%). For the period of 1982-86, both SPR and API have constantly shown a descending trend till recently when the epidemics in The Thar desert became widespread in the early ninetees ($P< 0.05$). Since 1987, *P. falciparum* proportion has significantly increased and by 1995 it has gone up by atleast 10 times (12.79%), with an escalation during the mid-eightees (average 19.8; range 6.83 – 13.18) when the *P. falciparum* shot up as high as 43.1% in 1983. It implies that under the changed malariological scenario, the SPR or API is not a good indicator of the true magnitude of the disease's impact. Instead, per cent *P. falciparum,* which is often related with deaths, seems to constitute a good basis for comparison vis-à-vis paddy cultivation or more correctly the paddy productivity.

Malaria epidemics and their pathways in the Thar Desert

Documentable instances of outbreaks of malaria epidemics in the Thar Desert are available since 1975, although there is every possibility that mortality might have also occurred in the desert during the nation-wide epidemic in 1908, and perhaps thereafter as well on other such occasions! As a consequence, at least 13 episode-years of malaria epidemics are identifiable since 1975. Save for the massive outbreaks of malaria in 1975, 1976, 1990 and 1994, most other epidemics were focal in nature covering often a part of a district or two. While the epidemics of 1975 and 1976 were predominantly characterized by a resurgence in *P. vivax* cases, the outbreaks of 1983-1984 onwards saw a phenomenal rise in *P. falciparum* (Tyagi, 2002).

(1) ***Malaria epidemic of 1975-77***: Even though no death was brought on record, it was crystal clear that, in view of the steep rise in the passive cases compared to the 1966 level of malaria cases, outbreaks of malaria had also struck the Thar Desert, along with other parts in the country.

(2) ***Malaria epidemic of 1983-1984***: Sri Ganganagar district experienced a focal *P. falciparum*-dominated malaria outbreak for the first time in 1983, with 4925 positive cases of which 43.4% were *P. falciparum* cases. In spite of the fact that in 1983 the annual parasite incidence (API) was but second lowest during the last few years, the malaria outbreak in the Nalibed area in the erstwhile Hanumangarh tehsil of the Sri Ganganagar caused high morbidity. Two primary health

centers (PHCs), Sangaria and Pilibanga, which were most severely affected, had jointly contributed 72.2% of the total positive cases in the district during the epidemic. It is noteworthy that *P. falciparum* had constantly increased for at least few years since 1987, while the API had kept a low ebb for the same period.

(3) ***Malaria epidemic of 1990***: In 1990, a widespread epidemic hit western Rajasthan districts, but particularly the Barmer district claiming 47 lives. A total of 19,322 positive cases (*P. falciparum* 66%) were recorded in Barmer district which was 2.5 times higher than what was reported in the preceding year. It was advocated that the epidemic in Barmer district occurred as a sequel to widespread floods in the region. It was demonstrated that the rise in the incidence of malaria was associated with the increase in the rainfall. During 1977, when annual rainfall was 13% above the normal, the annual parasite incidence (API) showed a conspicuous rise at 5.3 and annual *P. falciparum* incidence (AFI) between 0.03 and 0.1. However, 1985 was a year of exception when there was noticeable increase in cases, particularly of *P. falciparum*, despite low rainfall. This increase was, however, attributed to the migratory cases, since the cases did not proliferate locally due to non-conducive transmission conditions. This was substantiated by a significant decline in the incidence in the next 2-3 years

Both rural and urban areas were equally and simultaneously affected by the epidemic. Though all the eight PHCs presented a rising trend, the PHC Baitu was the worst affected with API touching a peak of 55.3 and AFI 19.6. The highest increase in AFI was in Mandli (4.5 times), Gudamalani (4 times), Ramsar (2.8 times), Ganga (2.3 times), Sindhari (1.75 times) and Chohtan (1.6 times). The Barmer town also recorded a 4-fold rise in API and 3-fold rise in AFI. The district overall API and AFI recorded 2.5 and 3-fold increases, respectively. The mass blood survey carried out during the epidemic showed that all age groups were affected. The slide positivity rate (SPR) was 61%. More than 93% of these had *P. falciparum* infection. A total of 122 infants were found positive of which less than 30% had *P. falciparum* infection. A total of 103 infant positives were from Baitu PHC only. About 88% of the positive infants were detected during September-November, 1990 only.

(4) ***Malaria epidemics of 1992-1993:*** During late 1992 and early 1993 epidemics of malaria broke out simultaneously in several districts, *i.e.*, Bikaner (Bajju and Kolayat PHCs), Jodhpur (Baap and Phalodi PHCs) and Jaisalmer (Pokaran, Nachna and Sangad PHCs). In Jaisalmer district alone a total of 2128 positive cases (*P. falciparum* 73.6%) were reported during the epidemic as against a combined total of of 1809 cases (*P. falciparum* 27.6%) in the two preceding years, *i.e.*, 1990 and 1991. If the malariometric data from the three PHCs are reviewed it becomes clear that the malaria scenario has drastically changed over the years. A

total of 2128 positive cases (*P. falciparum* 73.6%) were reported against a combined total for 1990, and 1991 of 1980 cases (*P. falciparum* 27.6%), clearly indicating a shift towards an ascending order.

During early 1993, the SPR was found to be as high as 85.0% and the *P. falciparum* proportion 88% in two desert villages near the Indira Gandhi canal, Madassar and Awai.

(5) ***Malaria epidemic in 1994*:** In mid-1994 a major epidemic of malaria broke out in whole of Rajasthan State taking a toll of 452 lives. The worst affected three districts included Barmer, Jaisalmer and Jodhpur, contributing nearly two-third of the total mortality. *Plasmodium falciparum* accounted for over 60% of cases in the desert and was considered as the main cause of death.

In a year-long pilot study during 1993-94, following a rapid mass blood survey immediately after the epidemic in the late autumn in 1992 in the two IGNP command villages (*P. vivax* 11.1%, *P. falciparum* 88.7%) and two unirrigated villages (*P. vivax* 17.1%, *P. falciparum* 82.7%), fever surveys conducted for one full year between April 1993 and March 1994 showed a slightly higher slide positivity rate in the IGNP villages (32.3%) than in the unirrigated villages (25.5%). The most important feature was, however, the higher proportion of *P. falciparum* in the IGNP villages (76.6%) compared to that in the unirrigated villages (16.6%). Influenced by the high percentage of *P. falciparum* in the study villages, the slide positivity rate maintained a high profile throughout. Mixed infections occurred in all villages, although more of them occurred in the unirrigated villages (11.5%) than the other two villages (3.2%). Interestingly, more gametocyte carriers were present in the IGNP non-command villages (16#, 17.8%) compared to the unirrigated villages (2#, 8.3%). Cumulatively, 73.7% of all the positive cases were found in children of less than 14 years of age. The male population (68.4%) suffered more than the females (31.5%).

(6) ***Malaria epidemic in 1995-1996*** : A total of 45 lives (60.7% of the total deaths in Rajasthan) were reported in the Thar Desert region alone, even though there is some debate whether the year of 1995 to be regarded as an year of epidemic or not. However, Najera *et al.* (1998) have rightly characterized it, along with 1996, as an epidemic year. The malariometric data of 1995 are comparable with some of the past epidemic years.

(7) ***Malaria epidemic of 1999:*** A focal outbreak of malaria occurred in the villages situated close to the main Indira Gandhi canal near Ramgarh in Jaisalmer district, western Rajasthan. Stagnation of water over a month's period in the main canal as well as long standing rain water in the form of large lake-like bodies near these villages formed vast breeding grounds for the vectors like *Anopheles culicifacies*, along with *A. stephensi* already breeding in the 'tanka' and 'beri' in the epidemic-hit

villages. A mass blood survey was conducted covering 380 persons with a history of fever in last one month in the epidemic hit villages, namely Seuva, Raghava, Raimala and Sadhana, all within a distance of 2 km from the main Indira Gandhi Cana. Of these 49 (12%) were *P. vivax* and 40 (10.5%) were *P. falciparum*, while 8 (2.1%) had mixed parasitism with both parasites. Results of the mass blood survey point out towards a high slide positivity rate (25.5%) and *P. falciparum* proportion (49.5%) in the villages hit by the epidemic. While Raghwa village has presented the highest SPR (30.4%), Raimala showed the lowest SPR (13.1%). Interestingly, however, while on one hand Raimala singled out with the highest % *P. falciparum* (60%), Seuva offered the lowest *P. falciparum* proportion (41.0%), on the other. Maximum *P. falciparum* cases were nevertheless originated from Sadhna village.

Possible factors responsible for conflagration of malaria, particularly *P. falciparum*-dominated malaria, in the Thar Desert

It is an enigma so far as to how *P. falciparum* ushered in the interior of the Thar Desert and manifested so often in the form of epidemics with high morbidity and mortality. A few possible interpretations are offered here:

(1) With the establishment of extensive canal-based irrigation in the interior parts of the Thar Desert, a far more serious vector of malaria, *An. culicifacies*, ushered in the xeric environments. This species being highly adaptive to new physiographic situations built up its density very fast under given circumstances of a large variety of breeding places in the IGNP area particularly the inevitable and numerous seepage water collections from the canals. What *Anopheles culicifacies* lacks in its susceptibility to infection by *P. falciparum*, it compensates by its relatively much higher density, hence also vigorous and prodigious biting on human host;

(2) *Anopheles stephensi*, chronologically the oldeset and most dominant vector of malaria in the typical sand-duned xeric environments of the Thar Desert, which traditionally bred only in the 'tanka', the only major source of reserved potable water invariably present in every desert village and but transmitted 'desert malaria' (Tyagi, 1996) at a low ebb, without epidemics, shifted its breeding habits with the availability of extensive and varied breeding sites due to extensive canalization to compete with that of *An. culicifacies;*

(3) The massive Indira Gandhi Nahar Pariyojana, in continuation with the Gang canal system and the Bhakra-Sirhind feeder canal system, together spanning over nearly 70 years of initiation, had attracted large labour forces from distant and often hyperendemic States like Bihar, Uttar Pradesh, Punjab, Haryana, Gujarat and Madhya Pradesh. Such labour forces quite possibly brought along

with them the parasite, possibly a resistant strain of, *P. falciparum*, which in the non-immune population had the potential to spread and manifest fast in suitable climatic situations of IGNP-induced environment in a part of the Thar Desert where it was pointed out that the 1994-utbreak of cerebral malaria in the Thar Desert proved fatal because the affected population had been exposed to the parasite for the first time; and,

(4) The resistant strain of *P. falciparum* inadvertently inveigled into the Thar interiors is uncontrollable by the normal regimen of the prescribed antimalarials.

The above hypothesis can be further aptly understood if the progress of pathways of malaria epidemics in the Thar Desert was traced out carefully (*cf*. Tyagi, 2002).

(2) *AEDES ALBOPICTUS* – TRANSMITTED DENGUE VIRUS IN KERALA STATE

The Kerala state has, in course of last couple of decades undergone an enormous change in respect of both its physiography/climate and risks of emergence or resurgence of several vector-borne diseases. Significantly Kerala, which remained virtually a terra incognito for dengue infection till mid-1990s, has been experiencing a string of widespread epidemics annually and has not attained hyperendemicity causing appearance of DHF and DSS cases as well. The recent emergence of dengue fever in Kerala during 2003 seems to have a definite correlation with the climatic change and the imposing anthropogenic stresses, highlighting the status of dengue fever in Kerala in the light of factors which have forced a sylvatic benign mosquito like *Ae. albopictus* to seek its new breeding and feeding possibilities near human habitation and facilitate transmission of dengue virus in a virtually non-immune population.

Kerala state is a unique ecosystem in view of its very strategic location. With a total area of 38,863 sq. km (urban 3365 sq. km and rural 35,498 sq. km), Kerala is one of the smaller states of the country. It nevertheless harbours a population of 3,18,38,619 (males 1,54,68,664; females 1,63,69,995) with a very high population density of 819 persons per sq. km., that is nearly three times the national average. Shaped as a high vertical slope with an average breadth of 50 km, it is endowed with diversified climatic richness, with a pinch of hydrological concern as 41 of its 44 rivers originating in the Western Ghat empty into the Arabian sea in less than 48 hours after a rain. The state's lowland region, which accounts for 10% of the total area, runs along the coastline and embodies beaches, swamps and lagoons, besides backwaters, paddy fields and coconut plantations. Kerala's midland (42% of the total landmass) is primarily made up of valleys with undulating small hills and meandering passages. A variety of seasonal, annual and perennial crops are grown in its mainly lateritic soil. Of all these crops, cocoa and pineapple plantations are of tremendous significance as dengue vectors have been found to breed there. The highland region (48% of the total state's area), with steep hills is rife with forests and small streams. Plantation crops like rubber, tea, coffee and cardamom are grown in this region (Fig. 3).

The diversity of its geographical features has resulted in a corresponding diversity in climate. While the high ranges along the eastern border have a cold and bracing climate throughout the year, the plains, invariably merging coastally into Arabian Sea along the western flank, are hot and humid. The average annual rainfall (3000mm) is quite high compared to other Indian states – almost three times higher than that in Karanataka while twice than in Tamil Nadu. The state basically enjoys four types of climate, viz., winter, summer, southwest monsoon and northeast monsoon.

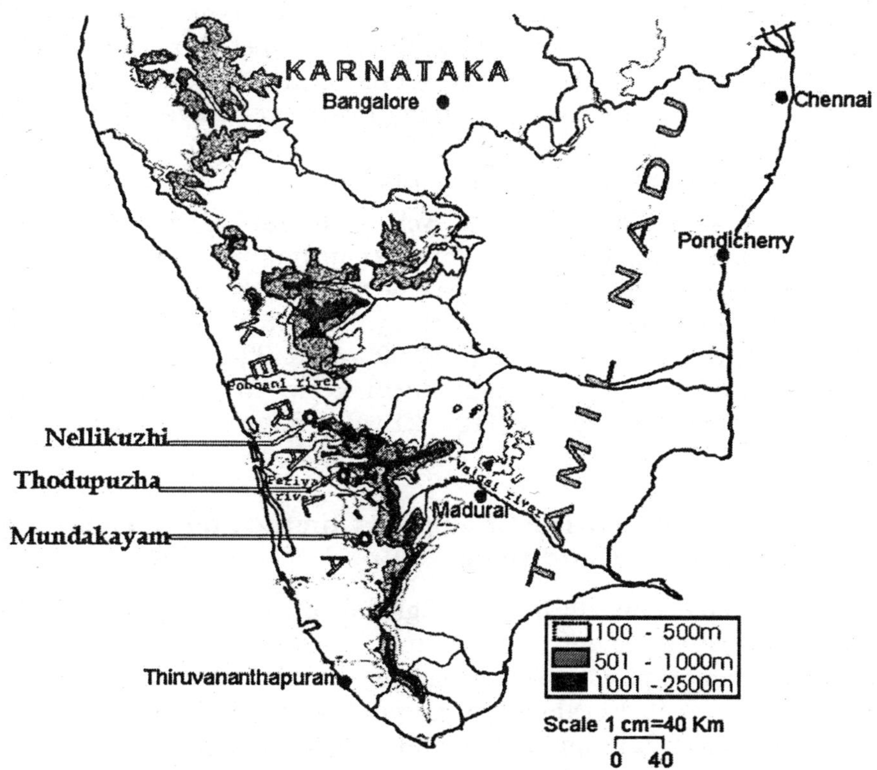

Fig. 3. Map of Kerala with neighbouring States, showing Western Ghats region embodying three study sites.

The winter season sets in during the month of December and continues till the end of February. During this season comparatively there is less rainfall, especially in the northern parts of the state. Winter is followed by the summer season. It starts in February and continues till May. Temperature is very high during this period. Occasional showers is a characteristic of this season. Kottyam district tucked in the western Ghat ranges receives the highest rainfall during this season. The southwest monsoon, accounting for 60-65% of the total rainfall, begins either in the end of May or in the beginning of June and fades out by September. The northeast monsoon, commencing in October, contributes to 25-30% of the total rainfall. The coastal areas record a maximum temperature of 32 °C. The interiors record a maximum of 37 °C during summer. In the coastal area it is hot and humid during April-May while cool during December-January.

A careful analysis of epidemics in Kerala illustrates that dengue not only seemed to have an epicenter in Kottayam before spreading over to entire Kerala, most of the cases had erupted in the mountainous and sylvan environs of the Western Ghat ranges on the south-western edge of the peninsular India facing Arabian sea coast in Kerala of which it covers nearly 90% of the total landmass. The Western Ghat ranges receive variable amount of rainfall from both the southwest (May-July) and the northeast monsoon (September – November), and is characterized with the Shola forest punctuated with man-growth extensive plantations of rubber (*Hevea brasiliensis*), cocoa (*Theobroma cacao*), cocunut (*Cocos nucifera*) and pineapple (*Ananans comosus*).

I have carried out studies in Kottayam, Idukki and Ernakulam, besides the coastal Thrissur and Thiruvananthapuram districts – all of which have had epidemics during past few years (Fig. 3). This region is fairly humid and warm throughout the year with the relative humidity and temperature varying between 70-90% and 22-34.5°C, respectively. The annual precipitation is high reaching up to 3000mm, with the maximum number of rainy days (18 – 21 rainy days/month) being in May to Agust.

Anthropocentric Activities: Change in Climate and Agriculture Versus Denguegenic Conditions

Anthropogenic impacts have brought about a discernible vicissitude in Kerala's forest cover, agricultural practices, traditional water harvesting, demography, urbanization and human mobility. It would be hard to believe that Kerala experienced a drought in 2003, with the average annual rainfall dipping to an all-time low of 2000 mm.

Although Kerala, which enjoys high rainfall, it still gets no more than 50-60 lit. of water for domestic use per person daily. Water allowance per capita has depleted miserably in past few years. The Kerala state has an estimated 77.35 billion cubic meters (BCM) of fresh water, but nearly 40% of the resources are lost as run off, meaning only 42 BCM of water being available for the whole pupulace though the state requires a minimum of 49.70 BCM annually for various indispensably vital activities such as irrigation, domestic use, industries and other miscellaneous use. This shortage has lead people to resort to exploiting ground water so severely that the water table in several districts has gone alarmingly low. The water scarcity also induces people to store water in a mélange of containers like pitchers, cement tanks, etc. as long as possible.

A cadence in both ambient temperature and relative humidity is considered highly supportive to a wide dissemination of the dengue vector in time and space. There are evidences to prove a decline in rainfall, on one hand, and a marginal yet perceptible rise in ambient temperature, on the other. Since 1980 Kerala has suffered seven years of deficient rainfall and another five years below normal rainfall. An analysis of the southwest monsoon since 1998 shows that except Thiruvananthapuram in 2001, all districts have been receiving highly deficient rainfall in most years. Kerala had 2270 mm rainfall in 2003, still about 33% lesser than average. It was amongst the few states in India declared drought-hit during 2003. It is noteworthy that since 1999 the annual rainfall has never touched the average value of 3107 mm; on an average, annual rainfall has dropped by 15-20%. The southwest monsoon has been particularly poor and in 2002 and 2003 it was deficient by as much as 33.3 and 26.4% respectively. The year 2003 was further impoverished due to failure of northeast monsoon, with no area getting any rain during January-March. A close observation of rainfall data in various districts since 1900 shows that the number of rainy days in Kerala has dropped drastically. A discernible climatic cadence in Kerala is possibly linked with the atmospheric EI Nino phenomenon which characterizes monsoon-related vagaries.

Soil in Kerala can retain water only for a maximum of three months. Unless it is replenished periodically through precipitation many parts of the state are subject to acute water shortage. For decades the marshy lands and wetlands, which recharge ground water and help retaining soil moisture, have been reclaimed for constructing houses and setting up industrial units including coconut and other plantations. Forest, which help in orographic or convectional rainfall (i.e., rain in hill during evening), play an integral role in holding soil together and retaining rainwater, but large scale deforestation has been occurring in the state since 1970s, mostly with a view to planting cash crops like rubber plantation. Kerala had an estimated forest cover of 44.45% in 1900 which, according to the recent satellite images, had reduced to 14.7% in 1983 and to a pathetic 9% at present.

Kerala originally had a total cropped area spanning over 2 million hectares, with paddy as the major crop covering about 35% of this area, followed by coconut (21%). By 1975 the paddy growing area reduced to 30% of the total cropped area, with only 0.881 million ha growing paddy. In 2003 the area under paddy cultivation nose-dived to a paltry 0.31 million ha. However, the coconut plantation sustained a good growth rate of 3% per annum through 1990s. It is noteworthy that, along with coconut, rubber plantation has de novo supported breeding of dengue vector. *Ae. albopictus*. At present while rubber plantations occupy the largest land area in Kerala, paddy fields have reduced and reclaimed largely for construction of human habitation.

Dengue Incidence in Kerala

Cyclic dengue epidemics in Kerala state have been occurring since 2001, even thought the first dengue report was brought on record from Kottayam district in 1997 with 14 cases and 4 deaths (Unpublished observation). This was followed by a more sever dengue outbreak implicating 67 cases (nearly 5-fold increase) and a toll of 13 human lives (3-fold increase) in 1998, again in the same district. The years of 1999 and 2000 were virtually free from any dengue cases. However, soon in 2001, epidemic dengue resurged mainly in Kottayam, idukki and Ernakulam reporting 70 cases, followed by 219 cases in 2002 with some deaths. The year 2003 experienced the severest epidemic till date yielding as many as 3546 confirmed cases (253-fold increase) and a toll of 68 human lives, spread for the first time all over the Kerala's fourteen districts. The occurrence of DHF and DSS was also clinically and serologically documented. In spite of this history of dengue-related malady during past few years, no mosquito species was ever incriminated with a dengue virus in nature in Kerala, which created a void in comprehending the disease epidemiology.

In view of a rather recent emergence of dengue infection in Kerala, the data on disease epidemiology are very scanty and inconclusive. During 2003, however, the dengue cases, which initially kept a low ebb, suddenly began to appear in large numbers from invariably all the fourteen districts soon after the southwest monsoon, reaching a maximum of 1337 in the month of July.

It can be judged from these observations that 75.5% of the total dengue cases in Kerala in 2003 occurred only during the southwest monsoon period when *Ae. albopictus* also prevailed in higher density and after that nine of the fourteen districts continued to yield dengue cases till the end of the year, with Thiruvananthapuram reporting the highest incidence of 789/3546 (22.2%). It is also noteworthy that 9 of the 14 districts, of which four are landlocked in the high ranges of the Western Ghat mountains, had reported dengue cases for the first time. However, all those districts (e.g., Kottayam, Idukki, Ernakulam, etc.) which had been reporting dengue earlier had shown an increase in the incidence.

Virus Detection and Isolation

Serologically, even though clinical cases had not appeared until late 1990s, the DEN-1 and DEN-4 viruses had been detected in 1970s from the human sera in a few districts of Kerala like Kozhikode, Kannur, Palakkad, Thriussur, Kottayam and Thiruvananthapuram. It was interesting that only *Ae. aegypti* without an instance of virus incrimination was recorded from all these districts covered under serological surveillance of local human population and *Ae. albopictus* was conspicuously absent. During the 2003outbreak, a total of 112 finger-pricked blood samples were collected from dengue suspected and/or confirmed cases. Of these, 74 (66.0%) sera originating from Pattanamthitta (22.9%), Kottayam (22.2%), Alleppy (Alappuzha) (24.3%), Ernakulam (6.7%) and Thiruvananthapuram (25.6%) were found positive in antigen capture MAC-ELISA using dengue positive human sera, the biological TOXO-IFA system was employed and DEN-3 (67%) was isolated from Kerala cases, implying introduction of a new dengue virus serotype. Additionally, a total of 37 human sera were tested by indigenously developed MAC-ELISA and 8 exhibited IgM antibodies against DEN-2 indicating recent dengue virus infection. These observations carry tremendous epidemiological significance from the view point of evolving hyperendemicity.

Virus detection from a dengue vector in Kerala had been elusive until 2002. A long term investigation on both *Ae. aegypti* and *Ae. albopictus* in Kerala during 2002-2004 yielded a total of 240 pools (5173 adults) of *Ae. albopictus* comprising 112 pools (2456 adults) of males and 128 pools (2717 adult of which 413 were wild-caught) of females which were screened for the presence of dengue virus through ELISA. Of these, one pool consisting of 20 females, reared from the field collected immatures from Mundakayam in Kottayam district, was positive for dengue virus, indicating the trans-ovarial mode of transmission. This phenomenon was further corroborated by a similar observation on *Ae. albopictus* in a nearby Nagercoil area (Kanyakumari district, Tamil Nadu) in the lower reaches of the Western Ghat ranges. Subsequently, during 2004, DEN-2 virus was finally isolated using the TOXO-IFA system and confirmed through ELISA from the two pools of wild caught *Ae. albopictus* female mosquitoes from Kottayam district, demonstrating with certainly the inevitable role of *Ae. albopictus* in the transmission of dengue in Kerala state. Ironically, so far *Ae. aegypti* has not been incriminated with a dengue virus in nature in Kerala.

Mosquito Fauna and Vectors of Dengue

A total of 9101 specimens belonging to 26 species under seven genera have been sampled in Kerala following various sampling methods. Among all the species, *Ae. albopictus* – the Asian tiger mosquito, was most dominant (69.7%), followed by *Armigeres subalbatus* (20.7%), *Ae. aegypti* (4.4%) and *Culex uniformis* (1.5%), with rest of the species accounting for less than 1% individually. *Aedes aegypti* was generally conspicuous by either a total absence or a very low density (Table II).

Sampled species exhibited good degree of flexibility in breeding adaptability to various kinds of habitats. *Aedes albopictus* surpassed all species by exhibiting a wide spectrum of breeding preferences with a clear cut predilection for coconut shells/plastic cups (79.7%) deployed in collection latex from the rubber trees. *Aedes albopictus* breeding in cocoa pods (*Theobroma cacao*) has been widespread in Western Ghat forest fringe areas and its significance has been duly emphasized in preserving the vector density during high monsoon times when the main breeding habitats such as the latex-collecting cups tethered along the main trunk of the rubber tree are rendered unsuitable for breeding due to latter being turned upside down or covered with polythene sheets to avoid water collection in cups. *Ae. albopictus* maintained two peaks of high density correlating with the two monsoon seasons. This observation was considerably substantiated by both the container larval positivity and the adult landing index for *Ae. albopictus*. While the container larval positivity was a high as 64.1-73.8% in June and 80-84.3% in December, the adult landing index too was correspon-dingly high in June (7.2-11) and December (7.1-7.8), respectively (Hiriyan & Tyagi, 2004; Hiriyan *et al.*, 2003).

Table III. Mosquito species in different collections from Kerala during March 2002 – November 2003.

Species	Landing collection (55 Man-hours)	Immature collection (64 Man-hours)	Egg collection (338 Trap-days)	Total	%
Ae. (Stg) aegypti	0	368	36	404	4.44
Ae. (Stg) albopictus	686	2356	3305	6344	69.68
Ar. (arm) subalbatus	60	1809	12	1881	20.66
Cx. (Lop) uniformis	0	137	0	137	1.5

High density of *Ae. albopictus* (Breteau Index about 20) compared to relatively very low BI for *Ae. aegypti* seemed to bear a direct impact on dengue transmission and dissemination since the highest cases seemingly arose during these two periods only; the more conspicuous peak of both cases and deaths being during the southwest monsoon.

CONCLUSIONS

The Thar Desert in northwestern India is today a hotspot for *P. falciparum* dominated malaria. Earlier malaria in Thar Desert was unstable and exclusively transmitted through *An. stephensi*. Today, not only is *P. falciparum* appreciably represented but also *An. culicifacies* becoming the major joint vector along with *An. stephensi*. But, in the true desert situations in the Thar Desert it is *An. stephensi* which solely transmits malaria, through sustenance of its populations in ageold "tanka" and "beri".

In Kerala, dengue is endemic today. My investigations have suggested that *Ae. albopictus*, prodigiously present in the Kerala's sylvan and mountainous Western Ghat ranges rife with rubber and cocoa plantations – the likely epicenter of the epidemic dengue episodes in the state, is effectively transmitting dengue virus even in the absence or insignificant presence of the principal vector, *Ae. aegypti*, alluding towards its primary, rather than secondary role in disease transmission. It is, therefore, clear that there is a direct correlation between the preponderance of *Ae. albopictus* and the conflagration of dengue cases in Kerala. While circulation of multiple strains (hyper-endemicity) in nature seems to be a comprehensible factor in conflagrating dengue outbreaks, discovery of DEN-3 virus in human sera in early 2003 and DEN-2 virus in *Ae. albopictus* soon after in 2004 lend a good support to the hypothesis that the dengue viruses are still in the process of dissemination and establishment across the length and width of the state. A constant occurrence of dengue cases from Kottayam district, adequately supported first by the detection of virus in Ae. albopictus in 2002 and subsequently its isolation from the same vector and site in 2004, allude towards a likely pathway of origin of dengue emergence in other neighbouring districts arising from Kottayam (Tyagi, 2003a,b, 2004; Tyagi *et al.*, 2003, 2004).

In view of these observations a clear parallelism can be viewed between the occurrence of dengue epidemic foci and the *Ae. albopictus* strongholds in Kerala, and the human-factor has obviously played an important role in dispersing the vector species in the hitherto uninvaded areas in the state and created conducive environment for man-*Ae. albopictus* contact. Human behaviour is often associated with the spatial and temporal distribution of *Ae. albopictus* in Kerala which has a tendency to displace *Ae. aegypti* from its habitats. At least three human activity-related reasons can be attributed to the spread of *Ae. albopictus* far from its original sylvatic abode in the Western Ghat in yore to the coastal plains harbouring congested human settlements at present: (i) massive deforestation during past three decades that forced *Ae. albopictus* to come out of its natural abode, (ii) development of human settlements along forest fringe areas where mosquito frequently fed on human blood peridomestically, and (iii) its potential and likeness for transportation through different modes. Although, nothing is known at present about zoonosis in Kerala, it has been clearly demonstrated that dengue is a zoonotic disease of monkeys in Malaysia maintained by *Ae. pseudoniveus / subniveus* at canopy level. Therefore, it would be of interest to know if monkey based zoonotic reservoir would be occurring in the

Western Ghat region abounding both in *Ae. albopictus* and different types of monkeys.

REFERENCES

Christophers, S.R., 1933., The fauna of British India. Diptera III. Family Culicidae, Tribe Anophelini. Taylor and Francis, London, pp. 371.

Covel, G., 1927. A critical review of the data recorded regarding transmission of malaria by different species of Anopheles, with notes on distribution, habits and breeding places. *Indian Medical Research Memoirs* 7: 87-99.

Hiriyan, J. Tewari, S.C. & Tyagi, B.K., 2003. Plastic cups inveigling *Aedes albopictus* (Skuse) breeding around tea vendor-spot in Ernakulam city, Kerala State. *Dengue Bull.* 27: 195.

Hiriyan, J. & Tyagi, B.K., 2004. Cocoa pod (*Theobroma cacao*) – A potential breeding habitat of *Aedes albopictus* (Skuse) in dengue sensitive Kerala state, *India. J. Am Mosquito Cont. Assoc* 20: 323, 2004.

Ijumba, J.N. & Lindsay, W., 2001. Impact of irrigation on malaria in Africa: paddies paradox. *Medical and Veterinary Entomology* 15: 1-11.

Nagpal, B.N. & V.P. Sharma, 1995. *Indian anophelines*. Oxford, New Delhi. 416 pp.

Puri, I.M., 1955. *The distribution of anopheline mosquitoes in India*. Health Bulletin No. 17; Malaria Bureau No. 7, Government of India, 104 pp.

Puri, I.M., 1960. *Synoptic keys for the identification of the full grown larvae of the Indian anopheline mosquitoes*. Health Bulletin No. 16, Malaria Bureau No. 7, Government of India Press, Calcutta, pp. 104.

Rao, A.S., 1996. Climatic changes in the irrigated tracts of Indira Gandhi Canal region of arid Western Rajasthan, India. *Annals of Arid Zone* 35: 111-116.

Rao, T.R., 1984. The anophelines of India. Malaria Research Centre, New Delhi. pp. 594.

Sikka, D.R. & S.M. Kulshrestha, 2001. Climate and Health studies: Status and scope in India. Centre for Ocean-Land-Atmosphere Studies/Centre for Application of Research on the Environment, Joint COLA/CARE Tech. Rep. No. 5; 66 pp.

Tyagi, B.K., 2002. Malaria in the Thar Desert: Facts, figures and future. Agrobios (India), Jodhpur. pp. 165.

Tyagi, B.K., 2004. A review of the emergence of *Plasmodium falciparum*-dominated malaria in irrigated areas of the Thar Desert, India. *Acta Tropica* 89: 227-239.

Tyagi, B.K., 2004. A review of the emergence of *Plasmodium falciparum*-dominated malaria in irrigated areas of the Thar Desert, India. *Acta Tropica* 89: 227-239.

Tyagi, B.K. & Chaudhary, R.C., 1997. Outbreak of falciparum malaria in the Thar Desert (India), with particular emphasis on physiographic changes brought about by extensive canalization and their impact on vector density and dissemination. *Journal of Arid Environments* 36: 541.

Tyagi, B.K. & S.P. Yadav, 1996. Malariological and sociological significance of 'tanka' and 'beri' in the Thar desert, western Rajasthan (India). *Journal of Arid Environments* 33: 497-501.

Tyagi, B.K., R.C. Chaudhary & S.P. Yadav, 1995. Epidemic malaria in Thar desert, India. Lancet 346: 634- 635.

Tyagi, B.K., Hiriyan, J. & Tewari, S.C., 2004. Dengue in India: A review, with special emphasis on the climate and environment variabilities. First International Seminar of Medical Entomology, Bhopal (Abstract) p. 33.

Tyagi, B.K., Hiriyan, J., Tewari, S.C. & Thenmozhi, V. 2003. Studies on dengue emergence in Kerala State, India. Annual Report 2002-2003. Centre for Research in Medical Entomology, Madurai, p. 49.

Tyagi, B.K., Hiriyan, J., Tewari, S.C., Philip Samuel, P. & Thenmozhi, V., 2005. Dengue in Kerala Disease epidemiology and vector population dynamics. 7th International Symposium on Vectors and Vector-borne Dieseases, Patiala, p. 34.

Tyagi, B.K., Hiriyan, J., Tewari, S.C., Thenmozhi, V. & Philip Samuel, P., 2004. Studies on dengue emergence in Kerala state, India. Annual Report. Centre for Research in Medical Entomology, Madurai, p. 48, 2003-2004.

Tyagi, B.K., 2003. Epidemic dengue/dengue haemorrhagic fever emergence in India, with special reference to Kerala State. National Symposium on Management of Aquatic Resources for Biodiversity Maintenance and Conservation, Jodhpur, (Abstract), p. 56.

Tyagi, B.K., 2003. Medical Entomology: A Handbook of Medically important Insects and Other Aarthropods. Scientific Publishers (India), Jodhpur, p.265.

Tyagi, B.K., 2003. Emergence of dengue in India as a major public health problem. National Symposium on Recent Trends in Zoological Sciences, Patiala, p. 34.

Tyagi, B.K., 2004. The Invincible Deadly Mosquitoes: India's health and Economy Enemy No.1, Scientific Publishers (India), Jodhpur, p. 276.

Chapter - 9

DIVERSITY OF BUTTERFLIES IN CHIDIYATAPU BIOLOGICAL PARK, ANDAMAN AND NICOBAR ISLANDS

C. Sivaperuman, C. Raghunathan and Suresh Kumar Shah

Zoological Survey of India, Andaman and Nicobar Regional Centre,
Port Blair- 744 102 Andaman and Nicobar Islands
E-mail: *c_sivaperuman@yahoo.co.in*

ABSTRACT

A study was conducted in the Chidiyatapu Biological Park and its adjoining areas to document the butterfly fauna along with species richness, abundance and diversity. The study areas were sampled using line transect method and observations made from 0600 to 1100 hours. The vegetation of this park could be classified as tropical semi evergreen forest, moist deciduous forest and littoral forest. A total of 49 species of butterflies belonging to 5 families and 41 genera were recorded during this study. The species richness and abundance was highest in semi evergreen forest followed by moist deciduous forest and littoral forest. Highest number of species were recorded from the family Nymphalidae. The highest diversity Index (H') was observed in the family Nymphalidae (2.03) and Pieridae (2.03) and was in Lycaenidae (1.44). Jaccard similarity index showed highest similarity between Semi evergreen forest and Moist deciduous forest (0.62). This study confirms the importance of butterfly conservation in Chidiyatapu Biological Park.

INTRODUCTION

Butterflies are essential part of any natural ecosystem as their adults perform pollination and larvae enact as primary herbivores thereby transferring radiant energy trapped by plants to the next trophic level; rendering dual roles as pollinators and as energy transferors. The holometabolus life history of butterflies exposes them to a wide range of environmental influences. They are highly sensitive to changes in temperature, humidity and light; parameters that are easily influenced by habitat deterioration (Murphy *et al.*, 1990). The lepidopterans, are also economically important as many of them are pests of a variety of crops.

Information on their host range, habitat preferences, distribution is very important in forest and agriculture canopy systems. Being closely associated with the vegetation, they are also important agents in nutrient cycling.

The butterfly fauna of the Andaman and Nicobar Islands is insular with its origins in the fauna of the Indo-Myanmar and Indo-Malayan regions. The Andaman elements have their closest affinities to Myanmar and mainland elements whereas the Nicobar elements appear most closely related to Malayan elements. The long isolation of these islands from the Asian continent and their undisturbed ecology provided optimal conditions for the evolution of many local and endemic taxa. Studies on butterflies of Andaman and Nicobar Island received attention after the publication of Wood Mason & de Nicebille form 1880-1882. Evans (1932) have worked on butterflies of this Islands. Later few researchers have been contributed on the fauna butterflies of Andaman & Nicobar Islands (Ferrar, 1948; Talbot, 1939, 1947; Arora & Nandhi, 1980, 1982; Khatri, 1989-1993, Chandra & Khatri, 1993; Devy *et al.,* 1994).

STUDY AREA

The archipelago comprises 572 islands known as the Andaman and Nicobar Islands extending over 800 km (Fig. 1). These islands were once a part of the Asian mainland but got detached some 100 million years ago during the Upper Mesozoic Period due to geological upheaval. The existing groups of islands constitute the physiographic continuation of the mountainous ranges of Naga and Lushai Hills and Arakan Yoma of Burma through Cape Negrais to the Andaman and Nicobar Islands and southeast of Sumatra (Achin Head). The chains of these islands are in fact the camel backs of the submerged mountain ranges projecting above the sea level running north to south between 6045′N and 13030′N latitudes and 90020′E and 93056′E longitudes.

Andamans and the Nicobars, the two groups are separated by the Ten-degree Channel which is about 150 km wide 400 fathoms deep. The Andaman group consists of 324 islands of which only 20 are inhabited. The main part of the group is collectively known as the Great Andamans comprising five closely adjoining islands, Baratang and Rutland Islands. All these are separated by narrow channels. The Great Andaman group covers a land area of 6408 km^2. This island is 48 km long and 27 km wide and its land area is about 960 km^2. The Nicobar group comprises 24 islands of which 13 are inhabited and others are of less significance.

Climate

These islands are tropical, that is, warm, moist and equable. The proximity of the sea and the abundant rainfall prevent extremes of heat.

The mountainous parts of the southern islands get about 300 cm of rain annually whereas the islands of north get lesser rainfall. Winter is practically unknown; the period from December to February is comparatively cool due to the effect of northeast monsoon. The warm weather extends from March to April when there is the least precipitation. In the May southwest monsoon breaks over the area and continues till October end. The variation of temperature over the islands is small (23°-31°C).

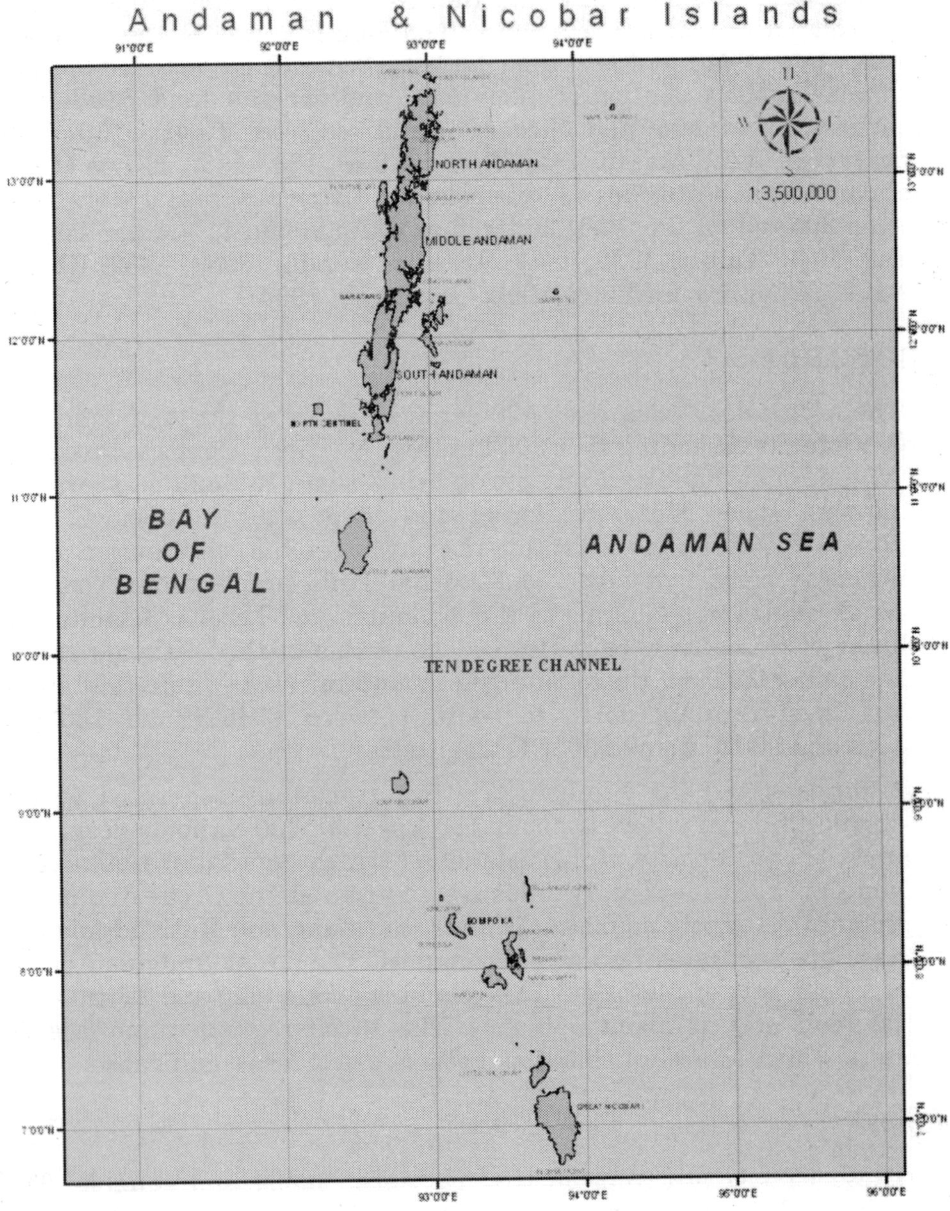

Fig. 1. Andaman and Nicobar Islands

Chidiyatapu Biological Park

This park is situated about 25km away from Port Blair and located in the southernmost tip of the South Andaman Islands. It is about 40 ha in area and is composted of littoral forests, most deciduous forest and semi evergreen forest. The major tree species recorded in the study area are *Eugenia jambulance*, *Pometia pinnata*, *Sageraea elliptica*, *Bombax insigne*, *Dipterocarpus* sp., *Terminalia baliata*, *Tetrameles nudiflora*.

Coordinates of transects in different habitats

Transect No.	Latitudes	Longitudes
1	11° 30.115' N	92°42.108' E
2	11° 38.571 N	92°42.298' E
3	11° 29.071 N	92°42.246 E
4	11° 30.369 N	92°42.072 E

METHODS

This study was conducted during May - June 2010. Butterfly species were estimated by 600 m line transect, traversed in one hour. Transects were enumerated between 0600 hours to 1100 hours. Butterflies were identified based on physical features with the help of field guides and reference books (Evans, 1932; Ferrar, 1948; Kehimkar, 2008). The Four transects were laid at different habitat *viz.* Semi evergreen, Moist deciduous, Littoral forests, and Sea side road. Observations were carried out in the morning between 0600-11.00h. Butterflies observed in each habitat were individually recorded. Unfamiliar species were collected for identification. Species observed outside the transects and forest edges were noted separately. Butterflies observed along the transects alone were considered for statistical analyses.

Species richness and abundance of butterflies

The total number of butterfly species and number of individuals seen in each transect and different habitats were calculated using the census data and field observations.

Diversity indices

Shannon index, Simpson index and Hill's diversity numbers N1 and N2 were calculated for different habitats using the programme SPDIVERS.BAS developed by Ludwig & Reynolds (1988).

Similarity measures

Jaccard similarity index between the different habitat were calculated using the formula of Magurran (1988).

Dominance Index

The dominance of the each butterfly species was calculated using the dominance index.

$$\text{Dominance Index} = n_i \text{ X } 100/ \text{ N}$$

where, n_i = Number of individuals

N = Total number of all the species seen during the study period

Rarefaction

The Rarefaction was calculated using the software BioDiversity (Lambshead *et al.,* 1997).

RESULTS

Occurrence of species

A total of 49 taxa of butterflies were recorded during the period, these belong to 5 families and 41 genera (Table 1). Of the recorded species highest number were recorded from family Nymphalidae followed by Pieridae.

Table 1. List of butterflies species recorded in and around the Chidiyatapu Biological Park

Sl. No	Family	Common Name	Scientific Name	Status*
1.	Hesperiidae	Brown Awl	*Badamia exclamationis*	Not rare, locally common
2.		Giant Red Eye	*Gangara thyrsis*	Not rare
3.		Common Snow Flat	*Tagiades japetus*	Not rare
4.		Plain Banded Awl	*Hasora vitta*	Not common
5.	Lycaenidae	Grams Blue	*Euchrysops cnejus*	Common
6.		Common Cerulean	*Jamides celeno blairana*	Common
7.		Yamfly	*Loxura atymnus*	Common
8.		Dingy Blue	*Petrelaea dana*	Not rare
9.		Pale Grass Blue	*Pseudozizeeria maha*	Common
10.		Common Onyx	*Horaga onyx*	Locally common
11.		Quaker	*Neopithecops zalmora*	Common
12.	Nymphalidae	Andaman Palmking	*Amathusia andamanensis*	Rare

13.	Leopard Lacewing	*Cethosia cyane*	Not rare
14.	Rustic	*Cupha erymanthis andamanica*	Common
15.	Andaman Map	*Cyrestis thyodamas andamanica*	Not common
16.	Autumn Leaf	*Doleschallia bisaltide*	Not rare
17.	Andaman Palmfly	*Elymnias cottonis*	Rare
18.	Andaman Crow	*Euploea core andamanensis*	Rare
19.	Pasha	*Herona marathus andamana*	Not rare
20.	Great Egg Fly	*Hypolimnas bolina jacintha*	Common
21.	Peacock Pansy	*Junonia almana*	Common
22.	Grey Pansy	*Junonia atlites*	Locally common
23.	Yellow Pansy	*Junonia hierta*	Common
24.	Commander	*Moduza procris*	Common
25.	Banded Bush Brown	*Mycalesis mineus*	Common
26.	Clear Sailor	*Neptis clina*	Rare
27.	Glassy Blue Tiger	*Parantica aglea*	Common
28.	Clipper	*Parthenos sylvia*	Rare
29.	Small Leopard	*Phalantha alcippe*	Locally common
30.	Cruiser	*Vindula erota*	Not rare
31.	White Commander	*parasarpa dudu*	Rare
32. Papilio-nidae	Common Mime	*Chilasa clytia*	Not rare
33.	Tailed Jay	*Graphium agamemnon*	Common
34.	Common Jay	*Graphium doson*	Locally common
35.	Andaman Swordtail	*Graphium epaminondas*	Rare
36.	Common Lime	*Papilio demoleus*	Very common
37.	Andaman Mormon	*Papilio mayo*	Not rare
38.	Common Mormon	*Papilio polytes*	Very common
39.	Andaman Helen	*Papilio prexaspes andamanicus*	Rare
40. Pieridae	Common Albatross	*Appias albina*	Common
41.	Common Emigrant	*Catopsilia pomona*	Common
42.	Lesser Gull	*Cepora nesissa lichenosa*	Locally common

43.	Three Spotted Grass Yellow	*Eurema blanda*	Common
44.	Common Grass Yellow	*Eurema hecabe*	Common
45.	Tree Yellow	*Gandaca harina andamana*	Not Common
46.	Great Orange Tip	*Hebomoia glaucippe*	Common
47.	Yellow Orange tip	*Ixias pyrene andamana*	Common
48.	Psyche	*Leptosia nina*	Common
49.	Common Wanderer	*Pareroni ceylanica naraka*	Common

* Kehimkar, 2008

Species composition in different habitats

Species of butterflies recorded in different habitat during the period of study is given in Table 2. Out of the forty nine species of butterflies 41 species were recorded from semi evergreen forest habitat followed by moist deciduous forest (35), Sea side road (28) and Littoral forest (27). Of the recorded butteflies species, seventeen species were recorded in all transects.

Table 2. Butterflies recorded in different transect in Chidiyatapu Biological Park

Sl. No.	Species name	Semi evergreen	Road side forest	Littoral forest	Sea side road
1.	*Badamia exclamationis*	√	√	√	-
2.	*Gangara thyrsis*	-	√	-	-
3.	*Tagiades japetus*	√	-		-
4.	*Euchrysops cnejus*	√	√	√	-
5.	*Jamides celeno blairana*	√	√	√	√
6.	*Loxura atymnus*	√	√	√	√
7.	*Petrelaea dana*	√	-	-	√
8.	*Pseudozizeeria maha*	√	-	√	√
9.	*Amathusia andamanensis*	-	√	-	√
10.	*Cethosia cyane*	√	√	√	-
11.	*Cupha erymanthis andamanica*	-	√	-	-
12.	*Cyrestis thyodamas andamanica*	√	-	-	-
13.	*Doleschallia bisaltide*	√	√	√	-
14.	*Elymnias cottonis*	√	√	√	√

15.	*Euploea core andamanensis*	√	√	√	√
16.	*Herona marathus andamana*	√	-	-	-
17.	*Hypolimnas bolina jacintha*	-	-	-	√
18.	*Junonia almana*	-	-	-	√
19.	*Junonia atlites*	√	√	-	√
20.	*Junonia hierta*	√	-	√	√
21.	*Moduza procris*	√	-		-
22.	*Mycalesis mineus*	√	√	√	√
23.	*Neptis clina*	√	-	-	-
24.	*Parantica aglea*	-	-	√	√
25.	*Parthenos sylvia*	√	√	√	√
26.	*Phalantha alcippe*	√	√	√	√
27.	*Vindula erota*	√	√	-	-
28.	*Chilasa clytia*	√	√	√	√
29.	*Graphium agamemnon*	√	√	√	√
30.	*Graphium doson*	√	√	√	√
31.	*Graphium epaminondas*	-	√	-	-
32.	*Papilio demoleus*	-	√	-	√
33.	*Papilio mayo*	√	√	√	√
34.	*Papilio polytes*	√	√	√	√
35.	*Papilio prexaspes andamanicus*	√	√	-	-
36.	*Appias albina*	√	√	√	√
37.	*Catopsilia pomona*	√	√	√	√
38.	*Cepora nesissa lichenosa*	√	√	√	√
39.	*Eurema blanda*	√	-	√	-
40.	*Eurema hecabe*	√	√	√	√
41.	*Gandaca harina andamana*	√	√	√	-
42.	*Hebomoia glaucippe*	√	√	-	√
43.	*Ixias pyrene andamana*	√	√	√	√
44.	*Leptosia nina*	√	√	-	-
45.	*Pareroni ceylanica naraka*	√	√	√	-
46.	*Parasarpa dudu*	√	-	-	-
47.	*Hasora vitta*	√	√	-	-
48.	*Horaga onyx*	√	-	-	-
49.	*Neopithecops zalmora*	√	√	-	√

Species richness and abundance of butterflies

Species richness and abundance of butterflies varied in the different families. Highest number of species richness was recorded from Nymphalidae followed by Lycaenidae, Papilionidae, Pieridae and Hesperidae (Fig. 2). Similarly species abundance also varies in different families. Highest number of individuals were recorded from Nymphalidae followed by Pieridae (Fig. 3).

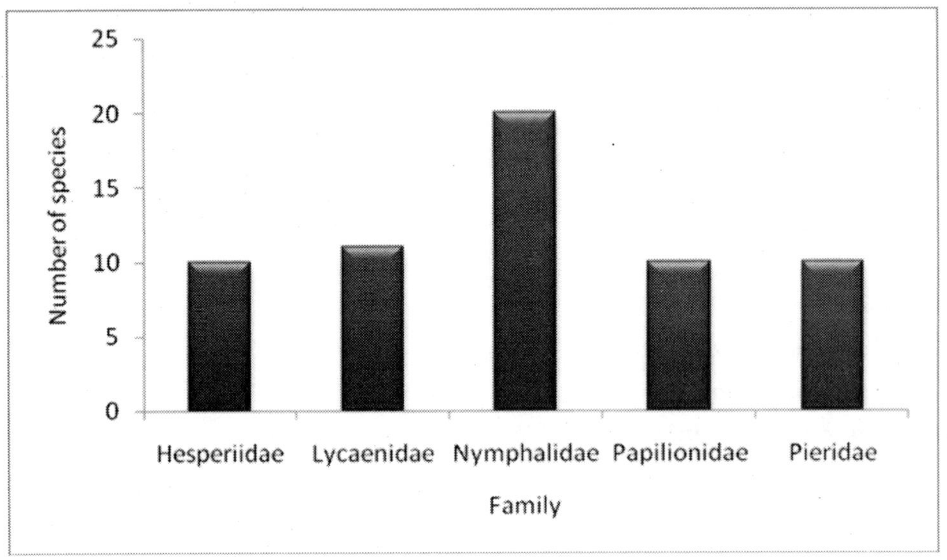

Fig. 2. Species richness of butterflies in different families

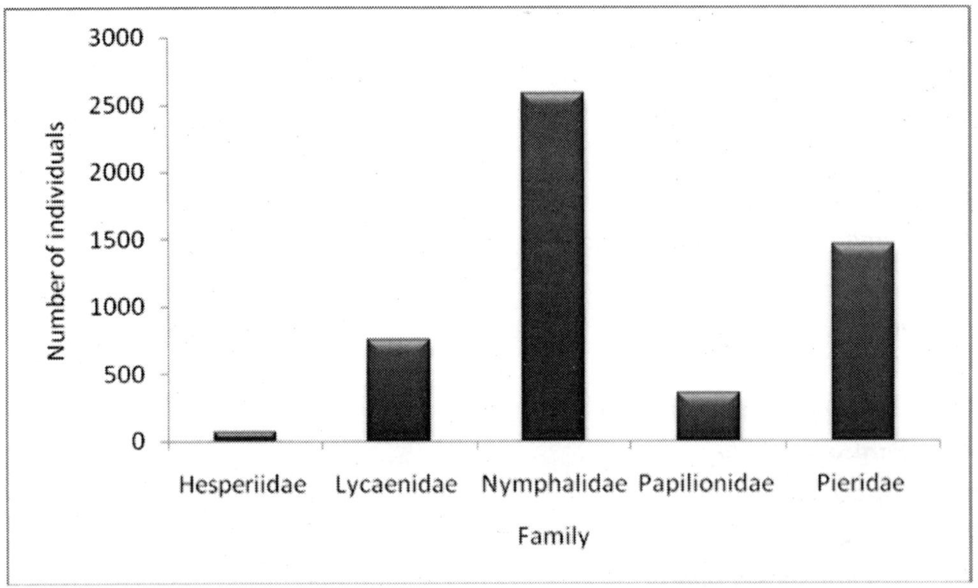

Fig. 3. Species abundance of butterflies in different families

Diversity

Most widely used diversity indices like Shannon Index of diversity, Simpson's Index of diversity and Hill's numbers N1 and N2 have been determined. Overall diversity index H' was 3.02. Similarly, the high values of richness indices also observed during the study (Table 3).

Table 3. Overall diversity indices of butterflies in the Chidiyatapu Biological Park

Richness		Diversity		Hills Numbers		Evenness	
R1	R2	λ	H'	N1	N2	E1	E2
7.01	0.85	0.08	3.02	20.49	13.24	0.73	0.34

Family wise diversity indices

The highest diversity Index (H') was observed in the family Nymphalidae (2.03) and Pieridae (2.03) and lowest diversity Index in Lycaenidae (1.44) (Table 4).

Table 4. Familywise diversity indices of butterflies in the Chidiyatapu Biological Park

Family	Richness		Diversity		Hills Numbers		Evenness	
	R1	R2	λ	H'	N1	N2	E1	E2
Hesperiidae	2.35	1.31	0.20	1.95	7.03	5.10	0.81	0.64
Lycaenidae	1.62	0.41	0.36	1.44	4.23	2.77	0.58	0.35
Nymphalidae	2.49	0.38	0.17	2.03	7.62	5.80	0.67	0.36
Papilionidae	1.70	0.58	0.18	1.88	6.58	5.71	0.79	0.60
Pieridae	1.33	0.25	0.16	2.03	7.58	6.30	0.84	0.69

Similarity indices between different habitats

Similarity indices between the different habitats were computed using qualitative data (Table 5). Jaccard similarity index showed highest similarity between Semi evergreen and Moist deciduous forest (0.62). Similarity between Litteral forest, sea side road was was less than fifty per cent.

Table 5. Jaccard similarity index between different habitats

Intensive study areas	Semi-evergreen	Moist deciduous	Littoral forest	Seaside
Semi-evergreen	0	0.62	0.49	0.47
Moist deciduous		0	0.51	0.53
Littoral forest			0	0.57
Seaside				0

Rarefaction

A comparison of number of species in different habitat for fixed number of individuals encountered is a useful exercise. For this purpose, expected number of species was calculated for varying number of individuals and presented in Fig. 4. The data shows that semi evergreen forest had more number of species followed by Sea side road.

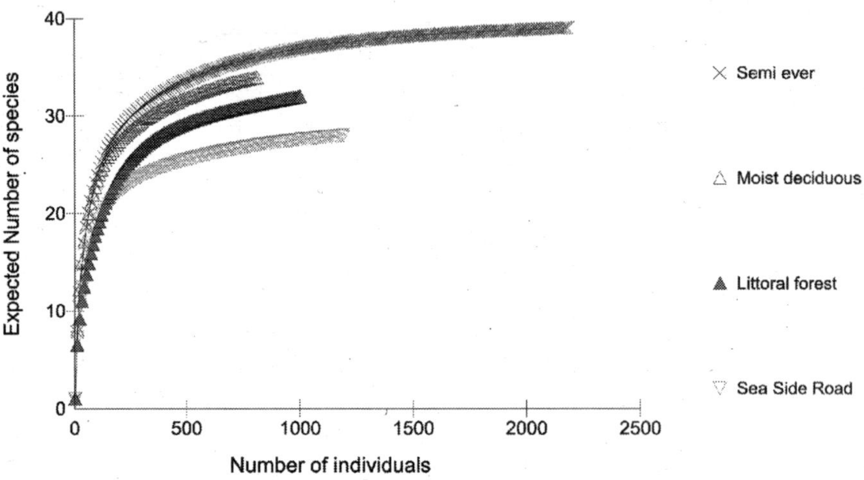

Fig. 4. Rarefaction for butterfly species in different habitats

Species composition in different habitat

Species richness of butterflies highest in the Semi evergreen forests, followed by Moist deciduous forest (Fig. 5). Similarly species abundance also was highest in Semi evergreen and lowest in Moist deciduous forests. (Fig. 6).

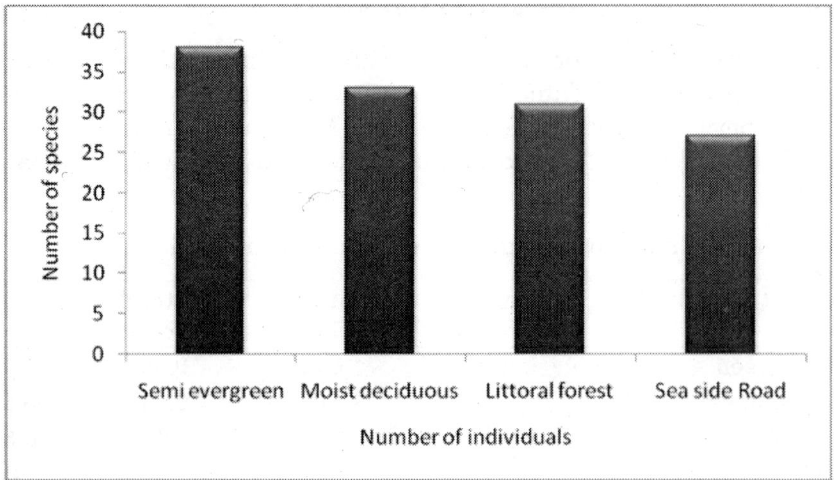

Fig. 5. Species richness in different habitat in Chidiyatapu Biological Park

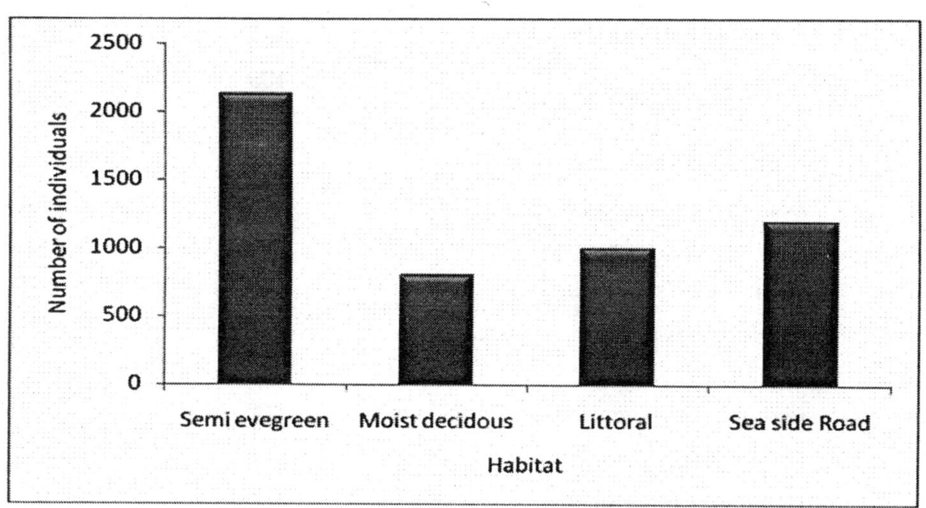

Fig. 6. Species abundance in different habitat in Chidiyatapu Biological Park
Diversity indices butterflies in different habitat

Species diversity index (H') was highest in Moist deciduous forest (2.90) and lowest in Littoral forest (2.41). Simpson Index was highest at Littoral forest (0.13), followed by sea side road (0.10) (Table 6).

Table 6. Diversity indices butterflies in different habitat

Habitat	Richness		Diversity		Hills Numbers		Evenness	
	R1	R2	λ	H'	N1	N2	E1	E2
Semi evergreen	4.83	0.82	0.08	2.87	17.65	11.94	0.79	0.46
Moist deciduous forest	4.78	1.16	0.07	2.90	18.19	13.55	0.83	0.55
Littoral forest	4.34	0.98	0.13	2.41	11.11	7.42	0.70	0.36
Sea side	3.67	0.78	0.10	2.63	13.85	10.37	0.80	0.51

Dominance of individual species

Of the recorded species of butterflies in the Chidyatappu Biological Park, *Junonia hierta* (15.22 per cent) was highest in dominance followed by *Phalantha alcippe* (13.64 per cent) and *Loxura atymnus* (9.62 per cent) (Table 7). Twenty three species were represented in less than 1 per cent.

Table 7. Abundance and dominance of butterfly species in the Chidyatapu Biological Park

Sl. No.	Species name	Abundance	Dominance
1.	*Junonia hierta*	780	15.22
2.	*Phalantha alcippe*	699	13.64
3.	*Loxura atymnus*	493	9.62

4.	*Euploea core andamanensis*	487	9.50
5.	*Appias albina*	439	8.56
6.	*Eurema hecabe*	269	5.25
7.	*Catopsilia pomona*	216	4.21
8.	*Ixias pyrene andamana*	216	4.21
9.	*Parthenos sylvia*	175	3.41
10.	*Jamides celeno blairana*	122	2.38
11.	*Papilio mayo*	99	1.93
12.	*Junonia atlites*	81	1.58
13.	*Gandaca harina andamana*	76	1.48
14.	*Neptis clinia*	76	1.48
15.	*Hebomoia glaucippe*	72	1.40
16.	*Graphium agamemnon*	71	1.39
17.	*Papilio polytes*	70	1.37
18.	*Discophora timora*	67	1.31
19.	*Cepora nesissa lichenosa*	66	1.29
20.	*Elymnias cottonis*	64	1.25
21.	*Graphium doson*	62	1.21
22.	*Pseudozizeeria maha*	59	1.15
23.	*Pareroni ceylanica naraka*	54	1.05
24.	*Mycalesis mineus*	49	0.96
25.	*Leptosia nina*	38	0.74
26.	*Chilasa clytia*	37	0.72
27.	*Badamia exclamationis*	28	0.55
28.	*Petrelaea dana*	26	0.51
29.	*Cethosia cyane*	22	0.43
30.	*Euchrysops cnejus*	16	0.31
31.	*Parantica aglea*	16	0.31
32.	*Vindula erota*	14	0.27
33.	*Junonia almana*	11	0.21
34.	*Neptis clina*	10	0.20
35.	*Eurema blanda*	9	0.18
36.	*Papilio prexaspes andamanicus*	8	0.16
37.	*Hypolimnas bolina jacintha*	6	0.12
38.	*Tagiades japetus*	5	0.10

39.	*Moduza procris*	4	0.08
40.	*Gangara thyrsis*	3	0.06
41.	*Papilio demoleus*	3	0.06
42.	*Amathusia andamanensis*	2	0.04
43.	*Cupha erymanthis andamanica*	2	0.04
44.	*Graphium epaminondas*	2	0.04
45.	*Cyrestis thyodamas andamanica*	1	0.02
46.	*Herona marathus andamana*	1	0.02
47.	Total	5126	100.00

DISCUSSION AND CONCLUSIONS

The study has generated data on the butterfly diversity in the Chidiyatapu Biological Park, South Andaman. The overall diversity index for the study area is (H') 3.02 and diversity in different habitats have been varied, semi evergreen forest (2.87), Moist deciduous forest (2.90), Littoral forest (2.41) and Sea side road (2.63). The evergreen forest is comparatively an undisturbed habitat in comparison with the other habitat.

The butterfly fauna of India is quite well known (Talbot, 1939, 1947; Wynter-Blyth, 1957; Larsen, 1987,1988). However, only very few studies were conducted with reference to Andaman & Nicobar Islands (Talbot, 1939, 1947; Ferrar, 1948; Arora & Nandhi, 1980, 1982; Khatri, 1989-1993, Chandra & Khatri, 1993; Devy *et al.,* 1994). In the present study 49 species of butterflies belonging to 41 genera and 5 families recorded. Most of the butterflies collected belongs to Nymphalidae, Pieridae, Lycaenidae, and Hesperidae.

The Chidiyatapu Biological Park consists a number of different habitats. This may be one of the reasons for the area exhibiting a high degree of species richness for butterflies. Tropical semi evergreen forests possessed the greatest diversity of butterfly in the Biological Park followed by Littoral and moist deciduous forest. From the conservation point of view, we found that our study area held several threatened species (Van Swaay & Warren, 1999). This confirms the importance of butterfly conservation in Biological Park.

Habitat association of butterflies can be directly related to the availability of food plants (Thomas, 1995). The criterion of representative diversity which is based on the recognition of assemblages of species that are typical for specific habitats seems more interesting (Dufrene & Legendre, 1997). A potentially useful tool is the single species approach based on focal or surrogate species, which can indicate ecological change, patterns of richness or habitat type (Caro & Doherty, 1999; Fleishman *et al.,* 2000). Each habitat has a specific set of micro environment suitable for a species. The overall species richness in the Chidiyatapu Biological Park is comparatively high. It might therefore, be expected that species with a smaller geographic range will thrive in restricted habitats having specific requirements with the Chidiyatapu Biological Park area. This emphasis their importance for maintaining biodiversity on a regional scale, the species with a restricted geographic distribution appear to be more sensitive human disturbance and forests structure changes than widespread species.

REFERENCES

Arora, G.S. & D.N. Nandi, 1980. On the butterfly fauna of Andaman and Nico-bar Islands (India). I. Papilionidae. *Records of Zoological Survey of India* 77: 141-151.

Arora, G.S. & D.N. Nandi, 1982. On the butterfly fauna of Andaman and Nicobar Islands India. II. Pieridae. *Ibid.* 80: 1-15.

Caro, T.M. & G.O. Doherty, 1999. On the use of surrogate species in conserv-ation biology. *Conserv. Biol.* 13: 805- 814.

Devy, M.S., T. Ganesh and P. Davidar, 1998. Patterns of butterfly distribution in the Andaman Islands: Implications for conservation. *Acta Oecologia* 19(6): 527-534.

Dufrene, M. & P. Legendre 1997. Species assemblages and indicator species: the need for a flexible asymmetrical approach. *Ecol. Monogr.* 67: 345-366.

Evans, W.H., 1932. *The Identification of Indian Butterflies.* Bombay Natural History Society, Bombay. 454 p.

Ferrar, M. L., 1951. On the butterflies of Andaman and Nicobar Islands. *Journal Bombay Natural History Society* 47: 470-491.

Fleishman, E., D.D. Murphy & P.F. Brussard, 2000. A new method for selection of umbrella species for conservation planning. *Ecol. Appl.*, 10: 569-579.

Gascon, C., T.E. Lovejoy, R.O. Bierregaard, J.R. Malcolm, P.C. Stouffer, H.L. Vasconcelos, W.F. Laurance, B. Zimmerman, M. Tocher & S. Borges, 1999. Matrix habitat and species richness in tropical forest remnants. *Biol. Conserv.* 91: 223–229.

Kehimkar, I., 2008. *The book of Indian Butterflies.* Bombay Natural History Society, Mumbai. 497 pp.

Khatri, T.C., 1989. A revised list of Butterflies from Bay Islands. *J. Andaman Sci. Assoc.* 5: 57-61.

Khatri, T.C., 1991. On some Nymphalidae (Rhopalocera: Lepidoptera) from the Andaman and Nicobar Islands. *Islands on March* 3: 82-94.

Khatri, T.C., 1992. On some Lycaenids (Rhopalcocera: Lepidoptera) from Andaman and Nicbar Islands. *Islands on March* 6: 8-16.

Kunte, J.K., 1997. Seasonal patterns in butterfly abundance and species diversity in four tropical habitats in the northern Western Ghats. *Journal of Bioscience* 22: 593-603.

Lambshead, P.J.D., G.L.J. Paterson & J.D. Gage, 1997. BioDiversity Professional Beta. The Natural History Museum and The Scottish Association for Marine Science.

Larsen, T.B., 1987. The butterflies of the Nilgiri mountains of South India (Lepidoptera: Rhopalocera). *Journal of the Bombay Natural History Society* 84: 291-316.

Larsen, T.B., 1988. The butterflies of the Nilgiri mountains of South India (Lepidoptera: Rhopalocera). *Journal of the Bombay Natural History Society* 85: 26-43.

Ludwig, J.A. & J.F. Reynolds, 1988. Statistical Ecology, A premier on Methods and Computing. A Wiley-Interscince publication. 337 pp.

Magurran, A.E., 1988. Ecological diversity and its measurement. Cambridge University Press, Cambridge: x + 179 pp.

Murphy, D.D., K.E. Freas & S.B. Weiss, 1990. An environment- metapopulation approach to population viability analysis for a threatened invertebrate. *Conservation Biology* 4: 41-51.

Talbot, G., 1939. The fauna of British India including Ceylon and Burma (Butterflies), Taylor and Francies, London.

Talbot, G., 1947. Fauna of British India. Butterflies II, Today & Romorrow Printers & Publishers.

Thomas, J.A., 1995. The ecology and conservation of *Maculinea arion* and other European species of large blue butterfly. pp. 180-210 *In:* A.S. Pullin (ed.) *Ecology and Conservation of Butterflies*. Chapman and Hall, London.

Van Swaay, C.A.M. & M.S. Warren, 1999. *Red data book of European butter-flies (Rhopalocera).* Nature and Environment, No. 99. Council of Europe, Strasbourg.

Wood-Mason, J. & L. de Niceville, 1980. List of diurnal Lepidoptera from Port Blair. Andaman Islands. *J. Asiat. Soc. Beng.* 49(2): 223-243.

Wood-Mason, J. & L. de Niceville, 1981b. Second list of Rhopalocerous Lepidoptera from Port Blair. Andaman Islands. *J. Asiat. Soc. Beng.*, 50(4): 243-262.

Wood-Mason, J. & L. de Niceville, 1982. Second list of Rhopalocerous Lepidoptera from Port Blair. Andaman Islands. *J. Asiat. Soc. Beng.*, 11: 14-20.

Wynter-Blyth, M.A., 1957. *Butterflies of Indian Region.* Bombay Natural History Society, Bombay.

Chapter - 10

STRUCTURE AND SPECIES COMPOSITION OF BUTTERFLIES IN GREAT NICOBAR BIOSPHERE RESERVE, ANDAMAN AND NICOBAR ISALNDS

C. Sivaperuman, Suresh Kumar Shah, C. Raghunathan and Ramakrishna*

Zoological Survey of India, Andaman and Nicobar Regional Station,
Port Blair - 744 102, Andaman and Nicobar Islands, India
E-mail: *c_sivaperuman@yahoo.co.in*
* Zoological Survey of India, Prani Vigyan Bhawan, M-Block,
New Alipore, Kolkata - 700 053, India

ABSTRACT

The Great Nicobar Biosphere Reserve (GNBR) is the southernmost Islands in the Andaman and Nicobar Islands. This study was conducted during 2008-2010. Line transect methods was used to assess the population of butterflies and 11 locations were surveyed during the period. A total of 65 species belonging to 5 families and 46 genera were recorded during the study. The family Nymphalidae and Lycaenidae were found to be the dominant in GNBR. Highest number of species was recorded from East-west road and lowest in Navi dera. The diversity in different locations was ranged from 3.09-3.74. High similarity was observed between Amphibian Road and B-quarry. The detailed conservation measures are discussed in this paper.

INTRODUCTION

Butterflies are our most popular insects and usually welcome visitors to gardens and parks with their brightly coloured wings. They are also excellent environmental indicators as their presence or absence provides useful information about the state of the local environment. They are essential part of any natural ecosystem as their adults perform pollination and larvae enact as primary herbivores thereby transferring radiant energy trapped by plants to the next trophic level; rendering dual roles as pollinators and as energy transferors. The holometabolus life history of butterflies exposes them to a wide range of environmental influences.

They are highly sensitive to changes in temperature, humidity and light; parameters that are easily influenced by habitat deterioration (Murphy *et al.,* 1990). Therefore, butterflies are good indicator species to monitor ecological changes in a habitat.

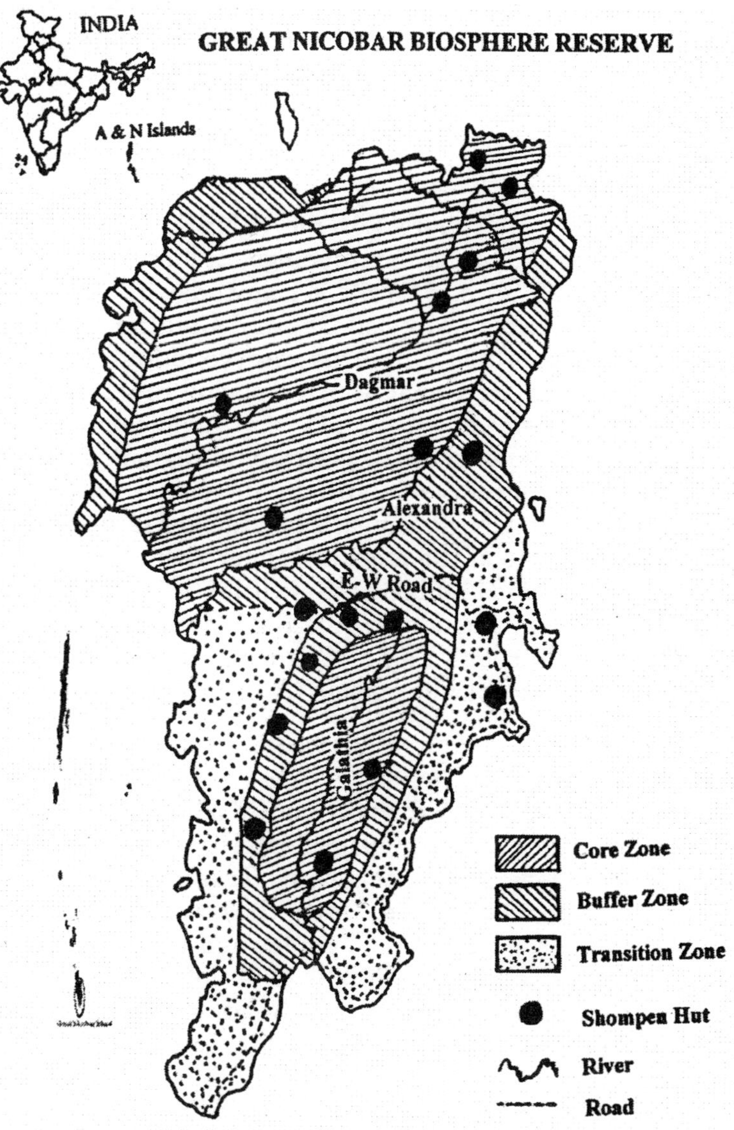

Fig. 1. Great Nicobar Biosphere Reserve

The butterflies of the Andaman and Nicobar Islands are insular with its origins in the fauna of the Indo-Myanmar and Indo-Malayan regions. The Andaman elements have their closest resemblance to Myanmar and mainland elements whereas the Nicobar elements appear most closely related to Malayan elements. Studies on butterflies of Andaman and

Nicobar Island received attention after the publication of Wood Mason & de Nicebille form 1880-1882. Evans (1932) have worked on butterflies of this Islands. Later few researchers have been contributed on the fauna butterflies of Andaman & Nicobar Islands (Talbot, 1939, 1947; Ferrar, 1948; Arora & Nandhi, 1980, 1982; Khatri, 1989-1993, Chandra & Khatri, 1993; Devy *et al.,* 1994). This study is aim to describe diversity and species abundance of butterflies in Great Nicobar Biosphere Reserve (GNBR).

STUDY AREA

The Great Nicobar Biosphere Reserve (GNBR) is the southernmost Island of Andaman and Nicobar archipelago. It lies between 6°45' and 7°15' N latitudes and 93°38' and 93°55' E longitudes in the Bay of Bengal (Fig. 1). The GNBR includes Campbell Bay and Galathea National Parks. It spreads over an area of 885 km^2. This reserve is known for its unique biodiversity and houses rich genetic germplasm resources. The Great Nicobar Biosphere Reserve represents the tropical rain forest in Andaman and Nicobar Islands of bio geographic region. The major forest area in this Biosphere Reserve is still in its virgin state and rich in species composition. The GNPR has rich heritage of faunal diversity and this area also inhabits one of the most endangered species *viz.*, Megapode (*Megapodius frycinet*) and Edible-nest swiftlet (*Collocalia fuciphaga*). Other important species include White Bellied Sea Eagle (*Haliaeetus leucogaster*), Nicobar Serpent Eagle (*Spilornis cheela*), Nicobar Pigeon (*Caloenas nicobarica*), Nicobar Parakeets (*Psittacula caniceps*).

METHODS

The study was conducted during month of August 2008, January 2009, August 2009, and February 2010. Butterfly species were estimated by 600 m transect, traversed in one hour. Transects were enumerated between 0600 hours to 1100 hours. Butterflies were identified based on physical features with the help of field guides and reference books (Evans, 1932; Ferrar, 1948; Kehimkar, 2008). The following locations were covered during the survey *i.e.* Amphibian Road, B Quarry, Gingen Basti, Gandhi Nagar, East West Road, Indira Point, Lakshman Beach, Mahar Nala, Navdara, Kopen heat and Kondul (Table 1). Species observed outside the transect and forest edges were noted separately. Butterflies observed along the transect alone were considered for statistical analyses.

Species richness and abundance of butterflies

The total number of butterfly species and number of individuals seen in each transect and different locations were calculated using the census data and field observations.

Table 1. Coordinates of the different locations in the study area

Location	Coordinates	
	Latitude	Longitude
East West Road	06° 52.531' N	93° 52.213' E
B-quary	06° 59.708' N	93° 56.680' E
Mahar Nala	06° 59.584' N	93° 55.079' E
Amphibian Road	07° 00.406' N	93° 54.788' E
Kondul	07° 13.027' N	93° 42.936' E
Lakshman Beach	07° 01.249' N	93° 54.321' E
Kopenheat	06° 50.922' N	93° 47.983' E
Indira Point	06° 45.822' N	93° 49.607' E
Navidara	07° 07.564' N	93° 53.144' E
Gandhi Nagar	06° 50.196' N	93° 53.680' E
Mahar Nala	06° 59.518' N	93° 55.104' E

Diversity indices: Shannon index, Simpson index and Hill's diversity numbers N1 and N2 were calculated using the programme SPDIVERS.BAS developed by Ludwig & Reynolds (1988).

Similarity measures: Similarity index between the different locations were calculated using the statistical software Biodiversity (Lambshead *et al.,* 1997).

Rarefaction

The Rarefaction was calculated using the software Biodiversity (Lambshead *et al.,* 1997).

RESULTS AND DISCUSSION

Occurrence of species

A total of 65 species of butterflies belong to 5 families and 46 genera were recorded during the period. Of the recorded species highest number were recorded from family Nymphalidae followed by Pieridae, Lycaenidae, Pieridae and hesperidae.

Distribution of butterfly species

Species of butterflies recorded in different locations during the period of study is given in Table 2. Out of sixty five species of butterflies 59 species were recorded from East-west Road, followed by Lakshman Beach (48), Mahar Nala & Kondul (47), Kopenheat (45), Gingem Basti (44),

Indira Point (42), Amphibian Road (40), B-quary (38), Gandhi Nagar (36) and Navidera (28).

Table 2. Distribution of butterfly species in different locations in GNBR

Sl. No.	Species name	Amphibian Road	B quarry	Gingen Basti	Gandhi Nagar	East West Raod	Indira Point	Lakshman Beach	Mahar Nala	Navidera	Kopen heat	Kondul
1.	*Atrophaneura aristolochiae*	√	√	√	√	√	√	√	√	√		
2.	*Papilio polytes*	√		√	√	√		√		√	√	√
3.	*Papilio mormon*	√	√	√	√	√	√	√	√			
4.	*Papilio polytes nicobarus*		√			√		√	√	√	√	√
5.	*Papilio demoleus*	√		√		√		√				
6.	*Graphium agamemnon decoratus*	√		√		√		√	√	√		√
7.	*Pachliopta aristolochiae kondulana*					√						√
8.	*Leptosia nina*		√		√	√	√		√	√		√
9.	*Cepora nerissa*			√		√		√	√	√	√	
10.	*Cepora nadina*	√		√	√	√		√				√
11.	*Leptosia nina nina*	√				√	√	√	√	√	√	√
12.	*Appias albino*	√		√	√	√		√	√		√	√
13.	*Appias lyncida*	√		√	√	√	√		√			
14.	*Catopsilia pomona*	√		√	√	√		√	√	√	√	√
15.	*Eurema blanda*	√	√	√		√	√	√	√		√	√
16.	*Hebomoia glaucippe*	√	√	√	√	√	√				√	√
17.	*Eurema blanda*		√		√	√	√	√	√		√	√
18.	*Eurema hecabe nicobarensis*		√					√	√	√	√	√
19.	*Spalgis epius*	√	√	√	√		√	√	√	√	√	
20.	*Discolampa ethion airavati*		√	√	√			√	√			√
21.	*Chilades parrhasius pila*					√	√		√		√	√
22.	*Lampides kankena*		√	√		√		√	√		√	√
23.	*Jamides bochus nicobaricus*					√	√	√			√	√
24.	*Jamides celeno kinkurka*	√		√		√						
25.	*Nacaduba hermus*	√	√		√	√		√	√		√	√
26.	*Prosotas nora dilate*		√	√	√		√	√	√			√
27.	*Curetis thetis obscura*		√	√			√	√	√		√	
28.	*Loxura atymnus nicobarica*	√	√		√		√				√	√

29.	*Hypolycaena thecloides nicobarica*	√	√	√	√	√		√				√
30.	*Deudorix epijarbas*	√		√		√	√	√	√		√	
31.	*Bindahara phocides areca*					√			√			√
32.	*Euchrysops pandava*		√	√	√	√	√	√	√		√	
33.	*Zizina otis*			√	√	√	√	√	√		√	√
34.	*Everes parrhasius pila*	√	√	√	√	√	√	√	√		√	√
35.	*Hypolimnas misippus*	√	√		√	√	√	√				
36.	*Hypolimnas bolina*		√	√	√	√	√		√		√	√
37.	*Precis atlites*	√	√	√		√	√	√	√	√	√	√
38.	*Junionia almana*		√	√	√	√	√	√	√	√	√	√
39.	*Athyma kanwa*				√	√	√	√		√	√	
40.	*Neptis hylas nicobarica*	√		√		√	√		√	√	√	√
41.	*Cirrochorea nicobaris*	√		√	√	√		√				√
42.	*Lethe rohria*					√					√	
43.	*Cethosia cyane*	√		√	√	√	√	√		√	√	√
44.	*Neptis hylas nicobarica*		√	√	√	√	√	√	√	√	√	√
45.	*Parthenos sylvia nila*	√	√	√		√	√	√	√		√	√
46.	*Cethosia biblis nicobarica*	√	√			√	√		√	√		√
47.	*Phalanta alcippe fratema*	√	√	√		√		√	√	√	√	√
48.	*Cirrochroa nicobarica*	√	√	√	√	√	√	√	√	√	√	
49.	*Hypolimnas antilope anomala*		√	√	√	√		√	√		√	√
50.	*Parantica aglea ageloides*	√	√			√	√	√	√	√		√
51.	*Danaus chrysippus*	√		√	√	√	√		√	√	√	√
52.	*Radena similis*	√		√	√	√	√	√			√	
53.	*Euploea crameri esperi*					√		√	√	√	√	√
54.	*Elymanias cottonis*	√	√	√	√	√	√		√	√	√	√
55.	*Tirumala limniace*	√	√	√		√	√	√				√
56.	*Danaus chrysippus*					√	√	√	√	√	√	
57.	*Euploea crameri*	√	√	√	√	√			√	√	√	√
58.	*Euploea scherzeri simulatrix*	√		√	√	√	√	√	√		√	√
59.	*Elymnias panthera*	√	√	√	√	√	√	√	√	√	√	√
60.	*Mycalesis anaxias manii*	√	√	√		√	√	√	√	√	√	
61.	*Hasora taminatus almea*					√				√		
62.	*Hasora badra*	√	√			√	√	√	√		√	√
63.	*Cephrenes chrysozona nicobarica*	√	√	√		√	√	√	√		√	√
64.	*Gangara thyrsis*	√	√	√	√	√	√				√	√
65.	*Tagiades japetus helferi*		√		√	√	√	√	√			√

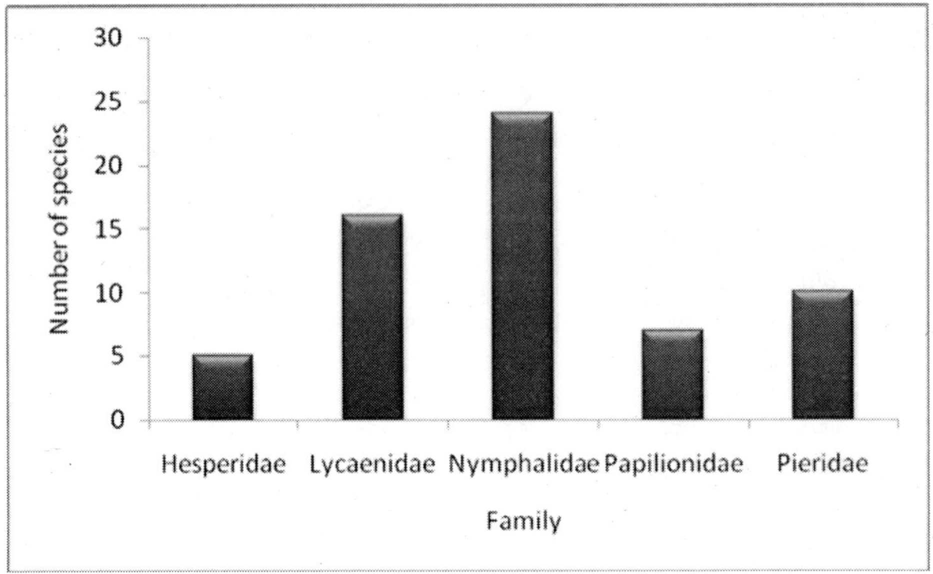

Fig. 2. Family wise species richness in GNBR

Familywise species richness and abundance of butterflies

Species richness and abundance of butterflies varied in the different families. Highest number of species richness was recorded from Nymphalidae (24) followed by Lycaenidae (16) and Pieridae (10) (Fig. 2). Similarly species abundance also varies in different families. Highest number of individuals was recorded from Nymphalidae (1012) and Papilionidae (400) (Fig. 3).

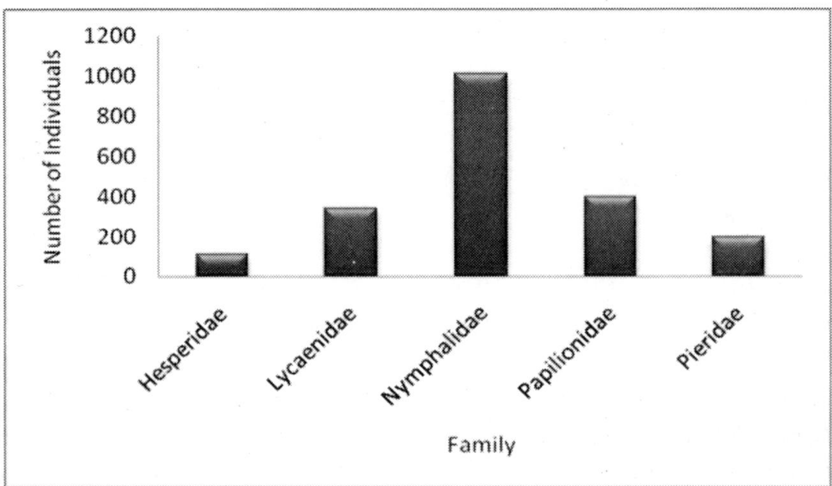

Fig. 3. Family wise abundance of butterflies in GNBR

The number of species recorded in different location with mean and standard deviation are present in Table 3.

Table 3. Number of species recorded in different locations with mean ± SD

Locations	Number of species	Mean ± SD
Amphibian Road	40	3.78 ± 4.05
B quarry	38	3.68 ± 3.43
Gingen Basti	44	5.82 ± 6.19
Gandhi Nagar	36	3.53 ± 3.68
East West Raod	59	4.49 ± 4.17
Indira Point	42	4.38 ± 4.61
Lakshman Beach	48	4.85 ± 4.17
Mahar Nala	47	3.38 ± 3.21
Navidera	28	4.50 ± 3.25
Kopen heat	44	4.50 ± 3.64
Kondul	47	4.64 ± 3.22

Rarefaction

A comparison of number of species in different locations for fixed number of individuals encountered is a useful exercise. For this purpose, expected number of species was calculated for varying number of individuals and presented in Fig. 4. The data shows that East-west Road had more number of species followed by Gingen Basti. The use of the rarefaction curve allows us to assess the differences in species richness of areas with different species accumulation rates.

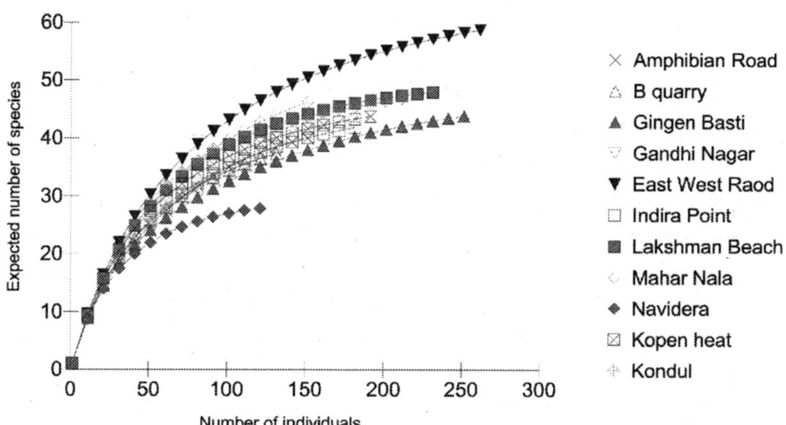

Fig. 4. Rarefaction for butterfly species in different locations

Diversity

Indices based on the proportional abundance of species are the best approach to measure diversity. Most widely used diversity indices like Shannon Index of diversity, Simpson's Index of diversity and Hill's numbers N1 and N2 have been determined. The diversity index (H') ranged from 3.09-3.74, with highest in East-west road (3.74) (Table 4).

Table 4. Diversity indices for different locations in GNBR

Location	Richness indices		Diversity indices		Hills Number		Evenness indices	
	R1	R2	LAMBDA	H'	N1	N2	E1	E2
Amphibian Road	7.77	3.26	0.05	3.27	26.36	21.51	0.89	0.66
B-quarry	7.49	3.21	0.04	3.31	27.28	24.20	0.91	0.72
Gingem Basti	7.75	2.75	0.04	3.32	27.73	22.76	0.88	0.63
Gandhi Nagar	7.23	3.19	0.05	3.21	24.84	20.26	0.90	0.69
East-West Road	10.39	3.62	0.03	3.74	42.03	36.31	0.92	0.71
Indira Point	7.86	3.10	0.04	3.36	28.66	22.69	0.90	0.68
Lakshman Beach	8.62	3.14	0.03	3.57	35.55	31.64	0.92	0.74
Mahar Nala	9.07	3.73	0.03	3.52	33.73	29.67	0.91	0.72
Navidera	5.58	2.49	0.05	3.09	22.05	21.87	0.93	0.79
Kopen heat	8.13	3.13	0.03	3.50	33.24	31.06	0.93	0.76
Kondul	8.54	3.18	0.03	3.62	37.52	37.42	0.94	0.80

Similarity indices between the different habitats

An alternative approach to measure the similarity of different sites is using similarity indices, using the proportional similarity measures. Similarity indices between the different habitats were computed using qualitative data (Table 5). The similarity index which is influenced by the most abundance species depicted the closet similarity of community between Amphibian Road and B-quarry, Kondul and Ginem Basti was the least similar to any other locations

During the study period total of 2057 individuals belonging to 5 families were recorded. Butterflies in Great Nicobar Biosphere Reserve represent five families, *i.e.,* Hesperidae, Lycaenidae, Nymphalidae, Papilionidae and Pieridae. The distribution of butterfly species showed that fifteen species were recorded in eight locations, twelve species in nine locations and one species only recorded in all the locations (Table 1). Diversity indices of Butterfly in Andaman and Nicobar Islands are meager, it is interesting to find that the diversity index of Great Nicobar Biosphere Reserve in this study. Changes in the diversity of butterfly in different locations in the study area are evident from the data, this because of the variation in the micro habitat, floristic structure and other habitat parameter (Plate 1).

Plate 1. Butterflies of Great Nicobar Biosphee Reserve:
(1) Parthenos demoleus, (2) Papilio polytes, (3) Atrophaneura aristolochiae, (4) Junonia almana, (5) Graphium agamemnon, (6) Eurema hecabe

Table 5. Similarity indices for different locations in GNBR

Location	Amphibian Road	B quarry	Gingen Basti	Gandhi Nagar	East West Raod	Indira Point	Lakshman Beach	Mahar Nala	Navidera	Kopenheat	Kondul
Amphibian Road	0.00	73.30	57.21	69.05	58.94	63.90	64.67	69.47	65.29	55.67	62.87
B quarry		0.00	54.75	67.52	63.00	65.07	60.24	75.81	68.03	61.37	65.91
Gingen Basti			0.00	55.02	54.62	51.30	47.34	47.76	48.00	46.17	44.16
Gandhi Nagar				0.00	56.55	58.78	59.38	62.74	64.07	58.08	58.66
East West Raod					0.00	54.54	56.66	60.00	60.29	59.84	54.84

Indira Point	0.00	53.84	63.16	57.67	54.57	57.88
Lakshman Beach		0.00	60.68	56.08	52.22	63.99
Mahar Nala			0.00	69.88	62.17	67.86
Navidera				0.00	59.53	60.30
Kopen heat					0.00	60.40
Kondul						0.00

The butterfly families site selection could be determine by the availability of some factors such as food, access to light to regulate their body temperature also open space to flight away from predators. The many of the species are localised to Great Nicobar Islands and few sub-species are intermixing in Andaman and Nicobar Islands. This indicates that the butterflies of Great Nicobar Islands are closely linked to that of Indo-Malayan than the Andaman Islands. Most of the species observed in this study are very specific to host plants. The study period was relatively short for an analysis of trends which may affect their conservation status. Long-term monitoring is needed to identify significant changes in diversity of butterflies, permitting the timely adjustment of management activities to reverse or prevent undesired trends.

ACKNOWLEDGEMENT

We are grateful to the Ministry of Environment and Forests, Government of India, for the support to this study, and Dr. Dinesh Kannan, Divisional Forest Officer, Department of Environment and Forests officials, Campbell Bay for providing necessary facilities to carry out this study.

REFERENCES

Arora, G.S. & D.N. Nandi, 1980. On the butterfly fauna of Andaman and Nico-bar Islands (India) I. Papilionidae. *Records of Zoological Survey of India* 77: 141-151.

Arora, G.S. & D.N. Nandi, 1982. On the butterfly fauna of Andaman and Nicobar Islands India II. Pieridae. *Ibid.* 80: 1-15.

Chandra, K. & Khatri, 1993. Butterflies of Great Nicobar Islands. *Indian Journal of Forestry* 18(4): 276-273.

Devy, M.S., T. Ganesh & P. Davidar, 1998. Patterns of butterfly distribution in the Andaman Islands: Implications for conservation. *Acta Oecologia* 19(6): 527-534.

Evans, W.H., 1932. *The Identification of Indian Butterflies.* Bombay Natural History Society, Bombay. 454 pp.

Ferrar, M.L., 1948. Butterflies of Andaman and Nicobar Islands. *Journal Bombay Natural History Society* 47(3): 470-491.

Kehimkar, I., 2008. *The book of Indian Butterflies.* Bombay Natural History Society, Mumbai. 497 pp.

Khatri, T.C., 1989. A revised list of Butterflies from Bay Islands. *J. Andaman Sci. Assoc.* 5: 57-61.

Khatri, T.C., 1991. On some Nymphalidae (Rhopalocera: Lepidoptera) from the Andaman and Nicobar Islands. *Islands on March* 3: 82-94.

Khatri, T.C., 1992. On some Lycaenids (Rhopalcocera: Lepidoptera) from Andaman and Nicbar Islands. *Islands on March* 6: 8-16.

Lambshead, P.J.D., G.L.J. Paterson & J.D. Gage, 1997. Biodiversity Profess-ional Beta. The Natural History Museum and The Scottish Association for Marine Science.

Ludwig, J.A. & J.F. Reynolds, 1988. Statistical Ecology, A premier on Methods and Computing. A Wiley-Interscince publication. 337 pp.

Murphy, D.D., K.E. Freas & S.B. Weiss, 1990. An environment- metapopulation approach to population viability analysis for a threatened invertebrate. *Conservation Biology* 4: 41-51.

Talbot, G., 1939. The fauna of British India including Ceylon and Burma (Butterflies), Taylor and Francies, London.

Talbot, G., 1947. Fauna of British India. Butterflies II, Today & Tomorrow Printers & Publishers.

Wood-Mason, J. & L. de Niceville, 1980. List of diurnal Lepidoptera from Port Blair. Andaman Islands. *J. Asiat. Soc. Beng.* 49(2): 223-243.

Wood-Mason, J. & L. de Niceville, 1981. Second list of Rhopalocerous Lepidoptera from Port Blair. Andaman Islands. *J. Asiat. Soc. Beng.* 50(4): 243-262.

Wood-Mason, J. & L. de Niceville, 1982. Second list of Rhopalocerous Lepidoptera from Port Blair. Andaman Islands. *J. Asiat. Soc. Beng.* 51(1): 14-20.

Chapter - 11

IMPACT OF *ENTEROBACTER* SP. INFECTION IN PHENOTYPIC CHANGES AND HAEMOLYMPH PROTEIN PROFILING IN THE SILKWORM, *BOMBYX MORI* L.

R. Chandrasekar[1] *, A. Vijay Ananad[2], P.G. Brintha[3] and B.K. Tyagi[4]

[1]Department of Entomology,
Agricultural Science Centre-North Building, USA
[2]Department of Biochemistry, M.I.E.T. Arts and Science College,
Tiruchirappalli, Tamilnadu, India
[3]Nachiyar Silk House, 1/4 NMK Colony, TVS Tollgate,
Tiruchirappalli 620 020, India
[4]Centre for Research in Medical Entomology (ICMR), 4-Sarojini St.,
Chinna Chokkikulam, Madurai 625002, TN, India
[1]Corresspondence E-mail: *rch225@uky.edu; koreanchandru@yahoo.com*

ABSTRACT

Silkworm larvae are often born to have bacterial or viral infections, causing heavy economic loss to silk industry. The present study is aimed to identify frequently infecting pathogen and its influence on storage proteins in the fifth larval instar of *Bombyx mori*. The pathogen was identified as *Enterobacter* sp. by Gram staining, cultural, biochemical tests and light microscopy. The phenotypic abnormalities of infected larvae noticed during metamorphosis were due to diarrhea, sluggish movement, less feeding, shrinkage of abdominal segments, and black colored thorax and anal region. Later in infection, large numbers of bacteria (>105 CFU per gram) are recoverable from the fat body, as well as the haemolymph and other tissues. SDS-PAGE protein profile of heamolymph showed several protein bands with molecular weight ranging between 14kDa - 100 kDa. Among these, the intensities of two major bands such as SP1 (82kDa) and SP2 (76 and 72 kDa) were found in infected haemolymph when compared to control. The next major group of peptides was localized at 30kDa region with 3 subunits of molecular weights 26, 29 and 31kDa, respectively. During infection the expression of 29kDa protein is believed to play a role in defense mechanism and hence it was gradually increased from 2nd to 6th day and subsequently declined on the day 7 when compared to control and further it was confirmed by immunoblot analysis. The 30kDa proteins may play an important role in the defense against pathogen.

INTRODUCTION

The domesticated silkworm, *Bombyx mori* has been the main target of many scientific studies due to its socio economic importance. The growth of the silkworm (Sericulture) industries in India especially in Tamilnadu is mainly affected by bacterial, fungal and viral diseases (Ponnuvel *et al.,* 2003). Insects are continuously exposed to potentially pathogenic microorganisms and other parasites, but only a few encounters and results in infection. Insects possess a mechanism to fight against microbial pathogens. Moreover, insects have an efficient innate immunity against microbial infection (Hoffmann *et al.*, 1997; De Gregorio *et al.*, 2001). Antibacterial peptides, a group of factors involved in insect humoral defense reactions, are rapidly secreted into the haemolymph against bacterial infection (Boman, 1991). To date, more than 150 antibacterial peptides and polypeptides have been isolated from different insect species. These antibacterial peptides and polypeptides are broadly classified into five major groups such as: cecropins, insect defensins, attacin-like (glycine-rich) proteins, proline-rich peptides and lysozymes (Hultmark, 1993; Wicker *et al.*, 1990). On the other hand, humoral reaction is supported by polyphenol oxidase activity, lectins, antibacterial proteins, storage protein and 30kDa serum proteins (Gillepic *et al.*, 1997; Daborn *et al.*, 2001; Chandrasekar, 2006).

The insect, hemolymph proteins are having diversified functions such as metabolic, endocrine, reproductive and immunological. In addition, hemolymph serves as the transport *milieu* for the exchange of certain essential macromolecules between cells, tissues and organs (Mullinas 1985; Lovallo *et al.* 2000). Interestingly a special class of hemolymph proteins known as storage proteins undergoes significant changes during infection (Levenbook, 1985; Nagata & Kobayashi, 1990). It has been reported that these larval storage proteins can accumulate in the organisms at the time of abundant dietary intake and they may act as either amino acid pool or protein reservoir during non-feeding phase of development (Levenbook, 1985; Tojo *et al.*, 1990; Chandrasekar, 2006; Chandrasekar *et al.,* 2009). According to Chandrasekar *et al.* (2007, 2008b, 2009), storage proteins serves as the main reservoirs for the supply of amino acids, both during metamorphosis and development. The protection of insect from infectious and parasitic agents is essential for its survival. Recently, several lines of evidence imply that arylphorin of the Lepidoptera members may be specifically induced by bacterial infections (Beresford *et al.,* 1997). Arylphorin suppresses hemocyte degradation and subsequent immune reactions that lead to the encapsulation of the parasites egg. Not only the storage protein but also some non-storage protein like 30kDa also plays an important role in the protective functions (Ujita *et al.*, 2005).

Antibacterial protein neither contributes significantly new or mass of protein to haemolymph nor requires compensatory increase in the synthesis of storage protein level. The increased 30kDa protein in the infected *B. mori* was reported by Janarthanan *et al.* (1995). Several authors reported that the antibacterial proteins are one of the main factors for protection against bacterial infection (Boman, 1991; Kimbrel, 1991), similarly the antibacterial proteins play a vital role during pathogen – host interactions. Recently, the molecular characterizations of storage protein and 30kDa proteins have been extensively analyzed in *Amsacta albistriga* (Chandrasekar *et al.*, 2008b). But, poorly understood is how infecting bacteria either overcome or avoid the peptide-mediated and cellular components of the invertebrate immune response (De Gregorio *et al.*, 2001). However, the data on immunological activity are still dispute. We are interested in using a model insect, the silkworm, *B. mori,* to describe in detail identification and the pattern of infection of an insect by an insect pathogenic bacterium, *Enterobacter* sp. The present work clearly addressed the phenotypic changes and expression of 30kDa proteins during bacterial infection.

MATERIALS AND METHODS

Experimental animal

The larvae of *B. mori* were collected from the State Sericulture Rearing Unit at Manikandam, Tiruchirappalli, India. The insect rearing was carried out according to the standard method of Chandrasekar et al. (2007). Fifth instar diseased silkworm larvae were used for the present investigation. Infected larvae were held individually with diet, and symptoms of toxicity were noted over time.

Isolation and identification of the pathogen

(a) Sterilization

Glassware was sterilized in hot air oven; the medium was autoclaved at 121°C for 15 minutes. The inoculation chamber was disinfected with 70% ethyl alcohol and UV–B radiation for 30 minutes. Surface sterilization of the diseased larvae was done with 70% ethanol and 0.1% $HgCl_2$ prior to dissection and further use.

(b) Preparation of Nutrient agar Plates and inoculums

Autoclaved nutrient agar was poured into sterile Petri plates and gently rotated to achieve uniform distribution of the medium. Each larva was surface sterilized with 70% ethanol and then bled to collect its total haemolymph fluid. After bleeding, each larva was then dissected to collect the internal organs. The cuticule and fat body were weighed individually and homogenized in PBS using a hand held Potter–Ehlvejen homogenizer. The rest of the insect body was defined as 'carcass', weighed and homogenized. To determine the number of recoverable bacteria (Colony Forming Unit per gram of wet weight of tissue or haemolymph), dilution series of tissue homogenates were prepared with PBS and serial dilutions were then plated in duplicate onto PP3-arabinose-ampicilin agar plates (Leuschner & Bew, 2004). Plates were incubated for 48 h at 30°C and individual colonies counted and maintained the stock culture throughout the experiments.

Biochemical Characterization

The following biochemical tests were carried out according to the method described by de Champes *et al.* (1993) to identify microorganisms- Catalase Test, Indole production test, Citrate utilization test, Voges proskauer test, Methyl red test, Mannitol salt agar test, Utilization of triple sugar iron test and Oxitase test.

Light microscopic analysis

For insect histopathology, sections of whole larvae or specific tissues from both infected and uninfected larvae were examined. For examination via light microscopy, whole larvae were fixed overnight in Bouin's solution after several incisions had been made in the cuticle to allow the fixative to permeate the cadaver. Then, the insects were washed twice with distilled water and subjected to dehydration in a graded series of 70-100% ethyl alcohol, after which they were embedded in paraffin wax. Sections were cut at 4–5 mm and then stained with haematoxylin/eosin (HE) and mounted on glass slides with DePeX mounting medium and they were examined under the Leitz Diaplass Light Microscope (Germany) as described by Chandrasekar *et al.* (2008b).

Sample collection and protein extraction

Each larva was surface sterilized with 70% ethanol and then haemolymph samples was collected by making a cut at the distal region of the prothoracic leg and draining the haemolymph into the small 1.5 ml microfuge tube with sufficient quantity of phenyl thio -urea (PTU) and 1mM phenyl methyl sulfonyl fluoride (PMSF). This mixture was centrifuged at 7,000g for 5 minutes to remove the haemocytes and other tissue debris was discarded. Haemolymph samples were stored at – 20ºC until use. The total protein concentration was determined by adapting the method of Bradford (1976).

SDS – PAGE

SDS-PAGE was performed by the method of Laemmli (1970) using a slap gel the separation of protein using 7% resolving gel. The molecular weight of the storage protein and 30kDa proteins were identified and separated by its relative mobility in SDS-PAGE and along with molecular marker. Protein bands were stained with Coomassie blue R- 250 (0.02% in 50% Methanol- 7.5% Acetic acid). Quantitative differences in the intensities of specific polypeptides were measured using a Densitometer (Hoefer, GS_{30} Transmittance/ Reflectance Scanning Densitometer).

Antibodies

Polyclonal antibodies to *Hyphantria cunea* SP1 were a gift from Prof. Seo Sook Jae, Geyongsang National University, Jinju, South Korea. This antibody has been shown to strongly cross-react with yolk protein (82 and 30kDa) from red hairy caterpillar (Chandrasekar *et al.,* 2007).

Immuno Blotting

The samples were resolved in the 10% SDS–PAGE, and electro-transferred to nitrocellulose membranes for Western blot following standard procedures (Bio-Rad Laboratories). The primary antibody was rabbit anti-IgG SP1 (1:1500 dilution) while the secondary antibody was horseradish peroxidase-conjugated goat anti-rabbit IgG (1:1000 dilution) visualization of any immunoreactions was carried out by incubating the membrane in a substrate solution of 30μg 4-Chlorol-n-napthol in 100ml and 30μl of H_2O_2 in 40 ml of TBS. The color reaction was stopped by transferring the membrane into Mill Q water and then photographed.

RESULTS

Symptoms of the infection

The present results presented in Fig. (1) showed that infected larvae were found to be inactive, including sluggish movement and stop feeding. The other symptoms include shrinkage of abdominal segments, diarrhea, blackening of thorax and anus regions with abdominal segments (Fig. 1 a, b, c) the blackening gradually reaches to entire region of the body exudes a foul – smelling, dark brown fluid protrude. However, very few of the infected larvae undergo irregular spinning to form deformed pupa, which shed old cuticle and leads to death (Fig. 1 g).

Identification of pathogen

Light microscopic study revealed that the pathogen is a Gram – negative, rod shaped bacterium (Fig. 2) colonies on nutrient agar plates appear as white. Bio- chemical tests (Table 1) and morphological examination helped to identify the organism up to generic level as *Enterobacter sp.* the organism showed catalase positive activity while oxitase negative test. It showed positive for citrate utilization, methyl red and mannitol test. Triple sugar iron agar test showed changes of colour from red to yellow, which infers acid slant, acid bud and H_2S production.

Fig. 1 Bacterial disease (*Enterobacter* sp) to induce the phenotypic abnormalities in the silkworm, *Bombyx mori.*

(a) Control larvae, (b) Infected larvae showing anus regions black in colour, (c) Symptoms of thoracic region in infected larvae, (d) The entire dark brownish appearance in the body of infected larvae, (e) Normal Pupa, (f) Prepupal incomplete metamorphosis, (g) Larval-pupal intermediate (LPI)

Table 1. Morphological and biochemical tests for identification of Entrobacter

Test	*Enterobacter* sp.
Colony morphology	
Configuration	Circular
Margin	Rhizoid
Elevation	Slightly raised
Surface	Shiny and smooth
Pigment	–
Opacity	semi-transparent
Gram's reaction	Negative
Cell shape	Rods
Size	Length: 2–3 µm; width: ≈1.0 µm
Arrangement	Singles
Spores (s)	Sporulating
Endospore	+
Position	Central
Shape	Ellipsoidal
Motility	+
Bio –Chemical test	
Catalase test	+
Formation of indole	-
Citrate utilization	+
Voges proskauer	+
Methyl red	+
Triple sugar iron test	Acid slant acid, but No H_2S
Oxitase	-
Mannitol	+
α-glucosidase	+
Arginine dehydrogenase	+
Sucrose	+

Growth of *Enterobactor* sp. in different insect tissues

The present authors examined the relative growth rates of *Enterobacter* sp. in different insect tissues by measuring the number of recoverable colony-forming unit (CFU) per unit weight of tissues dissected from injected hosts over time. Bacterial infected larva, the number of bacteria recoverable from either the haemolymph or the fat body increases steadily over time (Fig. 3A and B), with 10^8 CFU being recoverable per gram of infected tissue 24 h post infection. Within these tissues, the number of recoverable bacteria increased by six orders of magnitude within 24 h (24–36 h post infection). In contrast, although the final numbers of recoverable bacteria per gram are similar at 36h (Fig. 3B), there is a delay in colonization of, and multiplication within, the fat body. Similarly *Enterobacter* multiply slower in the remaining tissues of the insect cadaver and also yield a somewhat lower number (10^7 CFU per gram) of recoverable bacteria (Fig. 3D).

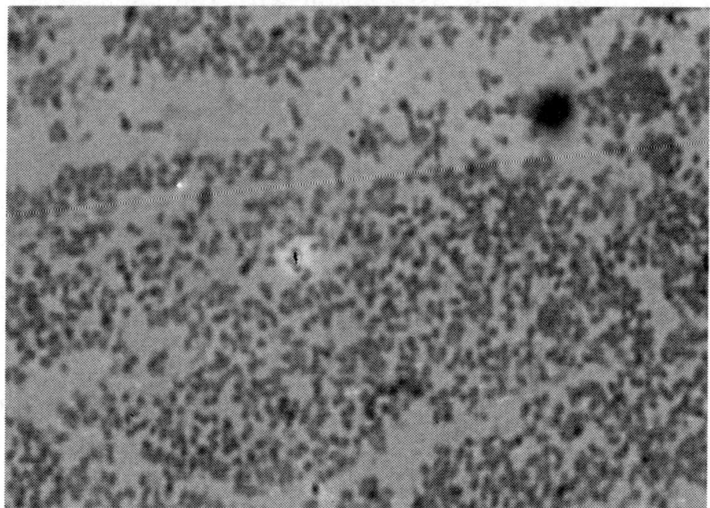

Fig. 2. Light microscopy view of pathogenic microorganism *Enterobacter* sp. (X20)

Effect on larval growth and metamorphosis

The larvae fed with normal leaf acquired increase in body weight (0.910 ± 0.13g), which started declining due to bacterial infection *Enterobacter* sp (0.49±1.01g) goading phenotypic abnormalities during the course of metamorphosis. Data in Figure 1 clearly showed the normal and affected larvae of *B. mori*. The following phenotypic changes shows reduced body length due to shrinkage of abdominal segment, delay of growth, blackening of anus and thoracic regions (Fig. 1b), growth abnormalities, inability of ecdysis resulting in frequent death with dark brownish body (Fig. 1d, f). These symptoms resemble those of silkworm disease were similar to flacherie in appearance. A very few of them were able to spin loosely woven bag of threads. The bacterial infected larvae, which failed in larval – pupal ecdysis hence, they become larval-pupal intermediate (LPI) (Fig.1 g). Hence the metamorphosis does not proceed further.

Light microscopic examination shows that histopathogenic

The ease of dissecting tissues from late instar *B. mori* larvae, and the well-characterized physiology of this insect model, allowed us to document the relative growth rate of bacteria in different insect tissues (Fig. 3). The colonization of the fat body proceeds in a specific anterior to posterior fashion with the bacteria occupying the highly folded fat body tissues, often next to muscles. The precise mechanism used by the bacteria to penetrate the cuticle matrix is not clear but we note that tracheae invade the extracellular matrix and that tracheae may therefore be one potential avenue of entry. Once beneath the extracellular matrix, the bacteria are in very close association with the fat body and muscle (Fig. 4A).

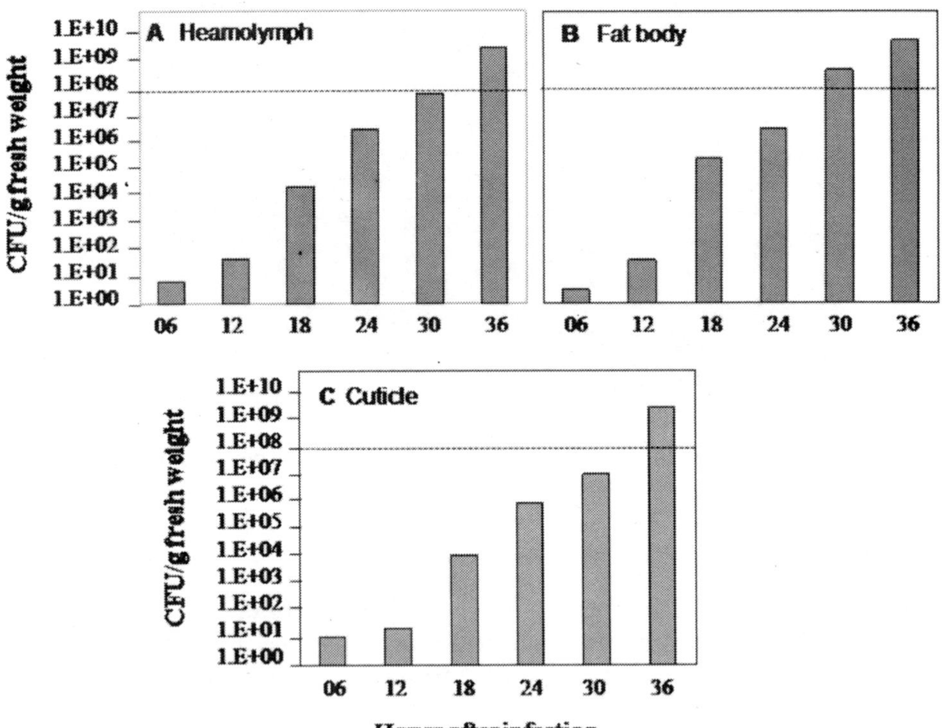

Fig. 3. Number of bacteria recoverable from different *Bombxy mori* tissues over time (h) after infection of *Enterobactor*. **A**, Hemolymph; **B**, fat body and **C**, Cuticle. Results are expressed as the mean number (± s.e. of at least three individual larvae) of colony-forming units (CFU) recoverable per gram of wet weight of insect tissue or haemolymph. Note that before 36 h most of the recoverable bacteria are found within the fat body and haemolymph. The line at 1.E + 08 is drawn to facilitate comparisons between the numbers of bacteria found in the different tissues over time.

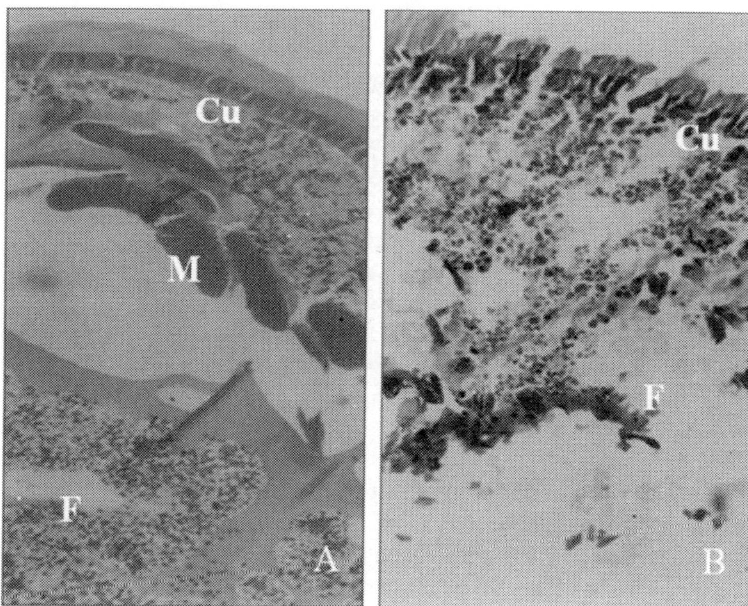

Fig. 4 Light microscope view (X20) of day 2 of the fifth Instar larva of control (**A**) and infected Silkworm (**B**). Cross section of paraffin-embedded anterior region of thoracic, stained with haematoxylin / eosin. Cu – cuticle, M- muscle, F- fat body tissue

Although there is no obvious reorganization of the fat body cells close to the sequestered bacteria, the cells of the cuticular epithelia are often detached from their basal lamina suggesting that apoptosis may already have been initiated. Later in infection, large numbers of bacteria ($>10^5$ CFU per gram) are recoverable from the fat body, as well as the muscle and the haemolymph itself. This suggests that colonization of the fat body tissues and haemolymph is initially more important than growth in other tissues and that early in infection *Entrobacter* may be actively destroying the muscle, fat body tissues and the haemolymph based immune system (Fig. 4B).

Effect on haemolymph protein profile

Protein profile of haemolymph from control and infected fifth instar larvae showed several bands with molecular weight ranging from 14 - 100kDa (Fig. 5A). Among these, two major peptides such as 80kDa and 30kDa concentration were found to be very high. The 80kDa had two polypeptides namely SP1 (82kDa) and SP2 (76kDa and 72kDa) an arylphorin were identified as storage protein and our results were confirmed with immuno blotting. The other major fraction of polypeptide was localized at 30kDa regions with a molecular weight 26kDa and 29kDa and 31kDa respectively and it was named as *Bm* LSP (*B. mori* larval serum protein).

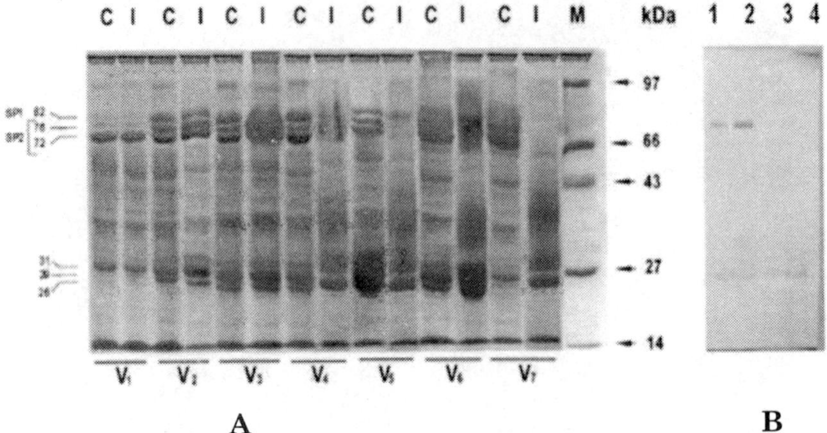

Fig. 5 SDS-PAGE (10%) of haemolymph protein profile of the fifth instar larvae of control and infected *Bombyx mori* (**A**) and immunoblot analysis (**B**) Infected (I), Control (C); V1 ~ V7 – respective development days of final instar larvae from day 1 -7; SP1 – Storage protein 1, SP2 – Storage protein 2; M – Molecular Weight Marker, kDa – Kilo Dalton.

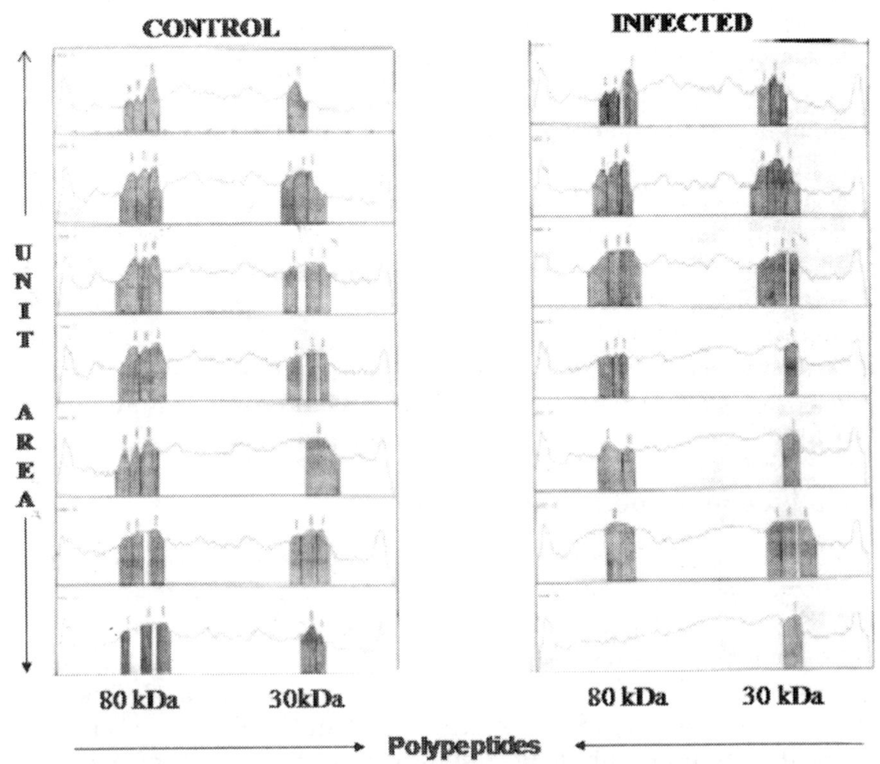

Fig. 6. Quantitative differences in the intensities of specific polypeptides (80kDa and 30kDa) were measured using a Densitometer (Hoefer,GS_{30} Transmittance/ Reflectance Scanning Densitometer).

Data presented here clearly demonstrated that there was a gradual increase of storage protein (SP1 and SP2) from day 1 to last day of fifth instar of control, where as in the infected larvae, sudden increase of storage protein on day 3, which drastically declined and finally disappeared at the end of the larval period. The appearance of 29kDa fractions in the bacterial infected larvae on 3 day (lane 6 Fig. 5A) could be considered as an evidence for the induction of protein synthesis in *B. mori.* The coomassie blue staining intensity of 80kDa proteins were fully disappeared however, 29kDa protein was slightly decreased during the end of the larval period. Immunobloting analysis of infected haemolymph sample revealed that the cross reactivity of SP1 antibody with 30kDa region was noticed (Fig. 5B). By contrast, there was no cross reactivity in the 80kDa regions, Thus the clearly emphasized densitometry scanning result revealed that SP1 amino acid pool was totally drained off during the infected stage (Fig. 6).

Discussion

The functions of haemolymph attributed to the appearance of selected proteins such as strorage proteins (Chandrasekar, 2006; Chandrasekar *et al.,* 2008a, b, 2009), immune proteins (Chandrasekar *et al.,* 2008b; Carlos *et al.,* 2002), lipophorins (Shapiro *et al.,* 1988) and hormone binding proteins (Jones *et al.,* 1990). Insect haemolymph proteins attracted a great deal of attention as biochemical model systems. The main haemolymph proteins lipophorin (LP), storage proteins (SP) and vitellogenin (Vg) are common in most of the insects and have special function for development, metamorphosis and reproduction (Wyatt & Pan, 1978; Chandrasekar *et al.,* 2007a,b, 2008a,b). Microorganisms induce drastic changes in the physiology and behavior of their host. These include transitory immunosuppression, induction of normal haemolymph proteins and metabolites arrest act.

The present results related to the infected larvae of *B. mori* showed that there may be a reduction in total food consumption, assimilation and other bioenergetic parameters. These presumably may be due to reduced rate of overall feeding activity. Reduced feeding activity after bacterial infection was also reported in *Cotesia kariyai.* The insect showed significant decrease of food utilization efficiency, suggesting nutrient usage for developing the parasitoid larvae (Nakamatsu *et al.*, 2001). Fluctuations in the numbers of recoverable bacteria from infected in *Manduca sexta* have been previously documented (Carlos *et al.,* 2002; Nakamatsu *et al.,* 2001). However, this is the first detailed numerical and spatial documentation of insect infection by this bacterium. Later in infection, large numbers of bacteria (>10^5 CFU per gram) are recoverable from the fat body and the haemolymph itself. This suggests that infected insect larvae, *Enterobacter* multiplies in both the fat body and hemolymph, only later colonizing the fat body and the remaining tissues of the cadaver. Our histopathological studies have revealed that destruction of fat body tissue and cuticle of host. Although the specific colonization events described may be *Enterobacter* or even specific strain, the dissection of infected insect demonstrates how bacterial infections can be readily detailed in insect models of sufficient size, such as *B. mori.*

The qualitative variations are also observed in this study, particularly in the haemolymph protein profile of the fifth instar larvae from day 1 to day 6, which indicates the accumulation of storage protein in response to bacterial infection when compared to control. Interestingly arylphorin, a storage protein, is a major target of some endoparasitoids; and hence it is inactivated or inhibits protein synthesis by teratocytes of *Microplitis croceipes* (Dong et al. 1996) or by polyadinovirus of *Campoletis sonorensis* (Shelby & Webb 1994). The inhibition of storage protein synthesis may be regarded as more evidence of nutrient redirection from host development to endoparasitoid development. Several authors hypothesized that the 30kDa protein is reported to be involved in multiplicity functions such as immunity (Tae-Yeon Kim *et al.,* 2002), reproduction and programmed cell death (Chandrasekar *et al.,* 2008ba, b, 2009).

Silkworm larvae exposed to bacteria show an increase in susceptibility to the infection. Aizawa (1991) reported the occurrence of anti-viral substance, having a molecular weight of 28kDa in the gut juice of *B. mori.* In *Sarcophaga peregrine,* two proteins are participating in the defense mechanism, because they are not only involved in the defense mechanism but also at specific developmental stage, specifically the early embryonic and pupal stage. Changes in protein pattern following the infection of cabbage moth larvae with nucleopolyhedrosis virus (Van der & Wassink 1969). Similarly Seiichi Hara & Minoru Yamakawatt (1995) demonstrated that some of synthetic peptides act as bactericidal substance during infection. Both quantitative and qualitative changes in haemolymph protein during the last larval instar and rarely in pupal stage have already been studied. In 1992, Slovak *et al.* had carried out the various antimicrobial proteins were inducible by the fat body of the flesh fly *Sarcophaga peregrine* during an acute-phase response to bacterial infection (Slovak *et al.,* 1992). The existing data authentically demonstrated that *B. mori* transferrin was found to strongly induce the fat body by the injection of *E coli,* suggesting that the fat body plays a vital role in self-defense. It is clear that insects have the ability to mount a potent defensive response against microbial parasites and this response is different from mammalian immunity (Yun *et al.*, 1999) and also some protein like moricin, cecropins, insect defensins, attacin-like proteins, proline-rich peptides and lysozymes, which contain very unique antibacterial peptide that shows strong antibacterial activity against both Gram-positive and Gram-negative bacteria (Hara & Yamakawa, 1995).

Moreover, reports on transgenic fruit fly (*Drosophila melanogaster*) suggest that insects harbour a local immune response system against microbial infection (Ferrandon *et al.*, 1998). Recently, the protein apoLp-III has been identified as an immune-stimulating molecule (Wiesner et al. 1997; Whitten *et al.*, 2004; Chandrasekar *et al.*, 2005). Injection of natively purified (Wiesner *et al.*, 1997) and recombinant apoLp-III from *Galleria mellonella* into the hemocoel of untreated larvae is followed by a strong increaser in antibacterial activity within the recipients' hemolymph. The induced antibacterial activity shows the same intensity, which can be maximally provoked by injecting bacteria (Wiesner *et al.*, 1997; Carlos *et al.,* 2002). It is interesting to speculate that 29kDa may have significant role in development of local immune response in the haemolymph, which is derived from the fat body. These results suggest that metabolism of the 30kDa proteins is a prerequisite for the normal immune development. Further proteomics work is in progress in our laboratory to clear these points.

In conclusion, the present study clearly demonstrates that antibacterial protein (29kDa) having important biological functions in role of both immunity and metamorphosis. The current study confirms that the storage protein is utilized to respond to pathological condition in *B. mori.* In India, the indigenous tropical polyvoltine races showed more resistance to diseases than temperate bivoltine races. In order to understand the differential response at molecular level, the present study may be an ideal approach to ken the physiological consequences. Further studies will be made on domesticated *B. mori*, and efforts in this direction will open new vistas of research in Seri-biotechnology, which will improve economy of the silk industry.

ACKNOWLEDGLEMENT

We are grateful to Prof. Michihiro Kobayashi, Nagoya University, Japan, for his continuous encouragement and for providing the Travel Grant for the APSERI-2008 (Asia Pacific Congress on Sericulture and Insect Biotechnology), held at Nagoya University, Japan, 21–25 March 2008, where this data was presented in the "poster session". This work was supported by grants from the ABRL program of Korea Science Engineering Foundation (R14-2002-056-01001-0) and BK 21 program from Geyongsang National University, South Korea and partial financial support from DST, New Delhi, India.

REFERENCES

Aziwa, K., 1991. Defense reactions of silkworm *Bombyx mori* against the nuclear polyhedrosis. *Sericologia* 31: 201.

Bradford, M., 1976. A rapid and sensitive method for quantitation of microgram quantities of protein utilizing the principle protein – dye binding, *Annl. Bio. Chem.* 72: 248-254.

Beresford, P.J., Basinki-Gray, J.M., Chiu, J.K., Chadwick, J.S. & Aston, W.P., 1997. Chacterization of hemolytic and cytotoxic Gallysins: a relationship with arylporins. *Dev. Comp. Immunol.* 21: 253-266.

Boman, H.G., 1991. Antibacterial peptides and key components needed in immunity. *Cell* 65: 205-207.

Carlos, P.S., Nicholas, R.W., Phillip, J.D., Paul, D., Timothy, C., Candy, PYA., Sadhana, S., Ursula, P., Stuart, E.R. & Richard, H.C., 2002. Bacterial infection of a model insect: *Photorhabdusluminescens* and *Manduca sexta. Cellular Microbiology* 4: 329-339.

Chandrasekar, R., Dhanalakshmi, R., Krishnan, M., Kim, H.J., Jeong, H.C. & Seo, S.J., 2005. Computational Analysis of Apolipophorin-III in *Hyphantria cunea. Int. J. Indust. Entomol.* 10: 25-33.

Chandrasekar, R., 2006. Expression and sequestration of storage protein 1 in differentiated fat body tissues of red hairy caterpillar, *Amsacta albistriga.* Ph.D thesis at Bharathidasan University, Tamilnadu, India. pp. 22-43.

Chandrasekar, R, Suganthi, L.M., Nirmala, X. & Krishnan, M., 2007. Expres-sion of silk gene with respect to the supplementation of hydrolyzed soybean protein (P-Soyatose) in the fifth instar male larvae of *Bombyx mori*. *J. Cell and Molecular Biology* 16: 163-174.

Chandrasekar, R., Seo, S.J. & Krishnan, M., 2008a. Expression and Localization of storage protein-1 (SP1) in differentiated fat body of red hairy caterpillar, *Amsacta albistriga. Archive. Insect Biochem. Physiol.* 69: 70-84.

Chandrasekar, R., Sumithira, P., Seo, S.J. & Krishnan, M., 2008b. Seque-stration of storage protein 1 (SP1) in differentiated fat body tissues of the female groundnut pest *Amsacta albistriga* (Lepidoptera: Arctiidae). *Inter. J. of Tropical* Insect *Science,* 28: 78-87.

Chandrasekar, R., Muthu Meenakshi, P. & Luiz Paulo, A.M., 2009. Hexamerin storage proteins: Biosynthesis, Utilization and Evolution. Short views on Insect Molecular Biology (Etd.) Chandrasekar R., International Book Mission Academic Press, India, pp. 49-68.

Daborn, P.J., Waterfield, N., Blight, M.A. & French-Constant, R.H., 2001. Measuring virulence factor expression by the pathogenic bacterium *Photorhabdus luminescens* in culture and during insect infection. *J Bacteriol.* 183: 5834-5839.

de Champes, Henquell, C., Guelon, D., Sirot, D., Gazuy, N. & Sirot, J., 1993. Clinical and bacteriological study of nosocomial infections due to *Enterobacter aerogenes* resistant to imipenem. *J. Clin. Microbiol.* 31: 123-127.

De Gregorio, E., Spellman, P.T, Rubin, G.M. & Lemaitre, B., 2001.Genome-wide analysis of the *Drosophila* immune response by using oligonucleotide microarrays. *Proc. Natl. Acad. Sci. USA* 98: 12590-12595.

Dong, K., Zhang, D. & Dahlman, D.L., 1996. Down-regulation of juvenile hormone esterase and arylphorin production in *Heliothis virescens* larval parasitized by *Microplitis croceipes. Arch. Insect Biochem. Physiol.* 32: 237-248.

Ferrandon, D., Jung, A.C. & Criqui, M.C., Lemaitre, B., Uttenweiler-Joseph, S., Michaut, L., Reichhart, J.M. & Hoffmann, J.A., 1998. A GFP-drosomycin reporter transgene reveals a local immune response in Drosophila that is not dependent on the Toll pathway, *EMBO J.* 17: 1217-1227.

Forst, S. & Clarke, D., 2001. Bacteria-nematode symbiosis. In: *Entomopathogenic Nematology*. Gaugler, R. (ed), London: CAB International, pp. 57–77.

Gillepic, J.P., Kanost, M.R. & Trcenezeek, T., 1997. Biological mediators of insect Immunity. *Annu. Rev. Entomol.* 42: 611-43.

Hara, S. & Yamakawa, M., 1995. A novel type of antibacterial peptide isolated from the silkworm, *Bombyx mori. J Biol. Chem.* 270: 29923-29927.

Hoffmann, J.A. & Reichhart, J.M., 1997. Drosophila immunity. *Trends Cell Biol.* 7: 309-316.

Hultmark, D., 1993. Immune reactions in Drosophila and other insects: a model for innate immunity. *Trends Genet* 9:178-183.

Janarthanan, S., Krishnan, M. & Ignacimuthu, S., 1995. Changes in protein profiles during development in silkworm *Bombyx mori* (L). *J. Exp. Biology.* 36:1158-1160.

Jones, G., Brown, N., Manczak, M. & Hiremath Kafatos, F.C., 1990. Molecular cloning, regulation, and complete sequence of a hemocyanin-related, JH-suppressible protein from insect hemolymph. *J. Biol. Chem.* 265: 8596-8602.

Kimbrel, D.A., 1991. Insect antibacterial proteins: not just for insects and against bacteria. *BioEssays.* 13: 657-663.

Laemmli, V.K., 1970. Cleavage of structural protein during the assembly of the head bactreiophage T4, *Nature* 227: 680-685.

Lovallo, N., Barratt, B. & Legge, M., 2000. Effect of Microtonus aethiopoides paras-itism on haemolymph protein composition across alternate hosts. *Entomol. Exp. Et. App.* 95: 213- 216.

Levenbook, L., 1985. Insect storage proteins. *In comprehensive insect biochemistry physiology and pharmacology*, Pergon Press Oxford 10: 307-313.

Leuschner, R.G. & Bew, J., 2004. A medium for the presumptive detection of *Enterobacter sakazakii* in infant formula: Inter laboratory study. *J AOAC Int.* 87: 604-613.

Mullinas, P.E., 1985. Chemistry and physiology of the haemolymph. In: *comprehensive insect Physiology, Biochemistry and Pharmacology,* kerkut 3: 355-400.

Nagata, M. & Kobayashi, M. 1990. Quantitative changes in storage protein during larval development of the silkworm, *Bombyx mori. J. Seric. Sci. Japan* 59: 461-468.

Nakamatsu, Y., Gyotoku. T. & Tanaka, 2001. The endoparasitoid *Cotesia kari-yai* (*Ck*) regulates the growth and metabolic efficiency of *Pseudaletia sepa-rata* larvae by venom and *Ck* polydnavirus. *J. Insect Physiol.* 47: 573-584.

Ponnuvel, K.M., Nakazawa, H., Furukawa, S., Asaoka, A., Ishibashi, J., Tanaka, H. & Yamakawa, M., 2003. A Lipaseisolated from the silkworm shows antiviral activity against NPV. *J. Virol.*77: 10725-10729.

Shapiro, J.P., Mayer, R.T. & Schroeder, W.J., 1988. Absorption and transport of natural and synthetic toxins mediated by hemolymph proteins, Sehnal, F., Zabza, A. and Denlinger, D.L. (ed), *Endocrinological Frontiers in Physiological Insect Ecology*. Wroclaw Technical University, Wroclaw, Poland, pp. 997-1005.

Shelby, K.S. & Webb BA., 1994. Polydnavirus infection inhibits synthesis of an insect plasma protein, arylphorin. *J. Gen. Virol.* 75: 2285-2292.

Seiichi Hara & Minoru Yamakawattt, 1995. A novel antibacterial peptide family isolated from the silkworm *Bombyx mori, Biochem. J.* 310: 651-656.

Slovak, M., Repka, V., Slovakova, L. & Kazimirova, M., 1992. Haemolymph proteins of the cabbage moth (*Mamestra brassicae* L) in some development stage: In Bennettova *B. etal* (eds), Advances in regulation of insect reproduction, Institute of Entomology (*Czech Acad. Sci Crech Budejovice)*, pp. 269-278.

Tae-Yeon Kim, Yun Choi & Hong-Seok Cheong, Joonho Choe, 2002. Identification of a cell surface 30kDa protein as a candidate receptor for Hantaan virus *J. Gen.Virol.* 83: 767–773.

Tojo, S., Nagata, M. & Kobayashi, M., 1990. Storage protein in the silkwor *Bombyx mori. Insect. Bio. Chem.* 10: 289-303.

Ujita M, Katsuno Y., Kawachi, I., Ueno, Y., Banno, Y., Fujii, H. & Hara, A., 2005. Glucan-binding activity of silkworm 30-kDa apolipoprotein and its involvement in defense against fungal infection. *Bioscience, Biotechnology, and Biochemistry,* 69:1178-1185.

Van der, G. & Wassink, H.J.M., 1969. Haemolymph protein of the cabbage army worm, *Mamestra brassicae*, after infection with a nucleopolyhedrosis virus, *J. Invertebr. Pathol.* 14: 419-420.

Yun, E.Y., Kang, S.W., Hwang, J.S., Goo, T.W., Kim, S.H., Jin B.R., Kwon, Y. & Kim, K.Y., 1999. Molecular cloning and characterization of a cDNA encoding a transferrin homolog from *Bombyx mori. Biol. Chem.* 380: 1455-1459.

Wiesner, A., Losen, S., Kopácek, P. & Götz, P., 1997. Isolated apolipophorin-III from *Galleria mellonella* stimulates the immune reactions of this insect. *J. Insect Physiol.* 43: 383-391.

Whitten, M.M.A., Tew, I.F., Lee, B.L. & Ratcliffe, N.A., 2004. A novel role for an insect apolipoprotein (apoLp-III) in -1,3-glucan pattern recognition and cellular encapsulation reactions. *J. Immunol.* 172: 2177-2185.

Wicker, C., Reichhart, J.M., Hoffmann, D., Hultmark, D., Samakovlis, C. & Hoffmann J.A., 1990. Insect immunity characterization of a *Drosophila* cDNA encoding a novel member of the diptericin family of immune peptides. *J. Biol. Chem.* 265: 22493-22498.

Wyatt, G. & Pan, M., 1978. Insect plasma proteins. *Ann. Rev. Biochem.* 47: 779-817.

Chapter - 12

REPRODUCTIVE BIOLOGY OF *TRIBOLIUM CASTANEUM*

R. Chandrasekar and S.R. Palli

Department of Entomology, Agricultural Science Centre,
University of Kentucky, Lexington – 40506, KY, USA.
E-mail: *koreanchandru@yahoo.com*

ABSTRACT

The red flour beetle *Tribolium castaneum* is emerging as a further standard insect model beside Drosophila. Its genome is fully sequenced and it is susceptible for genetic manipulations including RNA-interference. Tribolium flour beetles colonize patchily distributed grain stores, and juvenile and adult stages share the same food resources. Adults are highly promiscuous and female reproduction is distributed across an adult lifespan lasting approximately 1 year. Sexual selection is a major force driving the evolution of diverse reproductive traits. This evolutionary process is based on individual reproductive advantages that arise either through intrasexual competition or through intersexual choice and conflict. Sexual selection focuses mainly on differences in male mating success. More recent work focuses on the differences in paternity share, that may arise through sperm competition or cryptic female choice whenever females mate with multiple males. Thus, an integrative view of sexual selection needs to encompass processes that occur not only before copulation (pre-mating), but also during copulation (peri-mating), as well as after copulation (post-mating), all of which can generate differences in reproductive success. In Tribolium, seminal vesicles and spermathecae are the sperm storage organs of male and female, respectively, and sperm are transferred directly into the female. By encompassing mechanisms of sexual selection across all of these sequential reproductive stages, this review takes an integrative approach to sexual selection in Tribolium flour beetles, particularly well-studied and economically important model organism. Furthermore, recent advances suggest that sexual conflict may have shaped reproductive traits in this species.

INTRODUCTION

Insects are the most diverse of all groups of organisms, with over 800,000 species named and many thousands, probably millions, yet to be discovered. Insect diversity may be linked to their close association with the angiosperms (flowering plants). The Coleoptera (beetles) are the most diverse of all insect orders, with at least 350,000 species, representing one fourth of all known animal species. The body size- the largest beetle is 17cm long and the heaviest one weighs in at close to 100 grams, while the smallest species are well below 1 mm in size (Crowson, 1981). Key to this success may have been the most distinctive feature shared by all but a few beetles: the hardened forewings (elytra) that cover the hind wings used for flying. Insects and related arthropods are vectors of some of the world's most serious diseases, and are serious pests of food, fiber, and stored goods. The wide range of reproductive capacities among different kinds of organisms, striking occurrence of evolutionary novelties which pertain to mating success and the variety of genetic system have provided the basis for may current discussion pattern of reproduction. Reproduction is critical for maintenance of life on earth and sexual selection is a major force driving the evolution of diverse behavioral, morphological, anatomical, and physiological traits. This evolutionary process is based on reproductive advantages that arise either through the sexual competition or through interactions between the sexes (eg. choice, coercion, sensory exploitation). Insect mating systems present a wide variation of tactics to increase reproductive fitness. Charles Darwin stated that reproductive fitness was a test of an animal's ability to produce as many offspring as possible (Darwin, 1871). Whenever females mate with multiple males, the processes that take place during copulation (perimating), as well as after copulation (postmating), can produce difference in males paternity success. Paternity success refers to the proportion of a female's progeny sired by each mating male.

In insects, as in many mammals and other vertebrates, copulation brings male and female cells and molecules together in the female's reproductive tract. There, they can interact to facilitate post-copulatory events such as sperm along with seminal fluid proteins transport, storage, sperm maintenance, egg release and ultimately, fertilization. Much is known about the genes, cells and molecules that make contribute to post-copulatory processes (Chapman *et al.,* 2001; Wolfner *et al.,* 2005; Raviram and Wolfner, 2007). The chosen species whose genome sequence was described recently (NCBI, 2008) is the red flour beetle, *Tribolium castaneum*, is a well studied economically important model organism for sexual selection processes. Adults are highly promiscuous and employ several post-mating mechanisms to bias paternity. The factors influencing the male paternity share of female's progeny have been extensively studied (Schlager, 1960, Nilsson *et al.,* 2003; Fedina & Lewis, 2006). The purpose of this review is to provide a critical comprehensive view of various sexual selection processes in Tribolium flour beetles. In particular, we aim to summarize life history, behavioral, physiological, reproductive anatomy and post-copulation behavioral changes that may further our understanding of *Tribolium* Reproduction.

LIFE HISTORY

Habitate and food

Tribolium, primarily evolved as saprophylic insects and naturally occur under the bark of trees, feeding on decaying plant materials and occasionally in the nests of some Hymenoptera (Grimm, 2001; Halstead, 1969; Magis, 1954; Sokoloff, 1974; Goodnight, 1995). When ancient Egyptians started stockpiling grain to safeguard against periodic famines around 4.5 thousand years ago, many insects took advantage of this rich resource. Some attributes that may have facilitated movement into this newly created niche include tolerance for low water availability, polyphagy, and reproduction in constant low light environments (Levinson and Levison, 1995). Tribolium flour beetle have been a human commensal associated with stored food products since at least 2000 B.C. As a result of human commerce and transportation, *Tribolium* spp. are found worldwide, mainly inhabiting stored grain and cereal products, but also nuts, dried fruit, spices, chocolate and yeast (Pray & Goodnight, 1995).

Life history

Life history traits for *T. castaneum* has a generation time 3- 8 weeks depending on temperature and can reproduce for a long time. Adults are relatively long lived beetles with reproductive life spans of 0.5- 1 year (reviewed in Sokoloff 1974; Nilsson, *et al.,* 2002). *T. castaneum* males become sexually mature two days post-eclosion, while females, which can be fertilized as early as 3 h after eclosion, do not begin laying eggs until four days later (Dawson, 1964). Mated *T. castaneum* females lay on average 10 eggs per day for 3-6 months, after which oviposition rates decline; female lifetime fecundity ranges between about 500-1200 eggs (Sokoloff, 1974; Beranasconi *et al.,* 2006). Developmental period for *T. castaneum* in optimal conditions is about five days for eggs, 20 days for larvae, and 7 days for pupae (reviewed in Sokoloff, 1974). Most information concerning Tribolium life history traits and behaviors has been gathered from culture kept in the laboratory for many generations, and these may differ from natural population. Many Tribolium life history traits show considerable phenotypic plasticity in response to (abiotic & biotic) environmental variation (temperature, humidity, food). These beetles exhibit several life history high intrinsic population growth rate, fast maturation, rapid larval development, female high fecundity and good dispersal ability.

REPRODUCTIVE BIOLOGY

Male reproductive system /anatomy

The eight abdominal segment forms a complete ring. Enlarged sterna and tergal areas are hinged in the narrow pleurae. From Fig. 1 (A,B) the tergum is lightly sclerotized expect for the anterolateral corners where condyles articulate with those of the sterna apophyses. Within the eight abdominal segments lies the wishbone-shaped speculum gastrale. The united lower end of this structure lies ventral to the base of intromittent organ. The arms of the speculum curve laterally and posteriorly, the ends forming flattened, setae-bearing plates. These plates merge with a supra-anal region of sclerotization to form a ring which surrounds the median portion of the aedeagus and the anus. The aedeagus is composed a simple tubular penis (Fig. 2) and the articulated tegmen.

Male reproductive system consisted of a pair of multifollicular testes, tubular vas deferens, distinctly formed unichambered seminal vesicle, one pair of accessory glands, and highly muscular ejaculatory duct (Fig. 2). A pair of lobed testis occupies a greater space of the male abdomen. The median tube is the secretory ejaculatory duct. Testes are connected to ejaculatory duct through the distal thin walled tube, vas deferentia and proximal thick walled tube, seminal vesicle. There are two pairs of male accessory glands [one pair of coiled tubular accessory gland (TAG), and one pair of rod-shaped accessory gland (RAG)]. The duct of RAG and TAG and seminal vesicle terminate into ejaculatory duct at grand junction. The male accessory glands exist as rudimentary structures during the larval stage, proliferate and grow during the pupal stages. At imaginal ecdysis, the MAGs are fully formed. The growth of MAGs continues during adult stage, there is a significant growth 48-96 h after adult emergence which might mark the terminal differentiation and maturation of MAGs.

Male accessory gland substances

During copulation males transfer accessory gland substances to females in addition to sperm. In many insects, these substances have been shown to exert a profound influence on female reproductive physiology and behavior (Eberhard, 1996; Simmons, 2001). In Drosophila spp., male accessory gland proteins increase ovulation rate and suppress female receptivity (Wolfner, 2002). Tribolium spp. males possess two distinct parts (TAG and RAG) of male accessory gland.

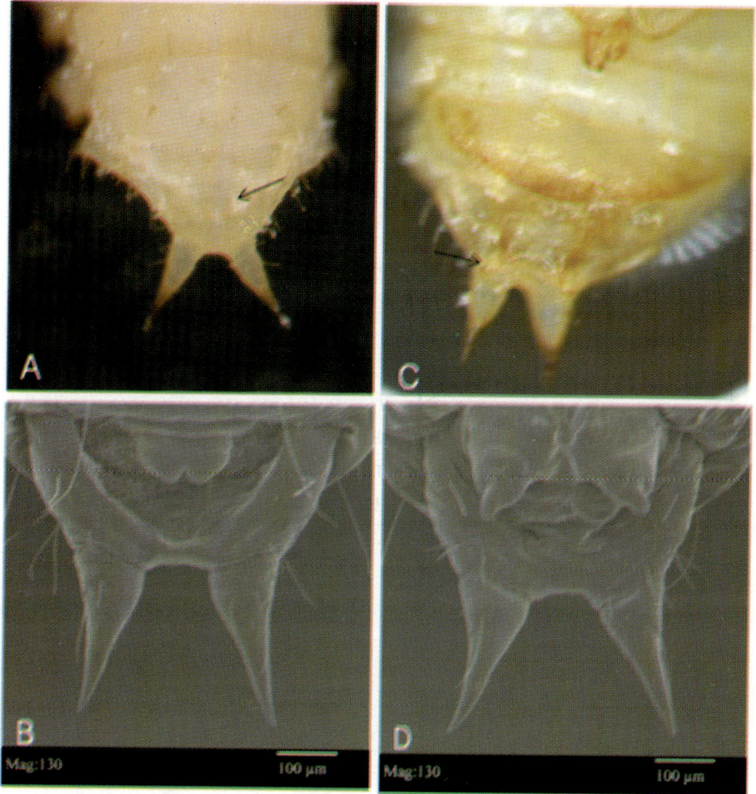

Fig. 1. Light and Scanning electron microscopy viewing male and female reproductive genital organ. A & B - Male genital organ; C & D – Female genital organ. Scale bar 100μm

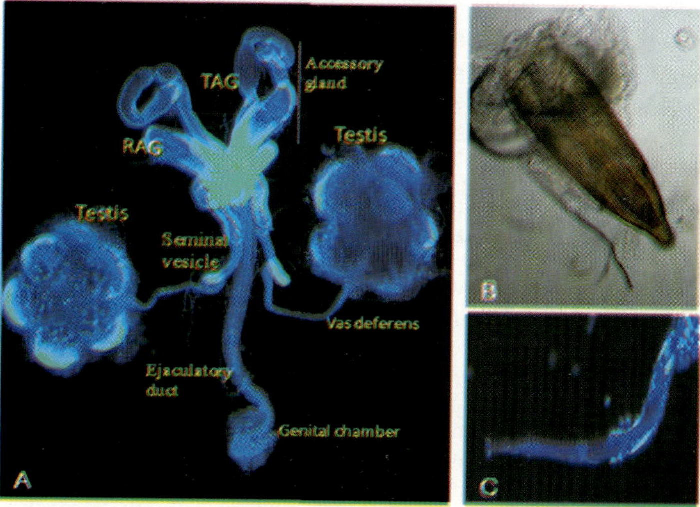

Fig. 2. Male reproductive system of Tribolium (A – male reproductive system stained with DAPI (pair of testis, accessory gland); B - ejaculatory pad; C – penis; TAG – Tubular accessory gland; RAG Rod shape accessory gland).

Table 1. Male accessory gland genes encode peptide and protein that can be categorized into a range of predicted molecular classes.

Accessory gland protein class	Number of molecules
Peptides / pheromones	44
Proteases	18
Protease inhibitors	7
Acid lipases	7
Lectins	8
Cystein rich secretory proteins	8
Defensin	1
RNAse Macroglobulin	1
Oxido reductases / Thioredoxin	3
Fascicilin	1
Alkaline phosphatase	1
Collagen like	3
Hydrolase	1
Chapeerones	4
Protein folding	3
Peroxidase	1
Total	**112**

Histological studies in *T. brevicornis*, *T. freeman,* and *T. anaphe* have identified at least five differentially staining cell types forming spatially distinct region within short male accessory gland (Sevener *et al.*, 1992; Novaczewski & Grimnes, 1996). Extensive studies have characterized male accessory gland proteins in *T. molitor*. Several proteins produced by long accessory glands may perform hormone-transport and sperm-capacitation functions, based on their molecular characteristics (Table 1) and sequence similarity with male proteins from Drosophila melanogaster (Paesen & Happ, 1994; Feng & Happ, 1996). In addition, the male bean-shaped accessory gland of *T. molitor* secrete trehalase into the male spermatopore; it serves as a structural component and may also provide sperm nutrition/activation in concert with trehalose found in females (Yaginuma & Happ, 1988; Yaginuma *et al.,* 1996).

Functional significance of the glandular secretions

Recent advances in bioinformatics, genomics and proteomics research have greatly helped in identifying Acps of non Drosophilids such as Ceratitis capital (Davies & Chapman, 2006), honey bees (Collins *et al.*, 2006), some orthopterans (Braswell *et al.,* 2006), Crickets (Andres *et al.*, 2006; Braswell *et al.,* 2006), medflies (Davies and Chapman, 2006), *Anopheles gambiae* (Dottorini *et al.,* 2007), *Aedes aegypti* (Sirot *et al.,*

2008), *Cimex lectularius* (Reinhardt *et al.*, 2007). There have been several significant reviews dealing with insect Acp gland structure, function and development (Happ, 1992; Gillott, 1988, 2003). They may stimulate an increase in (1) the number and rate of development of eggs, (2) influence sperm storage, (3) sperm organization, (4) modulate ovulation and oviposition. Antimicrobial agents in the secretions ensure that the female reproductive tract is a hospitable environment during sperm transfer. In a few species the secretions include noxious chemicals, which may protect it from predators and pathogens when laid. Though the morphology of MAGs has been studied in Tenebrionid beetles (Roberts and Grimnes, 1994; Novaczewski and Grimnes, 1996; Chen, 1984), the identification and functional characterization of Acps of beetles have not been reported so far. There are a few reports on secretory proteins from MAG of *Tenebrio molitor* (Paesen and Happ, 1994; Feng and Happ, 1996; Yaginuma *et al.*, 1996) and very few genetic studies *T. castanuem* showed influence of male gland substances of female physiology (Nilsson *et al.*, 2003; Attia & Tregenza, 2004). In our lab 112 genes that were highly expressed in male accessory glands (MAG) in the red flour beetles, *Tribolium castaneum* was identified. Out of these 112 identified genes, 59 of them contained sequences coding for signal peptide and cleavage site and the remaining 53 contained transmembrane domains (Parthasarathy et al., 2009) and further investigation under way. The 112 Acp genes encode peptide and protein that can be categorized into a range of predicted molecular classes. These classes include (1) small novel peptides or larger pheromone-like molecules that could be cleaved to release biologically active peptides, (2) predicted glycoproteins, and (3) proteins with sequence motifs predictive of either enzymatic functions (proteases and protease inhibitors, lipases), cellular or organismal protection (antimicrobial proteins, proteolysis) or sperm binding (lectin and cysteine rich secretory proteins. The conservation of seminal-fluid protein classes across a wide range of organisms with quite different reproductive physiologies is intriguing, given that several proteins in each of these conserved classes are known to have important reproductive functions in higher vertebrates (Malm *et al.*, 2000; Murer et al. 2001; Suarez 2002; Busso *et al.* 2005; Ellerman et al. 2006; Nixon et al. 2006) and, as we shall discuss in insects.

Female reproductive anatomy

The female insect's main reproductive function is to produce eggs, including the egg's protective coating, and to store the male spermatozoa until egg fertilization is ready. The female Tribolium reproductive organs include, paired ovaries which empty their eggs (oocytes) via the calyces into lateral oviducts, joining to form the common oviduct (Fig. 1C,D & 3). The opening (gonopore) of the common oviduct is concealed in a cavity called the genital chamber and this serves as a copulatory pouch (bursa copulatrix) when mating (Fig. 1C,D). The external opening to this is the vulva. Often in insects the vulva is narrow and the genital chamber becomes pouch or tube like and is called the vagina. Related to the vagina is a sac like structure, the spermatheca, where spermatozoa are stored ready for egg fertilization. A secretory gland nourishes the contained spermatozoa in the vagina. Egg development is mostly completed by the insect's adult stage and is controlled by hormones that control the initial stages of oogenesis and yolk deposition. Most insects are oviviparous, where the young ones hatch after the eggs have been laid.

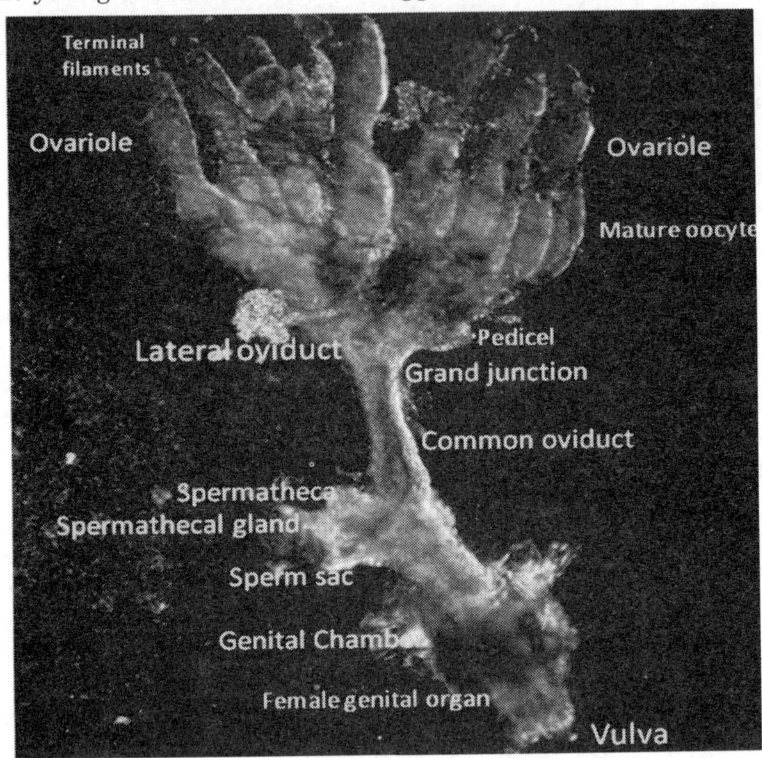

Fig. 3. Female reproductive system.

Female spermatheca

Spermatheca is female sperm storage organs which influence sperm displacement and the dynamics fertilization. In *T. castaneum*, a single copulation generally fills the female spermatheca to about two-thirds of its capacity (Lewis and Jutkiewicz, 1998), while second and further ejaculates result in displacement of previously stored sperm (Lewis *et al.*, 2005) by contrast, in *T. molitor* the spermatheca is not filled to capacity even after three mating (Drnevich, 2003). *T. castaneum* females dissected immediately after naturally terminated copulations always contained spermatophore that had already everted in the bursa copulatrix. This spermatophore appeared as a sperm-filled sac with posterior region (Fig.

4B). However, female may still use strategic sperm expulsion versus uptake to manipulate the relative representation of a male's sperm in storage. In view of their large mass, these spermatophores represent a substantial energy investment for the male, but whether this investment is intended to nourish females has not been chemically substantiated. Very few studies found a negative correlation between second-male paternity success and female's spermathecal volume (Fedina and Lewis, 2004). This negative relationship may reflect greater storage and/or reduced displacement of first-male sperm in females with larger storage organs. Future studies needed on these female processes might clarify widely reported variability in male paternity success in species with polyandrous females.

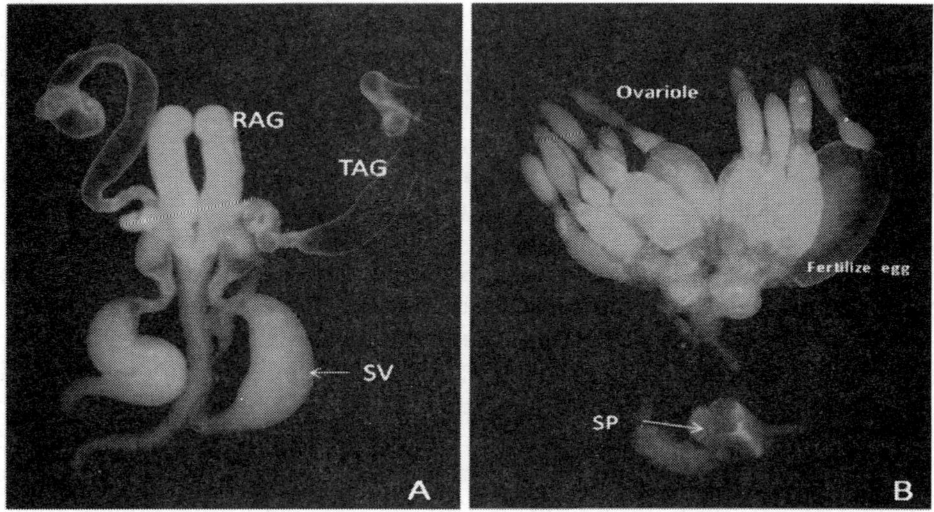

Fig. 4. Sperm storage organ of male (A) and female (B) stained with acriddin orange SV- seminal vesicle; SP – spermatheca

MATING BEHAVIOR

Characteristics of mating systems of Tribolium spp. are highly polygamous with both sexes mating frequently throughout their adult lives (Park, 1993; Sokoloff, 1974). Mating rates have been measured by observing beetles interacting in open arenas, and several studies indicate that males engage in copulations every 2-4 min (Shorde, 1960; Haubruge *et al.,* 1999; Lewis, 2004). Females in *T. castaneum* also mated with up to 12 different males per hour (Pai, Feil & Yan 2007). Given that females are able to lay over 700 fertile eggs over a span of 4 months following only a single mating (Bloch Qazi *et al.,* 1996). Polyandry (mating with multiple males) could provide females with an opportunity to obtain direct benefits from several males, and /or to obtain indirect genetic benefits by allowing sperm competition and exercising cryptic female choice. Polyandry may also lead to antagonistic coevolution between the sexes, and thus result in net costs to females (Rice, 1998; Arnqvist & Rowe 2005; Pai & Bernasconi, 2007). Peptide hormones play an important role in the control of mating behavior, yet understanding their in vivo role poses a number of very difficult experimental challenges.

Pre-mating

Many taxa rely on visual, acoustic, or chemical signals to attract mates over long distances. Because Tribolium inhabit grain storage areas with dim or no light, it is unlikely that visual signals play any major role in mate attraction or assessment in this group. In several Tribolium species, adult males produce a highly volatile pheromone, 4,8-dimethyldecanal (DMD), which is attractive to both sexes; this compound has been isolated from *T. castaneum, T. confusum, T. freeman* and *T. madens* (Arnaud and Haubruge, 2002; Suzuzki, 1980; Suzuki and Sugawara, 1979). Although DMD is commonly referred to as an aggregation pheromone, it may have a dual function as a sex attractant for females (Levinson and Levinson, 1995). Tribolium flour beetles show strong sexual dimorphism in the form of setiferous gland located on the prothoratic femora of males (absent in females) in *T. Castanuem, T.destructor, T. freeman, T. anaphe and T. amadens, T. confusum* and *T. alcine* (Faustini and Halsted, 1982). These glands produce a waxy secretion that accumulates on the gland surface as male age, and which is strongly attractive to both females and males in olfactometer tests (Faustini, Burkholder and Laub, 1981; Olsson *et al.,* 1006). Thus, the chemical signals active in Tribolium long-range mate attraction and possibly assessment are complex, and appear to involve several compounds that may work in opposition or synergistically. Once male and females reach close proximity, further assessment and choice among potential mates may occur based on different cues, including contact chemicals such as cuticular hydrocarbons (Howard and Blomquist, 2005), physical cues such as body size and shape, or courtship behaviors.

Peri-mating

Mating behavioral interactions during copulation have the potential to influence a male's subsequent paternity success by affection the processes of sperm transfer, sperm storage in spermatheca and use. *T. castaneum*, copulation durations are highly variable from 0.5 to 2 min or ~30 min (Sokoloff, 1974; Lewis, 2004). During copulation males rapidly rub their legs against the sides of the female's body interspersed with periods of inactivity (Bloch Qazi, 2003). In the case of female beetle usually exhibit quiescence behavior during which they stop moving and often assume a characteristics posture with extended forelegs and raised head (Bloch Qazi,, 2003). Female quiescence during copulation may therefore, indicate female cooperation in the insemination process successful.

Post-mating

Arthropods are unique in that females store sperm in specialized organ to fertilize their eggs later, sometimes several months after mating. Most arthropod females also mate with multiple males. These attributes select for traits promoting male sperm competition and cryptic female choice. Among them is female control over sperm storage and usage for fertilization, and male accessory gland substances influencing female physiology and incapacitating rival sperm (Eberhard, 1996; Birkhead and Pizzari 2002). In order to achieve fertilization, sperm must be passed to the female, enter sperm storage, survive in the sperm storage organs for long periods of time, exit storage, and finally enter an unfertilized egg. Moreover *T. castanuem* may mate with multiple male. There is an opportunity for differential use of sperm from different males by females, and consequently "competition" between sperm or ejaculates of different males, or "preference" by the females or sperm from different males. At 30 min post mating, spermatheca containing $>2.948 \pm 0.501 \times 10^5$ sperms/insect was observed in the female reproductive tract (Chandrasekar, unpublished data). Sperm introduced into mated females must reach sites of storage for efficient and long-term utilization. In insects, this process has been shown to require input from both females and males. In *T. castaneum* flour beetles (Bloch Qazi *et al.,* 1998; Fedina and Lewis, 2004; Ludovic Arnaud *et al.,* 2005) the ability of the female to undergo muscle contraction is necessary for sperm to be stored. In *D. melanogaster*, a feminized nervous system is necessary for proper accumulation of sperm into storage (Arthur *et al.,* 1998) and male-derived seminal proteins are needed for sperm storage and utilization in the female (Kalb *et al.,* 1993; Xue and Noll, 2000). In addition to the above-mentioned copulatory courtship behavior and olfactory attractiveness males may have evolved additional traits that are potential adaptations to increase their post-copulatory reproductive success. For instance, the brush-like structure on the male aedeagus (intromittent organ), or sperm size and number variation, may play a role in post-copulatory processes. Currently various advance in understanding post-mating sexual selection have been attained by using genetic markers to determine paternity of multiple males. Such markers can be morphological traits (eg. body colour), physiological traits (eg. allozymes, insecticide resistance), male sterility induced by irradiation, or molecular markers (Loxdale and Lushai, 1998; Behura, 2006).

PHENOMENOLOGY OF SPERM COMPETITION

In most insect species, seminal vesicles and spermathecae are the sperm storage organs of males and females, respectively. Sperm produced in the testes are transferred through the vas deferens to the seminal vesicles (SV) for storage until mating (Fig. 4A). At mating, once the male grasps the female head region with his abdominal terminalia, translocates sperm stored in the SV from the opening of the internal reproductive system, positioned near the abdominal tip, into the sperm vesicles (SV) of the secondary genitalia under abdominal segments 2 and 3 (Fig. 1A). After completing sperm translocation, the aedeagus of the secondary genitalia is inserted into the female genital opening, and sperm in the SPV are transferred along a slit of the aedeagus into the female storage organs (Fig. 2). Females usually store the received sperm in two storage organs, the bursa copulatrix (BC) and the spermatheca (SP). However, not all sperm survive in the storage organs. Sperm age after spermatogenesis, and the duration of sperm storage affects sperm quality through the aging process and chemical/physical damage to the cells (Reinhardt, 2007; Kaori Tsuchiya and Fumio Hayashi, 2010). Furthermore, sperm can also be damaged during transport between storage organs.

The protein composition of seminal fluid differs widely among closely related species, and clear evidence for species specificity in accessory gland activity is obtained from transplantation and injection studies (Chen, 1996). Robinson et al (1994), in double-mating studies with *T. castaneum* and *T. freeman*, found strong conspecific sperm precedence, which they characterized as post-copulatory prezygotic isolation. Price (1997) performed tests of sperm competition among *D. simulans, mauritiana* and *D. sechellia* by mating females with conspecific males followed by heterospecific males and vice versa. In all cases there was a strong tendency for females to use conspecific sperm, as though the foreign sperm were recognized and avoided. Recently Price et al. (2001) found that when the first male to mate is conspecific, the seminal fluid from this male incapacitates subsequent sperm from heterospecific males. However, when the first mating male is heterospecific and the second male is conspecific, then the heterospecific sperm is physically removed from storage. While these phenomena look similar to what happen with all conspecific mating, it may be premature to assume the underlying mechanism for inter- and intraspecific sperm competition.

FUTURE PROBLEM

The study of sperm competition seems to have no end to its ability to unearth new and unusual connections between biological processes and molecular reproduction. Most recently the remarkable effect of mating on immunocompetence in males (McKean and Nunney, 2001) has suggested a connection between mating and immune function. While there is no doubt that such discoveries will continue, three particular problems remain central to the evolutionary questions about sperm competition: (1) What is the association between molecular variation in Acps (accessory gland proteins) and sperm competitive success? (2) How relevant are the laboratory studies to natural conditions in the field? (3) How does the phenomenon of intraspecific sperm competition relate to conspecific precedence?

CONCLUSION

The production of spermatozoa in the male insect and transfer of the seminal mass to the female require significant investments of biochemical resources and the temporal coordination of development in several organs. Successful reproduction requires interaction between male and female cells, molecules, and genomes. In animals with internal fertilization, males transfer to females not only sperm but also seminal fluid that contains a cocktail of molecules important for reproductive success. The life history of *Tribolium* spp. flour beetles is representative of many coleopteran human commensals. Juveniles and adults live together in aggregation competing for the same food resources. Adults are long live, and female are highly polyandrous and lay eggs continually. These life history and mating system features are likely to influence mechanisms of sexual selection across three sequential stages of reproduction pre-mating, peri-mating and post-mating. Several mechanisms of paternity biasing may also operate after mating in Tribolium flour beetles, but more studies are needed to demonstrate this phenomenon clearly. Thus, movement of sperm into storage mainly under female control, but demonstration that females use this mechanism to favor certain males is lacking. Since accessory gland proteins affect processes that are involved in determining male post-copulatory reproductive successes, they may be subject to such evolutionary forces as sexual conflict or sexual selection through cryptic female choice or sperm competition.

Reproduction is correlated with both extrinsic environmental factors (food supply, tidal cycles, day length etc.,) and components within the population (density, social and age structure etc.). As our knowledge of the nature and role of male insect accessory gland products grow, so will the interest of wide range biological scientists. Biochemists will be able to compare similarly constructed molecules from other animal taxa in their search for structure-function relationship at the molecular mechanisms of reproductive physiology. Comparative physiologists, especially those curious to learn the ***"ins*** and ***outs***" of mating, will see parallels between insect and vertebrate mechanisms. Likewise, evolutionary biologists fascinated by animal mating strategies, sexual selection, and speciation will identify principles that are broadly applicable to other taxa, and applied entomologists/biotechnologist may evolve new ideas based Acps functions as they assist in designing potential strategies of management of insect pests.

ACKNOWLEDGEMENTS

We thank Dr. S. Perry, Plant Science and Dr. Dicky Yu of University of Kentucky for providing microscopy facility. This work was supported by National Science Foundation and this is contribution number 09-08-045 from the Kentucky Agriculture Experimental Station.

REFERENCES

Andrés, J.A., Maroja, L.S., Bogdanowicz, S.M., Swanson, W.J. & Harrison, R.G., 2006. Molecular evolution of seminal proteins in field crickets. *Molecular biology and evolution* 23(8):1574-84

Arnaud, L. & Haubruge, E., 2002. Insecticide resistance enhances male reproductive success in a beetle. *Evolution* 56: 2435 - 2444.

Arnqvist, G. & Rowe, L., 2005. Sexual conflict. Princeton University Press.

Aruthur, B.I.J., Hauschteck-Jungen, E., Nothiger, R. & Ward, P.I., 1998. A female nervous system is necessary for normal sperm storage in *Drosophila melanogaster*: a masculinized nervous system is as good as none. *Proc. Royal Society, London B: Biological Sci.* 265: 1749 – 1753.

Attia, F.A. & Tregenza, T., 2004. Divergence revealed by population crosses in the red flour beetle, *Tribolium castaneum*. *Evol. Ecol. Res.* 6: 927 – 935.

Behura, S.K., 2006. Molecular marker systems in insects: current trends and future avenues. Molecular Ecology 15: 3087 – 3113.

Bernasconi, G., Brostaux, Y., Meyer, E.P. & Arnaud, L., 2006. Do spermathecal morphology and inter-mating interval influence paternity in the polyandrous beetle *Tribolium castaneum*. *J. Evolutionary Biology* 14: 186-193.

Birkhead, T.R. & Pizzari, T., 2002. Post-copulatory sexual selection. *Nature Reviews Genetics* 3: 262 – 273.

Bloch Qazi, M.C., Boake, C.R.B. & Lewis, S.M., 1998. The femoral setiferous glands of *Tribolium castaneum* males and production of the pheromone 4,8-dimethyldecannal. *Entomologia Expt. Applicata* 89:313 – 317.

Bloch Qazi, M.C., Herbeck, J.T. & Lewis, S.M., 1996. Mechanisms of sperm transfer and storage in the red flour beetle (Coleoptera: Tenebrionidae). *Annals of the Entomological Society of America* 89: 892 897.

Bloch Qazi, M.C.B., 2003. A potential mechanisms for cryptic female choice in a flour beetle. *J. Evolutionary Biology* 16: 170 – 176.

Braswell, W.E., Andres, J.A., Maroja, L.S., Harrison, R.G., Howard, D.J. & Swanson, W.J., 2006. Identification and comparative analysis of accessory gland proteins in Orthoptera. *Genome* 49: 1069-1080.

Busso, D., Cohen, D.J., Hayashi, M., Kasahara, M. & Cuasnicu, P.S., 2005. Human testicular protein TPX1/CRISP-2: localization in spermatozoa, fate after capacitation and relevance for gamete interaction. *Mol. Hum. Reprod.* 11: 299 – 305.

Chapman, T., Herndon, L.A., Heifetz, Y., Partridge, L. & Wolfner, M.F., 2001. The acp-26Aa seminal fluid protein is a modulator of early egg hatachability in *Drosophila melanogaster*. *Proc. R. Soc. London Ser. B.* 268: 1647-1654.

Chen, P.S., 1984. The functional morphology and biochemistry of insect male accessory glands and their secretions. *Annu. Rev. Entomol.* 29: 233 – 255.

Chen, P.S., 1996. The accessory gland proteins in male Drosophila: structural, reproductive, and evolutionary aspects. *Experientia* 52: 503 -510.

Collins, A.M., Caperna, T.J., Williams, V., Garrett, W.M. & Evans, J.D., 2006. Proteomic analyses of male contributions to honey bee sperm storage and mating. *Insect Mol. Biol.* 15: 541-549.

Crowson, R.A., 1981. The Biology of Coleoptera (London: Academic Press).

Darwin, C., 1981. The descent of man and selection in relation to sex. London: John Murray.

Davies, S.J. & Chapman, T., 2006. Identification of gene expressed in the accessory glands of male Mediterranean fruit flies (Ceratitis capitata). Insect Biochem. Mol.Biol. 36, 846-856.

Dawson, P.S., 1964. Age at sexual maturity in female flour beetles, *Tribolium castaneum* and *T. confusum. Annals of the Entomological Society of America* 57: 1-3.

Dottorini, T., Nicolaides, L., Ranson, H., Rogers, D.W., Crisanti, A. & Catter-uccia, F., 2007. A genome-wide analysis in Anopheles gambiae mosquitoes reveals 46 male accessory gland genes, possible modulators of female behavior. Proc. Natl. Acad. Sci. USA, 104: 16215-16220.

Drnevich, J.M., 2003. Number of mating males and mating interval affect last-male sperm precedence in Tribolium molitor L. Animal Behavior 66: 349 – 357.

Eberhard, W. 1996. Female control: sexual selection by cryptic female choice. Princeton University Press, Princeton, N.J.

Ellerman, D.A., Cohen, D.J., Da Ros, V.G., Morgenfeld, M.M., Busso, D. & Cua-snicu, P.S., 2006. Sperm protein "DE" mediates gamete fusion through an evolutionarily conserved site of the CRISP family. *Dev. Biol.* 297: 228 – 237.

Faustini, D.L., Burkholder, W.E. & Laub, R.J., 1981. Sexually dimorphic setiferous sex patch in the male red flour beetle *Tribolium castaneum* site of aggregation pheromone production. J. Chemical Ecology 7: 465 – 480.

Faustini, D.L. & Halstead, D.G.H., 1982. Setiferous structures of male Coleoptera. *J. Morphology* 173: 43 – 72.

Fedina, T.Y. & Lewis, S.M., 2004. Female influence over offspring paternity in the red flour beetle *Tribolium castaneum. Proceedings of the Royal Society Biological Sciences Series* B 271: 1393 – 1399.

Fedina, T.Yu. & Lewis, S.M. 2006. Proximal traits and mechanisms for biasing paternity in the red floor beetle *Tribolium castaneum* (Coleoptera: Tenebronidae). *Behav. Ecol. Sociobiol.* 60: 844-853.

Feng, X. & Happ, G.M., 1996. Isolation and sequencing of the gene encoding Sp23, a structural protein of spermatophore of the mealworm beetle, *Tenebrio molitor. Gene* 179: 257-262.

Gillot, C., 2003. Male accessory gland secretions: modulators of female reproductive physiology and behavior. *Annu. Rev. Entomol.* 48: 163 – 184.

Gillott, C., 1988. Arthoropoda-Insecta. In Reproductive Biology of the Invertebrates, vol. III, Accessory Sex Glands (Adiyodi, K.G. and Adiyodi, R.G. edited), pp. 319 – 471. New York: John Wiley & Sons.

Grimm, R., 2001. Faunistik und taxonomie einiger arten der gattung *Tribolium* MacLay, 1825, mit beschreibung von drei neuen arten aus Africa (Coleoptera, Tenebrionidae). *Entomofauna* 22: 393-404.

Goodnight, C.J., 1995. Epistasis and the increase in additive genetic variance: Implications for phase 1 of Wright's shifting balance process. *Evolution* 49: 502- 511.

Happ, G.M., 1992. Maturation of the male reproductive system and its endocrine regulation. *Annu. Rev. Entomol.* 37: 303 – 320.

Haubruge, E.,Arnaud, L., Mignon, J. & Gage, M.J.G., 1999. Fertilization by proxy: Rival sperm removal and translocation in a beetle. *Proceeding of the Royal Society Biological Sciences Series B* 266: 1183 – 1187.

Halstead D.G.H., 1969. A new species of *Tribolium* from North America previously confused with *Tribolium madens* (Charp.) (Coleoptera: Tenebrionidae). *Journal of Stored Products Research* 4: 295-302.

Howard, R.W. & Blomquist, G.J. 2005. Ecological, behavioral, and biochemical aspects of insect hydrocarbons. *Annual Review of Entomology* 50: 371 – 393.

Kalb, J.M., DiBenedetto, A.J. & Wolfner, M.F., 1993. Probing the function of Drosophila melanogaster accessory glands by directed cell ablation. *Proc. Natl. Acad. Sci. USA* 90: 8093 -8097.

Kaori Tsuchiya & Fumio Hayashi, 2010. Factors Affecting Sperm Quality Before and After Mating of Calopterygid Damselflies. *PLoS ONE* 5 (3), e9904.

Levinson, A. & Levinson, H., 1995. Reflections on structure and function of pheromone glands in storage insect species. *Anzeiger fuer Schaedlingskunde Pflanzenschutz Umweltschutz* 68: 99-118.

Lewis, S.M., 2004. Multiple mating and repeated copulations: effect on male reproductive success in red flour beetle. *Animal Behavior* 67: 799 -804.

Lewis, S.M. & Jutkiewicz, E., 1998. Sperm precedence and sperm storage in multiply mated red flour beetles. Behav. Ecol. Sociobiol. 43: 365 – 369.

Lewis, S.M., Kobel, A., Fedina, T. & Beeman, R.W., 2005. Sperm stratification and paternity success in red flour beetles. *Physiological Entomology* 30: 303 – 307.

Ludovic Arnaud, Yves Brostaux, Stéphane Lallemand & Eric Haubruge, 2005. Reproductive strategies of Tribolium flour beetles. *J. Insect Science* 5:33, 12pp.

Loxdale, H.D. & Lushai, G., 1998. Molecular markers in entomology. *Bulletin of Entomological Research* 88: 577 – 600.

Malm, J., Hellman, J., Hogg, P. & Lilija, H., 2000. Enzymatic action of prostate-specific antigen (PSA or hK3): substrate specificity and regulation by Zn(2+), a tight-binding inhibitor. *Prostate* 45: 132 – 139.

Magis, N., 1954. Apercu de l'histoire naturelle des complexes d'especes du genre *Tribolium* (McLeay, 1925) (Coleoptera, Tenebrionidae), *Bull. Instr. r. Sci. nat. Belg.* **30** (1954), pp. 1–10.

Murer, V., Spetz, J.F., Hengst, U., Altrogge, L.M., de Agostini, A. & Monard, D., 2001. Male fertility defects in mice lacking the serine protease inhibitor protease nexin-1. *Proc. Natl. Acad. Sci. USA* 98: 3029-3033.

McKean, K.A. & Nunney, L., 2001. Increased sexual activity reduces male immune function in Drosophila mealnogaster. *Proc. Natl. Acad. Sci. USA* 98: 7904 -7909.

Nilsson, T., Fricke, C. & Arnovist, G., 2002. Pattern of divergence in the effects of mating on female reproductive performance in flour beetles. *Evolution* 56: 111-120.

Nilsson, T., Fricke, C. & Arnqvist, G., 2003. The effects of male and female genotype on variance in male fertilization success in the red flour beetle (*Tribolium castaneum*). *Behav. Ecol. Sociobiol.* 53: 227-233.

Nixon, B., MacIntyre, D.A., Mitchell, L.A., Gibbs, G.M., O'Bryan, M. & Aitken, R.J., 2006. The identification of mouse sperm-surface-associated proteins and characterization of their ability to act as decapacitation factors. *Biol. Reprod.* 74: 275 – 287.

Novaczewski, M. & Grimnes, K.A., 1996. Histological characterization of the reproductive accessory gland complex of *Tribolium anaphase* (Coleoptera: Tenebrionidae). *Behav. Ecol. Socibiol.* 53: 227 -233.

Olsson, P.O.C., Ryne, C., Wallen, R., Anderbrant, O. & Lofstedt, C. 2006. Male-produced sex pheromone in *Tribolium confusum*: Behavior and investigation of pheromone production locations. *J. Stored Products Research* 42: 173 – 182.

Paesen, G.C. & Happ, G.M., 1994. cDNA-inferred amino-acid sequence of a C protein, a heparin-binding, basic secretion product of the tubular accessory sex glands of the meal worm beetle, *Tenebrio molitor*. *Insect Biochem. Mole.Biology* 24: 21-27.

Pai, A. & Bernasconi, C., 2007. Polyandry and female control: the red flour beetle *Tribolium castaneum* as a case study. *J. Expt. Zoology* (Mol. Dev. Evol.) 308 D, DOI: 10.1002/lez.b.21146.

Pai, A., Feil, S. & Yan, G., 2007. Variation in polyandry and its fitness consequences among populations of the red flour beetle, *Tribolium castaneum*. *Evolutionary Ecology* DOI : 10.1007/s 10682-006-9146-4.

Park, T., 1993. Studies in population physiology. II. Factors regulating initial growth of *Tribolium confusum* populations. *J. Experimental Zoology* 65: 17 – 42.

Parthasarthy, R., Tan, A., Sun, Z., Chen, Z., Rankin, M. & Palli, S.R., 2009. Juvenile hormone regulation of male accessory gland activity in the red flour beetle, *Tribolium castaneum. Mechanisms of Development* 126: 563 – 579.

Price, C.S.C., 1997. Consepcific sperm precedence in *Drosophila. Nature* 338: 663 – 666.

Price, C.S.C., Kim, C.H., Gronlund, C.J. & Coyne, J.A., 2001. Cryptic reproduction isolation in the *Drosphila simulans* species complex. *Evolution* 55: 81 – 92.

Pray, L.A. & Goodnight, C.J., 1995. Genetic variation in inbreeding depression in the red flour beetle *Tribolium castaneum. Evolution* 49: 176-188.

Raviram, K. & Wolfner, M.F., 2007. Seminal influences: Drosophila Acps and the molecular interplay between males and females during reproduction. *Integrative and Comparative Biology* 47: 427-445.

Robinson, T., Johnson, N.A. & Wade, M. J., 1994. Postcopulatory, prezygotic isolation: intraspecific and interspecific sperm precedence in *Tribolium* spp., flour beetles. *Heredity* 73: 155-159.

Reinhardt, K., 2007. Biology of the bed begs (Cimicidae). *Annual Review of Entomology* 52: 351 -374.

Rice, W.R., 1998. Intergenomic conflict, interlocus antagonistic co-evolution, and the evolution of reproductive isolation. In Endless forms: *Species and Speciation* pp. 161 – 270.

Roberts, M.M. & Grimmes, K.A., 1994. Histological evidence for five cell types in the male accessory glands of *Tribolium freeman* (Coleoptera: Tenebrionidae). *Tribolium Information Bull.* 34: 72-74.

Schlager, G., 1960. Sperm precedence in the fertilization of eggs of *Tribolium castaneum*. Ann. Entomol. Soc. Am. 53: 557-560.

Sevener, J.D., Dennard, N.N. & Grimnes, K.A., 1992. Histological and hitochemical evidence for an additional cell type in the male accessory reproductive glands of *Tribolium brevicornis* (Coleoptera: Tenebrionidae). *Tribolium Information Bulletion* 32: 93 – 95.

Shrode, R.R., 1960. Evidence that mating is random in *Tribolium castaneum. Tribolium Information Bulletin* 2: 15 -16.

Simmons, L.W., 2001. Sperm competition and its evolutionary consequences in the insects. Princeton University Press.

Sirot, L.K., Poulson, R.L, McKenna, M.C., Girnary, H., Wolfner, M.F. & Harrington, L.C., 2008. Identity and transfer of male reproductive gland proteins of the dengue vector mosquito *Aedes aegypti*: potential tools for control of female feeding and reproduction. *Insect Biochem. Mol. Biol*. 38: 176 – 189.

Sokoloff, A., 1974. The biology of *Tribolium castaneum* with special emphasis on genetic aspects. Oxford University Press.

Suarez, S.S., 2002. Fromation of a Reservoir of sperm in the oviduct. *Reproduction Domestic Animal* 37: 140 – 143.

Suzuki, T., 1980. 4-8-dimethyldecanal the aggregation pheromone of the flour beetle *Tribolium castaneum* and *Tribolium confusum* (Coleoptera: Tenebrionidae). *Agricultural and Biological Chemistry* 44: 2519 -2520.

Suzuki, T. & Sugawara, R. 1979. Isolation of an aggregation pheromone from the flour beetles *Tribolium castaneum* and *Tribolium confusum*. *Applied Entomology and Zoology* 14: 228 – 230.

Wolfner, M.F., 2002. The gifts that keep on giving: physiological functions and evolutionary dynamics of male seminal proteins in Drosophila. *Heredity* 88: 85-93.

Wolfner, M.F., Applebaum, S. & Heifetz, Y., 2005. Insect gonadal glands and their gene products. In: Gilbert L, Iatrou K, Gills S, Editos. Comprehensive insect physiology, biochemistry, pharmacology and molecular biology. Amsterdam: Elsevier. Pp. 179-212.

Xue, L. & Noll, M., 2000. Drosophila female sexual behavior induced by sterile male showing copulation complementation. *Proc. Natl. Acad. Sci. USA* 97: 3272 – 3275.

Yaginuma, T. & Happ, G.M., 1988. Trehalase from the male accessory gland and the spermatophore of the mealworm beetle, *Tenebrio molitor*. *J. Comp. Physiol.* B, 157: 765-770.

Yaginuma, T., Mizuno, T., Mizuno, C., Ikeda, M., Wada, T., Hattori, K., Yamashita, O. & Happ, G.M., 1996. Trehalase in the spermatophore from the bean-shaped accessory gland of the male mealworm beetle, *Tenebrio molitor*: purification, kinetic properties and localization of the enzyme. *J. Comp. Physiol.* 166: 1-10.

Chapter - 13

ANOPHELINE SPECIES COMPLEXES IN SRI LANKA: A BIO-ECOLOGICAL PERSPECTIVE

S.N. Surendran

Department of Zoology, Faculty of Science, University of Jaffna, Jaffna 40000, Sri Lanka E-mail: *noble@jfn.ac.lk*

ABSTRACT

There are twenty-two anopheline mosquito species reported so far in Sri Lanka. Among the reported anopheline species complexes in South and Southeast Asia *Anopheles culicifacies, An. subpictus, An. annularis, An. barbirostric* and *An. maculates* are present in the country. The Culicifacies and Subpictus complexes have been studied in detail as their members play a role in malaria transmission. In the Culicifacies complex species B and E have been detected in the country. All four sibling species (A-D) have been reported in the Subpictus complex. Although its major role in malaria transmission has been reported, sibling species status is yet to be identified for *An. annularis*. Since malaria is a public health problem in the country a comprehensive knowledge on the bio-ecology of members of anopheline species complex is essential when one or more of their member(s) play a role in malaria transmission.

INTRODUCTION

Sri Lanka is an island nation located in the Indian Ocean and is well known for its biological diversity. The country has a wide range of climatic, ecological and geographical zones that influence the prevalence of anopheline mosquitoes (Ramasamy & Ramasamy, 1993). There are three climatic zones based on annual rainfall (Figure 1). The wet zone receives more than 2,500 mm of annual rainfall during the Southwest and Northeast monsoons. The dry zone receives 2,000 mm of rainfall per year mainly during the Northeast monsoon. An intermediate zone that lies between the other two zones receives annual rainfall between 2,000-2,500 mm.

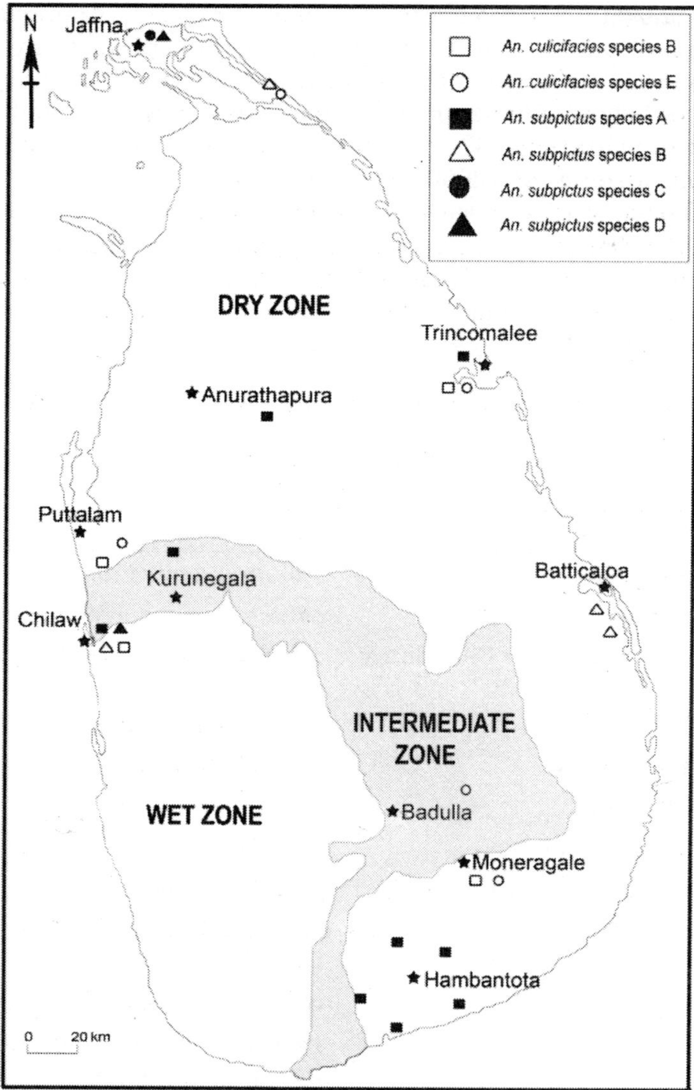

Figure 1. Climatic zones and distribution of sibling species in the Culicifacies and Subpictus complexes in Sri Lanka (Sources: Abhayawardana *et al.*, 1996; 1999 a&b; Kajatheepan and Surendran, 2006; Kannathasan *et al.,* 2008).

Anopheline mosquitoes are of prime importance among other mosquitoes as they act as vectors of human malaria. An update catalogue of the mosquitoes of Sri Lanka describes the presence of 140 mosquito species (Amerasinghe, 1991) with twenty-two belong to genes *Anopheles* (Amerasinghe, 1991) (Table 1). Density of anopheline mosquitoes were associated with monsoon rain (Ramasamy *et al.*, 1994; Konradsen *et al.,* 2000) and their prevalence is associated with rice fields, riverine system and irrigation projects (Ramasamy & Ramasamy, 1993; Konradsen *et al.,* 2000). There is a general tendency of increasing trend in the density of anopheline mosquitoes soon after the monsoon rain. Anopheline larval surveys carried out in association with irrigation projects in Sri Lanka revealed that the habitat change and niche reduction paved ways for the abundance of few anopheline species that found newly created conditions suitable for exploitations. This resulted in the immergence of new vector species (Konradsen *et al.*, 2000). Further, localized changes in the prevalence of different vector species sometimes occur consequent to ecological changes that accompany the rapid developmental changes in Sri Lanka. For example, *An. annularis* emerged as a potential vector species in newly built irrigation canals in the north-central province of the country towards the end of the last century (Ramasamy *et al.* 1992).

Table 1. Anopheline mosquitoes of Sri Lanka (Source: Amerasinghe, 1991)

Genus: *Anopheles*	
Sub genus: *Anopheles*	Sub genus: *Cellia*
aitkenii James (1903)	*aconitus* Donitz (1902)
barbirostris van der Wulp (1884)	*annularis* van der Wulp (1884)
barbumbrosus Strickland & Chowdhury (1927)	*culicifacies* Giles (1901)
gigas Giles (1901)	*elegans* James (1903)
interruptus Puri (1929)	*jamesii* Theobald (1901)
nigerrimus Giles (1900)	*karwari* James (1902)
peditaeniatus Leicester (1908)	*maculates* Theobald (1901)
peytoni Kulasekara et al. (1988)	*pallidus* Theobald (1901)
reidi Harrison (1973)	*pseudojamesi* Strickland & Chowdhury (1927)
	subpictus Grassi (1899)
	tessellates Theobald (1901)
	vagus Donitz (1902)
	varuna Iyengar (1924)

ANOPHELINE SPECIES COMPLEXES IN SRI LANKA

Mayr in 1942 coined the word sibling species to denote morphologically more or less similar and reproductively isolated populations within a taxon (Mayr, 1942). Sibling species of different Anopheline species complexes are known to exhibit different bio-ecological traits (Subbarao & Sharma, 1997; Coluzzi, 1992). In Southeast Asia 23 anopheline species have been identified as vectors of malaria. Among these nine species are reported to exist as species complexes (World Health Organization, 2007). Of these nine species *An. annularis*, *An.*

culicifacies, *An. subpictus*, *An. maculates* and *An barbirostris* are reported to be present in Sri Lanka.

(1) *Anopheles culicifacies* complex

Anopheles culicifacies Giles *s.l.* has a wide distribution in Asia (Subbarao, 1998; World Health Organization, 2007). This major vector species is widely distributed and present as a species complex comprising five members viz. A, B, C, D and E in the Indian subcontinent (World Health Organization, 2007). Species E is the new addition to this complex and has been reported in South India (Kar *et al.*, 1999). This species was first reported in 1999 in the Southern Indian island of Rameshwaram which is located ~32 km to the west of the coast of northern Sri Lanka and implicated as major vector of malaria (Kar *et al.* 1999). The different members of this complex show different bio-ecological characteristics, and species A, C, D and E are reported to be efficient vectors of malaria in the subcontinent (Subbarao & Sharma, 1997, Kar *et al.* 1999). In Sri Lanka only species B and E are reported to be present (Surendran *et al.* 2000).

Although species B and E of Sri Lanka appear to share considerable DNA sequence identity (Surendran *et al.* 2005), there is evidence to suggest that they show bio-ecological variations (Table 2). While species E appears to be numerically dominant in many parts of the country, both species are sympatric in all parts of Sri Lanka studied (Surendran *et al.* 2000). The collection of more species E than species B from cattle baited huts (a technique used to collect indoor feeding and resting mosquitoes), indicated that species E preferred to feed and rest indoors (Surendran *et al.* 2000). While this suggests that species E is likely to be anthropophagic, there is no definitive data yet on the relative anthrophagicity and zoophagicity of species B and E.

Limited studies on fecundity suggested that the mean fecundity of species E tended to be higher than that of species B (Surendran *et al.* 2002; Surendran *et al.* 2006). Investigations on longevity suggested that species E tended to survive longer than species B with nearly 16% of the field caught species E surviving long enough to support the extrinsic developmental cycle of *Plasmodium vivax* and *P. falciparum* (Surendran *et al.* 2006). The causes for the possible greater longevity of species E compared to species B are presently unknown.

Studies on the relative infectivity of human malaria parasites to species B and E in Sri Lanka showed that while species E could support the extrinsic cycle of *P. vivax* and *P. falciparum*, no species B isolated in Sri Lanka could be infected with either parasite through feeding on infected blood (Surendran *et al.* 2006). These observations are consistent with the poor vector capability of species B reported in India (Subbarao & Sharma, 19997) and Pakistan (Mahmood *et al.* 1984). Therefore, species E, and not species B, is the major malaria vector in Sri Lanka.

Both species B and E are completely susceptible to the pyrethroids λ-cyhalothrin and deltamethrin, and totally resistant to DDT in standard tests (Surendran *et al.* 2006). However, species E was found to be significantly more resistant to malathion than species B (Surendran *et al.* 2006). Specific causes for the different degrees for malathion resistance in species B and E have not been investigated, largely because it has not yet been possible to establish colonies of the two species or to obtain large enough numbers from field collections.

A recent study based on biochemical analysis aiming at determining resistance mechanisms among *An. culicifacies s.l.* populations collected from different parts of the country suggested that there may be different resistance mechanisms to the various common insecticides. A plausible explanation for the complexity of the resistance variation is the presence of sibling species with different resting and breeding preferences (Perera *et al.* 2008). Indoor resting sibling species will be under greater selection pressure due to indoor residual spray (IRS) and may develop resistance more rapidly than outdoor resting species.

Table 2. Bio-ecological characteristics of members of Culicifacies complex in Sri Lanka.

Characteristics	*An. culicifacies*	
	Species B	Species E
Prevalence (Surendran et al. 2000)	Sympatric	Sympatric
Breeding sites (Surendran and Ramasamy, 2005; Jude et al. 2010)	Rock pools, sand pools, quarries	Rock pools, sand pools, quarries, irrigation channels, brackish water pools, wells,
Resting and feeding places (Surendran *et al.*, 2000, 2006)	Indoor and outdoor	Indoor and outdoor
Vector potential (Surendran *et al.*, 2006)	Poor vector or non-vector	Vector of *P. vivax* and *P. falciparum*
Longevity (Surendran *et al.*, 2006)	Tendency to survive less than 3 ovipositions	Tendency to survive more than 3 ovipositions

Vector control strategies through water management were proposed based on the assumption that *An. culicifacies s.l.* only breeds in freshwater (van der Hoek *et al.* 1998). An investigation on larval breeding sites suggest that species E is able to exploit a wide range of breeding habitats with different limnological characteristics (Surendran & Ramasamy, 2005). This may be considered to be an adaptive variation in species E. A high density of species E was also found in open bound wells and unbound wells used for domestic purposes, and in irrigation channels (Surendran & Ramasamy, 2005). Species B was isolated from quarries created for domestic or economic purposes, and rock pools and sand pools in the river margins. However, none of the collected water samples contained both species B and E larvae suggesting the existence of local inter-species competition or niche segregation due to unknown environmental factors. A recent study from eastern coast of the country, which was badly affected by 2004 Asian tsunami, reveals the breeding ability of species E in brackish water (Jude *et al.* 2010). This adaptive ecological feature should be taken in to consideration as global warming is expected to produce more brackish waters in coastal area and thus poses a threat of malaria transmission.

(2) *Anopheles subpictus* complex

Anopheles subpictus Grassi *s.l.* is relatively important among the many secondary vectors of malaria in Sri Lanka, since its role in the transmission of *P. vivax* and *P. falciparum* has been reported from many parts of the country (Amerasinghe *et al.* 1992; Amerasinghe *et al.* 1999).

The taxon *An. subpictus* is reported to exist as a species complex comprising four members A, B, C and D in South India (Suguna, 1994; World Health organization, 2007). All four members can be distinguished by polytene chromosome banding patterns (Suguna *et al.,* 1994, World Health Organization, 2007) and stage specific morphometric characteristics (Suguna *et al.,* 1994). Initially, a study based on polytene chromosome banding sequence revealed the presence of species A and B in Sri Lanka (Abhayawardana *et al.,* 1996). Later studies based on morphometric diagnostic characteristics revealed the presence of all four sibling species in the country (Kannathasan *et al.* 2008; Kajatheepan & Surendran, 2006; Abhayawardana *et al.* 1999a).

The members of the Subpictus complex show different bio-ecological characteristics (Table 3). While species B, C and D were detected in the Jaffna district of the Northern Province (Kannathasan *et al.* 2008; Kajatheepan & Surendran, 2006), the presence of all four members was reported from North-Western Province of the country (Abhayawardana et al. 1999a; Abhayawardana *et al.* 1999b). Species B was predominant in East coast of the Jaffna district and the West coast of the Chilaw district in the North Western Province. Species C was predominant in inland localities of Jaffna and Chilaw districts (Abhayawardana *et al.* 1999a&b; Kannathasan *et al.* 2008; Kajatheepan & Surendran, 2006). Density of species A and D in the coastal areas were found to be low in comparison with the inland areas (Abhayawardana *et al.* 1999a&b).

Among the siblings, species B prefers to feed and rest outdoors while species A and C prefers indoors (Abhayawardana *et al.* 1999b). However species D does not show any significant difference in feeding and resting preference (Abhayawardana *et al.* 1999a&b). Peak abundance of sibling species C in inland has been observed in November and January while species B is abundant in January, December, April and July in coastal areas of the Chilaw district (Abhayawardana *et al.* 199a). Although indoor resting siblings can be inferred to prefer feeding on humans (anthropophagic) and similarly outdoor resting sibling species to prefer feeding on animals (zoophagic), the feeding preferences of the different *An. subpictus* sibling species remain to be definitively established.

Although *An. subpictus s.l.* has been incriminated as a vector of *P. vivax* and *P. falciparum* on many occasions (Konradsen *et al.* 2000), there is no confirmatory information on which sibling species is involved. A limited study from North-Western Province (Chilaw district) identified a single species C mosquito with sporozoites (Amerasinghe, 2001). Hence there is a need to establish the relative potential of the different sibling species of *An. subpictus* to transmit malaria in Sri Lanka.

Table 3. Bio-ecological characteristics of members of Subpictus complex in Sri Lanka

Characteristics	*An. subpictus*			
	Species A	Species B	Species C	Species D
Prevalence (Abhayawardana *et al.* 1999a,b; Kannathasan *et al.* 2008)	Sympatric	Sympatric	Sympatric	Sympatric
Breeding sites (Abhayawardana *et al.* 1999a,b; Kajatheepan and Surendran, 2006)	Inland	Coastal	Inland	Inland
Resting and feeding places (Abhayawardana *et al.* 1999a,b)	Indoor	Outdoor	Indoor	Indoor and outdoor
Peak abundance (Abhayawardana *et al.* 1999a,b)	Not known	January, April, December and July	November and January	Not known
Vector potential (Amerasinghe, 2001)	Not known	Not known	Vector	Not known

In general, *An. subpictus s.l.* populations are reported to be highly resistant to DDT and malathion throughout the country. A detailed analysis based on resistance data reported from many parts of Sri Lanka, during the period 1991–2003, shows a stratified (coastal and inland) resistance pattern for *An. subpictus s.l.* that could be attributed to the distribution of sibling species (Kelly-Hope *et al.,* 2005). A study in the Jaffna district revealed that species B, C and D were highly susceptible to malathion and that species B and D were highly resistant to DDT (Kannathasan *et al.,* 2008). The results of a recent study based on biochemical analysis for resistance mechanisms of *An. subpictus s.l.* populations collected from different malarious areas revealed the presence of heterogeneity of resistance to common insecticides except DDT and malathion to which all the populations are resistant (Perera *et al.* 2008).

(3) *An. annularis* complex

The taxon *An. annularis* van der Wulp comprises two members namely A and B in India. Although both species are sympatric in many parts of India, sibling species A is in association with malaria transmission (Atrie *et al.* 1999; Tauqeer *et al.* 2007). *Anopheles annularis,* which breeds in the newly constructed irrigation canals in the dry zone of north-central Sri Lanka, functioned locally as the most important malaria vector with a very high entomological inoculation rate of 0.12 infective bites per hour during a peak transmission season (Ramasamy *et al.,* 1992). However, sibling species status of this taxon in Sri Lanka is unknown.

(4) *An. barbirostris* complex

The taxon *An. barbirostris* van der Wulp is reported to exist as three cytotypes viz. A, B and C in Southeast Asia. Species B and C are highly anthropophagic and associated with filarial and malaria transmission (World Health Organization, 2007). Although specimens identified as *An. barbirostris* in the dry zone of Sri Lanka were detected with sporozoites (Konradsen, *et al.* 2000), sibling species status of this taxon is unknown

(5) *An. maculates* complex

There are nine members with differential vector potential and ecological characteristics present in this complex. *Anopheles maculates* Theobald *s.s.* is composed of three chromosomal forms namely B, E and F. The Sri Lankan populations are reported to *An. maculates s.s.* (B form) which is reported to be a non vector (World Health Organization, 2007). So far no specimens of *An. maculates* in the country have been detected for the presence of malaria parasites

CONCLUSIONS

The available evidence suggests the presence of sibling species with different eco-biological traits in the *An. culicifacies* and *An. subpictus* complexes of Sri Lanka. Therefore knowing the distribution of the different sibling species in different parts of the country is important for formulating effective malaria control measures based on vector control. Other biological properties of the different sibling species need to be more firmly established as these also have an impact on the adoption of appropriate vector control measures.

Apart from the two major malaria vector species, identified secondary vectors such as *An. annularis* and *An. barbirostris*, and *An. maculates* also exist as species complexes, and similar considerations apply to distinguishing the sibling species and studying the bio-ecological parameters of their sibling species.

ACKNOWLEDGEMENT

Assistance of N. Prasanthan and P.J. Jude (University of Jaffna) is gratefully acknowledged.

REFERENCES

Abhayawardana, T.A, M.B. Wickramasinghe & P.H. Amerasinghe, 1999a. Indoor resting behaviour of different sibling species of *Anopheles subpictus*. *Proc. Sri Lanka Assoc. Advan. Sci.* 55: 18.

Abhayawardana, T.A., M.B. Wickramasinghe & F.P. Amerasinghe, 1999b. Sibling species of *Anopheles subpictus* and their seasonal abundance in Chilaw area. *Proc. Sri Lanka Assoc. Advan. Sci.* 55: 17.

Abhayawardana, T.A., S.R.E. Wijesuria, R.R.K.C. Dilrukshi, 1996. *Anopheles subpictus* complex: distribution of sibling species in Sri Lanka. *Indian J. Malariol*. 33: 53–60.

Amerasinghe, P.H., 2001. Malaria vectors in Sri Lanka. *Proceedings of the Workshop on Malaria Risk Mapping-Implications for its use in control*, pp. 20-23. International Water Management Institute, Sri Lanka.

Amerasinghe, P.H., F.P. Amerasinghe, F. Kondradsen, K.T. Fonseka & R.A. Wirtz, 1999. Malaria vectors in a traditional dry zone village in Sri Lanka. *Am. J. Trop. Med. Hyg*. 60: 421-29.

Amerasinghe, P.H., F.P. Amerasinghe, R.A. Wirtz, N.G. Indrajith, W. Somapala, L.R. Preira & A.M.C. Rathnayake, 1992. Malaria transmission by *Anopheles subpictus* Grassi in a new irrigation project in Sri Lanka. *J. Med. Entomol*. 29: 577-81.

Amerasinghe, F.P., 1991. A catalogue of the mosquitoes of Sri Lanka. Natural Resources, Energy and Science Authority of Sri Lanka, Colombo.

Atrie, B., S.K. Subbarao, M.K.K. Pillai, S.R.V. Rao & V.P. Sharma, 1999. Population cytogenetic evidence for sibling species in *Anopheles annularis* (Diptera: Culicidae). *Ann. Entomol. Soc. Am*. 92: 243-249.

Coluzzi, M., 1992. Malaria vector analysis and control. *Parasitol Today*.8:113-18.

Hackett, L.W., 1937. *Malaria in Europe*. pp. 47-84. Oxford University Press. Humphrey Milford, London.

Jude, P.J., S. Dharshini, M. Vinobaba, S.N. Surendran & R. Ramasamy, 2010. *Anopheles culicifacies* breeding in brackish waters in Sri Lanka and implications for malaria control. *Mal. J*. 9:106

Kajatheepan, A. & S.N. Surendran, 2006. Prevalence and insecticide resistance of sibling species B in the *Anopheles subpictus* complex in tsunami affected Vadamarachchi-east. *Proc. Jaffna. Sci. Assoc.* 14: 55

Kannathasan, S., A. Antonyrajan, K.A. Srikrishnaraj *et al.,* 2008. Studies on prevalence of anopheline species and community perception of malaria in Jaffna district, Sri Lanka. *J. Vector Borne Dis.* 45: 231-239.

Kar I., S.K. Subbarao, A. Eapen, J. Ravindran *et al.,* 1999. Evidence for a new vector species E within the *Anopheles culicifacies* complex (Diptera: Culicidae). *J. Med. Entomol.* 36: 595-600.

Kelly-Hope L.A., A.M.G.M. Yapabandara, M.B. Wickramasinghe, M.D.B. Perera and S.H.P.P. Karunaratne *et al.,* 2005. Spatiotemporal distribution of insecticide resistance in *Anopheles culicifacies* and *Anopheles subpictus* in Sri Lanka. *Trans. R. Soc. Trop. Med. Hyg.* 99: 751–61.

Konradsen, F., F.P. Amerasinghe, W. van der Hoek & P.H. Amerasinghe, 2000. Malaria in Sri Lanka: Current knowledge on transmission and control. Colombo, Sri Lanka. International Water Management Institute.

Mahmood F., R.K. Sakai & K. Akhtar, 1984. Vector incrimination studies and observations of species A and B of the taxon *Anopheles culicifacies* in Pakistan. *Trans. R. Soc. Trop. Med. Hyg.* 78: 607-16.

Mayr, E. 1948. The bearing of the new systematics on genetical problems the nature of species. *Advanced Genetics* 2: 205-36.

Perera, M.D.B., J. Hemingway & S.H.P.P. Karunaratne, 2008. Multiple insecticide resistance mechanisms involving metabolic changes and insensitive target sites selected in anopheline vectors of malaria in Sri Lanka. *Mal. J.* 7: 168

Ramasamy, R., R. de Alwis, A. Wijesundera & M.S. Ramasamy, 1992. Malaria transmission at a new irrigation project in Sri Lanka: The emergence of *Anopheles annularis* as a major vector. *Am. J. Trop. Med. Hyg.* 47: 547-553.

Ramasamy, R. & M.S. Ramasamy, 1993. Changes in malaria vector prevalence in Sri Lanka: the need for local surveillance. *Ceylon Med. J.* 38, 43-46.

Ramasamy, M.S., R. Kulasekara, K.A. Srikrishnaraj & R. Ramasamy, 1994. Population dynamics of anthropophilic mosquitoes during the northeast monsoon season in the malaria epidemic zone of Sri Lanka. *Med. Vet. Entomol.* 8: 265-274.

Subbarao, S.K. & V.P. Sharma, 1997. Anopheline species complexes and malaria control. *Indian J. Med. Res.* 106: 164-73.

Suguna, S.G., K.G. Rathinam, A.R. Rajavel & V. Dhanda, 1994. Morphological and chromosomal descriptions of new species in the *Anopheles subpictus* complex. *Med. Vet. Entomol.* 8: 88–94.

Surendran, S.N. & R. Ramasamy, 2005. Some characteristics of the larval breeding sites of *Anopheles culicifacies* species B and E in Sri Lanka. *J. Vector Borne Dis.* 42: 39-44.

Surendran, S.N., T.A. Abhayawardana, B.G.D.N.K de Silva, M.S. Ramasamy & R. Ramasamy, 2000. *Anopheles culicifacies* Y chromosome dimorphism indicates the presence of sibling species (B and E) with different malaria vector potential in Sri Lanka. *Med. Vet. Entomol.* 14: 437–40.

Surendran, S.N., B.G.D.N.K. de Silva, M.S. Ramasamy & R. Ramasamy, 2002. Comparative fecundity and age composition of two sibling specie of *Anopheles culicifacies* (Diptera: Culicidae) complex in Sri Lanka (abstract). *Proc. Sri Lanka Assoc. Advan. Sci.* 58:154.

Surendran, S.N., N.J. Hawkes, A. Steven, J. Hemingway & R. Ramasamy, 2006. Molecular studies of *Anopheles culicifacies* (Diptera: Culicidae) in Sri Lanka: sibling species B and E show sequence identity at multiple loci. *Eur J. Entomol.* 103: 233-37.

Surendran, S.N., M.S. Ramasamy, B.G.D.N.K. De Silva & R. Ramasamy, 2006. *Anopheles culicifacies* sibling species B and E in Sri Lanka differ in longevity and in their susceptibility to malaria parasite infection and common insecticides. *Med. Vet. Entomol.* 20: 153–6.

van der Hoek, W., F.P. Amerasinghe, F. Konradsen & P.H. Amerasinghe, 1998. Characteristics of malaria vector breeding habitats in Sri Lanka: Relevance for environmental management. *Southeast Asian J. Trop. Med. Public Health.* 29: 168-72.

World Health Organization (2007). Anopheline species complexes in South and South-east Asia. SEARO Technical Publication No. 57.

Chapter - 14

ECOBIOLOGY AND MALE DIMORPHISM IN RHINOCEROS BEETLE *ORYCTES RHINOCEROS* L. (COLEOPTERA: CARABAEIDAE)

Mariamma Jacob and Mathew M. Oommen

Department of Zoology, University of Kerala,
Kariavattom P.O. Thiruvananthapuram - 695581, Kerala, India
E-mail: mjacob1953@yahoo.com; Mathew_M_Oommen@hotmail.com

ABSTRACT

Oryctes rhinoceros L. popularly known as the coconut rhinoceros beetle, is a major pest of coconut palm *Cocos nucifera* L. The adults pierce the crown of the palm and eat away the growing meristem and cause damage to the palm. The breeding sites in which the immature stages develop are widespread, and often difficult to remove or destroy. The larvae are harmless and usually seen in decaying palm tops, on decaying logs on ground, in cow dung pits or compost pits, feeding on the decaying organic matter. There are three larval instars, and the last (III instar) stage show wide variation in time of molting (45-130 days). The male and female adults are of varying sizes. In the big sized males (major males) a prominent horn is present, with its tip directed backward, whereas in females the horn is not significant as observed in males. Medium sized males (minor males) having small or no horn are also present in a population and they mimic similar sized females. Laboratory studies conducted by the first author give evidence that minor males emerge from type I third instar larva which molt after 45 days of development and the major males emerge from type II third instar larva which transforms to pupa after 130 days. The I and II types of larvae have two timings of endocrine mechanism for molting. The ecological conditions, their biology, the laboratory observations of factors influencing male dimorphism, the significance of male dimorphism in a colony, the role of dimorphic forms in reproductive strategies and the evolutionary significance are discussed in the present review.

INTRODUCTION

Arthropods are ubiquitous in distribution and are most common than any other animals. It is not surprising, therefore, that more than three fourth of animal species come under the phylum Arthropoda. They encompass a great diversity of animal taxa known from the Cambrian to the present day (Budd & Telford 2009). The five living groups are Onychophora, Myriapoda, Insecta, Crustacea, and Arachnida. They share a set of distinctive features, more notably the clear segmentation of their bodies, a sclerotised cuticle and joined appendages. Among arthropods, insects occupy a dominant position, outnumbering all present day living organisms. Their tenacity for life is amazing, and their capacity for multiplication and adaptation have made them to occupy the top level in animal kingdom. This has enabled them to migrate and disperse, apart from being able to discover their mates, seek food and escape from enemies. Their ability to live in extremes of environmental conditions is most noteworthy, and it is this ability for adaptation that has made it possible for them to invade and colonize every nook and corner of the world. Special adaptations for reproduction are present in most insects, and the embryonic and postembryonic stages of insects are also well adapted to their environments.

The environment of an insect population naturally includes the abiotic and biotic factors, as in the case of any other animals or plants. Natural populations have well defined properties, which include their density or degree of crowding in a unit of space, birth rate or natality, or population decline factor, genetic make up, dispersion or distribution, and space range or territoriality. Smith (1935) introduced the term density - dependent and density-independent factors for the stability of a population. The combined action of density independent factors like weather, imperfectly density dependent factors like interspecific competition, and perfectly density dependent factors including intraspecific competition for food and space influence the stability of a population. Individuals in a population vary in size, structure, physiology, and fecundity and behavior. As these variations are genetically induced, the actions of environmental factors tend to be selective, and in course the population gets adapted to the environmental stresses. It is this adaptation that provides a mechanism tending to maintain equilibrium between the population and the environment. Reproductive potential influences natality and the average life span of individuals of a population under ideal conditions. Intra- and interspecific competition in populations play an important role in determining the kind and number of organisms. Some restrict their area of search for food, the home range and territoriality, though common in birds, also occur in insects. Such behavioral patterns serve to regulate population (Nayar *et al.,* 1976).

Among insects, Coleoptera comprises beetles and weevils having extremely varied habitats, wide distribution and countless adaptations, and have occupied a dominant position in animal kingdom. The heavily sclerotized head and appendages present several variations in shape and size. They are very large to medium sized, smooth, shiny, brilliantly coloured, and sometimes sculptured. Horns and tubercles are present on head and pronotum which exhibit sexual dimorphism (Nayar *et al.*, 1976).

Male dimorphism and trimorphism are reported for dung beetles of the family Scarabaeidae (Emlen, 1997; Rowland & Emlen, 2009). It is interesting to notice that many of the dung beetles have two or three male phenotypes, and they can adopt alternative reproductive tactics. In most of these species males are of two types. One is of large size and horns (major males) that guard females, and the other is of smaller size (minor males) that lacks horns, sneaks copulation. Major males use their large body size and large horn size to guard entrances to burrows containing females from rival males. The minor males sneak into these tunnels either by slipping past the guarding males or by digging side tunnels that intercept tunnels beneath the guarding males (Emlen, 2000; Moczek & Emlen, 2000; Rowland & Emlen, 2009).

The Rhinoceros beetle, Oryctes *rhinoceros* Linnaeus (Coleoptera: Scarabaeidae: Dynastinae) is one of the serious and important pests of coconut palm *Cocos nucifera* Linnaeus, and has a wide distribution in all coconut growing countries. The adult beetles cause severe damage by feeding away tender parts of the palms. They live in crevices between leaf sheaths near the crown and burrows into softer portions. The tender fronds in the folded condition are cut across and when unfolded they bear characteristic clippings or holes in the leaflets. Such repeated attacks cause stunting of palms. Injury to spathe is also common resulting in loss of nuts. The larvae spend their lifetime in decaying organic matter either in the ground or on dead standing palm. *Oryctes rhinoceros* exhibits sexual dimorphism but studies on male dimorphism are scanty and the investigations for the causes of male phenotypes are less. This review describes the ecology and biology of *O. rhinoceros*, and the probable factors influencing dimorphism in males as observed through laboratory investigations in the light of endocrinological, genetic aspects, and evolutionary significance.

ECOLOGY

According to Nirula (1955) the earliest record of damage to coconut palm by *O. rhinoceros* was in 1889 by Riddley in the Straits Settlements. Wide ecological distribution is reported for *O. rhinoceros* (Bedford 1976, 1980). *Oryctes rhinoceros* as a pest occur in coconut growing regions of Asia (Catley, 1969) spread from West Pakisthan, through India, the Maldive Islands, Ceylon, Hainan, Taiwan, Hong Kong, Thailand, Vietnam, the Malayan peninsula, the islands of Java, Sumatra, Bali, Lombok, Kalimantan, Celebes, Ceram and Amboina in Indonesia, to the Philippine islands (Zelazny, 1977; Zelazny & Alfiler, 1991). In Burma, the pest first appeared in the extreme south of the peninsula. It probably entered from Malaysia, spread to North throughout the coconut growing areas of lower Burma (McKenna & Shroff, 1911). It was accidentally introduced to a number of copra-producing areas of the Pacific and Indian Oceans. It is believed to have been introduced in rubber seedling Pot plants from Ceylon to the Pacific islands of Upolu, Western Samoa in 1909 (Jepson, 1912) from where it moved to the neighbouring island in American Samoa. In 1921 the beetle was recorded in Keppel Island in the Kingdom of Tonga, but found to be eradicated in 1930 but again reported from Tonga in 1952, (Dumbleton, 1952). Wallis Island about 320 km west of Samoa became infested in 1931 (Cohic, 1950).

Oryctes rhinoceros was introduced to the Palau Islands by 1942 (Gressitt, 1953). It reached New Britain by 1942, and subsequently spread to New Ireland in 1952, West Iran, Pak Island in 1960, and Manus Island (New Guinea) in 1960 (Catley, 1969). The insect was found in main islands of Fiji early in 1953 and spread in many places in Fiji (Bedford, 1980). In the Indian Ocean, the island of Diego Garcia became infested in the First World War (Orian, 1959). The beetle was found in Mauritius (Vinson, 1963) and in The Reunion in 1978 (Monty, 1978).

In India, the pest occurrence was reported from coconut growing centers in South India especially in Kerala (Cherian & Ananthanarayanan, 1939, Kurian & Pillai, 1964; Catley, 1969). They are now present in all coconut growing places in the country, like Tamil Nadu, Karnataka, Maharashtra, West Bengal, Assam, Orissa and Madhya Pradesh (Khare, 1919), from West Punjab (Ahmad & Ullah, 1951) and from Andhra Pradesh (Indiravathi *et al.*, 2001). *Oryctes rhinoceros* was reported as a serious pest of coconut palm in the Andaman, Nicobar and Laccadive Islands (Nirula, 1955; Jacob & Bhumannavar, 1991).

Adults of *O. rhinoceros* are found to occupy two kinds of habitats namely crown of the palms or rarely stems of other crop plants for feeding and dumps of decaying organic matter for breeding purposes. Decaying trunks of coconut palm and heaps of cattle dung (undisturbed and semi-dry) are the major centers of breeding activity in India (Cherian & Ananthanarayanan, 1939; Nirula, 1955; Kurian & Pillai, 1964; Catley, 1969; Monty, 1978). Dead standing coconut palms destroyed either by beetles or by lightening of felled coconut logs are found to be major breeding centers, *e.g.,* decaying piles of coconut leaves (Nirula, 1955), rotting heaps of coconut husks (Goonewardena, 1958), damp decomposing choir refuse (Cherian & Ananthanarayanan, 1939), decaying trunks of banana (Nirula, 1955; Sison, 1957), rotting rubber stumps (Corbett, 1938), rotting heaps of saw dust (Sison, 1957; Goonewardena, 1958; Catley, 1969), sugarcane thrash (Cherian & Ananthanarayanan 1939; Monty 1978).

Eggs are laid in the breeding medium and the larvae feed on the decaying organic matter and pupate in the medium in a small case and the young adults after two weeks will come to the surface and fly to palms for feeding. They reach a decaying medium for breeding and after egg laying they move away for feeding. They move between habitats by flight at night.

Although coconut palm is the principal host species, a wide variety of other palms and crop plants are also found to be infested by *O. rhinoceros.* According to Goonewardena (1958) there are 29 host genera for *O. rhinoceros* and 16 among them are palms like, *e.g.*, Royal palm, Buri palm, Latanier palm, the Raphia palm, Fishtail palm, Oil palm, the Palmyra palm and the Toddy palm. *Oryctes rhinoceros* commonly attacks them. This beetle rarely attacks Talipot palm, Nipa palm, Nibong palm and Sago palm. The Arecanut palm has also been reported as an alternative host of this pest in Malaya (Corbett, 1938).

BIOLOGY

Life history of *O. rhinoceros* has been reported by many authors (Cherian & Ananthanatrayanan, 1939; Gressitt, 1953; Nirula, 1955; Goonewaerdena, 1958; Kurian & Pillai, 1964; Catley, 1969; Bedford, 1976, 1980) who have shown that the size and duration of life stages are liable to variation and is related to its environment. Climatic and nutritional conditions affect larval duration and adult size. Factors like temperature, humidity and moisture content in the substratum where they grow play a major role in life cycle. In our laboratory studies (1984 to 2007), larvae were collected mainly from local manure pits, decaying vegetable debris, man made vermicompost pits, decaying standing coconut palm and from palm fallen on ground. They were washed in running tap water and dipped in mild alcohol (to remove mites and other parasites) and were reared in sterilized cow dung. Individual plastic containers (8 cm diameter x 9 cm height, closed with lids having holes) were used because the larvae showed cannibalism. Adults were reared in bigger glass containers (17 cm diameter and 30 cm height) with holes on lid, half filled with cow dung and decaying leaves and food given was ripe banana or sugarcane bits. Cow dung is collected from 1 to 1.5 meter deep pits as the larvae in natural condition occupy decaying cow dung. Usually larval stages of *O. rhinoceros* are occupying this level, third instar larvae towards pupation go deep and finds a place where sand and cow dung mix, form a small case by the fast movement of non-feeding larva and prepupa and pupate inside this case. Details of eggs and other stages from laboratory studies are reported. The adults and larvae were reared under controlled laboratory conditions like 75-90% relative humidity, temperature 26-32°C and approximately 12:12 photo-period (Jacob, 1989a, 1989c; Priyaranjini, 2003; Salini, 2003).

Eggs (Fig. 1)

These are creamy white when freshly laid, oblong, measuring 2- 2.3 mm x 1-1.5 mm size (Table 1). Incubation period lasted for 8-10 days, as the days are advancing colour changes to light brown. At the time of hatching, the eggs are about 3-3.4 mm x 2-3 mm with almost an oval shape. According to Goonewardena (1958) the freshly laid eggs are of still bigger size (3-4mm x 2-3mm) and incubation period was 8.72 days and Kurian and Pillai (1964) Nirula (1955) recorded it as 10 days. However, Catley (1969) reported as 12 days whereas Cherian & Ananthanarayanan (1939) as 9-14 days and Monty (1978) and Sison (1957) observed it as 15 days. Bedford (1980) gave a conclusive report from observation of many authors regarding the incubation period of eggs of *O. rhinoceros* as 8-12 days.

There are three larval instars for *O. rhinoceros*. The larvae are found to dig and burrow interior of any medium to avoid light. This is very specific with last (III instar larva). Bedford (1980) has reported that there exists negative phototaxis to larvae of Oryctes species and this is an adaptation against desiccation and predation.

I instar larva (Fig. 2)

Body is too delicate, creamy white in colour including head at hatching, but within 3-4 hours of hatching head turns red. The body weight increases from 0.103 to 0.412 g from 1day to five day, length from 10 to 25 mm, body width 2-2.5mm with head capsule width 1.5 -2 mm, respectively. Finally they reach about 2 g weight, body size 37 - 40 mm x 5-6mm and 3mm head width (Priyaranjini, 2003). They continue in first instar for 20 to 21 days and after one day's non-feeding stage moults to 2nd instar larva. Catley (1969) reported that first instar lasted for 12-18 days while Bedford (1980) recorded the duration as 12-21 days. The variation can be due to the changes in the environment where they are living.

II instar larva (Fig. 3)

Body is slender, having a dull white colour longer than first instar larva as they grow, the abdominal region become curved and darker as the cow dung is accumulating in the pouch of hindgut. Newly moulted larvae are about 2 g weight, 40 mm x 6 mm size and head width is 3.5mm, which reached 5mm by 10 days. They continue feeding for 19 –20 days, undergo non-feeding stage for a day and then moult to third instar on 20th or 21st day reaching almost 3 g weight and 53-55 mm x10-12 mm size (Priyarenjini, 2003).

Figure **1.** Eggs of *Oryctes rhinoceros;* **2.** I Instar larva (1day old); **3.** II instar larva (10 days old); **4.** III instar larva (70 days old); **5.** Prepupa (2 days old); **6.** Pupa, female and male (10 days old)

III instar larva (Fig. 4)

This is the final instar stage and they are very active feeders. 1-5 day old larva has 3-3.9 gram weight, 54 -68 mm length and width of 10- 11 mm and head width increased from 6- 9mm (Salini, 2003) (Table I). One-month-old larva has about 9 g weight, 100-110mm length, 15-18mm width and 10mm head width. It is noticed that 35-40 days old ones which crossed 9 g weight, undergo non-feeding stage for 5 to 6 days and start pre-pupation (Type I as described below). However, others feed voraciously and increase in size. The body weight reaches to more than 16g so also the body width to 20-25mm, whereas the length of the body (110-112 mm) has no significant increase and the head width (10 mm) remained the same. These larvae are very active feeders till 90 days. They live for a total of 130 days (Siny, 2009) in third instar stage. Feeding rate is reduced from 90 to 100 days, and lead a non-feeding stage for 20 to 30 days and then transform to pre-pupa (Type II as described below). A full-grown larva is creamy white due to the massive accumulation of fat body. The fat body shows specific arrangement in the body and is visible due to the presence of ingested cow dung in the abdomen. Wide variability in the duration of third instar larvae is noticed in earlier studies also. Goonewardena (1958) reported it as 85 days, Hurpin & Fresneau (1967) as 167 days, Catley (1969) as 90-120 days and Bedford (1976) as 60 days. According to Catley (1969) and Bedford (1980) unfavorable and nutritional conditions delayed larval development and smaller adults are resulted.

Third instar larvae are different from the I and II instar larvae for its voracious feeding rate, its size, head width, distinct bean shaped chitinous plate around spiracles, very specific wing disc just above the first spiracle (only one thoracic spiracle seen on the first thoracic segment is bigger than the rest 8 abdominal spiracles) and male genital disc. Male larvae can be clearly identified from female by the sclerotized triangular mark at mid-ventral region on the 9th sternite;- a chitinous plate in male genital disc inside the body (Jacob, 1989a). In the last (III) instar larvae, the mark is very clear to the naked eyes because of the size of the larva and also due to the accumulation of creamy white fat more at the ventral side of abdomen. The chitinous plate in the genital disc is visible in the I and II instar but needs magnification. During pupal moult, this sclerotised plate on the genital disc will be expelled along with the larval exuvium, cast off by prepupa.

Pre pupa (Fig. 5)

During the non-feeding stage, the III instar larvae which are located at the bottom of the rearing bottle will move around a specific site in the cow dung. Continuous movement in the cow dung hardens that particular region. In addition to this, larva ejects a blackish secretion that comes from the mouth; act as sticky substance to hold the cow dung together in the form of a case around the larva. Further development of the larva is inside this case, away from external pressures especially from enemies. Towards the close of these movements, the weight and the length of the body is reduced (Table 1) and the body segments showed wrinkles. Thus the larva enters a prepupal stage. They have a rhythmic up and down motion initially with the head region, while abdomen is free. But later the head and the abdomen will move together and the body curls and then relaxes. The movements by prepupa give a polished appearance to the interior of the case. If disturbed, they show fast movements with head and abdomen. Prepupal stage lasts for 6-8 days (Jacob, 1989a). In the initial stage, they are dull creamish in color but towards pupal emergence they appeared dirty brown as the pharate pupa is growing inside. The movement of the larvae and succeeding case formation, and prepupal and pupal life can be clearly seen when they are reared in glass bottles kept in the dark. If they are kept in dry cow dung or in sand or among dry leaf litter during wandering stage, they abstain from making a case, and continue normal development, but if disturbed they moult as larval-pupal intermediates.

Pupa (Fig. 6)

It is wheatish when comes out of prepupal case turns to a light brick coloured one entirely different from the larva or adult. Within four to five hours, colour changes to brick red. They live in the cow dung case created by prepupa and live up to 16-18 days, the colour changes to brownish red (Jacob, 1989a,c). The weight of the pupa at emergence is 5-6 g reaching 9-10 g towards adult emergence, length varied from 34 mm to 51mm and width15-25 mm. Usually they are immobile but at times they show jerking movements at the abdomen. Male pupae of the second type of III instar larvae have a specific horn. Male and female pupae are identified by the arrangement of last abdomen sternites (Mini & Prabhu, 1988).

Adult (Fig. 7)

Adult beetles are 30-55 mm long 15-22 mm in width with 3-4 mm head width. At emergence the head is reddish brown, but the body is wheatish and delicate. Within 3-4 hours the body turns reddish brown and after three to four weeks, to blackish brown and stout. Males and females clearly show sexual dimorphism. Among big sized adults with 40-55 mm in length, males (major males) possess a characteristic horn on head projecting dorsally upward from the head. The horn is 5-8 mm long and the tip is directed backwards in male but is short in female. Females usually have only a triangular extension on head but if they are bigger in size (40-50mm long), akin to a hose, tip is straight and narrow and not significant as in males. The second visible character is the shape of pygidium (last abdominal segment) and the arrangement of hairs on it. In males, pygidium is almost round and clearly devoid of hairs. The hairs are arranged around the genital opening/anus on ventral side. However, in females, pygidium is of conical shape and is clothed with a lot of hairs. It is of interest to notice that there are medium sized adults (minor adults, 3- 3.8 cm long) males and females (Fig. 8, 9) having a small or no horn. The pygidium is either round or conical but devoid of hairs. Thus the minor males mimic females in their size, nature of horn and also in some, and the shape of pygidium. Both types of adults (minor and major) breed in manure pits, undisturbed decaying organic debris, and dead coconut logs etc. Females lay eggs at the bottom of the substratum. A single female lays 20-24 eggs at one time. The emerged adults break open the case and move at the bottom of the bottle and move to the surface of the cow dung after two weeks, if the bottle is kept open they will fly away at night probably to coconut palm for feeding.

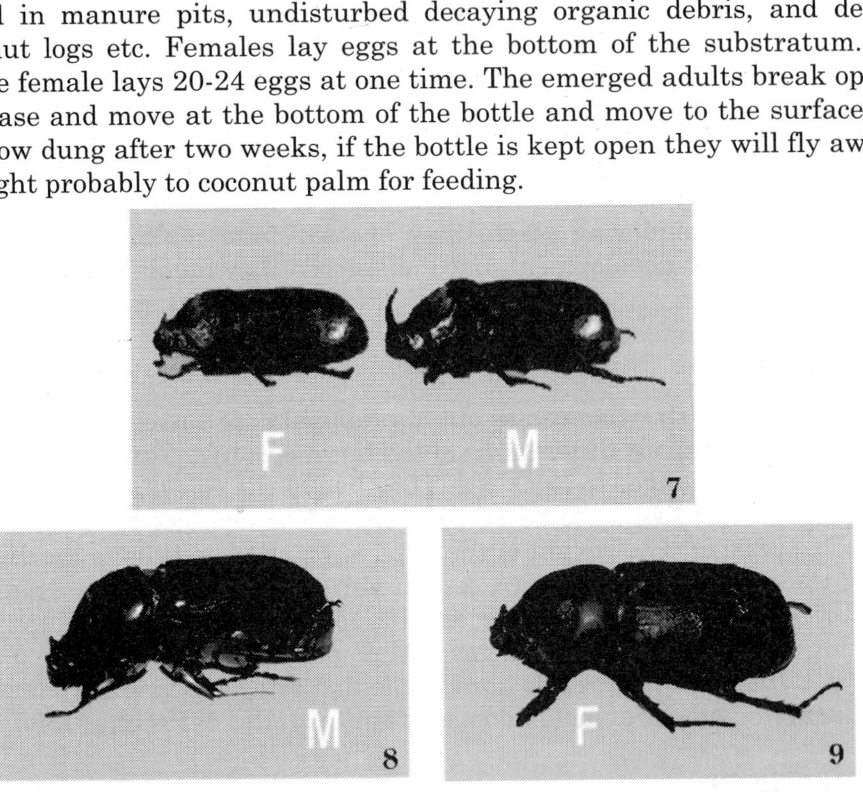

Figure **7.** Adults, major female & male; **8.** Minor male (magnified to show small horn); **9.** Minor female (magnified to show horn)

In our laboratory studies, the adults are found to take food after two weeks and are reproductively mature after one month. Medium sized adults (minor group) live up to 3 months, whereas the bigger sized males and females (major group) are active for 6months and may continue 20-25 days more but most of them lead an inactive period, may be because they are under captivity. Most of the early reports show that the adult life span is 4-7 months (Cherian & Ananthanarayanan, 1939; Goonewardena, 1958; Kurian & Pillai, 1964; Catley, 1969). The healthy adults were found to live up to 9 months (Bedford, 1980). Regression of reproductive system was noticed in many field collected (either in light traps or people catching from coconut fronds) blackish brown major males (probably aged ones) brought to our laboratory (Jacob, 1989c). They were very active with an intact copulatory organ and swollen spermatophore sac with live spermatozoa, but they have regressed testes and associated ducts, narrow digestive tract and with reduced body weight.

Table 1. Life stages of *O. rhinoceros* with morphometry (1) Early 1-5 days, in I, II and III instar larvae (2) Late 20-21 in I and II instar, and 36-40 day old III instar (3) 90-100 days old III instar larva

Stages	Duration (days)	Body weight (gm)	Body length (mm)	Body width (mm)	Head width (mm)
Eggs	8-10		0 day 2- 2. 3	0 day 1- 1.5	
			8 day 3-3.4	8 day 2- 3	
LI	20-21	(1) 0.103-0.412	(1) 10-25	(1) 2-2.5	(1) 1.5 -2
		(2) 1.543-1.980	(2) 37- 40	(2) 5- 6	(2) 3
LII	20-21	(1) 2.033- 2.244	(1) 40 – 45	(1) 6-6.7	(1) 3.5-3.9
		(2) 2.903-2.975	(2) 53 - 55	(2) 10-12	(2) 4.7-5
LIII	45-130	(1) 3.108-3.942	(1) 54 – 68	(1) 10-10.8	(1) 6 – 9
		(2) 9.708-9.979	(2) 100-110	(2) 15 - 18	(2) 10
		(3) 16.661>16.130	(3) 110-112	(3) 20 -25	(3) 10
Pre-pupa	6-8	7- 11	80 - 85	15 -18	
Pupa	16-18	6 - 10	34-51	15- 25	
Adult	3-7 months	6-8 > 4- 4.5	30- 55	15-22	3- 4

LABORATORY STUDIES ON MALE DIMORPHISM

1. Juvenile hormone analogue Methoprene

It was topically applied to just moulted pupae as a part of the studies on endocrine influence on male reproductive physiology in *O. rhinoceros* (Jacob, 1989b). The topical application of methoprene (dissolved in acetone/50 and 100μl) to just moulted pupae, they all emerged as adultoids (characterized by adult head, pupal body, pupal wing buds and protruded intermittent organs) after 16-18 days while controls which received acetone only moulted as normal adults. During the initial phase, 35 days old III instar male larvae were used for the experiments. Even though some of the controls moulted after 45 days, the experimental larvae continued feeding and after almost two months, most of them died while a few molted as larval-pupal intermediates. At that time the first author had no patience to search the reason for such a result and was not interested to wait for a long time to get a result after hormone application, hence switched on to topical application of the JH analogue to pupae which usually undergoes adult emergence after 16-18 days. When a number of experiments were conducted in the following years, then only the author could understand that the delaying of pupal emergence in 35 days old III instar larvae used for topical application of methoprene is due to the increase in the titer of JH in the body of all experimental larvae. They continued in the III instar stage for a long time and then died or moulted as abnormal.

2. Diflubenzuron (Dimilin)

A benzoylphenylurea compound, it could suppress metamorphosis in insects, as it is a good moulting disrupter (Retnakaran *et al.*, 1985; Bowers, 1990). Oral application of diflubenzuron to I, II, and III instar larvae of *O. rhinoceros* gave interesting results (Jacob, 1994). Diflubenzuron (25% wetted powder) diluted in distilled water and various concentrations were mixed with 20 gm cow dung and orally fed to I, II and III instar larvae of different ages. The first instar larvae and early second instar larvae were sensitive and died in the medium. Treated late (17 day old) II instar larvae molted to III instar and they gradually stopped feeding as there was no digestion. No fecal pellets were seen in the rearing bottle, later entering into a small pupa and an adult, which was abnormal without proper folding of wings, the last abdomen segments protruded along with aedeagus and died after few hours. The third instar larvae 0 day old, 7 days old and 1 month old also were treated and found to be insensitive to the compound (Jacob, 1994). However, it was a surprise to the first author when 35 days old III instar larvae were fed with the compound, some of them showed non-feeding stage and entered prepupal stage after 5 days and moulted as small pupae and subsequently as normal but small adults both males and females with a small or no horn. Others went on feeding and they moulted as normal big adults after 70-80 days and the males with prominent horn. Controls also showed similar result. As the treated larvae moulted as normal adults like controls, these larvae were recorded as insensitive to the compound. However, the behavior of III instar larvae with regard to their moulting at two stages as observed in the experiment provoked the author to do a feeding experiment for the third instar larva alone, and monitor the changes at each level.

3. Feeding experiments with III instar larvae (Table 2)

Eggs collected from adults reared in the laboratory, I and II instar larvae collected from cow dung pits were used for experiments. Third instar larvae (10 numbers repeated more than 10 times) on emergence were kept in individual plastic containers on 20 gm sterilized cow dung and renewed after 5 days. It has been observed that third instar larvae of

Oryctes rhinoceros after 35-40 days, which attained about 9 gram, weight clearly showed two different life pattern (Table 2). Almost 60% (from total experiments) started non-feeding stage even though food was renewed. They are designated as Type I larva. They underwent pupation on 45 or 46th day and molted as adults about 30-38mm long. These are minor adults, which emerge from the Type I larva. Male and female (Figs. 8,9) are almost alike with a small/no horn (just an extension of exoskeleton), while 40% of the larvae went on feeding, increasing the size reaching about 16 gm weight by 90 days and then feeding is reduced, enter non-feeding stage at 100-130 days and pupated. These larvae are designated as Type II larva. These on emergence were of bigger sized male and female, male 40-55 mm long having a prominent horn (5 to 8 mm length, with a sharp curve backwards). Females are about 40-50 mm long with a horn on head having a narrow tip, which did not reach the size of horn in males. It appears that major adults emerged from Type II larvae. Minor males emerged from type I third instar larvae are medium sized and without a prominent horn and mimic females, their pygidium is round or have a conical shape as that of females but it is devoid of hairs while in females, pygidium is clothed with hairs. Type I and type II larvae are having same head width and similar length but they differ in width and in body weight.

Table 2. Observations on III instar larvae reared in the laboratory after feeding with same quantity (20gm) of sterilized cow dung renewing after five days providing same laboratory conditions.

S. No	Duration (days)	Body weight (gm)	Body length (mm)	Body width (mm)	Head width (mm)
1	1-5	3.108 – 3.942	54 - 68	10 - 10.8	6 - 9
2	6-10	4.54 - 4.740	63 - 78	10.5 - 10.8	9
3	11-15	4.910 - 5.947	80 - 90	10.5 - 10.8	9
4	16-20	6.180 - 6.860	90 - 100	10.8 - 11	10
5	21-25	6.874 - 7.883	92 - 100	10.5 - 11	10
6	26-30	8.250 - 8.924	95 - 100	11 - 11.5	10
7	31-35	9.098 - 9.663	98 - 106	15 - 18	10
8	36-40	9.708 - 9.979	100 - 110	15 - 18	10
9 type I	41-45	9.806 > 8.609 (pupate)	100 - 110	15 - 18	10
9 type II	41-45	10.944 - 11.725	110	15 - 18	10
10	46-50	11.854 - 12.429	110	20 - 22	10
11	51-55	12.680 - 13.842	110	20 - 22	10

12	56-60	13.865 - 14.321	110 - 112	20 - 22	10
13	70-80	14.865 - 15.697	110	20 - 25	10
14	90-100	16.661 > 16.130	110 - 112	20 - 25	10
15	120-130	14.101 > 11.564 (pupate)	110 > 90	20 > 15	10

When larvae were collected from the field, III instar larvae outnumbered the other two larval stages. Type I and type II third instar larvae could be clearly identified in the field collected ones by the measurements of body and their early and late molting behavior as observed in the feeding experiments. However, it should be admitted that the percentage of type I third instar larvae exceeded in the collection from cow dung pits and those reared in the laboratory on cow dung.

4. Experiments by starving the III instar larvae

After one month feeding, III instar larvae (10 to 20 numbers, experiments repeated with several generation), which attained almost 9 gm weight, were kept plastic container individually half- filled with clean dry sand. After 3-5 days 60% showed pre pupal movements and entered into pupal stage. They seem to be Type I larvae. They molted as small sized adults. Others went on moving inside search of food, then gradually slowed down motion, got reduced in size (2 cm), showed movements of pre pupa for more than one month and became an ugly small stumpy red coloured organism which died inside the larval covering. These are probably Type II larvae from which bigger adults are supposed to be emerged. Here again two types of larvae having two ways of moulting become evident. Precocious metamorphosis is already noticed in *O. rhinoceros* followed by starvation (Mini & Prabhu, 1986). However, they have not mentioned the two types of larvae. Starvation or any unfavourable conditions in the environment may lead to precocious metamorphosis in insects. It is a common phenomenon in insects and it is to overcome unfavourable situation (Novak, 1975).

Neem oil (EC) dissolved in warm water (1 ml + 99 ml water) and 5 ml of the diluted compound was mixed with cow dung (20 gm) and one month old third instar larvae were introduced. Sudden pre-pupation movements were shown by a few and molted to normal but small pupa while others showed speedy movements inside the container for two to three days and then died. Here those moulted to pupa appeared to be type I larva. They were trying to escape from the unfavourable condition by pupation as they have reached the competence for pupation. However, second ones, which died, seemed to be type II larvae. They could not moult, as they have not reached the competence for moulting. No detailed experiments were carried out but this study confirming neem oil could produce antifeedant effect on *O. rhinoceros*.

ENDOCRINE AND GENE REGULATION

Topical application of juvenile hormone analogue methoprene on larvae described above, clearly show that the moulting in two types of larvae can be equalized by providing extra juvenile hormone, that the experimental larvae continue in the same stage for more than two months. Type I larvae usually moult after 40-45 days as they reach competence for pupal moult by 30-35 days of development. By methoprene application, juvenile hormone titre in the body is increased, hence they continue in larval stage. Juvenile hormone is the growth hormone and larval size increases by feeding. Type II third instar larvae, which usually feed and grow in size is due to the continued juvenile hormone activity up to 90-100 days and then they reach the competence for pupal moult. In the experiments, some treated larvae died after two months may be type I larva and those moulted as larval-pupal intermediates may be type II as they could not moult normally as the JH is in increased state. In contrast to this, starvation experiments and neem oil treatment bring forth the fact that unfavourable conditions can induce precocious moulting in Type I third instar larvae of *O. rhinoceros* they moulted to small pupa, as they reached the competence for moulting, whereas the Type II larvae died in the rearing bottle as they had a delayed moulting mechanism. Hormones controlling insect metamorphosis are well established (Wigglesworth, 1934). Moulting takes place in insects when Juvenile Hormone (JH) decreases and Ecdysone or Moulting Hormone (MH) increases in the body. Truman & Riddiford (1999, 2002) explored the roles of ecdysone and a juvenile hormone in the evolution of complete metamorphosis. The authors stress that JH is playing a key role in the evolution of insect metamorphosis.

Juvenile hormone is the growth-promoting (growth of larval parts by acquiring food) hormone synthesized in corpora allata and circulated through haemolymph. Biologically active JH will be hydrolyzed to JH-acid by the enzyme JH esterase (JHE) (Hammock *et al.*, 1995). A non-feeding stage or wandering stage usually refers to reduced juvenile hormone in the haemolymph. Shinoda and Itoyama (2003) are giving much importance to the key enzyme JH acid methyl transferase (JHAMT) that mediates synthesis of active JH from JH acid has a role in initiating pupal molt. The authors suggested that the suppression of transcription of JHAMT gene bring down the active JH in corpus allatum and this initiated non feeding stage in *Bombyx mori*. Juvenile hormone esterase (JHE) increase in haemolymph and subsequent increase in ecdysone take place leading to pupal molting. From the various experiments described above it appears that in type I larvae of *O. rhinoceros*, JH synthesis is suppressed and suddenly enters the non-feeding stage (40-45 days) and undergo pupal molt. However, in the type II larvae, active JH is continued till 90 days and gradually reduced and moult to pupa by 130 days. In both types of III instar larvae of *O. rhinoceros*, the availability of active JH and its degradation control-feeding rate at two different times. However, an interaction of genes and hormones, *viz.*, neurohormones controlling feeding and moulting, enzymes, receptors, moulting hormones (JH and MH) and genes controlling them are involved. The role of genes and hormones in insect metamorphosis are reported in lepidopteran insects (Riddiford *et al.*, 1999; Zhou & Riddiford, 2001).

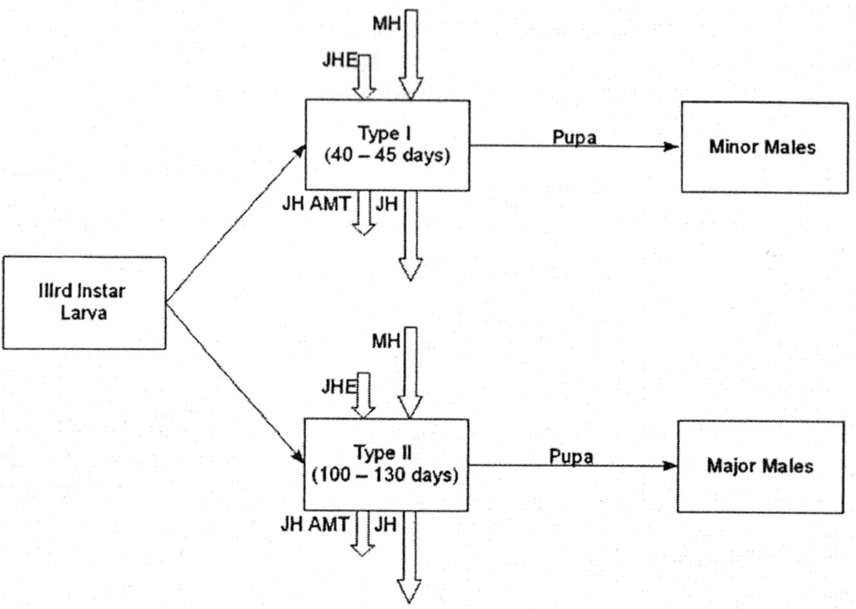

A proposed simple model – for the endocrine mechanism in Type I and Type II third instar larvae of *Oryctes rhinoceros*, resulting in minor and major males. Type I resumes non-feeding at 40-45 days and Type II at 100-130 days. When the enzyme JHAMT (Juvenile hormone Acid Methyl transferase) decreases in corpus allatum, active JH is reduced. Meanwhile JHE (Juvenile hormone esterase) increases in haemolymph hydrolyzing JH to JH acid and a subsequent increase in MH mediating ecdysis.

EVOLUTIONARY SIGNIFICANCE

The studies explained above suggest that male dimorphism exists in *Oryctes rhinoceros,* as there are minor males with small /no horn and major males having a prominent horn. This phenotypic variation is arising at last larval instar towards their pupal molt. The III instar larvae of *Oryctes rhinoceros* when they reach about 40-45 days (type I) molt to pupa and subsequently emerge as small adults (minor males/ females). Minor males have small horn and mimic that of females. There is another set of larvae (type II), which continued in III instar for more than 3 months and undergo metamorphosis to major males/females. When the type I third instar larva pupates early, they are leaving space and food for the type II larva, thus reducing competition for food and shelter in the place of their occurrence. The small males, which metamorphose early, will fly away to palms, feed and breed, and die after three months of emergence. In the laboratory, the minor males are found to prefer similar sized females for mating. The type II larvae emerge as major males only after 3-4 months. The larger size of such males and their prominent horn make them major males having high reproductive potential and live up to nine months. However, larval mortality is high in Type II larvae when unfavorable conditions arise in nature.

Emlen (1997) observed male dimorphism and the alternative reproductive tactics in the dung beetles *Onthophagus acuminatus.* Detailed studies of horns in dung beetles and the alternative phenotypes show evolutionary significance (Emlen, 2000; Moczek & Emlen, 2000). Male trimorphism (alpha, beta and gamma males) was reported in the scarabaeid dung beetles *Oxyștemon conspicillatum* (Rowland & Emlen 2009). The authors regard that at least two qualitatively different threshold mechanisms are regulating horn expression in dung beetles. The three males take three different reproductive tactics, a dominant male, a subordinate (sneak) tactic, and a female mimic tactic. Hartfelder and Emlen (2005) reported that hormones control polyphenism in dung beetles.

Many authors refer to the larval feeding or larval environment in which temperature variation has a crucial impact on phenotypic variation. Brakefield (2006) reported that adult wing pattern in the butterfly *Bicyclus anynana* is dependent on the temperature under which larvae are reared. According to Hodin & Riddiford (2000), different mechanisms underlie phenotypic variation and interspecific changes in Drosophilids.

In *O. rhinoceros*, the laboratory reared and the field collected larvae exhibited nature of molting at two different time. Eggs of the females laid at the same time, also exhibited the variation among larvae in the late third instar stage. When the same quantity of food was supplied to both larvae, the type I larvae initiated wandering or non-feeding stage after 40 days. Hence in *O. rhinoceros*, it is not the scarcity of food that induce phenotypic variation in adults, but a varied endocrine mechanism arising at the time of late III instar larval stage. Genes and hormones in turn control this behavior. It appears that decrease in JHAMT (Shinoda & Itoyama, 2003) brings down active JH in the type I larvae of *O. rhinoceros* at an early time, leading them to a non-feeding stage. The type II larvae, which continue feeding for at least three months, give evidence that active JH is present in their haemolymph and then the JH titer is reduced and enter a non-feeing stage after 100 days. If type II larvae are starved when they are 35 days old, they are unable to transform to pupa because they have not attained the competence for molting, whereas type I larvae can overcome an unfavorable condition if they are 30-35 days old, by immediate pupation. It cannot be overruled that in the evolutionary time scale the minor and major adults (emerged from Type I and II third instar larvae respectively) may diverge to two species.

ACKNOWLEDGEMENT

The first author is indebted to the great teacher and scientist, the late Prof. K.K. Nayar (Former Head, Department of Zoology, University of Kerala) for introducing her into the field of Entomology, and offers deep gratitude to Dr. V.K.K. Prabhu (Retd. Professor, Department of Zoology, University of Kerala) for helping to grow as a researcher in Entomology. She acknowledges the award of Research Associateship and Pool officership during her postdoctoral research period by the Council of Scientific and Industrial Research (CSIR), New Delhi and to all colleagues who associated with her in the Department of Zoology, University of Kerala.

REFERENCES

Ahmad, M.N & M.H. Ullah, 1951. Date cultivation in the Punjab (Pakistan). *Punjab Fruit Journal* - January-April.

Bedford, G.O., 1976. Observations on the biology and ecology of *Oryctes rhinoceros* and *Scapanes australis*; pests of coconut palm in Melanesia. *J. Aust. Entomol. Soc.* 15: 241-251.

Bedford, G.O., 1980. Biology, Ecology and control of Palm Rhinoceros beetle. *Annu. Rev. Entomol.* 25: 309-339.

Bowers, W.S., 1990. Prospects for the use of Insect Growth Regulators in Agriculture. In. *Advances in Invertebrate Reproduction*. Hoshi, M and O. Yamashita (Eds) Elsevier Science Pub. Amsterdam, pp. 365-382.

Brakefield, P.M., 2006. Physiology, Development, Genetics, and the evolution of phenotypic plasticity: Studies with butterfly eyespots. In: *Comparative Developmental Physiology Contributions Tools and Trends*. Ed Stephen J. Warburton, Warren W. Burggren, Berud Pelster, Carl, L. Reiber and John Spicer. Oxford University Press, Oxford, New York, pp. 58-69.

Budd, G.E. & M.J. Telford, 2009. The origin and evolution of arthropods. *Nature* 457: 812-817.

Catley, A., 1969. The coconut rhinoceros beetle *Oryctes rhinoceros* (L) (Coleoptera Scarabaeidae: Dynastinae). *PNAS* 15: 18-30.

Cherian, M.C., K.P. Ananthanarayanan, 1939. Studies on the coconut palm beetle (*Oryctes rhinoceros* Linn.) in south India. *Indian J. Agri. Sci.* 9: 541-559.

Cohic, F., 1950. Insect pests in the Wallis Islands and Futuna. *South Pac. Comm. Tech. Pap.* No.8, pp. 25.

Corbett, G.H., 1938. *Oryctes rhinoceros* L. and oil palms. *Malayan Agri. J.* XXVI.

Dumbleton, F.C., 1952. Rhinoceros beetle in the Kingdom of Tonga. *South Pac. Comm. Tech. Pap.* No. 34, pp. 7.

Emlen, D.J., 1997. Alternative reproductive tactics and male dimorphism in the horned beetles *Onthophagus acuminatus* (Coleoptera: Scarabaeidae) *Behav. Ecol. Sociobiol.* 41: 335-341.

Emlen, D.J., 2000. Integrating development with evolution: a case study with beetle horns. *Bioscience* 50: 403-418.

Goonewardena, H.F. 1958. The rhinoceros beetle (*Oryctes rhinoceros* L.) in Ceylon. Introduction, distribution, and life history. *Trop. Agric.* Ceylon 114, 39-60.

Gressitt, J.L., 1953. The coconut rhinoceros beetle (*Oryctes rhinoceros*) with particular reference to the Palau Islands. *Bull. Bernice P Bishop Mus.* No. 212, pp. 157.

Hammock, B.D., K. Touhara, B.C. Bonning & G.D. Prestwich, 1995. Action of Juvenile hormone (JH) esterase on JH-JH binding protein complex. An *in vitro* model of JH metabolism in a caterpillar. *Insect Biochem. Mol. Biol.* 25: 727-734.

Hartfelder, K. & Emlen D.J., 2005. Endocrine control of insect polyphenism. *Comprehensive Molecular Insect Science* 3: 651-703.

Hodin, J. and L.M. Riddiford, 2000. Different mechanism underlie phenotypic plasticity and inter specific variation for a reproductive character in Drosophilids (Insecta: Diptera). *Evolution* 54: 1638-1653.

Hurpin, B. & Fresneau, M., 1967. Contribution a lalutte contre les Oryctes nuisibles aux palmiers. Elevage en laboratoire de *O. rhinoceros* L. *Oleagineuz* 22: 1-6.

Indiravathi, G., M. Rajamannar & A Sujatha, 2001. Biology of rhinoceros bee-tle, *Oryctes rhinoceros* L. in coastal Andhra Pradesh. *J. Appl. Zool. Res.* 12: 14-18.

Jacob, M. 1989a. Differentiation of male reproductive system in *Oryctes rhinoceros* (Coleoptera : Scarabaeidae). *Proc. Indian Acad. Sci. (Anim. Sci)* 98:(4) 233-242.

Jacob, M., 1989b. Influence of Methoprene on the male reproductive system of *Oryctes rhinoceros* (Coleoptera: Scarabaeidae) *Current Science* 58: (8) 469-471.

Jacob, M., 1989c. *In vivo* and *in vitro* studies of the male reproductive physiology of *Oryctes rhinoceros* (Coleoptera:Scarabaeidae), a major pest of coconut palm. *Proc. Kerala Sci. Cong.* 1: 30-39.

Jacob, M., 1994. Effect of Diflubenzuron on the development of rhinoceros beetle *Oryctes rhinoceros* L. *J. Appl. Zool. Res.* 5: (2): 81-85.

Jacob T. K. & B. S. Bhumannavar, 1991. The coconut rhinoceros beetle *Oryctes rhinoceros* L. Its incidence and extent of palm damage in the Andaman and Nicobar Islands (India). *International J. Pest Management* 27: 80-84.

Jepson, F.W. 1912. The rhinoceros beetle (*Oryctes rhinoceros* Linn.) in Samoa. *Fiji Dep. Agri. Bull.* No.3, pp. 25.

Khare, J.L., 1919. *Oryctes rhinoceros* L. Proc. III Ent. Mtg., Pusa, Delhi.

Kurian, C. & G.B., Pillai 1964. Rhinoceros beetles: a major menace to coconut cultivation. *World Crops* 16: 20-24.

McKenna, J. & K.D. Shroff, 1911. The rhinoceros beetle (*Oryctes rhinoceros* Linn.) and its ravages in Burma. *Burma Dept Agri. Bull.* No.4, pp. 6.

Mini A. & V.K.K. Prabhu, 1986. Significance of critical developmental stage on starvation induced endocrine mediated precocious metamorphosis in *Oryctes rhinoceros* (Coleoptera Scarabaeidae). *Proc. Indian Acad. Sci. (Anim. Sci.)* 95: 379-385.

Mini, A & V.K.K. Prabhu, 1988. Sexual dimorphism in the larva and pupa of coconut rhinoceros beetles, *Oryctes rhinoceros* (Coleoptera: Scarabaeidae). *Current Science* 57: 686-687.

Moczek A.P. & D.J. Emlen, 2000. Male horn dimorphism in scarab beetle *Onthophagus taurus*: Do alternative reproductive tactics favor alternative phenotypes. *Anim. Behav*. 59: 459-466.

Monty, J., 1978. The coconut palm rhinoceros beetle *Oryctes rhinoceros* (L) in Mauritius and its control. *Rev. Agr. Sucr. Ile Maurice* 57: 60-76.

Nayar, K.K., T.N. Ananthakrishnan & B.V. David 1976. *General and Applied Entomology*, Tata McGraw-Hill Publishing Company Limited, New Delhi.

Nirula, K.K., 1955. Investigations on the pests of the coconut palm. Part II. *Oryctes rhinoceros* L. *Indian Coconut J*. 8: 161-180.

Novak V.J.A., 1975. *Insect Hormones*. Chapman and Hall London John Wiley & Sons New York. Pp. 600.

Orian, A.J.E., 1959. Report on a visit to Diego Garcia *Rev. Agr. Sucr. Ile Maurice* 38:127-143.

Priyaranjini, G.K., 2003. Localization of brain neurosecretory cells and neuropeptides of I and II instar larvae of *Oryctes rhinoceros* (Coleoptera: Scarabaeidae). Dissertation submitted in partial fulfillment of the requirements for the Degree of Master of Science in General and Applied Zoology, Department of Zoology, Univ. Kerala, Trivandrum, Kerala, pp. 57.

Retnakaran, A. J. Granett & T. Ennis, 1985. Insect Growth Regulators. In. *Comprehensive Insect Physiology, Biochemistry and Pharmacology* 12:

Kerkut G.A. and L.I. Gilbert (Eds). Pergamon Press. Pp. 529-601.

Riddiford L.M., K. Hiruma, Q. Lan & B. Zhou, 1999. Regulation and role of nuclear hormone receptors during larval molting and metamorphosis of Lepidoptera *Amer. Zool*. 39: 736-746.

Rowland, J. M. & D.J. Emlen, 2009. Two thresholds, three male forms result in Facltative male Trimorphism in Beetles. *Science* 323: 773-776.

Salini, V.S., 2003. Localization of brain neurosecretory cells and neuropeptides of III instar larvae of *Oryctes rhinoceros* (Coleoptera: Scarabaeidae) Dissertation submitted in partial fulfillment of the requirements for the Degree of Master of Science in General and Applied Zoology, Department of Zoology, University of Kerala, Trivandrum, Kerala. Pp. 60.

Shinoda, T. & K. Itoyama, 2003. Juvenile hormone acid methyltransferase: A key regulatory enzyme for insect metamorphosis. *PNAS* 100: 11986-11991.

Siny, G.B., 2009. Characterization and molecular cloning of bug corazonin neuropeptide gene. PhD thesis, Department of Zoology, University of Kerala, Kerala. Pp. 130.

Sison, P. 1957. Further studies on the biology ecology and control of the black beetles of coconut *Oryctes rhinoceros* Linn. *Proc. 8th Pac. Sci Congr.* 1953 3A 1299-1318.

Smith, H.S., 1935. The role of a biotic factors in the determination of population densities. *J. Econ. Entomol.* 28: 873-898.

Truman, J.W. & L.M. Riddiford, 1999. The origin of Insect Metamorphosis. *Nature* 401:447-452.

Truman J.W. and L.M. Riddiford, 2002. Endocrine insights into the Evolution of Metamorphosis in Insects. *Annu. Rev. Entomol.* 47: 467-500.

Vinson, J. 1963. Sur l'apparition a Maurice d'un scarabee nuisible,*Oryctes rhinoceros* Linne. *Rev. Agri. Sucr. Ile Maurice* 42: 173-80.

Wigglesworth, V.B., 1934. The physiology of ecdysis in *Rhodnius prolixus* (Hemip-tera) II Factors controlling molting and metamorphosis. *Quart. J. Micr. Sci.* 77:191-222.

Zelazny B., 1977. Occurrence of the baculovirus disease of the cocnut palm rhinoceros beetle in the Philippines and in Indonesia. *FAO Plant Prot. Bull.* 25: 73-77.

Zelazny, B. & A. Alfiler, 1991. Ecology of infected and healthy adults of *Oryctes rhinoceros* (Coleoptera Scarabaeidae) on coconut palms in the Philippines. *Ecological Entomology* 16 (2): 253-259.

Zhou, B. & L.M. Riddiford, 2001. Hormonal regulation and patterning of the Broad Complex in the epidermis and wing discs of the tobacco horn worm *Manduca sexta. Dev. Biol.* 231: 125-137.

Chapter - 15

DIVERSITY OF APHIDIINAE PARASITOIDS (HYMENOPTERA: BRACONIDAE) OF APHIDS (HOMOPTERA: APHIDIIDAE) - AN INDIAN PERSPECTIVE

Debjani Dey

Division of Entomology, Indian Agricultural Research Institute, New Delhi – 110012, E-mail: *ddeyiari@hotmail.com*

INTRODUCTION

Aphids are well known pests throughout the world. The biological peculiarities of the aphids along with extensive monoculture, injudicious and indiscriminate use of agricultural chemicals etc. have made aphids' control difficult. This has stimulated research on their biological control within the frame of IPM. Most parasitoids associated with aphids belong to the subfamily Aphidiinae (Hymenoptera: Braconidae). Perusal of all the available literature reveals a basic lack of information on the faunal composition, distribution and biology of this important group of natural enemies of aphids. Although members of Aphidiinae are known to occur throughout the world in all major habitats, except Arctic and Antarctic, yet consolidated studies on them are limited perhaps due to their minute size.

The prevalent climatic condition in India allows the occurrence of several generations of the pest and its associated parasitoids in a year indicating the suitability of this area for biological control activities. Perusal of literature indicates the presence of a very rich aphid as well as its parasitic fauna in India. Hence an attempt was made at a very critical analysis of the prevalent aphidiinae parasitoid fauna in India with an aim to create a base for future research and its possible utilization in biological control of aphids.

Wasps of the subfamily Aphidiinae can be easily distinguished by their yellowish brown to brownish black colouration, small size usually about 1.5-3.5 mm long, antenna with 18 or fewer segments, presence of occipital carina at least laterally, weakly sclerotized metasomal tergum, smooth scutellar sulcus, reduced wing venation to the extent that many species have only a single large, median cell in the fore wing, absence of cross vein *cu-a* in the hind wing, presence of a short ovipositor with a modified sheath or hypopygium in females etc.

Around 400 species under 60 genera of Aphidiinae have been described (Stary, 1988; Dolphin & Quicke, 2001) worldwide. Four tribes, Ephederini, Praini, Aphidiini and Trioxini (Mackauer, 1968; Belshaw & Quicke, 1997) have been recognized within Aphidiinae based on adult and larval morphology (Mackauer, 1968; O'Donnell, 1989; Finlayson, 1990). As with its position within the Braconidae, the tribal relationships within the Aphidiinae have proved intractable (Mackauer, 1961; Cho, 1984; Gardenfors, 1986). Two of the tribes, the Aphidiini and Trioxini, are united by the possession of a higher derived female reproductive system (LaRalec, 1993) and embryology (Tremblay & Calvert, 1971). Relationships with the other two tribes are problematic as these appear to be plesiomorphic in different characters: the Ephedrini in wing venation, and the Praini in venom gland apparatus (Edson & Vinson, 1979) and pupation habit. Modern molecular phylogenetic analysis are also inconclusive, indicating there are three, *viz.*, Ephederini, Praini and Aphidiini (Smith *et al.,* 1999; Sanchis *et al.,* 2000) or four, *viz.*, Ephederini, Praini, Aphidiini and Trioxini (Belshaw & Quicke, 1997) or five, *viz.*, Ephederini, Praini, Monoctonini, Trioxini and Aphidiini (Sanchis *et al.,* 2000) tribes wherein all recognized tribes are monophyletic (Sanchis *et al.,* 2000; Smith *et al.,* 1999).

Significant contributions to the knowledge of the aphidiine fauna on the ubiquitous basis has been made from the world (Mackauer & Stary, 1967; Stary, 1970), Mediterranean (Stary, 1976), Central Asia (Stary, 1979), Far East (Takada, 1968; Stary, 1975), India and adjacent countries (Stary & Ghosh, 1983), Africa south of Sahara (Stary *et al.,* 1984), Nearctic America (Marsh, 1979). Studies on the systematics of Aphidiine fauna include contributions by Nees (1818), Haliday (1833, 1834), Ratzeburg (1844, 1848, 1852), Foerster (1862), Marshall (1896), Dalla Torre (1898), Szepligeti (1904), Gahan (1911), Watanbe and Takada (1964, 1964a), Stary (1966, 1970), Stary & Remaudiere (1982), Stary & Schlinger (1967), Mackauer (1967) and Marsh (1979) from around the world.

In India, research on members of Aphidiinae have been restricted to description of new species or scattered isolated records (Subba Rao & Sharma, 1958; Shuja-Uddin, 1973, 1974, 1975, 1978, 1982, 1983; Stary and Ghosh, 1975, 1979; Stary & Raychaudhuri, 1977, 1978; Stary and Bhagat, 1978; Stary 1979; Takada & Rishi, 1980; Agarwala *et al.,* 1980, 1981; Bhagat, 1981, 1981a, 1982, 1982a; Saha *et al.,* 1982; Samanta *et al.,* 1983, 1983a, 1985; Das & Chakrabarti, 1989, 1989a, 1989b, 1990, 1991, 1991a); compilation of checklists (Narayanan *et al.,* 1960, 1962; Stary &

Ghosh, 1983); studies on population dynamics and field parasitism (Singh & Upadhyay, 1999; Bhagat, 2000; Men & Kandalkar, 2001; Dogra *et al.*, 2003).

DISTRIBUTION

Aphidiids are cosmopolitan as a group, but their distribution in the different climactic zones and continents is rather uneven, closely following that of their hosts. Most of the species occur in the temperate and subtropical belts of the northern hemisphere. The tropical fauna consists of relatively few endemic groups and of several species that have become pan-tropical or cosmopolitan. Several endemic groups, most of them rather different from the northern elements, live in the subtropical and temperate zones of the southern hemisphere. Apart from this, widely distributed species have also penetrated into this area. Some species occur even in the cold climatic zones of the subantartics and subarctics. Many a times, accidental immigrants as also deliberate introduction of species obscure the relationships of the original faunas (Stary, 1970).

STATUS OF APHIDIINE PARASITOIDS IN INDIA

The Indian fauna is represented by around 123 species under 23 genera which constitute around 31% of the world fauna, belonging to four tribes - Ephederini, Praini, Aphidiini and Trioxini with 4, 3, 12 and 4 genera, respectively, comprising about 15, 13, 44 and 28 percent of the total aphidiine fauna in India. The 123 recorded Indian species are placed under the following 23 genera, *viz.*, *Adialytus*, *Aphidius*, *Archaphidus*, *Areopraon*, *Betuloxys*, *Cristicaudus*, *Diaeretiella*, *Diaeretus*, *Ephedrus*, *Indaphidius*, *Indoephedrus*, *Kashmiria*, *Lipolexis*, *Lysaphidus*, *Lysiphlebia*, *Lysiphlebus*, *Monoctonus*, *Neoephedrus*, *Parapraon*, *Pauesia*, *Praon*, *Toxares* and *Trioxys* occurring from the plains to very high altitudes (Dey & Akhtar, 2007). The maximum number of species, *i.e.,* 83 have been reported from North India from the state of Delhi, Himachal Pradesh, Jammu and Kashmir, Punjab and Uttar Pradesh followed by 55 species from Eastern India from the state of Assam, Bihar, Manipur, Meghalaya, Nagaland, Sikkim, Tripura and West Bengal and 13 species from South India from the state of Karnataka and Tamil Nadu (Table 3). The genus *Aphidius* is the most abundant with maximum number of species on record followed by genus *Binodoxys*. The only aphidiine with a truly cosmopolitan distribution is *Diaeretiella rapae*.

SEASONAL HISTORY

The seasonal history of aphidiines is primarily dependent on the occurrence of the host species, but some responses to environmental effects are different in the parasitoids. Adaptations of the parasitoids to the life cycles and seasonal history of their host aphids are numerous and often rather specific and peculiar. The number of parasitoid generations per season depends on weather conditions as well as on the adaptation to the life cycle of the host. On the other hand, the number of generations can vary due to seasonal effects on the aphid-parasitoid interaction. On the other hand, the number of generations in parasitoids is strictly determined by that of the aphid host as close adaptation (diapause) occurs (Hafez, 1961; Stary, 1970; Hight *et al.*, 1972; Cierniewska, 1976; Chua, 1977; Agarwala & Raychaudhuri, 1981; Subba Rao & Sharma, 1962; Rao *et al.*, 1969, Sithumadhavan & Dharmadikari, 1969; Dharmadikari & Ramaseshiah, 1970; Das & Chakrabarti, 1990; Saha *et al.*, 1982; Stary & Ghosh, 1979).

HOST RANGE

The aphidiinae are parasitoids of the Aphidoidea only; the Adelgoidea are not parasitized. Host range patterns in particular species are relatively easy to define, although the taxonomy and ecology of the hosts indicate various significances. The host range patterns may differ in the various parts of the distribution range. However, several host range patterns can be distinguished based on Stary (1970) and Stary & Rejmanek (1981) (Table 1).

(i) The host range is restricted to a single host species (strict monop-hagy) as can be seen in the following species, *viz.*, *Adialytus salicaphis, Aphidius gifuensis, A. polycostulari, A. salignae, A. smithi, A. staryi, A. transcapious, Betuloxys assamensis, B. takecallis, Binodoxys basicurvus, B. ceratovacunae, B. equates, B. eutrichosiphi, B. jaii, B. joshimathensis, B. kashmirensis, B. kumaonensis, B. longispinus, B. mackaueri, B. oregama, B. shillongensis, B. tomentosae, B. uroleucon, Cristicaudus garhwalensis, C. nepalensis, C. indicus, Diaeretus leucopterus, Ephedrus laevicollis, E. cerasicola, E. urticae, Indoephedrus neoficicola, I. reticulata, Lipolexis myzakkaiae, Lysiphlebia rugosa, Lysiphlebus confuses, Neoephedrus helichrysi, N. kalimpongensis, Pauesia antennata, P. arcuate, P. gulmargensis, P. hazratbalensis, P. himalayensis, P. indica, P. lachniella, P. laricis, P. mashobrica, P. orientalis, Praon dorsal, P. himalayensis, P. hyperomyzus, P. kashmirensis, P. mollitrichosiphi, P. necans, P. orientale, P. pubescens, P. stagona, Toxares zakai, Trioxys areolatus, T. auctus, T. pallidus, T. peniculatus, T. rishii, T. rosaecola,* and *T. soporensis.*

(ii) The host range is restricted to two or more species of the same aphid genus *viz., Aphidius absinthii, A. cingulatus, Archaphi-*

dius greenidae, Betuloxys intermedius, Binodoxys rubicola, B. sinensis and *Pausia pini.*

(iii) The host range is restricted to species of two or more genera within the same aphid subfamily, *viz., Aphidius colemani, Aphidius eglanteriae A. hortensis, A. picipes, A. rosae, A. similes, A. urticae; A. uzbekistanicus; Areopraon lepelleyi; Binodoxys acalephae, B. brevicornis, B. centaureae, B. indicus, Diaeretiella rapae, Ephedrus dioscoreae, E. nacheri, E. niger, E. persicae, Kashmiria aphidis, Lipolexis gracilis, L. pseudoscutellaris, Lysiphlebia mirzai, Lysiphlebus testaceipes, Lysaphidus delhiensis, Praon abjectum, P. absinthii, P. myzophagum, P. volucre, Toxares deltiger* and *T. macrosiphophagum.*

(iv) The host range is restricted to species of two or more genera of two or more subfamilies of the same aphid family as in *Aphidius matricariae, A. qadri, Ephedrus lacertosus, E. plagiator, E. srinagarensis* and *Toxares shigai.*

(v) The host range is restricted to species of several genera of two or more aphid families as in *Adialytus ambiguous, Indaphidius curvicaudatus* and *Lipolexis scutellaris.*

The evolutionary trend for the parasitic Hymenoptera is apparently towards oligophagy, *i.e.* parasitization of several species of an aphid genus occurs most, followed by parasitization of species of several aphid genera of the same aphid subfamily. In the general classification, presumably less than one third of the total number of species is monophagous. However, the classification of monophagy may often be due to incomplete knowledge of the host range.

The different host species for a particular parasitoid are not all equal within the range as the parasitoids have certain preferences. Correspondingly, the hosts can also be divided into primary and secondary species.

PARASITOID SPECTRUM

The parasitoid spectrum for the aphid species of the Indian subcontinent comprising all the associated aphidiine parasitoid species was compiled from many primary and secondary sources of information material, both electronic and non-electronic, *viz.*, Review of Applied Entomology (Series A); CAB Pest CD 1973-2006; Zoological Records; Thompson Catalogue (1953), Hymenopterorum Catalogus (Shenefelt, 1957) and various other related books, journals, bulletins, newsletters etc. This parasitoid spectrum has been subdivided according to the host range of the parasitoids (Table 2). Similarly, the spectrum can also be described with respect to a certain area, distribution, habitat, plant species, season etc.

Table 2. The aphidiine parasitoid spectrum of various aphid species in the Indian subcontinent

Sl. No.	Aphids	Family / Tribe	Parasitoid/s
1.	*Acyrthosiphon gossypii*	Aphidinae : Macrosiphini	*Praon dorsale*
2.	*Acyrthosiphon pisum*	"	*Aphidius smithi, Aphidius* sp., *Ephedrus dioscoreae, E. plagiator*
3.	*Acyrthosiphon* sp.	"	*Aphidius urticae, A. uzbekistanicus*
4.	*Aphis citricola*	Aphidinae : Aphidini	*Binodoxys indicus, B. sinensis, Ephedrus plagiator, Ephedrus* sp., *Lipolexis oregamae, Praon myzophagum, Toxares macrosiphophagum, Trioxys* sp., *Cristicaudus nepalensis*
5.	*Aphis craccivora*	"	*Aphidius matricariae, Aphidius* sp., *Binodoxys acalephae, B. indicus, Diaeretiella rapae, Ephedrus persicae, Lipolexis gracilis, Lipolexis oregamae, Lysaphidus testaceipes, Praon abjectum, Praon* sp., *Trioxys* sp.
6.	*Aphis fabae*	"	*Aphidius matricariae, Binodoxys indicus, Ephedrus plagiator, Lipolexis oregamae*
7.	*Aphis fabae solanella*	"	*Binodoxys indicus*
8.	*Aphis farinosa*	"	*Binodoxys longispinus, Toxares shigai*
9.	*Aphis gossypii*	"	*Aphidius colemani, A. gifuensis, A. matricariae, A. picipes, A. similes, Aphidius* sp., *Binodoxys acalephae, B. basicrvus, B. equatus, B. indicus, B. rubicola, B. sinensis, Diaeretiella rapae, Ephedrus cerasicola, E. persicae, E. plagiator, E. srinagarensis, Lipolexis oregamae, Lipolexis pseudoscutellaris, Praon absinthii, P. myzophagum,*
10.	*Aphis nasturtii*	"	*Aphidius matricariae, Binodoxys indicus, Lipolexis oregamae*
11.	*Aphis nerii*	"	*Aphidius colemani, Binodoxys indicus, Lipolexis oregamae,*

12. *Aphis pisum*	"	*Ephedrus plagiator*
13. *Aphis ruborum longisetosus*	"	*Aphidius* sp., *Binodoxys indicus, B. rubicola, Diaeretiella rapae, Ephedrus persicae, Ephedrus* sp., *Lipolexis oregamae,*
14. *Aphis solanella*	"	*Aphidius colemani, A. matricariae, Aphidius* sp., *Ephedrus* sp. *Lipolexis oregamae, Toxares acrosiphophagum*
15. *Aphis* spiraecola	"	*Aphidius matricariae, A. similis, Aphidius* sp., *Binodoxys indicus, Ephedrus plagiator, Ephedrus* sp., *Lipolexis oregamae, Toxares deltiger*
16. *Aphis* sp.	"	*Aphidius colemani, A. transcaspious, Binodoxys indicus, Cristicaudus indicus, Ephedrus dioscoreae, Ephedrus* sp. *Lipolexis oregamae, Trioxys* sp.
17. *Aulacorthum magnoliae*	Aphidinae : Macrosiphini	*Binodoxys indicus*
18. *Aulacorthum* sp.	"	*Aphidius* sp., *Binodoxys indicus, Ephedrus* sp.
19. *Betacallis querciphaga*	Drepanosiphinae : Phyllaphidini	*Betuloxys assamensis*
20. *Brachycaudus cardui*	Aphidinae : Macrosiphini	*Aphidius matricariae*
21. *Brachycaudus helichrysi*	"	*Aphidius matricariae, A. qadrii, A. similes, Aphidius* sp., *Binodoxys indicus, Diaeretiella rapae, Ephedrus plagiator, Lysiphlebia rugosa, Lysiphlebia* sp., *Neoephedrus helichrysi, Praon* sp., *T. macrosiphophagum, T. shigai, Toxares deltiger*
22. *Brachycaudus* sp	"	*Aphidius matricariae, A. qadrii, T. macrosiphophagum, Toxares deltiger, Trioxys rishii, Trioxys* sp.
23. *Brevicoryne brassicae*	Aphidinae : Macrosiphini	*Aphidius matricariae, Diaeretiella rapae*
24. *Callaphis juglandis*	Drepanosiphinae : Phyllaphidini	*Praon himalayensis*
25. *Capitophorus carduinus*	Aphidinae : Macrosiphini	*Aphidius matricariae*
26. *Capitophorus formosartemisiae*	"	*Aphidius matricariae, Betuloxys intermedius, Cristicaudus garhwalensis*

27. *Capitophorus hippophaes indicus*	"	*Aphidius matricariae, Toxares deltiger, T. macrosiphophagum,*
28. *Capitophorus* sp.	"	*Aphidius qadrii, Betuloxys intermedius*
29. *Cavariella egopodii*	"	*Ephedrus laevicollis*
30. *Cavariella* sp.	"	*Ephedrus laevicollis*
31. *Ceratovacuna silvestri*	Hormaphidinae : Ceratophidini	*Binodoxys ceratovacunae, B. indicus, Ephedrus plagiator*
32. *Ceratovacuna* sp.? *perglandulosa*	"	*Binodoxys oregama*
33. *Cervaphis schouteniae*	Greenideinae : Cervaphidini	*Trioxys peniculatus*
34. *Chaetosiphon gracilicorne*	Aphidinae : Macrosiphini	*Aphidius* sp.
35. *Chaitophorus leucomelas*	Chaitophorinae : Chaitophorini	*Toxares shigai*
36. *Chaitophorus pakistanicus*	"	*Praon kashmirensis*
37. *Chaitophorus* sp.	"	*Adialytus salicaphis, Aphidious qadrii*
38. *Chromaphis juglandicola*	Drepanosiphinae: Phyllaphidini	*Trioxys pallidus*
39. *Cinara abieticola tenuipes*	Lachninae : Cinarini	*Pauesia arcuata*
40. *Cinara atrotibialis*	"	*Pauesia laricis*
41. *Cinara confinis tenuipes*	"	*Pauesia mashobrica*
42. *Cinara lachnirostris*	"	*Pauesia pini*
43. *Cinara maculipes*	"	*Pauesia orientalis*
44. *Cinara* sp.	"	*Pauesia gulmargensis, P. pini*
45. *Dactynotus minatii*		*Ephedrus lacertosus*
46. *Diphorodon cannabis*		*Aphidius matricariae*
47. *Diphorodon* sp.		*Aphidius matricariae*
48. *Dysaphis* sp.	Aphidinae : Macrosiphini	*Aphidius colemani*
49. *Eriosoma ulmi*	Eriosomatinae : Eriosomatini	*Areopraon lepellepi.*
50. *Eriosoma* sp.		*Areopraon lepellepi, Areopraon* sp

51. *Eulachnus thunbergii*	Lachninae : Cinarini	*Diaeretus leucopterus*
52. *Eutrichosiphum* sp.	Greenideinae : Greenideini	*Binodoxys eutrichosiphi*
53. *Greenidea ficicola*	Greenideinae : Greenideini	*Archaphidus greenideae*, *Trioxys areolatus*
54. *Greenidea formosana*	"	*Archaphidus greenideae*, *Lipolexis oregamae*
55. *Greenidea kumaoni*	"	*Binodoxys kumaonensis*
56. *Greenidea neoficicola*	"	*Indoephedrus neoficicola, I. reticulate*
57. *Hayhurstia atriplicis*		*Diaeretiella rapae*, *Ephedrus nacheri*
58. *Hysteroneura setariae*	Aphidinae : Aphidini	*Binodoxys indicus*,
59. *Hyadaphis coriandri*	Aphidinae : Macrosiphini	*Aphidius picipes*
60. *Hyalopterus pruni*	Aphidinae : Aphidini	*Aphidius colemani*, *Aphidius* sp., *Lysiphlebia mirzai*
61. *Hypermyzus carduellinus*	Aphidinae : Macrosiphini	*Ephedrus plagiator*
62. *Hypermyzus lactucae*	"	*Ephedrus persicae*
63. *Hypermyzus* sp.	"	*Toxares deltiger*, *T. macrosiphophagum*
64. *Hyperomyzus carduellinus*	"	*Praon hyperomyzus*
65. *Lachnus tropicalis*	Lachninae : Lachnini	*Aphidius* sp., *Pauesia himalayensis*, *Pauesia indica*, *Pauesia lachniella*, *Toxares deltiger*, *T. macrosiphophagum*
66. *Lachnus* sp.	"	*Pauesia hazratbalensis*
67. *Liosomaphis himalayensis*		*Aphidius hortensis*, *Lipolexis gracilis*
68. *Liosomaphis* sp.		*Monoctonus* sp.
69. *Lipaphis erysimi*	Aphidinae : Macrosiphini	*Aphidius colemani*, *A. hortensis*, *A. matricariae*, *Aphidius* sp., *Binodoxys brevicornis*, *Binodoxys indicus*, *Diaeretiella rapae*, *Ephedrus plagiator*, *Ephedrus* sp., *Lipolexis gracilis*,

			Lysiphlebus testaceipes, Praon volucre
70.	*Longiunguis sacchari*		*Lysiphlebia mirzai, Lysiphlebus delhiensis*
71.	*Lonicera quinqueocularis*		*Ephedrus srinagarensis*
72.	*Macrosiphoniella kikungshana*	Aphidinae : Macrosiphini	*Aphidius absinthii*
73.	*Macrosiphoniella oblonga*	"	*Monoctonus* sp.
74.	*Macrosiphoniella pseudoartemisae*	"	*Aphidius absinthii, A. matricariae, Diaeretiella. rapae*
75.	*Macrosiphoniella pseudoartemisiae*	"	*Aphidius eglanteriae, A. matricariae*
76.	*Macrosiphoniella sanborni*	"	*Aphidius absinthii, Aphidius* sp., *Ephedrus niger, Toxares macrosiphophagum, T. shigai*
77.	*Macrosiphoniella spinipes*	"	*Praon* sp., *Trioxys* sp.
78.	*Macrosiphoniella yomogifoliae*		*Aphidius absinthii*
79.	*Macrosiphoniella* sp.	"	*Aphidius absinthii, Binodoxys centaureae, Toxares deltiger, T. macrosiphophagum, Praon* sp.,
80.	*Macrosiphum miscanthi*	Aphidinae : Macrosiphini	*Aphidius uzbekistanicus, Binodoxys indicus, Binodoxys joshimathensis, Ephedrus lacertosus*
81.	*Macrosiphum rosae*	"	*Aphidius rosae, Aphidius* sp., *Ephedrus plagiator, Indaphidius curvicaudatus*
82.	*Macrosiphum* sp.	"	*Adialytus ambiguus*
83.	*Melanaphis arundinariae*	Aphidinae : Aphidini	*Ephedrus nacheri*
84.	*Melanaphis donacis*	"	*Trioxys* sp.
85.	*Melanaphis indosacchari*	"	*Lysiphlebus delhiensis*
86.	*Melanaphis megha-layensis meghalayensis*	"	*Binodoxys mackaueri*
87.	*Melanaphis sacchari*	"	*Lysiphlebia mirzai, Lysiphlebus delhiensis, Lysiphlebus* sp.

88. *Melanaphis* sp.	"	*Lysiphlebus delhiensis*
89. *Metopolophium drihodun*	Aphidinae : Macrosiphini	*Diaeretiella rapae*
90. *Metopolophium euryae*	Aphidinae : Macrosiphini	*Toxares deltiger, Toxares macrosiphophagum*
91. *Microlophium carnosum*		*Ephedrus urticae*
92. *Microlophium* sp.		*Binodoxys kashmirensis*
93. *Mollitrichosiphum (Metatrichosiphon) nandii*	Greenideinae : Greenideini	*Indaphidius curvicaudatus*
94. *Mollitrichosiphum alni*	"	*Indaphidius curvicaudatus*
95. *Mollitrichosiphum* sp	"	*Indaphidius curvicaudatus*
96. *Mollitrichosiphum tenuicorpus*	"	*Praon mollitrichosiphi, Trioxys* sp.
97. *Myzakkaia verbasci*		*Lipolexis myzakkaiae*
98. *Myzaphis rosarum*	Aphidinae : Macrosiphini	*Trioxys rosaecola*
99. *Myzus boehmerica*	Aphidinae : Macrosiphini	*Toxares deltiger, T. macrosiphophagum*
100. *Myzus cerasi*	"	*Aphidius matricariae*
101. *Myzus cornutus*	"	*Aphidius staryi*
102. *Myzus dycei*	"	*Aphidius matricariae, Ephedrus* sp., *Lysiphlebus confusus*
103. *Myzus numecola*	"	*Aphidius qadrii*
104. *Myzus obtusirosris*	"	*Toxares deltiger, Toxares macrosiphophagum*
105. *Myzus ornatus*	"	*Aphidius matricariae, A. qadrii, Aphidius* sp., *Praon* sp., *Toxares deltiger, T. macrosiphophagum,*
106. *Myzus persicae*	"	*Aphidius colemani, A. hortensis, A. matricariae, A. similes, Aphidius* sp., *Binodoxys brevicornis, B. indicus, Diaeretiella rapae, Ephedrus lacertosus, E. persicae, E. plagiator, Lipolexis oregamae, Neoephedrus kalimpongensis, Praon myzophagum, Praon* sp., *Toxares deltiger, T. macro-siphophagum, T. shigai, T. zakai,*

107. *Myzus sorbi*	"	*Aphidius eglanteriae, Binodoxys tomentosae, Ephedrus persicae,*
108. *Myzus umefoiae*	"	*Aphidius matricariae*
109. *Myzus varians*	"	*Aphidius colemani*
110. *Myzus* sp.	"	*Aphidius matricariae, Ephedrus srinagarensis, Kashmiria aphidis, Praon volucre, Praon* sp.
111. *Nasonovia* sp.	Aphidinae : Macrosiphini	*Praon pubescens*
112. *Neomyzus circumflexus*	Aphidinae : Macrosiphini	*Aphidius* sp., *Ephedrus lacertosus*
113. *Phorodon cannabis*	Aphidinae : Macrosiphini	*Aphidius matricariae*
114. *Phorodon* sp.	"	*Aphidius matricariae, Binodoxys acalephae, Ephedrus plagiator, E. srinagarensis*
115. *Prociphilus* sp.	Eriosomatinae : Pemphigini	*Binodoxys indicus, Ephedrus srinagarensis, Kashmiria aphidis*
116. *Pseudoacyrthosiphon holsti*		*Aphidius urticae*
117. *Pterocomma populeum*	Aphidinae : Pterocommatinae	*Aphidius cingulatus*
118. *Pterocomma* sp.	"	*Aphidius cingulatus*
119. *Rhopalosiphum maidis*	Aphidinae : Aphidini	*Aphidius colemani, A. uzbekistanicus, Aphidius* sp., *Lipolexis oregamae, Lysiphlebia mirzai, Lysiphlebus delhiensis*
120. *Rhopalosiphum nympheae*	"	*Aphidius colemani, Pauesia antennata, Praon abjectum, Toxares deltiger, T. macrosiphophagum, Praon necans, Praon* sp.
121. *Schizaphis rotundiventris*	Aphidinae : Aphidini	*Aphidius* sp., *Toxares deltiger, T. macrosiphophagum*
122. *Schzaphis* sp.	"	*Aphidius uzbekistanicus*
123. *Schizoneurella indica*	Eriosomatinae : Eriosomatini	*Areopraon lepellepi*
124. *Shinjia orientalis*		*Toxares* sp.
125. *Shinjia pteridifoliae*		*Aphidius matricariae*
126. *Shivaphis celti*	Drepanosiphinae: Phyllaphidini	*Binodoxys jaii, Trioxys auctus*

127. *Shivaphis* sp.	"	*Trioxys soporensis*
128. *Sinomegoura citricola*	Aphidinae : Macrosiphini	*Binodoxys indicus*
129. *Sinomegoura pyri*	"	*Binodoxys shillongensis*
130. *Sipha burakowskii*	Chaitophorinae : Siphini	*Aphidius matricariae*
131. *Sipha maydis*	"	*Adialytus ambiguus*
132. *Sitobion avenae eleusinae*	Aphidinae : Macrosiphini	*Diaeretiella rapae*
133. *Sitobion pachysiphon*	"	*Praon* sp.
134. *Sitobion rosaeformis*	"	*Aphidius rosae, Aphidius* sp, *Diaeretiella rapae, Ephedrus plagiator, Indaphidius curvicaudatus, Praon* sp.
135. *Sitobion* sp.	"	*Aphidius polycostulari, A. uzbekistanicus, Aphidius* sp., *Praon volucre,*
136. *Sitobion? alopecuri*	"	*Aphidius uzbekistanicus*
137. *Stagona* sp.		*Praon stagona,*
138. *Takecallis arundinariae*	Drepanosiphinae: Phyllaphidini	*Betuloxys takecallis*
139. *Takecallis himalayensis*	"	*Ephedrus lacertosus*
140. *Tetraneura* sp.	Eriosomatinae : Tetraneurini	*Ephedrus plagiator*
141. *Therioaphis trifolii*	Calaphidinae : Panaphidini	*Lipolexis gracilis*
142. *Toxoptera aurantii*	Aphidinae : Aphidini	*Binodoxys indicus, Ephedrus* sp., *Lipolexis oregamae, Lipolexis pseudoscutellaris, Trioxys* sp.
143. *Toxoptera citricidus*	"	*Binodoxys indicus, Lipolexis oregamae, Lipolexis pseudoscutellaris, Trioxys* sp.
144. *Toxoptera odinae*	"	*Lipolexis oregamae*
145. *Tuberculatus indicus*	Drepanosiphinae : Phyllaphidini	*Trioxys* sp.
146. *Tuberolachnus salignus*	Lachninae : Lachnini	*Aphidius salignae, Lipolexis oregamae*
147. *Uroleucon pseudotanaceti*	Aphidinae : Macrosiphini	*Ephedrus niger*
148. *Uroleucon tanaceti*	"	*Toxares shigai*

149. *Uroleucon* sp.	"	*Binodoxys centaureae, B. uroleucon, Ephedrus niger, Praon absinthii, Praon orientale*
150. Unknown aphids		*Aphidius absinthii, A. erysimi, A. urticae, Betuloxys hortorum, Binodoxys indicus, Diaeretiella rapae, Ephedrus niger, E. persicae, E. srinagarensis, Ephedrus* sp., *Lipolexis gracilis, Lysiphlebus confusus, Lysiphlebus* sp., *Monoctonus nervosus, Monoctonus* sp., *Praon* sp., *Toxares deltiger, T. macrosiphophagum, T. shigai, Trioxys* sp.
151. Without host records		*Aphidius areolatus, A. erysimi, A. setiger*

An in-depth analysis shows that the number of parasitoid species in the spectrum are rather unequal. Significant differences occur among the different aphid groups, which might be due to the interaction of various evolutionary trends of the parasitoids. In principle, the numbers of parasitoid species in the spectrum per aphid species appears to be directly proportional to the number of species in that particular aphid group. Correspondingly, the largest number of parasitoid species has been determined in the Aphididae (Table 3) followed by the other aphid groups. Within the spectrum, a particular parasitoid species can also be subdivided according to its host range patterns, and a certain hierarchy can be distinguished.

REFERENCES

Agarwala, B.K. & D.N. Raychaudhuri, 1981. Observations in the population trends of rose aphids and their Hymenopterous parasites in Kalimpong, West Bengal (India). *Entomon* 6: 211-214.

Agarwala, B.K., D. Ghosh, S.K. Das, S.C. Poddar & D.N. Raychaudhuri, 1981. Parasites and predators of aphids (Homoptera: Aphididae) from India V. New records of two aphidiid parasites, nine arachnids and one dipteran predator from India. *Entomon* 6: 233-238.

Agarwala, B.K., D. Raychaudhuri & Raychaudhuri, D.N. 1980. Parasites and predators of aphids in Sikkim and Manipur (Northeast India). III. *Entomon* 5: 39-42.

Belshaw, R. & D.L.J. Quicke, 1997. A molecular phylogeny of the Aphidi-inae (Hymenoptera: Braconidae). *Molecular Phylogenetics and Evolution* 7: 281-293.

Bhagat, K.C., 2000. Natural parasitism of wheat aphid *Macrosiphum (Sitobion) avenae* Fab. by *Lysiphlebia mirzai* Shuja-Uddin (Hymenoptera: Braconidae). *Insect Environment* 6: 9.

Bhagat, R.C., 1981. *Areopraon lepelleyi* (Hymenoptera: Aphidiidae), a new record form India. Jodhpur University *Geobios* 8: 177-178.

Bhagat, R.C., 1981a. On some new aphid parasitoids from Kashmir, India (Hymenoptera, Aphidiidae). *Oriental Insects* 15: 195-200.

Bhagat, R.C. ,1982. Aphid galls and their parasitoids from Kashmir. *Entomon* 7: 103-105.

Bhagat, R.C., 1982a. New aphid parasitoids (Hymenoptera: Aphidiidae) from Kashmir, India. *Oriental Insects* 16: 113-117.

Cho, L., 1984. The phylogeny of Aphidiidae (Hymenoptera) of Taiwan. *Journal of Agricultural Research (China)* 33: 437-446. [In Chinese]

Chua,T.H., 1977. Population studies of *Brevicoryne brassicae* (L.), its parasites and hyperparasites in England. *Researches on population ecology* 19 : 125 – 139.

Cierniewska, B., 1976. Studies on the ecology of Ephedrus persicae Frog. (Hymenoptera : Aphidiidae), a parasite of the rosy apple aphid, Dysaphis plantaginea (Pass) (Homoptera : Aphididae). *Roczniki Nauk Rolnicych, Ser. E*, 6: 59-75.

Dalla Torre, K.W.V., 1898. Catalogus Hymenopterorum huiusque descriptorum systematicus et synonymicus. Vol. IV. Braconidae. Lipsiae.

Das, B.C. & S. Chakrabarti, 1989. Aphidiid parasitoids (Hymenoptera; Aphidiidae) of graminaceous aphids in Garhwal, of Western Himalayas. *Oriental Insects* 24: 265-372.

Das, B.C. & S. Chakrabarti, 1989a. Two new aphid parasitoids (Hymenoptera: Aphidiidae) from Garhwal range of Western Himalaya, India. *Entomon* 14: 339-343.

Das, B.C. & S. Chakrabarti, 1989b. *Praon himalayensis*, a new walnut aphid parasitoid (Hymenoptera: Aphidiidae) in Garhwal range of Western Himalaya. *Entomon* 14: 345-347.

Das, B.C. & S. Chakrabarti, 1990. New and little known aphidiid parasitoids (Hymenoptera; Aphidiidae) of gall forming aphids in Western Himalaya, with notes on their seasonal history. *Oriental Insects* 24: 399-414.

Das, B.C. & S. Chakrabarti, 1991. Aphidiid (Hymenoptera: Aphidiidae) fauna of Garhwal, Western Himalaya. *Journal of the Bombay Natural History Society* 88: 409-413.

Das, B.C. & S. Chakrabarti, 1991a. Two new *Cristicaudus* (Hymenoptera: Aphidiidae) as parasites of aphids in Garhwal range of Himalaya, India. *Entomon* 16: 305-309.

Dey, D. & M.S. Akhtar, 2007. Diversity of natural enemies of aphids belonging to Aphidiinae (Hymenoptera : Braconidae) in India. *Journal of Asia Pacific Entomology* 10: 281-296.

Dharmadhikari, P.R. & G. Ramaseshiah, 1970. Recent records of aphidiids (Hymenoptera: Aphidiidae) in India. *Technical Bulletin of Commonwealth Institute of Biological Control* 13: 83-89.

Dogra, I., N. Devi & Desh-Raj, 2003. Parasitization of mustard aphid, *Lipaphis erysimi* Kalt. by *Diaeretiella rapae* M'Intosh in the mid-hill zone of Himachal Pradesh (India). *Journal of Entomological Research* 27: 145-149.

Dolphin, K. & D.L.J. Quicke, 2001. Estimating the global species richness of an incompletely described taxon: an example using parasitoid wasps (Hymenoptera: Braconidae). *Biological Journal of the Linnean Society* 73, 279-286.

Edson, K.M. & S.B. Vinson, 1979. A comparative morphology of the vemon apparatus of the female braconids (Hymenoptera: Braconidae). *Canadian Entomologist* 111: 1013-1024.

Finlayson, T., 1990. The systematics and taxonomy of finalinstar larvae of the family Aphidiinae (Hymenoptera). *Memories of Entomological Society of Canada* 152: 3-74.

Foerster, A., 1862. Synopsis der Familien und Gattungen der Braconen. Verhandlungen des Naturhistorischen Vereins der Preussischen Rheinlande und Westfalens. 19: 225-250.

Gahan, A.B., 1911. Aphidiinae of North America. *Maryland Agricultural Experiment Station. Bulletin* 152: 147-200.

Gardenfors, U., 1986. Taxonomic and biological revision of palearctic *Ephedrus* Haliday (Hymenoptera: Braconidae, Ahpidiinae). *Entomologica Scandinavica,* Supplement, 27: 1-95.

Hafez, M. 1961.Seasonal fluctuations of population density of the cabbage aphid, *Brevicoryne brassicae* (L.) in the Netherlands, and the role of its parasite, *Aphidius (Diaeretiella) rapae* (Curtis). *Tijdschrift voor Plantenziekten* 67 : 445-548.

Haliday, A.D., 1833. An essay on the classification of parasitic Hymenoptera of Britain, which correspond with the *Ichneumones minuti* of Linnaeus. *Entomological Magazine,* 1: 259-276 and 480-491.

Haliday, A.D., 1834. Essay on the classification of parasitic Hymenoptera of Britain, which correspond with the *Ichneumones minuti* of Linnaeus, *Entomological Magazine* 2: 93-106.

Hight. S.C., R.D. Eikenbary, R.J. Miller & Starks, K.J., 1972. The greenbug and Lysiphlebus testaceipes. *Environmental Entomology* 1 : 205-209.

LeRalec, A., 1993. Les hymenopteres parasitoides: Adaptations de l'appareil reproducteur femelle. Morphologie et ultrastructure de l'ovaire, de l'oeuf et de l'ovipositeur. These doctorat de l'Universitaire de Rennes I.

Mackauer, M. & P. Stary, 1967. Hym. Ichneumonoidea, World Aphidiidae. in: Deluechi V. and Remaudiere G. (Edts.): Index of entomophagous insects. Le Francois, Paris, pp. 195.

Mackauer, M., 1961. Die Typen der Unterfamilie Aphidiinae des Britischen Museums London (Hymenoptera: Braconidae). *Beitrage zur Entomologie* 11: 96-154.

Mackauer, M., 1967. A new genus and several new species of aphid parasites (Hymenoptera: Aphidiidae). *Entomophaga* 12: 139-147.

Mackauer, M., 1968. Aphidiidae. Pars 3. Hymenopterorum Catalogus (n. ed.). edited by Ferriere G. and v. d. Vecht J. Dr. W. Junk N. V., The Hague, pp. 103.

Marsh, P.M., 1979. *Aphidiidae* in Krombein, K. V., Hurd, P. D. Jr., Smith, D. R. and Durks, B. D. *Catalog of hymenoptera in America North of Mexico. Vol. I.* Smithsonian Institution Press, Washington, 295-313.

Marshall, T.A., 1896. Les Braconides. In: Andre E. (ed.) 'Species des Hymenopteres d'Europe et d'Algerie.' Tome 5. pp. 635.

Men, U.B. & H.G. Kandalkar, 2001. Parasitization of mustard aphids *Lipaphis erysimi* Kaltebach, by *Diaretiella rapae* M'Intosh (Hymenoptera: Braconidae). *Insect Environment* 6: 176.

Narayanan, E.S., B.R. Subba Rao, and A.K. Sharma, 1960. A catalogue of the Baluchiknown species of the world belonging to the subfamily Aphidiinae (Hymenoptera: Braconidae). *Beitrage zur Entomologie* 10: 545-581.

Narayanan, E.S., B.R., Subba Rao, A.K. Sharma & P. Stary, 1962. Revision of a catalogue of the known species of the world belonging to the subfamily Aphidiinae (Hymenoptera : Braconidae). *Beitrage zur Entomologie* 12: 662-720.

Nees, v. E.C.G., 1818. Genera et Families Ichneumonidium Adscitorum Exhibens. *Nova Acta Physico Medica Acad. Ceasar. Leop. Carol. Erlangen* 299-310.

O'Donnell, D.J., 1989. A morphological and taxonomic study of first instar larvae of Aphidiinae (Hymenoptera: Braconidae). *Systematic Entomology* 14: 197–219.

Rao, V.P., P.R., Dharmadhikari, G., Ramaseshiah, V.R. Phalak & T.V. Sethu Madhavan, 1969. Study of natural enemies of aphids (for the U.S.A.). *Report of Commonwalth Institute of Biological Control* 1968: 45-47.

Ratzeburg, J.T.C., 1844. Die Ichneumonen der Forstinsecten in forstlicher und entomologischer Beziehung. Berlin. pp. 224.

Ratzeburg, J.T.C., 1848. Die Ichneumonen der Forstinsecten in forstlicher und entomologischer Beziehung. Zweiter Band. Berlin. pp. 238.

Ratzeburg, J.T.C., 1852. Die Ichneumonen der Forstinsecten in forstlicher und entomologischer Beziehung. Dritter Band. Berlin. pp. 272.

Saha, J.L., S.C. Poddar, S.K. Das, B.K. Agarwala & D. Raychaudhuri, 1982. Studies on the aphid parasites (Hymenoptera: Aphidiidae) from Himachal Pradesh, India. *Akitu* 44: 12.

Samanta, A.K., D.R. Pramanik & D. Raychaudhuri, 1983. Some new aphid parasitoids (Hymenoptera: Aphidiidae) from Meghalaya, North East India. *Akitu* 54: 8.

Sanchis, A., A. Latorre, F. Gonza lez-Candelas & J.M. Michelena, 2000. An 18S rDNA-Based Molecular Phylogeny of Aphidiinae (Hymenoptera: Braconidae). *Molecular Phylogenetics and Evolution* 2: 180–194.

Sethumadhavan, T.V. & P.R. Dharmadhikari, 1969. Notes on *Diaeretiella rapae* (M'Intosh) (Hymenoptera : Aphidiidae) parasitic on aphids on crucifers in

India. *Technical Bulliten, Commonwalth Institute of Biological Control* 11: 173-177.

Shuja-Uddin, 1973. Three species of *Trioxys* Haliday (Hymenoptera: Aphidiiae) recorded from India. *Indian Journal of Entomology* 35: 9-14.

Shuja-Uddin, 1974. Two new species of the genus *Toxares* Westwood (Aphidiiae: Hymenoptera) from India with a note on the genus. *Indian Journal of Entomology* 36: 268-274.

Shuja-Uddin, 1975. Two new species of Aphidiidae (Hymenoptera) from India. *Records of the Zoological Survey of India* 68: 415-420.

Shuja-Uddin, 1978. New record of *Adialytus* Forster (Hymenoptera; Aphidiidae) from India with comments on the validity of the genus. *Journal of Entomological Research* 2: 160-162.

Shuja-Uddin, 1982. Description of new species of genus *Trioxys* Haliday (Hymenoptera: Aphidiidae) and new record of *Trioxys* (*Trioxys*) *pallidus* (Hal.) from Kashmir (India). *Journal of Entomological Research* 6: 146-149.

Shuja-Uddin, 1983. A new species of genus *Trioxys* Haliday (Hymenoptera: Aphidiidae) form Kashmir, India. *Journal of Entomological Research* 7: 36-38.

Singh, R. & B.S. Upadhyay, 1999. Parasitism of *Aphis gossypii* (Homoptera: Aphididae) by *Binodoxys indicus* (Hymenoptera: Braconidae) at varying parasitoid-host ratios. *Entomologia Generalis* 24: 271-276.

Smith, P.T., S. Kambhampati, W. Voelkl & M. Mackauer, 1999. A phylogeny of aphid parasitoids (Hymenoptera: Braconidae: Aphidiinae) inferred from mitochondrial NADH 1 dehydrogenase gene sequence. *Molecular Phylogenetics and Evolution* 11: 236-245.

Stary, P. & M. Rejmanek, 1981. Number of aphid parasitoids per host in different systematic groups of aphids: the implications for introduction styrategy in biological control (Homoptera: Aphidoidea; Hymenoptera: Aphidiidae). *EntomologiaScandinavic*, Supplementum 15: 341-351

Stary, P., G. Remaudiere & A. Autrique, 1984. Les Aphidiides parasites de pucerons en region ethiopienne. **In**: G. Remaudiere and A. Autrique (ed.), *Contribution a l'Ecologie des Aphides Africains. Production Vegetale et Protection des Plantes*, Etude FAO, Rome, 64: 95-102.

Stary, P. & R.C. Bhagat, 1978. Records and descriptions of aphid parasitoids from Kashmir, India (Hymenoptera: Aphidiidae). *Acta Entomologia Bohemoslov* 75: 387-393.

Stary, P. & A.K. Ghosh, 1975. Aphid parasites (Hymenoptera: Aphidiidae) from Meghalaya, India. *Oriental Insects* 9: 343-349.

Stary, P. & A.K. Ghosh, 1979. Redescription of *Trioxys nepalensis* (Takada) and new host records of some aphid parasitoids (Hymenoptera: Aphidiidae) from Meghalaya, India. *Oriental Insects* 13: 41-45.

Stary, P. & A.K. Ghosh, 1983. *Aphid parasitoids of India and adjacent countries (Hymenoptera: Aphidiidae).* Technical Monograph, Zool. Survey of India 7:1-96.

Stary, P. & D.N. Raychaudhuri, 1977. A new species of *Pauesia* (Hymenoptera: Aphidiidae) parasitic on *Lachnus tropicalis* (Homoptera: Aphididae) in India. *Oriental Insects* 11: 233-235.

Stary, P. & D.N. Raychaudhuri, 1978. *Trioxys (Betuloxys) takecallis*, sp. nov. from India (Hymenoptera: Aphidiidae). *Oriental Insects* 12: 365-368.

Stary, P. & G. Remaudiere, 1982. New genera, new species and host records of aphid parasitoids (Hymenoptera, Aphidiidae) from Mexico. *Annales de la Societe Entomologique de France* 18: 107-127.

Stary, P. & E.S. Schlinger, 1967. *A revision of Far East Asian Aphidiidae (Hymenoptera).* Dr. W. Junk N. V., Publishers, The Hague. pp. 204.

Stary, P., 1966. Aphid parasites of Czechoslovakia. A review of the Czechoslovak Aphidiidae (Hymenoptera). Dr. W. Junk N. V., Publishers, The Hague. pp. 142.

Stary, P., 1970. Biology of Aphid parasites (Hymenoptera: Aphidiidae) with respect to integrated control. Dr. W. Junk N. V., Publishers, The Hague. pp. 643.

Stary, P., 1975. *Aphidius colemani* Viereck: its taxonomy, distribution and host range (Hymenoptera: Aphidiidae). *Acta Entomologia Bohemoslov* 72: 156-163.

Stary, P., 1976. *Aphid parasites (Hymenoptera: Aphidiidae) of the Mediterranean Area.* Dr. W. Junk N. V., Publishers, The Hague. pp. 95.

Stary, P., 1979. *Indaphidius curvicaudatus* gen. n., sp. n. (Hymenoptera: Aphidiidae), an aphid parasitoid from India. *Acta Entomologia Bohemoslov* 76: 34-37.

Stary, P., 1988. Chapter 9. Natural Enemies. 9.1. Parasites 9.1.1 Aphidiidae. In Minks, A. K. and Harrewijn, P. Aphids, Their biology, Natural Enemies and Control. World crop Pests. 2B, 171-184. Elsevier, Amsterdam.

Subba Rao, B.R. & A.K. Sharma, 1958. *Trioxys (Trioxys) indicus*, new species, a parasite of *Aphis gossypii* Glover. *Indian Journal of Entomology* 20: 199-202.

Subba Rao, B.R. & A.K. Sharma, 1962. Studies on the biology of *Trioxys indicus* Subba Rao and Sharma 1958, a parasite of *Aphis gossypii* Glover. *Proceedings of National Institute of Sciences of India* (B) 28: 164-182.

Szepligeti, G.V., 1904. Braconidae, in Wytsman Genera insectorum Fasc. 22.

Takada, H. & N.D. Rishi, 1980. Records of fifteen species of Aphidiidae (Hymenoptera) from Kashmir, India with descriptions of three new species. *Kontyu* 48: 234-240.

Takada, H., 1968. Aphidiidae of Japan (Hymenoptera). *Insecta Matsumurana* 30: 67-124.

Tremblay, E. & D. Calvert, 1971. Embryosystematics in the Aphidiinae (Hymenoptera: Braconidae). *Bollettino del Laboratorio di Entomologia Agraria Filippo Silvestri* 29: 223–249.

Watanabe, C. & H. Takada, 1964. A note on *Pauesia konoi* (Watanabe) (Hymenoptera: Aphidiidae). *Insecta Matsumurana* 27: 11.

Watanabe, C. & H. Takada, 1964a. Occurrence of two species of the genus *Praon* Haliday Quilis in Japan (Hymenoptera: Aphidiidae). *Insecta Matsumurana* 27: 8-11.

Chapter - 16

PLANT PROTEINASE INHIBITORS AND THEIR POTENTIAL IN INSECT PEST MANAGEMENT

E.R. Prasad[1], Vijayaprasad R. Komera[2], K. Padmasree[1] and A. Dutta-Gupta[2]

[1]Department of Plant Sciences, School of Life Sciences, University of Hyderabad, Hyderabad 500 046, AP, India
[2]Department of Animal Sciences, School of Life Sciences, University of Hyderabad, Hyderabad – 500 046, AP, India
E-mail: *prasader@rediffmail.com*; *kvpr5@yahoo.com*; *kpssl@uohyd.ernet.in*; *apdgsl@uohyd.ernet.in*

ABSTRACT

Plant proteinase inhibitors (PIs) are the natural defense related proteins, which play a potent defensive role against predators and pathogens. They are constitutively present in seeds and induced in certain plant tissues by herbivory or wounding. This chapter describes the classification of PIs, their physiological roles, localization and mechanism of action, as well as applications of plant PIs against insect pests and other applications.

PLANT DEFENSE PROTEINS

Plants have evolved certain defense mechanisms most of which are concentrated in the seeds since these are the vehicles for propagation and survival of the species. Seed tissues may accumulate a wide array of defensive compounds constitutively or after induction that confer resistance against phytophagous predators and infection by viruses, bacteria, fungi, nematodes, etc. (Ryan, 1990; Carlini & Grossi-de-Sá, 2002). The plant proteins that are involved in defense mechanisms are lectins, ribosome-inactivating proteins, inhibitors of proteolytic enzymes and glycohydrolases. The class of inhibitors of proteolytic enzymes comprises proteinase and α-amylase inhibitors. Other plant proteins involved in the complex mechanisms of defense are the arcelins, chitinases, canatoxin and modified forms of storage proteins (Sales *et al.*, 2000; Carlini & Grossi-de-Sá, 2002). However, the components of defense system of some plants might have been lost during the artificial selection imposed by domestication. For instance, arcelins, a family of proteins displaying insecticidal properties are now found only in wild accessions of the *Phaseolus vulgaris*, and coding genes have somehow been silenced in the domesticated plant. Similarly, the susceptibility of the *Vigna unguiculata* to predation by *Callosobruchus maculatus* is supposed to be related to an increased digestibility of vicilins (a major class of storage proteins) by the insect's enzymes, as compared to the protein found in resistant seeds of Nigerian cowpea (Sales *et al.,* 2000).

PROTEINASES

Proteinases or proteolytic enzymes catalyze the cleavage of peptide bonds in proteins. The term "protease" includes both "endopeptidases" and "exopeptidases" whereas, the term "proteinase" is used to describe only "endopeptidases" (Ryan, 1990; Fan & Wu, 2005). Since most of the plant inhibitors target mainly endopeptidases, the term 'proteinase inhibitor' is used. Proteinases are classified according to their mechanism of catalysis and the amino acid present in the active center as serine proteinases, with a serine and histidine; cysteine proteinases, with a cysteine; aspartic proteinases, with an aspartate group and metalloproteinases, with a metallic ion (Fan and Wu, 2005). Proteolysis is a key process in all living organisms, but must be highly regulated which otherwise could be very hazardous to their natural environment (Bode & Huber, 2000). Hence, it is not surprising that a large number of naturally occurring PIs have been described in animals, plants and microorganisms and have been extensively studied in order to elucidate their structural and functional properties.

PROTEINASE INHIBITORS

The proteinases inhibitors (PIs) are the natural antagonists of proteinases. They are small proteins which are quite common in nature and also present in all life forms. Most of the PIs interact with their target proteinases by contact with the active (catalytic) site of the proteinase, resulting in the formation of a stable proteinase-inhibitor complex and loss of enzymatic activity (Bode & Huber, 2000). In plants, PIs are present as one of the most abundant class of proteins. Most storage organs such as seeds and tubers contain PIs, which inhibits different types of enzymes. Seeds usually accumulate PIs during maturation and the concentrations of these inhibitors vary between 1-10% of the total seed proteins. In addition PI gene expression was also detected in leaves in response to wounding caused by herbivory (Ryan, 1990). Because of their ability to inhibit the enzymes involved in the digestive processes of humans and animals, plants PIs have been referred to as "antinutritional compounds" so far. Some of them have now been favorably reconsidered in view of the potential exploitation of their biological properties in pharmacological and medical applications (Scarafoni *et al.,* 2007). Majority of PIs studied in plant kingdom originates from three main families namely Leguminoceae, Solanaceae and Gramineae.

CLASSIFICATION OF PLANT PIs

Proteinases are classified based on the active site residues involved in hydrolysis, as serine, cysteine, aspartic and metallo proteinases. Hence, the classification of PIs generally follows the proteinase catalytic type. Thus there are many families of the inhibitors that inhibit serine proteinases, several families that inhibit cysteine proteinases and others that inhibit aspartate and metallo-proteinases (Table 1). In addition, there are multi-headed inhibitors which have more than one active site on one PI peptide chain and PIs that consists of multiple peptide chains with different specificities, and as well as multifunctional PIs that have only one active site but inhibit more than one PI class (Bode and Huber, 2000).

Table 1. Classification of plant proteinse inhibitors

Proteinase inhibited	Name of the PI family	Molecular mass	Distribution
Serine	Kunitz inhibitor	19-24 kDa	Legumes, potato, cereals and *Arabidopsis*.
	Bowman-Birk inhibitor	6-9.5 kDa (cereals 14-15 kDa)	Legumes, maize, rice, wheat and barley
	Cereal trypsin and α-amylase inhibitor	11.5-14 kDa	Barley, wheat, sorghum, ragi, rice and rye
	Proteinase inhibitor 1	7.2-9.1 kDa	Solanaceous species, cereals, squash (seed and phloem), *Arabidopsis* and legumes
	Proteinase inhibitor 2	5-6 kDa	Solanaceous species. Found in tubers, fruits, seeds, leaves and flowers
	Mustard trypsin inhibitor	6.6-7.1 kDa	Brassicaceae, mustard, rape and *Arabidopsis*
	Squash inhibitor	3-3.5 kDa	Cucurbit seeds
	Serpins	4.2-4.4 kDa	Wheat, rice, barley, squash, *Arabidopsis*, soybean and cotton
Cysteine	Cystatins	10-16 kDa	Rice, maize, wheat, potatoes, soybean and *Arabidopsis* Found in seeds, tubers, pollen, flowers and leaves

	Multicystatin	6-8.7 kDa	Potato (8 active sites), sunflower (3 active sites) and other species
	Kunitz inhibitor	20 kDa	Potato
	Pineapple bromel-lain inhibitor	5.8-5.9 kDa	Pineapple
Aspartic	Kunitz	20-21 kDa	Potato
	SQAPI	10.5 kDa	Squash
	Wheat inhibitor	58 kDa	Wheat
Metallo	Carboxy-Peptidase inhibitor	4.1-4.3 kDa	Solanum

(i) Serine PIs

Serine PIs are universal throughout the plant kingdom and have been described in many plant species and the number of known and partially characterized serine PIs are enormous (Haq *et al.,* 2004). All the serine PI families shown in table 1 are competitive inhibitors and inhibit the proteinases with a similar standard mechanism (Bode & Huber, 2000). However, some non-competitive type of inhibitors are also reported (Macedo *et al.,* 2003; Araújo *et al.,* 2005; Prasad *et al.,* 2010 a,b). In general, serine PIs behave as pseudo-substrates, with the amino acid at position P_1 of the inhibitor determining the specificity for the enzyme, either trypsin or chymotrypsin (Bode & Huber, 2000). Some of the plant serine PIs are bifunctional molecules being able to inhibit trypsin as well as α-amylase (Haq *et al.,* 2005). The different families of serine PIs are as follows,

Kunitz-inhibitor family: The Kunitz inhibitor family is a widespread family first described in legumes, but also found in wide range of species including *Arabidopsis,* cereals, and solanaceous species. A protein exhibiting the activity against trypsin was isolated from soybean (*Glycine max*) and crystallized in 1945 (Kunitz, 1945) and was later named as the soybean Kunitz trypsin inhibitor (SKTI). This was the first PI to be purified from a plant source and characterized in detail. The SKTI contains 181 amino acid residues and has two disulfide bonds. The active site of the inhibitor, localized inside a peptide loop (limited by the disulfide bond between Cys39 and Cys86 residues) contains amino acid residues Arg63-Ile64 in the position P1-P1'.

Further large number of Kunitz type inhibitors was detected in leguminous plants, mostly belonging to subfamilies Mimosoidae and Caesalpinoidae, which comprise more ancient representatives of the family (Norioka *et al.,* 1988). The majority of the inhibitors appear as proteins with a molecular mass of about 20 kDa, containing 160-200 amino acid residues joined together into a single polypeptide chain and two disulfide bonds (Macedo *et al.,* 2003; Garcia *et al.,* 2004; Araújo *et al.,* 2005; Liao *et al.,* 2007; Oliveira *et al.,* 2007; Bhattacharyya & Babu, 2009). However, some Kunitz PIs are also composed of two polypeptide chains of unequal size, which are linked by a disulfide bond (Zhou *et al.,* 2008; Lopes *et al.,* 2009). The two chains are formed as a result of proteolytic cleavage of the original single chain molecule and it is not clear why the presence of two-chain inhibitors is limited to the species of the subfamily Mimosoidae in leguminous plants.

The majority of inhibitors from the Kunitz family have one active site. However, the inhibitor from asparagus pea (*Psophocarpus tetragonolobus* L.) is an exception, since it is capable of binding two chymotrypsin molecules at the same time. In general, the PIs contain several isoforms of an inhibitor. Thus, at least ten SKTI genes are found in the soybean genome, which are expressed in distinct parts of the plant or at different stages of its development (Krishnan, 2001). Individual isoforms may differ considerably in their effects on the enzymes.

Bowman-Birk Inhibitors (BBIs) family: The BBI family is one of the most widespread groups of serine PIs, with members found in many botanical families. The family is named after Bowman (1946) and Birk (Birk *et al.,* 1963), who were the first to identify and characterize a member of this family from soybean. The soybean BBI is now the most-well studied member of this family and is often referred as the classic BBI. These are particularly abundant in legume seeds compared with their concentrations in other legume organs or in other plant families. The BBI type PIs have been mainly found in legumes and cereals (Laing & McManus, 2002).

BBI from legumes such as soybean and pea, are proteins consist of a single polypeptide chain with molecular masses in the range of 6-9 kDa (Clemente & Domoney, 2006). They usually comprise two distinct binding loops, responsible for the inhibition of two enzyme molecules that may be same or different enzymes (Singh & Appu Rao, 2002). Such inhibitors can interact simultaneously and independently with two target proteinases without any substantial conformational change. BBI type PIs interact with the enzymes *via* their binding loops that adopt a characteristic canonical proteinase inhibitory conformation, similar to that of a bound substrate and the resulting non-covalent complex inactivate the proteinase (Qi *et al.,* 2005; Clemente & Domoney, 2006). The first reactive site in these inhibitors is usually inhibits trypsin and second reactive site inhibits either trypsin or chymotrypsin or elastase (Qi *et al.,* 2005).

The importance of these inhibitory sites differs from each other with respect to biological significance (Clemente & Domoney, 2006). The trypsin inhibitory site has been associated with protection of plants against insects and microorganisms, and also showed negative effect on bioavailability of dietary proteins. Conversely, the chymotrypsin inhibitory site has been implicated in cancer chemopreventive effects (Kennedy, 1998). The active site configuration in these inhibitors is stabilized by the presence of seven conserved disulfide bonds (Qi *et al.,* 2005).

BBIs from monocotyledonous plants are of two types. One group consists of a single polypeptide chain with a molecular mass of about 8 kDa and consists of single reactive site. Another group has a molecular mass of 16 kDa with two reactive sites (Qi *et al.,* 2005). In the case of double-headed BBIs, it has been found that the relative affinity of binding of proteases is altered when one site is already occupied.

The presence of different isoforms among BBI type PIs within the species is well-documented (Prasad *et al.,* 2010 a,b). This polymorphism can be attributed both to the expression of distinct genes and to post-translational modifications of the primary gene products (Domoney *et al.,* 2002). Recently cyclic peptides based on inhibitory loops have been used as molecular tools with which the functional and biological properties of BBI and its variants can be investigated (Losso, 2008). Interestingly, a naturally occurring cyclic peptide has been isolated from sunflower (*Helianthus annuus*) that inhibits trypsin strongly (K_i = 100 pM) and shows homology to sequence within the inhibitory domain of BBI-like proteins, although derived apparently from a precursor that is unrelated to BBI (Mulvenna *et al.,* 2005). Many aspects of the biology of BBI proteins have been investigated, ranging from structural features to functional and biological properties (Losso, 2008).

Cereal trypsin/α-amylase inhibitor family: The members of this family have serine proteinase inhibitory activity and/or α-amylase inhibitory activity. A large number of inhibitors in this family have only α-amylase inhibitory activity, however inhibitors from barley (*Hordeum vulgare*), rye (*Secale cereale*), tall fescue (*Festuca arundinacea*), maize (*Zea mays*) and ragi (*Elusine coracana*) are active against trypsin. The inhibitors of this family consist of a single polypeptide chain with a molecular mass of about 13 kDa, and possessing five disulfide bonds (Christeller and Laing, 2005).

Potato inhibitor I (PI-1) family: The inhibitors of this family are widespread in plants and have been described in many species, including potato tubers, tomato fruit, bitter gourd seeds, squash phloem exudates and in tomato leaves in response to wounding (Habib & Fazili, 2007). These inhibitors have the molecular mass of 8 kDa and are generally monomeric. However isoinhibitors from potato tubers form an oligomeric structure consistent with a pentamer on molecular sieving chromatography. PI-1 family inhibitors are unusual in that they generally lack disulphide bridges, which are typical of most inhibitor families, acting to assist in stabilizing the inhibitor structure. However, Cucurbit and potato tuber inhibitors possess a single disulphide bond, but this is not involved in stabilizing the reactive site loop. The inhibitory mechanism in this family is considered to fit the standard model (Christeller and Laing, 2005).

Potato inhibitor II (PI-2) family: The inhibitors of this group have been reported only from the members of Solanaceae family. These inhibitors have been found in leaves, flowers, fruit and phloem of other solanaceaous species (Christeller & Laing, 2005). A low molecular-mass inhibitor of this family has been found to be constitutively present in Jasmme tobacco (*Nicotiana alata*) flowers. Six small wound-inducible PIs of this family have been reported from tobacco leaves. An analysis of isolated inhibitor II proteins and genes has shown that they are composed of multiple repeat units, varying from one to eight (Christeller & Laing, 2005). Inhibitors in this family have been reported to inhibit chymotrypsin, trypsin, elastase, oryzin, Pronase E and subtilisin.

Mustard trypsin inhibitor (MSI) family: These are small single polypeptide chain inhibitors with the molecular mass of about 7 kDa, found in the family Brassicaceae, which includes *Arabidopsis* (Laing & McManus, 2002). These inhibitors have been isolated and characterized from a number of species including white mustard (*Sinapis alba*) and rape (*Brassica napus*). These inhibitors are expressed in seeds during their development and are also wound-inducible. There is a possibility that these inhibitors evolved from BBIs as the two inhibitors show about a 27% similarity of amino acids over 63 residues that make up the mature mustard seed inhibitor (Christeller & Laing, 2005). The inhibitors followed the standard model in forming a tight binding complex with trypsin, and being cleaved by trypsin (Habib & Fazili, 2007).

Squash inhibitors family: The squash inhibitor family is a plant unique inhibitor family, have been described only in plants, and particularly form the seeds of Cucurbitaceae. It is the smallest inhibitor family known, consist of a small single peptide chain containing between 28 and 30 amino acids with molecular mass of 3-3.5 kDa, and P_1 at ~5th amino acid from the N-terminal side. The squash inhibitors also have three disulphide bridges and fold in a novel knottin structure with similarity to potato carboxypeptidase inhibitor. Recently two different but inter-convertible (cis-trans isomers) inhibitors have been isolated and characterized from seeds of wax gourd (*Benincasa hispida*) (Atiwetin *et al.,* 2006). The small size of these inhibitors, combined with potential activity against important biological molecules such as Hageman factor, human leucocyte elastase and cathepsin G, has made them particularly attractive for studying proteinase and inhibitor interactions (Habib & Fazili, 2007).

Serpin family: Plant serpins have been identified, purified and characterized from cereal seeds, from the phloem exudate of squash, and possibly from various pollens (Christeller & Laing, 2005). Analysis of the *A. thaliana* genome indicates at least seven serpin genes, similar to the number found in cereals, suggesting that serpins may be more widespread in plants than is currently confirmed (Laing & McManus, 2002). Plant serpins appear to be exclusively inhibitors of serine proteinases, including trypsin, elastase and chymotrypsin. They have a molecular mass around 39-43 kDa, with amino acid and nucleotide sequence homology to other well characterized serpins. Plant serpins show a similar suicide substrate mechanism to other serpins, with essent-ially irreversible inhibition and the inhibitor-proteinase complex stable to boiling in SDS, indicative of covalent acyl bond formation (Christeller & Laing, 2005).

(ii) Cysteine PIs

Plant cystatins or phytocystatins are the second most studied class of inhibitors and have been identified and characterized from several plants like sunflower, rice, wheat, maize, soybean, sugarcane, cowpea, potato, cabbage, ragweed, carrot, papaya, apple fruit, avocado, chestnut, Job's tears, etc (Haq *et al.,* 2004). The transgenic plants expressing cysteine PIs have shown the most promising results, probably because most phytophagous insects employ these proteinases in their digestive mechanism. The rice cysteine PIs are well studied of all the cysteine PIs and are highly heat stable.

(iii) Aspartic PIs

Aspartic PIs are a relatively less studied class partly due to their rarity of occurrence. Potato tubers possess an aspartic proteinase (cathepsin D) inhibitor that shares considerable amino acid sequence identity with SKTI (Haq *et al.,* 2004). Cathepsin D inhibitor is the only well-characterized aspartic PI. The cathepsin D inhibitor (27 kDa) is unusual as it inhibits trypsin and chymotrypsin as well as cathepsin D, but does not inhibit aspartyl proteases such as pepsin, rennin or cathepsin E. Aspartic PIs have also been isolated from sunflower, barley and cardoon (*Cyanara cardunculus*) flowers named as cardosin A. The squash (*Cucurbita maxima*) phloem exudates expressed a novel aspartic PI (SQAPI), constituting a fifth family of aspartic PIs. However, a comparison of the SQAPI sequence to the phytocystatin (a cysteine PI) family sequences showed 30% identity (Christeller *et al.,* 2006).

(iv) Metallo PIs

The metallo-proteinase inhibitors in plants are represented by the metallo-carboxypeptidase inhibitor family in tomato and potato plants (Haq *et al.,* 2004). The inhibitors of the metallo-carboxypeptidase from tissue of tomato and potato are polypeptides (4 kDa) that strongly and competitively inhibit a broad spectrum of carboxypeptidases from both animals and microorganisms, but not the serine carboxypeptidases from yeast and plants (Habib & Fazili, 2007). The inhibitor is found in tissues of potato tubers where it accumulates during tuber development along with potato inhibitor I and II families of serine PI. The inhibitor also accumulates in potato leaf tissues along with inhibitor I and II proteins in response to wounding (Ryan, 1990).

PHYSIOLOGICAL ROLES OF PIs

Besides their protective role against insects and microbial pests, there are other important physiological roles for PIs in seeds, which are summarized below:

Endogenous function: PIs regulate endogenous proteinase levels before and during seed germination for storage protein digestion and to control protein turnover. Arg-1, a proteinase isolated from *Ipomoea batatas* was completely inhibited by sweet potato trypsin inhibitor (Hou and Lin, 2002). Proteinase inhibitor II (PIN2) from *Solanum americanum* not only enhanced resistance to caterpillars when expressed exogenously, but also inhibited endogenous proteases that are expressed during seed development (Chye *et al.,* 2006).

As storage proteins: The PIs serving as storage proteins in plants was identified by the presence of high content of inhibitors in seeds and other storage organs of plants and of their dynamics in the course of seed maturation and germination (Mosolov & Valueva, 2005). It was also reported that certain PIs and storage proteins had common origin, and belong to the same protein family. Furthermore, certain storage proteins of plants exhibit activities of PIs. A 2S storage protein from the seed of *Brassica juncea* is formed via processing of a precursor exhibiting the activity of a trypsin inhibitor (Mandal *et al.,* 2002).

Modulation of apoptosis: A new role for PI in the modulation of apoptosis or programmed cell death (PCD) has been identified. Cysteine proteinase plays an important role in the regulation of programmed cell death leading to hypersensitive (HR) reaction, following pathogen attack. It has been shown that the ectopic expression of cystatin inhibits the induced cysteine proteinase activity, which in turn blocks PCD (Solomon *et al.,* 1999). Sin and Chye (2004) demonstrated that *S. americanum* PI-II proteins, SaPIN2a and SaPIN2b are expressed in floral tissues that are destined to undergo developmental PCD, suggesting possible endogenous roles in inhibiting trypsin- and chymotrypsin-like activities during flower development.

LOCALIZATION OF PIs

In plants, PIs are produced either as prepro-proteins or pre-proteins, which are subsequently subjected to *in vivo* processing. Furthermore, they have been subcellularly localized to various compartments in the plant cell. Similar to other storage proteins, inhibitors are located in vacuoles and protein bodies. The SKTI was reported to be associated with the cell walls and to a lesser extent with protein bodies, cytosol, and nuclei, while the BBI has been localized in protein bodies, nuclei, and cytosol (Fan and Wu, 2005). However, wound-induced inhibitors of tomato, potato, and *S. americanum* (Sin and Chye, 2004) are located in vacuoles. Xu et al. (2004) described the expression of a PIN2 protein of *S. americanum* in phloem of stems, roots, and leaves suggesting a novel endogenous role for PIN2 in phloem. A chickpea Kunitz trypsin inhibitor is located in the cell wall of different organs (Hernández-Nistal *et al.,* 2009).

MECHANISM OF ACTION OF PIs AGAINST INSECT PESTS

The mechanism of binding of the plant PIs to the insect proteinases appears to be similar with all the four classes of inhibitors. The inhibitor binds to the active site of the enzyme to form a complex with a very low dissociation constant (10^{-7} to 10^{-14} M at neutral pH values), thus effectively blocking the active site (Lawrence and Koundal, 2002). A binding loop conformation on the inhibitor formed by a disulphide bond, projects from the surface of the molecule and contains a peptide bond (reactive site) cleavable by the enzyme. This peptide bond will be cleaved in the enzyme inhibitor complex, but the cleavage does not affect the interaction. The inhibitor thus directly mimics a normal substrate for the enzyme, but does not allow the normal enzyme mechanism of peptide bond cleavage to proceed to completion i.e., dissociation of the product. Thus PIs inhibit the proteinase activity of insect gut proteolytic enzymes and reduce the quantity of proteins that can be digested, and also cause hyper production of the digestive enzymes which enhances the loss of sulfur amino acids, and as a result of which, the insects become weak with stunted growth and ultimately die (Ramos *et al.,* 2009). Retardation of insect development, slower rate of growth and reduced fitness for survival would allow a much wider window within which the other pesticides including biopesticides could be successfully employed for the management of insects.

APPLICATIONS OF PIs IN COMBATING INSECT PESTS

The possible role of PIs in plant protection was well documented. Following the early studies, there have been many examples of PIs active against certain insect species, proved by both *in vitro* assays against insect gut proteases (Giri *et al.,* 2003; Oliveira *et al.,* 2007; Pereira *et al.,* 2007; Prasad *et al.,* 2009) and in *in vivo* artificial diet bioassays (Bhattacharyya *et al.,* 2007; Ramos *et al.,* 2009; Telang *et al.,* 2009; Prasad *et al.,* 2010c). On the other hand, the PI gene expression has been detected in leaves of several species following wounding caused by herbivory. These studies suggest their role in protecting plants from insect attack and microbial infection (Ryan, 1990).

Plant PIs as inhibitors of gut proteinases: In vitro assays are the initial assessment to check the potentiality of plant PIs as insecticidal agents against insect pests. PIs isolated from most of the plants show inhibitory activity against various lepidopteran, coleopteran and dipteran insect gut proteinases. Information regarding the *in vitro* effect of plant PIs against gut proteinases of different insect pests is provided in the Table 2 and these inhibition studies are further substantiated using the casein-SDS-PAGE. For instance, the zymogram of midgut homogenate of *Achaea janata* shows the inhibition of higher molecular mass proteinases in presence of red gram PI (Fig. 1).

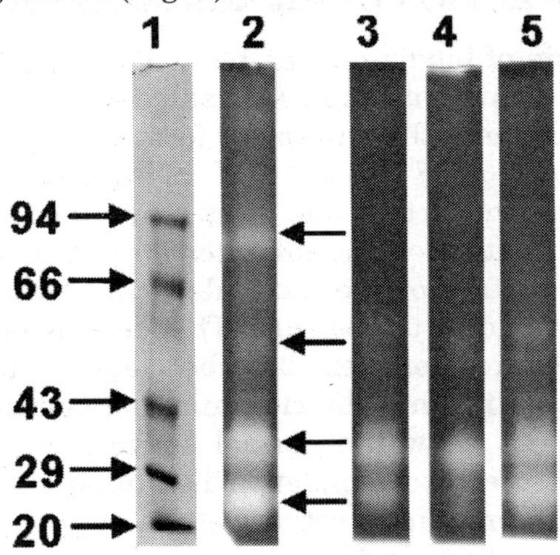

Fig. 1. Casein-SDS-PAGE (7.5%) showing zymogram of larval midgut proteinases of *Achaea janata*. Lane 1, marker; lanes 2-5 loaded with midgut proteinases. After electrophoresis, gel was incubated with 100 mM glycine-NaOH, pH 10.5 (lane 2, control); SKTI (lane 3); BBI (lane 4); and PI purified from *C. cajan* (lane 5). The inhibitors (100 µg/ml) were dissolved in above buffer and used.

Table 2. *In vitro* effect of plant PIs against gut trypsin-like proteinases

PI source	Insect	Order	Reference
Cajanus cajan	*Manduca sexta*	Lepidoptera	Prasad *et al.*, 2010a
	A. janata, Spodoptera litura	Lepidoptera	Prasad *et al.*, 2009; Prasad *et al.*, 2010c
	Helicoverpa armigera	Lepidoptera	Chougule *et al.*, 2003; Prasad *et al.*, 2010c
	Papilio demoleus, Amsacta albistriga, Daphnis nerii	Lepidoptera	Prasad *et al.*, 2010c
Vigna mungo	*A. janata, S. litura, H.*	Lepidoptera	Prasad *et al.*, 2010c

	armigera, P. demoleus, A. albistriga, D. nerii		
Glycine max	*Mamestra brassicae*	Lepidoptera	Chougule *et al.*, 2008
	Anthonomus grandis	Coleoptera	Franco *et al.*, 2004
	Tribolium castaneum	Coleoptera	Oppert *et al.*, 2003
Cicer arietinum	*Spodoptera frugiperda*	Lepidoptera	Gomes *et al.*, 2005a
	A. grandis	Coleoptera	
	H. armigera	Lepidoptera	Srinivasan *et al.*, 2005
Prosopis julifora	*Acanthoscelides obtectus C. maculatus*	Coleoptera	Oliveira *et al.*, 2002
	T. castaneum, C. maculatus	Coleoptera	Sivakumar *et al.*, 2005
Momordica charantia	*H. armigera, S. litura*	Lepidoptera	Telang *et al.*, 2003
Vigna unguiculata	*A. grandis*	Coleoptera	Franco *et al.*, 2003
Psophocarpus tetragonolobus	*H. armigera*	Lepidoptera	Giri *et al.*, 2003; Telang *et al.*, 2009; Harsulkar *et al.*, 1999
Lycopersicon esculentum	*H. armigera*	Lepidoptera	Damle *et al.*, 2005
Hyptis suaveolens	*Prostephanus truncatus*	Coleoptera	Aguirre *et al.*, 2004
Crotalaria pallida	*S. frugiperda*	Lepidoptera	Gomes *et al.*, 2005b
	C. maculatus	Coleoptera	
Pithecellobium dumosum	*Zabrotes subfasciatus,*	Coleoptera	Oliveira *et al.*, 2007
	Ceratitis capitata	Diptera	
Archidendron ellipticum	*S. litura*	Lepidoptera	Bhattacharyya *et al.*, 2007
Dimorphandra mollis	*C. maculatus*	Coleoptera	Macedo *et al.*, 2002
Phaseolus coccineus	*Hypothenemus hampei*	Coleoptera	Pereira *et al.*, 2007
Plathymenia foliolosa	*Anagasta kuehniell*	Lepidoptera	Ramos *et al.*, 2009
Capsicum annum	*H. armigera*	Lepidoptera	Tamhane *et al.*, 2005; Tamhane *et al.*, 2007
Albizzia kalkora	*H. armigera, S. exigua, Pieris rapae*	Lepidoptera	Zhou *et al.*, 2008

Tamarindus indica	*C. maculatus*	Coleoptera	Araújo *et al.*, 2005
	Rhyzopertha dominica		
	S. frugiperda	Lepidoptera	
	Alabama argillacea		
	C. capitata	Diptera	
Poecilanthe parviflora	*Diatraea saccharalis, Corcyra cephalonica, S. frugiperda*	Lepidoptera	Garcia *et al.*, 2004
Cassia obtusifolia	*Pieris rapae*	Lepidoptera	Liao *et al.*, 2007
Peltophorum dubium	*Anagasta kuehniella*	Lepidoptera	Macedo *et al.*, 2003
Adenanthera pavonina	*C. maculatus*	Coleoptera	Macedo *et al.*, 2004

The feeding experiments conducted on whole leaves of wild groundnut species, *Arachis correntina*, with *Spodoptera litura* larvae showed that the feeding affects the larval development, storage protein levels in the haemolymph and physiological processes like molting and emergence of pupa from larvae (unpublished data). Recently, three proteinase inhibitors have been purified from *A. correntina* leaves, which showed PI activity in gelatin-zymogram assay (data not shown).

In vivo bioassays for insecticidal activity of plant PIs: Based on the of *in vitro* activity reports, the capability of plants PIs was further checked on the growth and development of various insect pests. The *in vivo* bioassays were done by artificial diet assays or leaf coating assays and the reports of various research groups are summarized in Table 3.

Table 3. *In vivo* bioassays of plant PIs against growth and development of insect pests

PI source	Insect	Order	Reference
Cajanus cajan	*M. sexta*	Lepidoptera	Prasad *et al.*, 2010a
	A. janata, S. litura	Lepidoptera	Prasad *et al.*, 2010c
Vigna mungo	*A. janata, S. litura*	Lepidoptera	Prasad *et al.*, 2010c
Glycine max	*Mamestra brassicae*	Lepidoptera	Chougule *et al.*, 2008
	A. grandis	Coleoptera	Franco *et al.*, 2004
	T. castaneum	Coleoptera	Oppert *et al.*, 2003
	Ceratitis capitata	Diptera	Silva *et al.*, 2006
Cicer arietinum	*S. frugiperda*	Lepidoptera	Gomes *et al.*, 2005a
	A. grandis	Coleoptera	
	H. armigera	Lepidoptera	Srinivasan *et al.*, 2005

Momordica charantia	*H. armigera, S. litura*	Lepidoptera	Telang *et al.*, 2003
Vigna unguiculata	*A. grandis*	Coleoptera	Franco *et al.*, 2003
Psophocarpus tetragonolobus	*H. armigera*	Lepidoptera	Telang *et al.*, 2009; Harsulkar *et al.*, 1999
Lycopersicon esculentum	*H. armigera*	Lepidoptera	Damle *et al.*, 2005
Crotalaria pallida	*S. frugiperda*	Lepidoptera	Gomes *et al.*, 2005b
	C. maculatus	Coleoptera	
Archidendron ellipticum	*S. litura*	Lepidoptera	Bhattacharyya *et al.*, 2007
Dimorphandra mollis	*C. maculatus*	Coleoptera	Macedo *et al.*, 2002
Plathymenia foliolosa	*Anagasta kuehniell*	Lepidoptera	Ramos *et al.*, 2009
Capsicum annum	*H. armigera*	Lepidoptera	Tamhane *et al.*, 2005; Tamhane *et al.*, 2007
Tamarindus indica	*C. maculatus*	Coleoptera	Araújo *et al.*, 2005
	C. capitata	Diptera	
Cassia obtusifolia	*Pieris rapae*	Lepidoptera	Liao *et al.*, 2007
Peltophorum dubium	*Anagasta kuehniella*	Lepidoptera	Macedo *et al.*, 2003
Pisum sativum	*Acyrthosiphon pisum*	Thysanoptera	Rahbé *et al.*, 2003
Adenanthera pavonina	*C. maculatus*	Coleoptera	Macedo *et al.*, 2004

Transgenic plants against insect resistance: The economically important orders of insect pests namely Lepidoptera, Diptera and Coleoptera, use serine and cysteine proteinases in their digestive system to degrade proteins in the ingested food, efforts have generally been directed at genes encoding PIs active against these mechanistic classes of proteinases for developing transgenic plants. The PI genes have been particularly useful in developing transgenic plants resistant to insect pests, as they require the transfer of a single defensive gene, and can be expressed from the wound-inducible or constitutive promoters of the host (Boulter, 1993). The first PI gene successfully transferred was cowpea trypsin inhibitor (CpTI) and the transgenic tobacco produced showed significant resistance against *Heliothis virescens* (Hilder *et al.*, 1987). Later, several transgenic plants expressing PIs produced in the last 20 years have been reviewed extensively (Carlini and Grossi-de-Sá, 2002; Lawrence & Koundal, 2002; Haq *et al.*, 2004; Fan & Wu, 2005). Further, the recent PI genes transferred to various plant species and the resulting transgenic plants were summarized in Table 4.

Table 4. Transgenic plants expressing proteinase inhibitors

Plant transformed	Gene used for transgenesis	Pest targeted	Reference
Tobacco	Cowpea trypsin inhibitor (CpTI)	*H. virescens, S. frugiperda*	Pujol *et al.,* 2005
	SKTI, C-II and PI-IV inhibitor	*Spodoptera littoralis*	Marchetti *et al.,* 2000
	Solanum americanum PI (*SaPIN2a*)	*H. armigera, S. litura*	Luo *et al.,* 2009
	Nicotiana tobacum PI	*H. armigera, S. litura*	Srinivasan *et al.,* 2009
Potato	SKTI, C-II and PI-IV inhibitor	*S. littoralis*	Marchetti *et al.,* 2000
Rice	Maize PI (mpi)	*C. suppressalis*	Vila *et al.,* 2005
	BTI-CMe	*Sitophilus oryzae*	Alfonso-Rubí *et al.,* 2003
Sugarcane	SKTI and soybean BBI	*Diatraea saccharalis*	Falco and Silva-Filho, 2003
Pigeonpea	CpTI	*H. armigera*	Lawrence and Koundal, 2001
Apple	*Nicotiana alata* protease inhibitor	*Epiphyas postvittana*	Maheswaran *et al.,* 2007
Oil seed rape	Oryzacystatin I	*Myzus persicae*	Rahbé *et al.,* 2003
Poplar	*Arabidopsis thaliana* cysteine PI (*Atcys*)	*Chrysomela populi*	Delledonne *et al.,* 2001

The co-evolution of plants and their predators has led to an adaptation of phytophagous insects to the deterrent compounds present in the plants they feed on. There are several reports in the literature describing the lack of effect of plant proteinase and α-amylase inhibitors upon the digestive enzymes of the insects adapted to a particular plant species (Jongsma & Bolter, 1997). However the same inhibitors on the other hand are very efficient in blocking enzymes from mammalian source or from other insects that do not feed on that plant. Thus, the genetic engineering of plants aiming to increase resistance to insect predation or to phytopathogens may rely exclusively on the repertoire of genes found in plants (Carlini & Grossi-de-Sá, 2002). Hence it is necessary to screen continuously for potential PIs from non-host plants and wild types to circumvent the adaptation of insect pests to PIs (Harsulkar *et al.,* 1999; Prasad *et al.,* 2009; Prasad *et al.,* 2010c).

OTHER APPLICATIONS OF PIs

PIs are well established as a class of cancer chemopreventive agents (Kennedy, 1998). Whereas PIs from different families have been shown to prevent the carcinogenic process, the most potent of the known anticarcinogenic PIs are those with the ability to inhibit chymotrypsin-like proteinases. The BBI from soybean is until now the PI that has been most studied as an anticarcinogenic agent. The soybean derived BBI with a well characterized ability to inhibit trypsin and chymotrypsin is particularly effective in suppressing carcinogenesis in a variety of *in vivo* and *in vitro* systems (Clemente *et al.*, 2005; Kennedy *et al.*, 2003). A crude BBI concentrate (BBIC) was developed by Kennedy (1998) to allow clinical studies based on the properties of BBI. The BBIC has undergone preclinical, phase I, and II of clinical trials (Losso, 2008).

Epidemiological studies have suggested that people who consume soybean or pulses regularly have a reduced rate of some degenerative diseases that strike later in life. The BBI had more successful stories in the medical or paramedical scientific community than in the nutrition scientific community. The BBI showed the property of prevention of hair and weight loss in animal models, increased life span, ability to potentiate the cytotoxicity of cis-platinum in lung cancer treatment, ability to prevent and not cause some birth defects (Losso, 2008). Kennedy (1998) pointed out that people from Asia who are regularly on a high soybean diet reproduce themselves without any sign of BBI toxicity.

Chymase is a chymotrypsin-like serine proteinase found in mast cells of many mammalian species. It is a chemoattractant that may catalyze the accumulation of inflammatory cells in the development of chronic inflammatory response of allergic and non-allergic diseases (Tani *et al.*, 2000). The chemotactic activity of chymase was inhibited by BBI, suggesting that the hydrolytic activity of chymase may be responsible for its chemotactic activity (Tani *et al.*, 2000). Inflammation associated with tumor promotion was inhibited by BBI (Kennedy, 1998). US Patent 6,767,564 covers the use of BBI for the treatment of multiple sclerosis and other autoimmune diseases (Morris *et al.*, 2005). Serine PIs have also been shown to control abnormal exocytosis and secretion of endogenous proteinases, which is characteristic of a number of diseases (Fook *et al.*, 2005).

Diet has shown the ability to affect the onset or progression of chronic diseases, but only recently that the molecular basis of diseases prevention by dietary involvement has been under scientific consideration (Losso, 2008). Functional foods are foods consumed as part of a normal diet rather than pills. PIs, especially BBIs fit the definition of a functional food. Proteinase levels in the body are biomarkers of health and performance. Furthermore, on ingestion of PIs they form complexes with digestive proteinases in the intestinal tract, which are subsequently excreted. The deficiency of active digestive proteinases in the small intestine induces the further synthesis of proteolytic enzymes by the pancreas, as a result of the suppression of the negative feedback regulation of pancreatic secretion through increased release of the hormone cholecystokinin (CCK) from the intestinal mucosa (Losso, 2008). The CCK release has been associated with the feeling of satiety or fullness and reduced hunger in humans and animal models. If CCK production has to be increased in humans as a result of PIs in the diet without the enlargement of the pancreas, then intake of those PIs may also be good for obesity control as they provide the feeling of fullness and satiety (Losso, 2008).

ACKNOWLEDGMENTS

This work was supported by grants from Institute of Life Sciences, University of Hyderabad (No. EFL/II/CS-MoU/112/1866/14-12-2004), UGC, New Delhi, India (No. F.31-166/2005 (SR) to K.P.S and CSIR, New Delhi, India (No. 38(1009)/01/EMR-II) to A.D.G., and Junior/Senior Research Fellowship from UGC to E.R.P. and CSIR to K.V.P. The facilities generated under DST-FIST-II, DBT-CREBB and UGC-CAS-I were used for the present study.

REFERENCES

Aguirre, C., S. Valdés-Rodríguez, G. Mendoza-Hernández, A. Rojo-Domínguez & A. Blanco-Labra, 2004. A novel 8.7 kDa protease inhibitor from chan seeds (*Hyptis suaveolens* L.) inhibits proteases from the larger grain borer *Prostephanus truncatus* (Coleoptera: Bostrichidae). *Comp. Biochem. Physiol. B* 138: 81-89.

Alfonso-Rubí, J., F. Ortego, P. Castañera, P. Carbonero & I. Díaz, 2003. Transgenic expression of trypsin inhibitor CMe from barley in *indica* and *japonica* rice, confers resistance to the rice weevil *Sitophilus oryzae*. *Transgenic Res.* 12: 23-31.

Araújo, C.L., I.W.L. Bezerra, A.S. Oliveira, F.T. Moura, L.L.P. Macedo, C.E.M. Gomes, A.E.A.D. Barbosa, F.P. Macedo, T.M.S. Souza, O.L. Franco, C. Bloch-J & M.P. Sales, 2005. In vivo bioinsecticidal activity towards *Ceratitis capitata* (fruit fly) and *Callosobruchus maculatus* (cowpea weevil) and in vitro bioinsecticidal activity toward different orders of insect pests of a trypsin inhibitor purified from tamarind tree (*Tamarindus indica*) seeds. *J. Agric. Food Chem.* 53: 4381-4387.

Atiwetin, P., S. Harada & K. Kamei, 2006. Serine proteinase inhibitor from wax gourd (*Benincasa hispida* [Thunb] Cogn.) seeds. *Biosci. Biotechnol. Biochem.* 70: 743-745.

Bhattacharyya, A. & C.R. Babu, 2009. Purification and biochemical characterization of a serine proteinase inhibitor from *Derris trifoliata* Lour.

seeds: insight into structural and antimalarial features. *Phytochemistry* 70: 703-712.

Bhattacharyya, A., S.M. Leighton & C.R. Babu, 2007. Bioinsecticidal activity of *Archidendron ellipticum* trypsin inhibitor on growth and serine digestive enzymes during larval development of *Spodoptera litura*. *Comp. Biochem. Physiol. C* 145: 669-677.

Birk, Y., A. Gertler & S. Khalef, 1963. A pure trypsin inhibitor from soya beans. *Biochem. J.* 87: 281-284.

Bode, W. & R. Huber, 2000. Structural basis of the endoproteinase-protein inhibitor interaction. *Biochim. Biophys. Acta* 1477: 241-252.

Boulter, D., 1993. Insect pest control by copying nature using genetically engineered crops. *Phytochemistry* 34: 1453-1466.

Bowman, D.E., 1946. Differentiation of soy bean anti-tryptic factors. *Proc. Soc. Exp. Biol. Med.* 63: 547-550.

Carlini, C.R. & M.F. Grossi-de-Sá, 2002. Plant toxic proteins with insecticidal properties. A review on their potentialities as bioinsecticides. *Toxicon* 40: 1515-1539.

Chougule, N.P., E. Doyle, E. Fitches & J.A. Gatehouse, 2008. Biochemical characterization of midgut digestive proteases from *Mamestra brassicae* (cabbage moth; Lepidoptera: Noctuidae) and effect of soybean Kunitz inhibitor (SKTI) in feeding assays. *J. Insect Physiol.* 54: 563-572.

Chougule, N.P., V.K. Hivrale, P.J. Chhabda, A.P. Giri & M.S. Kachole, 2003. Differential inhibition of *Helicoverpa armigera* gut proteinases by proteinase inhibitors of pigeonpea (*Cajanus cajan*) and its wild relatives. *Phytochemistry* 64: 681-687.

Christeller, J. & W. Laing, 2005. Plant serine proteinase inhibitors. *Protein Pept. Lett.* 12: 439-447.

Christeller, J.T., P.C. Farley, R.K. Marshall, A. Anandan, M.M. Wright, R.D. Newcomb & W.A. Laing, 2006. The squash aspartic proteinase inhibitor SQAPI is widely present in the cucurbitales, comprises a small multigene family, and is a member of the phytocystatin family. *J. Mol. Evol.* 63: 747-757.

Chye, M.L., S.F. Sin, Z.F. Xu & E.C. Yeung, 2006. Serine proteinase inhibitor proteins: exogenous and endogenous functions. *In Vitro Cell. Dev. Biol. Plant* 42: 100-108.

Clemente, A. & C. Domoney, 2006. Biological significance of polymorphism in legume protease inhibitors from the Bowman-Birk family. *Curr. Protein Pept. Sci.* 7: 201-216.

Clemente, A., J.M. Gee, I.T. Johnson, D.A. Mac Kenzie & C. Domoney, 2005. Pea (*Pisum sativum* L.) protease inhibitors from the Bowman-Birk class influence the growth of human colorectal adenocarcinoma HT29 cells in vitro. *J. Agric. Food Chem.* 53: 8979-8986.

Damle, M.S., A.P. Giri, M.N. Sainani & V.S. Gupta, 2005. Higher accumulation of proteinase inhibitors in flowers than leaves and fruits as a possible basis for differential feeding preference of *Helicoverpa armigera* on tomato (*Lycopersicon esculentum* Mill, Cv. Dhanashree). *Phytochemistry* 66: 2659-2667.

Delledonne, M., G. Allegro, B. Belenghi, A. Balestrazzi, F. Picco, A. Levine, S. Zelasco, P. Calligari & M. Confalonieri, 2001. Transformation of white poplar (*Populus alba* L.) with a novel *Arabidopsis thaliana* cysteine proteinase inhibitor and analysis of insect pest resistance. *Mol. Breed.* 7: 35-42.

Domoney, C., T. Welham, N. Ellis, P. Mozzanega & L. Turner, 2002. Three classes of proteinase inhibitor gene have distinct but overlapping patterns of expression in *Pisum sativum* plants. *Plant Mol. Biol.* 48: 319-329.

Falco, M.C. & M.C. Silva-Filho, 2003. Expression of soybean proteinase inhibitors in transgenic sugarcane plants: effects on natural defense against *Diatraea saccharalis*. *Plant Physiol. Biochem.* 41: 761-766.

Fan, S.G. & G.J. Wu, 2005. Characteristics of plant proteinase inhibitors and their applications in combating phytophagous insects. *Bot. Bull. Acad. Sin.* 46: 273-292.

Fook, J.M.S.L.L., L.L.P. Macedo, G.E.D.D. Moura, F.M. Teixeira, A.S. Oliveira, A.F.S. Queiroz & M.P. Sales, 2005. A serine proteinase inhibitor isolated from *Tamarindus indica* seeds and its effects on the release of human neutrophil elastase. *Life Sci.* 76: 2881-2891.

Franco, O. L., S.C. Dias, C.P. Magalhães, A.C.S. Monteiro, C. Bloch Jr., F.R. Melo, O.B. Oliveira-Neto, R.G. Monnerat & M.F. Grossi-de-Sá, 2004. Effects of soybean Kunitz trypsin inhibitor on the cotton boll weevil (*Anthonomus grandis*). *Phytochemistry* 65: 81-89.

Franco, O.L., R.C. Santos, J.A.N. Batista, A.C.M. Mendes, M.A.M. Araújo, R.G. Monnerat, M.F. Grossi-de-Sá & S.M. Freitas, 2003. Effects of black-eyed pea trypsin/chymotrypsin inhibitor on proteolytic activity and on development of *Anthonomus grandis*. *Phytochemistry* 63: 343-349.

Garcia, V.A., M.G.M. Freire, J.C. Novello, S. Marangoni & M.L.R. Macedo, 2004. Trypsin inhibitor from *Poecilanthe parviflora* seeds: purification, characterization, and activity against pest proteases. *Protein J.* 23: 343-350.

Giri, A.P., A.M. Harsulkar, M.S.B. Ku, V.S. Gupta, V.V. Deshpande, P.K. Ranjekar & V.R. Franceschi, 2003. Identification of potent inhibitors of *Helicoverpa armigera* gut proteinases from winged bean seeds. *Phytochemistry* 63: 523-532.

Gomes, A.P.G., S.C. Dias, C. Bloch Jr., F.R. Melo, J.R. Furtado Jr., R.G. Monnerat, M.F. Grossi-de-Sá & O.L. Franco, 2005a. Toxicity to cotton boll weevil *Anthonomous grandis* of a trypsin inhibitor from chickpea seeds. *Comp. Biochem. Physiol. B* 140: 313-319.

Gomes, C.E.M., A.E.A.D. Barbosa, L.L.P. Macedo, J.C.M. Pitanga, F.T. Moura, A.S. Oliveira, R.M. Moura, A.F.S. Queiroz, F.P. Macedo, L.B.S. Andrade, M.S. Vidal & M.P. Sales, 2005b. Effect of trypsin inhibitor from *Crotalaria pallida* seeds on *Callosobruchus maculatus* (cowpea weevil) and *Ceratitis capitata* (fruit fly). *Plant Physiol. Biochem.* 43: 1095-1102.

Habib, H. & K.M. Fazili, 2007. Plant protease inhibitors: a defense strategy in plants. *Biotechnol. Mol. Biol. Rev.* 2: 68-85.

Haq, S.K., S.M. Atif & R.H. Khan, 2004. Protein proteinase inhibitor genes in combat against insects, pests, and pathogens: natural and engineered phytoprotection. *Arch. Biochem. Biophys.* 431: 145-159.

Haq, S.K., S.M. Atif & R.H. Khan, 2005. Biochemical characterization, stability studies and N-terminal sequence of a bi-functional inhibitor from *Phaseolus aureus* Roxb. (Mung bean). *Biochimie* 87: 1127-1136.

Harsulkar, A.M., A.P. Giri, A.G. Patankar, V.S. Gupta, M.N. Sainani, P.K. Ranjekar & V.V. Deshpande, 1999. Successive use of non-host plant proteinase inhibitors required for effective inhibition of *Helicoverpa armigera* gut proteinases and larval growth. *Plant Physiol.* 121: 497-506.

Hernández-Nistal, J., I. Martín, T. Jiménez, B. Dopico & E. Labrador, 2009. Two cell wall Kunitz trypsin inhibitors in chickpea during seed germination and seedling growth. *Plant Physiol. Biochem.* 47: 181-187.

Hilder, V.A., A.M.R. Gatehouse, S.E. Sheerman, R.F. Barker & D. Boulter, 1987. A novel mechanism of insect resistance engineered into tobacco. *Nature* 330: 160-163.

Hou, W.C. & Y.H. Lin, 2002. Sweet potato (*Ipomoea batatas* (L.) Lam) trypsin inhibitors, the major root storage proteins, inhibit one endogenous serine protease activity. *Plant Sci.* 163: 733-739.

Jongsma, M.A. & C. Bolter, 1997. The adaptation of insects to plant protease inhibitors. *J. Insect Physiol.* 43: 885-895.

Kennedy, A.R., 1998. Chemopreventive agents: protease inhibitors. *Pharmacol. Ther.* 78: 167-209.

Kennedy, A.R., D. Kritchevsky & W.C. Shen, 2003. Effects of spermine-conjugated Bowman-Birk inhibitor (spermine-BBI) on carcinogenesis and cholesterol biosynthesis in mice. *Pharm. Res.* 20: 1908-1910.

Krishnan, H.B., 2001. Characterization of a soybean [*Glycine max* (L.) Merr.] mutant with reduced levels of Kunitz trypsin inhibitor. *Plant Sci.* 160: 979-986.

Kunitz, M., 1945. Crystallization of a trypsin inhibitor from soybean. *Science* 101: 668-669.

Laing, W.A. & M.T. McManus, 2002. *In: Protein protein interactions in plants*, (Mc Manus, M.T., W.A. Laing and A.C. Allan) (Eds.). Sheffield Academic Press, Sheffield, *Annual Plant Reviews*, Vol. 7: pp. 77-119.

Lawrence, P.K. & K.R. Koundal, 2001. *Agrobacterium tumefacians*-mediated transformation of pigeonpea (*Cajanus cajan* L. Millsp.) and molecular analysis of regenerated plants. *Current Sci.* 80: 1428-1432.

Lawrence, P.K. & K.R. Koundal, 2002. Plant protease inhibitors in control of phytophagous insects. *Electron. J. Biotechnol.* 5: 93-109.

Liao, H., W. Ren, Z. Kang, J.H. Jiang, X.J. Zhao & L.F. Du, 2007. A trypsin inhibitor from *Cassia obtusifolia* seeds: isolation, characterization and activity against *Pieris rapae*. *Biotechnol. Lett.* 29: 653-658.

Lopes, J.L.S., N.F. Valadares, D.I. Moraes, J.C. Rosa, H.S.S. Araújo & L.M. Beltramini, 2009. Physico-chemical and antifungal properties of protease inhibitors from *Acacia plumosa*. *Phytochemistry* 70: 871-879.

Losso, J.N., 2008. The biochemical and functional food properties of the Bowman-Birk inhibitor. *Crit. Rev. Food Sci. Nutr.* 48: 94-118.

Luo, M., Z. Wang, H. Li, K.F. Xia, Y. Cai & Z.F. Xu, 2009. Overexpression of a weed (*Solanum americanum*) proteinase inhibitor in transgenic tobacco results in increased glandular trichome density and enhanced resistance to *Helicoverpa armigera* and *Spodoptera litura*. *Int. J. Mol. Sci.* 10: 1896-1910.

Macedo, M.L.R., G.C. Mello, M.G.M. Freire, J.C. Novello, S. Marangoni & D.G.G. Matos, 2002. Effect of a trypsin inhibitor from *Dimorphandra mollis* seeds on the development of *Callosobruchus maculatus*. *Plant Physiol. Biochem.* 40: 891-898.

Macedo, M.L.R., C.M. Sa, M.G.M. Freire & J.R.P. Parra, 2004. A Kunitz-type inhibitor of coleopteran proteases, isolated from *Adenanthera pavonina* L. seeds and its effect on *Callosobruchus maculatus*. *J. Agric. Food Chem.* 52: 2533-2540.

Macedo, M.L.R., M.G.M. Freire, E.C. Cabrini, M.H. Toyama, J.C. Novello & S. Marangoni, 2003. A trypsin inhibitor from *Peltophorum dubium* seeds active against pest proteases and its effect on the survival of *Anagasta kuehniella* (Lepidoptera: Pyralidae). *Biochim. Biophys. Acta* 1621: 170-182.

Maheswaran, G., L. Pridmore, P. Franz & M.A. Anderson, 2007. A proteinase inhibitor from *Nicotiana alata* inhibits the normal development of light-brown apple moth, *Epiphyas postvittana* in transgenic apple plants. *Plant Cell Rep.* 26: 773-782.

Mandal S, P. Kundu, B. Roy & R.K. Mandal, 2002. Precursor of the inactivate 2S seed storage protein from the Indian mustard *Brassica juncea* is a novel trypsin inhibitor. *J. Biol. Chem.* 277: 37161-37168.

Marchetti, S., M. Delledonne, C. Fogher, C. Chiabà, F. Chiesa, F. Savazzini & A. Giordano, 2000. Soybean Kunitz, C-II and PI-IV inhibitor genes confer different levels of insect resistance to tobacco and potato transgenic plants. *Theor. Appl. Genet.* 101: 519-526.

Morris, C.A., L.D. Morris, A.R. Kennedy & H.L. Sweeney, 2005. Attenuation of skeletal muscle atrophy via protease inhibition. *J. Appl. Physiol.* 99: 1719–1727.

Mosolov, V.V. & T.A. Valueva, 2005. Proteinase inhibitors and their function in plants: A review. *Appl. Biochem. Microbiol.* 41: 227-246.

Mulvenna, J.P., F.M. Foley & D.J. Craik, 2005. Discovery, structural determination, and putative processing of the precursor protein that produces the cyclic trypsin inhibitor sunflower trypsin inhibitor 1. *J. Biol. Chem.* 280: 32245-32253.

Norioka, N., S. Hara, T. Ikenaka & J. Abe, 1988. Distribution of the Kunitz and the Bowman-Birk family proteinase inhibitors in leguminous seeds. *Agric. Biol. Chem.* 52: 1245-1252.

Oliveira, A.S., L. Migliolo, R.O. Aquino, J.K.C. Ribeiro, L.L.P. Macedo, L.B.S. Andrade, M.P. Bemquerer, E.A. Santos, S. Kiyota & M.P. Sales, 2007. Purification and characterization of a trypsin-papain inhibitor from *Pithecelobium dumosum* seeds and its *in vitro* effects towards digestive enzymes from insect pests. *Plant Physiol. Biochem.* 45: 858-865.

Oliveira, A.S., R.A. Pereira, L.M. Lima, A.H.A. Morais, F.R. Melo, O.L. Franco, C. Bloch Jr., M.F. Grossi-de-Sá & M.P. Sales, 2002. Activity toward bruchid pest of a Kunitz-type inhibitor from seeds of the algaroba tree (*Prosopis juliflora* D.C.). *Pesti. Biochem. Physiol.* 72: 122-132.

Oppert, B., T.D. Morgan, K. Hartzer, B. Lenarcic, K. Galesa, J. Brzin, V. Turk, K. Yoza, K. Ohtsubo & K.J. Kramer, 2003. Effects of proteinase inhibitors on digestive proteinases and growth of the red flour beetle, *Tribolium castaneum* (Herbst) (Coleoptera: Tenebrionidae). *Comp. Biochem. Physiol. C* 134: 481-490.

Pereira, R.A., A. Valencia-Jiménez, C.P. Magalhães, M.V. Prates, J.A.T. Melo, L.M. Lima, M.P. Sales, E.Y.T. Nakasu, M.C.M. Silva & M.F. Grossi-de-Sá, 2007. Effect of a Bowman-Birk proteinase inhibitor from *Phaseolus coccineus* on *Hypothenemus hampei* gut proteinases *in vitro*. *J. Agric. Food Chem.* 55: 10714-10719.

Prasad, E.R., A. Dutta-Gupta & K. Padmasree, 2009. Inhibitors from pigeonpea active against lepidopteran gut proteinases. *J. Econ. Entomol.* 102: 2343-2349.

Prasad, E.R., A. Dutta-Gupta & K. Padmasree, 2010b. Purification and characterization of a Bowman-Birk proteinase inhibitor from the seeds of black gram (*Vigna mungo*). *Phytochemistry* 71: 363-372.

Prasad, E.R., A. Dutta-Gupta & K. Padmasree, 2010c. Insecticidal potential of Bowman-Birk proteinase inhibitors from red gram (*Cajanus cajan*) and black gram (*Vigna mungo*) against lepidopteran insect pests. *Pesti. Biochem. Physiol.* In press, doi:10.1016/j.pestbp.2010.05.003.

Prasad, E.R., H. Merzendorfer, C. Madhurarekha, A. Dutta-Gupta & K. Padmasree, 2010a. Bowman-Birk proteinase inhibitor from *Cajanus cajan* seeds: purification, characterization, and insecticidal properties. *J. Agric. Food Chem.* 58: 2838-2847.

Pujol, M., C.A. Hernández, R. Armas, Y. Coll, J. Alfanso-Rubí, M. Pérez, C. Ayra & A. González, 2005. Inhibition of *Heliothis virescens* larvae growth in transgenic tobacco plants expressing cowpea trypsin inhibitor. *Biotechnologia Aplicada* 22: 127-130.

Qi, R.F., Z.W. Song & C.W. Chi, 2005. Structural features and molecular evolution of Bowman-Birk protease inhibitors and their potential application. *Acta Biochim. Biophys. Sin.* 37: 283-292.

Rahbé, Y., E. Ferrasson, H. Rabesona & L. Quillien, 2003. Toxicity to the pea aphid *Acyrthosiphon pisum* of anti-chymotrypsin isoforms and fragments of Bowman-Birk protease inhibitors from pea seeds. *Insect Biochem. Mol. Biol.* 33: 299-306.

Ramos, V.S., M.G.M. Freire, J.R.P. Parra & M.L.R. Macedo, 2009. Regulatory effects of an inhibitor from *Plathymenia foliolosa* seeds on the larval development of *Anagasta kuehniella* (Lepidoptera). *Comp. Biochem. Physiol. A* 152: 255-261.

Ryan, C.A., 1990. Protease inhibitors in plants: genes for improving defenses against insects and pathogens. *Annu. Rev. Phytopathol.* 28: 425-449.

Sales, M.P., I.R. Gerhardt, M.F. Grossi-de-Sá & J. Xavier-Filho, 2000. Do legume storage proteins play a role in defending seeds against bruchids? *Plant Physiol.* 124: 515-522.

Scarafoni, A, C. Magni & M. Duranti, 2007. Molecular nutraceutics as a mean to investigate the positive effects of legume seed proteins on human health. *Trends Food Sci. Tech.* 18: 454-463.

Silva, F.C.B.L., A. Alcazar, L.L.P. Macedo, A.S. Oliveira, F.P. Macedo, L.R.D. Abreu, E.A. Santos & M.P. Sales, 2006. Digestive enzymes during development of *Ceratitis capitata* (Diptera: Tephritidae) and effects of SBTI on its digestive serine proteinase targets. *Insect Biochem. Mol. Biol.* 36: 561-569.

Sin, S.F. & M.L. Chye, 2004. Expression of proteinase inhibitor II proteins during floral development in *Solanum americanum*. *Planta* 219: 1010-1022.

Singh, R.R. & A.G. Appu Rao, 2002. Reductive unfolding and oxidative refolding of a Bowman-Birk inhibitor from horsegram seeds (*Dolichos biflorus*): evidence for 'hyperreactive' disulfide bonds and rate-limiting nature of disulfide isomerization in folding. *Biochim. Biophys. Acta* 1597: 280-291.

Sivakumar, S., O.L. Franco, P.D. Tagliari, C. Bloch Jr., M. Mohan & B. Thayumanavan, 2005. Screening and purification of a novel trypsin inhibitor from *Prosopis juliflora* seeds with activity toward pest digestive enzymes. *Protein Pept. Lett.* 12: 561-565.

Solomon, M., B. Belenghi, M. Delledonne, E. Menachem & A. Levine, 1999. The involvement of cysteine proteases and protease inhibitor genes in the regulation of programmed cell death in plants. *Plant Cell* 11: 431-443.

Srinivasan, A., A.P. Giri, A.M. Harsulkar, J.A. Gatehouse & V.S. Gupta, 2005. A Kunitz trypsin inhibitor from chickpea (*Cicer arietinum* L.) that exerts anti-metabolic effect on podborer (*Helicoverpa armigera*) larvae. *Plant Mol. Biol.* 57: 359-374.

Srinivasan, T., K.R.R. Kumar & P.B. Kirti, 2009. Constitutive expression of a trypsin protease inhibitor confers multiple stress tolerance in transgenic tobacco. *Plant Cell Physiol.* 50: 541-553.

Tamhane, V.A., A.P Giri, M.N. Sainani & V.S. Gupta, 2007. Diverse forms of Pin-II family proteinase inhibitors from *Capsicum annuum* adversely affect the growth and development of *Helicoverpa armigera*. *Gene* 403, 29–38.

Tamhane, V.A., N.P. Chougule, A.P. Giri, A.R. Dixit, M.N. Sainani & V.S. Gupta, 2005. In vivo and in vitro effect of *Capsicum annum* proteinase inhibitors on *Helicoverpa armigera* gut proteinases. *Biochim. Biophys. Acta* 1722: 156-167.

Tani, K., F. Ogushi, H.Kido, T. Kawano, Y. Kunori, T. Kamimura, P. Cui & S. Sone, 2000. Chymase is a potent chemoatractant for human monocytes and neutrophils. *J. Leukoc. Biol.* 67: 585–589.

Telang, M., A. Srinivasan, A. Patankar, A. Harsulkar, V. Joshi, A. Damle, V. Deshpande, M. Sainani, P. Ranjekar, G. Gupta, A. Birah, S. Rani, M. Kachole, A. Giri & V. Gupta, 2003. Bitter gourd proteinase inhibitors: potential growth inhibitors of *Helicoverpa armigera* and *Spodoptera litura*. *Phytochemistry* 63: 643-652.

Telang, M.A., A.P. Giri, P.S. Pyati, V.S. Gupta, M. Tegeder & V.R. Franceschi, 2009. Winged bean chymotrypsin inhibitors retard growth of *Helicoverpa armigera*. *Gene* 431: 80-85.

Vila, L., J. Quilis, D. Meynard, J.C. Breitler, V. Marfà, I. Murillo, J.M. Vassal, J. Messeguer, E. Guiderdoni & B.S. Segundo, 2005. Expression of the maize proteinase inhibitor (*mpi*) gene in rice plants enhances resistance against the striped stem borer (*Chilo suppressalis*): effects on larval growth and insect gut proteinases. *Plant Biotechnol. J.* 3:187-202.

Xu, Z.F., W.L. Teng & M.L. Chye, 2004. Inhibition of endogenous trypsin- and chymotrypsin-like activities in transgenic lettuce expressing heterogeneous proteinase inhibitor SaPIN2a. *Planta* 218: 623-629.

Zhou, J.Y., H. Lia, N.H. Zhang, L. Tang, Y. Xu & F. Chen, 2008. Identification of a Kunitz inhibitor from *Albizzia kalkora* and its inhibitory effect against pest midgut proteases. *Biotechnol. Lett.* 30: 1495-1499.

Chapter - 17

EMERGING TRENDS IN FORENSIC ENTOMOLOGY: AN EYE TOWARDS THE FUTURE

Devinder Singh and Bhanvi Wadhawan

Department of Zoology, Punjabi University, Patiala - 147 002, Punjab, India
E-mail: *devinder.ss.61@gmail.com*

INTRODUCTION

Post Mortem Interval (PMI) or time elapsed since death can be determined by various methods including the physical and chemical changes of the corpse (Henssge *et al.*, 1995). These include *livor mortis*, *algor mortis*, *rigor mortis*, post mortem cooling, changes in the chemical constituents of the body, autolysis and putrefaction, which can be detected during 2-3 days after death (Kashyap & Pillay, 1989; Gennard, 2007). The carrion passes through a series of decomposition stages that include: fresh, bloated, active decay, advanced decay, dry decay and remains (Payne, 1965). These stages are recognizable but the boundaries between them are diffuse and overlap considerably (Campobasso *et al.*, 2001). As a result, the accurate determination of the onset and duration of these stages may be considerably difficult, thus hindering the accurate and precise estimation of the PMI. On the other hand, insects offer a much longer time scale measured in hours, days, weeks, months or even years (Greenberg & Kunich, 2002). Forensic entomology is the most well researched method of determining time since death in the later post mortem period (Buchan & Anderson 2001) and can be defined as the application of the study of insects and other arthropods to investigations of violent crimes such as rape, physical abuse and murder (Keh, 1985; Catts & Goff, 1992).

At present there are two approaches available for estimating the PMI using insect evidence, and the application of either one largely depends on the state of decomposition of the corpse at the time of discovery. The first method involves the analysis of the pattern of colonization of the carrion by successive waves of insects and other arthropods (Schoenly & Reid, 1987; Haskell *et al.*, 1997). Insect succession can generally be used from about a month to a year or until the carrion is completely skeletonised

(Anderson, 1999). The second method relies on the rate of development of immature stages of flies that are available on the carrion after death. Of the many species of Diptera, the blow flies (Calliphoridae) are usually the first colonizers of carrion and the flesh flies (Sarcophagidae) arrive later (Smith, 1986). Development data for primary carrion flies (blow flies) provide the most accurate means of estimating the minimum PMI using arthropod information (Greenberg, 1991).

Forensic entomology has an interesting history spanning over several decades (Nuorteva, 1977; Smith, 1986; Erzinçlioglu, 1990; Marchenko, 2001; Amendt *et al.*, 2004). Its main application has so far been the estimation of the minimum PMI, though there is much more potential for its use in crime investigation. There are many variables that affect the rate of insect development and thus interfere in the calculation of PMI. A great deal of research has already been undertaken to help the investigators, but there are many questions yet to be answered and there is still room for experimentation. Some of the areas where gaps exist and which have potential for research include:

Insect succession

A corpse is colonized by insects in a predictable pattern, thus the presence or absence of a taxon can help in the calculation of PMI (Hewadikaram & Goff, 1991; Anderson, 2001; Grassberger & Frank, 2003a; Tabor *et al.*, 2004; De Jong & Hoback, 2006). The research is methodically site and season specific. There are many factors which can influence succession pattern including the amount of light the corpse is exposed to, its size, type of habitat, temperature, humidity etc. (Amendt *et al.*, 2004; Hobischak *et al.*, 2006; Joy *et al.*, 2006). Succession research is extremely labour intensive. So it becomes an onerous job to produce replicated data to obtain a statistical confidence interval of a PMI estimate (LaMotte & Wells, 2000). Resampling by means of computers can overcome this limitation (Schoenly, 1992). There is a great need to produce and apply such computer generated algorithms to insect succession research. It has been observed by several workers that the adult flies of different species exhibit preferences for differing habitat types (Norris, 1965). So there arises a possibility of determining post mortem corpse movement if the immature insects found on the corpse are not characteristic of the site where the body has been discovered (Amendt *et al.*, 2004). However, one must be prudent while basing this solely on entomological evidence. Densely populated landscapes exhibit little separation of habitats and thus an assortment of insect communities can be found on corpses discovered there. There is an urgent need to document the carrion communities in different regions and the pattern of their succession so that the data can be used for future interpretations.

Aquatic entomology

Research on freshwater and marine fauna found associated with submerged corpses has been sporadic (Nuorteva *et al.*, 1974; Goff & Odom, 1987; Haskell *et al.*, 1989; Vance *et al.*, 1995; Sorg *et al.*, 1997; Davis & Goff, 2000). Awareness of the role of aquatic arthropods like those belonging to orders Ephemeroptera (mayflies), Trichoptera (caddisflies) and families Chironomidae (midges) and Simuliidae (black flies) of order Diptera during decomposition is inadequate (Keiper *et al.*, 1997; Anderson, 2001; Merrit & Wallace, 2001; Hobischak & Anderson, 2002). Decomposition in an aquatic environment takes place at half the rate as on land, mainly due to the lack of access for many insects and more or less constant cooler temperatures. Aquatic insects are not obligatory sarcophagous, hence the use of these insects for estimating time of death is difficult. Hobischak and Anderson (2002) tried to assess if arthropod succession in freshwater environments could be used to estimate the PMI. There was a predictable succession of invertebrates but further studies are necessary to confirm if the results are carrion or season specific. Research done in a marine environment (Anderson & Hobischak, 2004) showed distinct stages of decomposition dissimilar with patterns observed on land or in freshwater. No successional insect model has been developed which might describe the different waves of colonization of a corpse in aquatic habitats and thus help in deciphering PMI (Merrit & Wallace, 2001). Scant research has focussed on corpses that decompose in a terrestrial environment and are then transferred to an aquatic environment. Singh and Greenberg (1994) ventured in this area by studying the effect of submergence on the survival of blow fly pupae. Valuable data from this type of studies can be garnered and then applied in cases where immature stages are discovered on a submerged corpse.

Identification, phylogenetics and population genetics

Accurate identification of the insect specimen is of vital importance for a forensic entomologist (Smith, 1986). Immature stages of several forensically relevant taxa are difficult to identify morphologically and have to be reared to the adult stage for accurate identification. For some groups, like the sarcophagids (flesh flies), the situation is even worse where identification of adults is very cumbersome and requires the skills of an experienced taxonomist. Under these circumstances, identification by focussing on molecular methods can be used. However, the research done so far has been disproportionate in its treatment of forensically significant insect taxa. Majority of research papers have been published on calliphorids (Harvey *et al.*, 2003; Ratcliffe *et al.*, 2003; Chen *et al.*, 2004; Wells & Williams, 2007) which indicates their significance as indicator species for PMI estimation. Lesser data exists for sarcophagids (Wells *et al.*, 2001; Ratcliffe *et al.*, 2003; Zehner *et al.*, 2004b). The main limitation to this method is that only a small number of individuals from each species can be analyzed. Replications in species diagnostic testing also need to be addressed to ensure reliability (Wells & Williams, 2007).

Species determination of forensically related fly species takes into account several loci. Mitochondrial DNA markers used for this objective include the cytochrome oxidase subunit one (COI) gene (Wells & Sperling, 2001; Zehner *et al.*, 2004a, b; Wells & Williams, 2007) and the cytochrome oxidase subunit two (COII) gene (Sperling *et al.*, 1994; Wells & Sperling, 2001; Wallman & Donnellan, 2001). Other mitochondrial markers that have been used consist of the control region (Thyssen *et al.*, 2005), subunit 5 of the NAD dehydrogenase (ND5) gene (Zehner *et al.*, 2004a, b) and the tRNA-leucine gene (Wells & Sperling, 2001). Nuclear markers have been exploited to a lesser extent as well, such as the internal transcribed spacer regions (Ratcliffe *et al.*, 2003) and the gene for 28S ribosomal RNA (Stevens & Wall, 2001). Some closely-related species cannot be differentiated in this manner using COI and COII (Wallman & Donnellan, 2001; Wells *et al.*, 2004), so future research should examine other markers as well. The methodologies are varied and comprise of direct sequencing (Harvey *et al.*, 2003; Chen *et al.*, 2004; Zehner *et al.*, 2004b; Ames *et al.*, 2006), polymerase chain reaction-restriction fragment length polymorphism (PCR-RFLP) analysis (Sperling *et al.*, 1994; Malgorn & Coquoz, 1999; Thyssen *et al.*, 2005), random amplified polymorphic DNA (RAPD) analysis (Benecke 1998), and allozyme electrophoresis (Wallman & Adams, 2001). These relatively lesser utilized methodologies can be prompt, inexpensive and highly discerning. Therefore, they warrant further evaluation as forensic tools. Most of the genetic research in forensic entomology has focused on individual or species identification. Limited investigations have been carried on phylogenetic patterns (Stevens & Wall, 1996; Stevens & Wall, 1997; Wells & Sperling, 1999; Bernasconi *et al.*, 2000; Stevens & Wall, 2001; Stevens *et al.*, 2002; Wells *et al.*, 2004, 2007). Phylogenetic analysis requires a greater amount of genetic information than what is normally used for simple identification. One strategy is to use multiple mitochondrial genes (Wallman *et al.* 2005). Junqueira *et al.* (2004) even sequenced the entire mitochondrial DNA of a blow fly. Combination of mitochondrial and nuclear loci shows a lot of promise as a new approach e.g. Stevens *et al.* (2002) used both COI and 28S. Close attention should be paid to newer protocols developed for other fly taxa, because that technology might be applicable on forensically important flies with little modification.

The information about genetic investigation of forensically important insects can be relevant to forensic entomologists (Böhme, 2006). These might hint at regional genetic differences which exist in a forensically important species, *e.g.* development rate. Regional genetic differences could also make it possible to infer the post-mortem movement of a corpse if a larva discovered on it has a genotype typical of another area. Thus an expansion in population genetic studies is the need of the hour. Especially well suited for this purpose are the STR genotypes though their development for forensically important insects has been quiet difficult because these loci are less widespread in insects as compared to vertebrates (Ji *et al.,* 2003). Still, some work has been initiated in this direction for calliphorid flies (Florin & Gyllenstrand, 2002; Torres *et al.*, 2004; Torres & Azeredo-Espin, 2005). However, validation studies are necessary before the aforementioned methods are actually used for solving criminal cases. Wells & Williams (2007) tested the suitability of a published method (Wells & Sperling, 2001) for identifying individuals in the blow fly subfamily Chrysomyinae in North America. They obtained genetic data for already identified specimens numbering in hundreds and scrutinized them as unknowns. All of them were correctly identified by the previously published method. There is a great need of such validation studies taking into account many specimens of a species to authenticate the published results based on a few specimens.

Maggot gut content studies

Studies on the DNA of human blood or tissues obtained from the alimentary canal of maggots and blood feeding insects is an area demanding immediate attention. If a corpse is moved from one place to another, the maggots recovered from the first area might help in the potential identification of the corpse by the utilization of gut contents in DNA studies. Campobasso *et al.* (2005) reviewed these and other potential applications of this analysis. Most of the published research has focussed on blow flies only (Linville *et al.*, 2004; Linville & Wells, 2002; Zehner *et al.*, 2004b) while little attention has been paid to sarcophagids and muscids (Wells *et al.*, 2001). Rest of the literature deals with mosquitoes (Kreike & Kampfer 1999; Mukabana *et al.*, 2002), lice (Replogle *et al.,* 1994; Lord *et al.,* 1998; Mumcuoglu *et al.,* 2004), and a sap beetle (DiZinno *et al.,* 2002). The benchmark for profiling human DNA is Short tandem repeat (STR) analysis. Gut content studies have similarly utilized STR loci as markers (Schiro, 2001; Linville *et al.,* 2004; Mumcuoglu *et al.* 2004; Zehner *et al.,* 2004a). Constant upgradation in methods which help in DNA profiling is urgently required. Stress should be on developing methods which help in better discernment and exploit such tissues which were previously unsuitable for analysis. Undoubtedly, these will then be used for forensic insect gut content analysis. Limited work has been done to detect semen in the gut content of maggots (Clery, 2001). Campobasso *et al.* (2005) pointed out that actively feeding stages are required for the successful isolation of host DNA from the alimentary tract of the larvae. Nevertheless, fragments of 197 base pairs (bp) and 87 bp of host sheep DNA (satellite I region) have been reported in all stages of immatures including pupae of *Calliphora dubia* (Carvalho *et al.,* 2005). Studies on insects feeding on humans to determine if there were species specific salivary proteins left behind by them is yet another area where information is greatly needed.

Weather and temperature data collection

Inaccurate weather data collection from the site of collection is detrimental to a forensic investigation (Haskell & Williams, 1990; Haskell *et al.,* 2000; Amendt *et al.,* 2007). Archer (2004) concentrated on perfecting estimates of temperature prior to corpse recovery based on data collected from meteorological station obtained after corpse recovery. Variations in microclimate can play havoc on the estimates of environmental conditions thus directly affecting PMI estimation. Hunt (2005) found in a prelimi-nary study that the weather data obtained from satellites correlates with actual data on the ground. Thus, forensic entomologists must show an increased dependence on weather satellites for data collection. It is now well known that the temperature inside the maggot masses is much higher than the ambient temperature (Turner and Howard, 1992; Ireland and Turner, 2006). Development is a function of temperature for forens-ically important flies, so the temperature of the maggot mass should be considered in future studies.

Parasitoids

The necrophagous insects can suffer from an infestation by parasitoids, mainly hymenopterans. Fabritius & Klunker (1991) enlisted 83 parasitoid species in Europe which attack the larval and pupal stages of synanthropic Diptera. There are few reports on using parasitoids for the benefit of forensic investigation (Smith, 1986; Haskell *et al.,* 1997; Amendt *et al.* 2000; Anderson & Cervenka, 2002; Grassberger & Frank, 2003b; Voss *et al.,* 2009). Pupal parasitoids can play an especially important role in the estimation of the PMI (Anderson & Cervenka, 2002). The estimated developmental time of the parasitoid when added to the time of development of the host fly can provide an extended time frame for PMI estimation (Grassberger & Frank, 2003b). However, parasitoids might also create significant problems in forensic investigations. Holda-way and Evans (1930) explained the alteration of developmental times of *Lucilia sericata* after an attack by its parasitoid *Alysia manducator* which resulted in a premature pupariation. There is a need to better understand the life cycles of various parasitoids and the effects they have on the development of forensically important flies and consequently on the PMI estimation.

Multivariate approaches

Forensic entomology is a highly specialised science and there have been few studies enumerating the combined use of entomological and non- entomological evidences for PMI investigation (Schoenly *et al.*, 1991). Non-entomological post-mortem phenomena which can potentially be examined in addition to entomological data include rate of DNA decay (Anderson *et al.*, 2005), variations in the volatile organic products formed during the decomposition of corpses (Statheropoulos *et al.*, 2005), and growth rate of algae on corpses found in aquatic locations (Haefner *et al.*, 2004). The mathematical models and computer algorithms designed for calculation of PMI on the basis of insect evidence (Schoenly, 1992; Wells & LaMotte, 2001; Byrd & Allen, 2001) can also be used for non-entomological phenomena. A combined investigation spanning the fields of Anthrop-ology, Botany and Entomology is a highly efficient way of estimating PMI of buried bodies (Galloway *et al.*, 2001). This combined outlook can enrich and further corroborate the data based solely on insect investigations.

Wildlife and veterinary forensic entomology

Forensic entomology is usually employed in human death investigations. However, it can also be applicable to crimes related with wildlife (Merck, 2007). Anderson (1999) described a case of killing of two young bear cubs illegally in Canada where entomological data were used to estimate the PMI. There is a lot of scope for research in this field which is almost virgin so far.

Time of oviposition

There are various factors which might affect the time of oviposition e.g. covering corpses with leaf litter or wrapping them in blankets, carpets or plastic bags and placement of the body in freezers or indoors (Higley & Haskell, 2001). Cold and rainy weather and other seasonal fluctuations may limit fly activity and delay oviposition (Erzinclioglu, 1996). However, Faucherre *et al.* (1999) were able to observe flying as well as ovipositing *Calliphora vicina* on a corpse under extreme environ-mental conditions (about 5°C). Before this study the commonly acknowle-dged notion was that fly activity ceases below an air temperature of 10°C (Williams, 1984) or even 12°C (Smith 1986; Erzinclioglu 1996). The effects of various other climatic factors on oviposition and fly activity need to be elucidated.

Nocturnal oviposition

Peaks in oviposition activity by blow flies can be observed in the early afternoon (Nuorteva, 1959; Baumgartner & Greeenberg, 1984, 1985). Earlier it was believed that these do not lay eggs during night time and if a corpse is discovered around noon and is teeming with recently hatched blow fly larvae, we might assume that the eggs might have been laid in the evening the previous day. Also, if fly eggs are detected on a corpse during the night, we might deduce that death occurred on the previous day (Nuorteva, 1977). Greenberg (1990) published the first experimental evidence of nocturnal oviposition by three forensically important blow flies. However, the study by Tessmer *et al.*, (1995) was at odds with the aforementioned evidence and they reported that blow flies fail to lay eggs at night both in urban (with lighting) and rural dark habitats. On the other hand, studies by Singh and Bharti (2001, 2008) conformed with those of Greenberg (1990). Hence nocturnal oviposition is a potential confounding factor for PMI estimation and it should be taken into account. There is scope for further research work in this direction.

Diapause

Diapause, the period during which development of insects is discontinued, presents a challenge for forensic entomologists because it directly affects PMI estimation. Onset and duration of diapause is generally dependent on temperature and photoperiod. Waning day length and/or dwindling temperatures can induce diapause thus suspending development till the return of the favourable conditions. Diapause, in case of blow flies, is usually under maternal control and if females are exposed to short day lengths, diapause occurs in the subsequent generation (Vinogradova, 1991). Species spread over a large geographical range are less susceptible because they have to endure alterations in day length throughout the year. McWatters and Saunders (1998) observed that the threshold day length which induces diapause will be longer for populations having a northern distribution as compared to those in the south. Temperature may also influences the occurrence of diapause (Vinogradova & Zinovjeva, 1972) and can have considerable variation both within and between years (McWatters & Saunders, 1998). Rising temperatures are known to diminish the prevalence of diapause in forensically important dipterans, *Liopygia argyrostoma* (Saunders, 1975), *Protophormia terraenovae* (Vinogradova, 1986) and *Calliphora vicina* (McWatters & Saunders 1998). The length of diapause is another important factor. It was reported that in case of *C. vicina* kept at temperatures of 15°C and 20°C, diapause was terminated in most larvae within 30 days (McWatters & Saunders, 1998). However, the diapause terminated earlier in the larvae whose parents had been kept at 20°C than those whose parents had been kept at 15°C. Thus, the forensic entomologists need to be cautious about PMI estimation when temperatures are low and/or day length is declining.

Competition

The field of population ecology studies of carrion insects needs to be explored further. Additional research should focus on insect abundance and colonization and growth rates. Intra- and interspecific competition affects development and growth of the larvae. Smith & Wall (1997a, 1997b) published data which showed that the larvae of *Lucilia sericata* in carcasses are subject to considerable levels of competition and that it affects the feeding and development of the larvae thus indirectly affecting PMI estimation.

Precocious egg development

One more parameter which might complicate matters is precocious egg development in flies (Reiter, 1984; Smith 1986; Erzinclioglu, 1990). Fertilized eggs may be detained in the oviduct in some female flies ahead of oviposition (Wells & King, 2001). If appropriate oviposition site is not available, the eggs may be kept inside the fly's oviduct till they complete their embryonic development. It has been testified for several species of blow flies that the larva hatches from such eggs immediately following oviposition (Erzinclioglu, 1990; Wells & King, 2001). The proportion of wild flies carrying an egg that is about to hatch can be quite high (Wells & King, 2001). This may affect the PMI estimation by several hours and is a subject potentially important for further investigations.

Neglect

Forensic entomology may also help in investigations dealing with living, but ill, people by revealing neglect. The occurrence of maggots in the wounds or natural orifices of living persons may indicate such neglect. Estimating the age of these maggots can reveal how long the neglect has been happening (Benecke, 2003). This approach needs further standardization before it could be put to proper use.

Entomotoxicology

Entomotoxicology deals with the analysis of toxins in arthropods (mainly immature stages of blow flies) that feed on carrion. Using arthropods feeding upon a corpse or found at the crime scene, investigators can determine whether toxins were present in a body at the time of death. Previously, such deductions were quiet difficult in the case of severely decomposed bodies devoid of soft tissues and body fluids. The qualitative determination of toxic substances and drugs can augment information on both the cause of death and the geographical area where the cadaver had been lying. The xenobiotics present in decomposing tissues might alter the development rate or the succession of arthropods on the cadaver, so, their qualitative and quantitative determination can provide a more accurate calculation of PMI (Gagliano-Candela & Aventaggiato, 2001). Entomotoxicology in India has been unable to gain a firm foothold so far. Sufficient data need to be generated in the fields of toxin bioaccumulation, insect metabolism of drugs and quantitative analyses of insect evidence. Moreover, previous research done on the foreign shores has remained concentrated on heroin, cocaine and other drugs of abuse (Goff *et al.*, 1997; Sadler *et al.*, 1997; Carvalho *et al.*, 2001; Introna *et al.*, 2001), whereas, the thrust area of research now need to concentrate on drugs and toxins specific to India. Also, the research done is species specific and species of flies commonly found in India need to be investigated further. Given its wide range of prospective applications and its ability to supplement established forensic techniques, entomotoxicology is the science of the future.

CONCLUSION

The forensic entomologists are privileged that meticulous and species specific research was conducted during the last century on fly biology, behaviour, growth and development. Recent forensic entomology studies have commenced in novel areas for investigation such as: aquatic insects, succession of carrion insects from geographic areas not previously analysed, fly DNA cataloguing, human DNA in maggot gut contents, effect of climate change, entomotoxicology and recovery of insect remains on prehistoric humans to name just a few. However, there is a dire need of broadening the areas of research with studies focussing on varied death scenarios and geographical locations. It is also essential that previously published results be validated and replicated. Only then, we can realise the full potential of insect evidence in crime investigations.

REFERENCES

Amendt, J., C. P. Campobasso, E. Gaudry, C. Reiter, H. N. LeBlanc & M. J. R. Hall, 2007. Best practice in forensic entomology–standards and guidelines. *Int. J. Legal Med.* 121: 90-104.

Amendt, J., R. Krettek, C. Niess, R. Zehner & H. Bratzke, 2000. Forensic entomology in Germany. *Forensic Sci. Int.* 113: 309-314.

Amendt, J., R. Krettek & R. Zehner, 2004. Forensic entomology. *Naturwissenschaften* 91: 51-65.

Ames, C., B. Turner & B. Daniel, 2006. The use of mitochondrial cytochrome oxidase I gene (COI) to differentiate two UK *blow fly* species – *Calliphora vicina* and *Calliphora vomitoria*: *Forensic Sci. Int.* 164: 179-182.

Anderson, G. S., 1999. Wildlife forensic entomology: determining time of death in two illegally killed black bear cubs. *J. Forensic Sci.* 44: 856-859.

Anderson, G. S., 2001. Succession on carrion and its relationship to determining time of death. In *Forensic entomology: the utility of arthropods in legal investigations*. Byrd, J.H. and J. L. Castner (eds.), Boca Raton, CRC, pp. 143-175.

Anderson, G. S. & V. J. Cervenka, 2002. Insects associated with the body: their use and analyses. In: *Advances in forensic taphonomy: method, theory and archaeological perspectives*. Haglund, W. D. and M. H. Sorg (eds) CRC, Boca Raton, Fla., pp. 173-200.

Anderson, G.S. & N.R. Hobischak, 2004. Decomposition of carrion in the marine environment in British Columbia Canada. *Int. J. Legal Med.* 118: 206-209.

Anderson, S., B. Howard, G. R. Hobbs & C. P. Bishop, 2005. A method for determining the age of a bloodstain. *Forensic Sci. Int.* 148: 37-45.

Archer, M.S. 2004. The effect of time after body discovery on the accuracy of retrospective weather station ambient temperature corrections in forensic entomology. *J. Forensic Sci.* 49: 553-559.

Baumgartner, D.L. & B. Greenberg, 1984. The genus *Chrysomya* (Diptera: Calliphoridae) in the New World. *J. Med. Ent.* 21: 105-113.

Baumgartner, D. L. & B. Greenberg, 1985. Distribution and medical ecology of the blow flies (Diptera: Calliphoridae) of Peru. *Ann. Entomol. Soc. Am.* 78: 565-578.

Benecke, M., 1998. Random amplified polymorphic DNA (RAPD) typing of necrophageous insects (Diptera: Coleoptera) in criminal forensic studies: validation and use in practice. *Forensic Sci. Int.* 98: 157-168.

Benecke, M., 2003. Neglect of the elderly: cases and considerations. Proceedings of the first meeting of the European Association for Forensic Entomology. pp. 29-30.

Bernasconi, M. V., C. Valsangiacomo, J. C. Piffaretti & P. I. Ward 2000. Phylogenetic relationships among Muscoidea (Diptera: Calyptratae) based on mitochondrial DNA sequences. *Insect Mol. Biol.* 9: 67-74.

Böhme, P. 2006. Population genetics of forensically important North American blow flies (Diptera: Calliphoridae) using the A+T-Rich region of mitochondrial DNA. Diploma Thesis, Üniversität Bonn.

Buchan, M. & G. S. Anderson, 2001. A review of the current status of methods used in the later *post-mortem* interval to determine time since death. *Canadian Soc. Forensic Sci. J.* 34: 1-22.

Byrd, J. H. & J. C. Allen, 2001. Computer modelling of insect growth and its application to forensic entomology. In: *Forensic entomology: the utility of arthropods in legal investigations.* Byrd, J. H. and J. L. Castner (eds) CRC, Boca Raton, FL, pp. 303-330.

Campobasso, C. P., G. DiVella & F. Introna, 2001. Factors affecting decomposition and Diptera colonization. *Forensic Sci. Int.* 120: 18-27.

Campobasso, C. P., J. G. Linville, J. D. Wells & F. Introna, 2005. Forensic genetic analysis on insect gut contents. *Am. J. Forensic Med. Path.* 26: 161-165.

Carvalho, F., I. R. Dadour, D. M. Groth & M. L. Harvey, 2005. Isolation and detection of ingested DNA from the immature stages of *Calliphora dubia* (Diptera: Calliphoridae): a forensically important *blow fly*. *Forensic Sci. Med. Path.* 1: 261-265.

Carvalho, L. M. L., A. X. Linhares, & J. R. Trigo, 2001. Determination of drug levels and the effect of diazepam on the growth of necrophagous flies of forensic importance in southeastern Brazil. *Forensic Sci. Int.* 120: 140-144.

Catts, E. P. and M. L. Goff, 1992. Forensic entomology in criminal investigations. *Annu. Rev. Ent.* 37: 253-272.

Chen, W. Y., T. H. Hung & S. F. Shiao, 2004. Molecular identification of forensically important blow fly species (Diptera: Calliphoridae) in Taiwan. *J. Med. Ent.* 41: 47-57.

Clery, J. M., 2001. Stability of prostate specific antigen (PSA) and subsequent Y-STR typing of *Lucilia (Phaenicia) sericata* (Meigen) (Diptera: Calliphoridae) maggots reared from a simulated *post-mortem* sexual assault. *Forensic Sci. Int.* 120: 72-76.

Davis, J. B. & M. L. Goff, 2000. Decomposition patterns in terrestrial and intertidal habitats on Oahu Island and Coconut Island Hawaii. *J. Forensic Sci.* 45: 836-842.

De Jong, G. D. & W. W. Hoback, 2006. Effect of investigator disturbance in experimental forensic entomology: succession and community composition. *Med. Vet. Ent.* 20(2): 248-258.

DiZinno, J. A., W.D. Lord, M. B. Collins-Morton, M. R. Wilson & M. L. Goff, 2002. Mitochondrial DNA sequencing of beetle larvae (Nitidulidae: Omosita) recovered from human bone. *J. Forensic Sci.* 47: 1337-1339.

Erzinclioglu, Y. Z., 1990. On the interpretation of maggot evidence in forensic cases. *Med. Sci. Law* 30: 65-66.

Erzinclioglu, Y. Z., 1996. *Blow flies*, Naturalists' Handbooks No. 23. Richmond Publishing, Slough, UK, pp. 1-71.

Fabritius, K. & R. Klunker, 1991. Die Larven- und Puparienparasitoide von synanthropen Fliegen in Europa. *Merkbl Angew Parasitenkd Schadlingsbekampf* 32: 1-24.

Faucherre, J., D. Cherix & C. Wyss, 1999. Behavior of *Calliphora vicina* (Diptera, Calliphoridae) under extreme conditions. *J. Insect Behav.* 12: 687-690.

Florin, A. B. and N. Gyllenstrand, 2002. Isolation and characterization of polymorphic microsatellite markers in the *blow flies Lucilia illustris* and *Lucilia sericata*. *Mol. Ecol. Notes* 2: 113-116.

Gagliano-Candela, R., & L. Aventaggiato, 2001. The detection of toxic substances in entomological specimens. *Int. J. Legal Med.* 114: 197-203.

Galloway, A., H. Walsh-Haney & J. H. Byrd, 2001. Recovering buried bodies and surface scatter: the associated anthropological botanical and entomo-logical evidence In: *Forensic entomology: the utility of arthropods in legal investigations*, Byrd, J. H. and J. L. Castner (eds) CRC, Boca Raton, pp. 223-262.

Gennard, D. E., 2007. *Forensic entomology: An introduction*. Wiley & Sons Publications, England, pp. 224.

Goff, M. L., M. L. Miller, J. D. Paulson, W. D. Lord, E. Richards & A. I. Omori, 1997. Effects of 3,4-methylendioxymethamphetamine in decomposing tissues on the development of *Parasarcophaga ruficornis* (Diptera: Sarcophagidae) and detection of the drug in *post-mortem* blood, liver tissue, larvae and puparia. *J. Forensic Sci.* 42: 276-280.

Goff, M. L. & C. B. Odom, 1987. Forensic entomology in the Hawaiian Islands: three case studies. *Am. J. Forensic Med. Path.* 8: 45–50.

Grassberger, M. & C. Frank, 2003a. Initial study of arthropod succession on pig carrion in a central European urban habitat. *J. Med. Ent.* 41(3): 511-523.

Grassberger M. & C. Frank, 2003b. Temperature-dependent development of the parasitic wasp *Nasonia vitripennis* and its forensic implications. *Med. Vet. Ent.* 17: 257-262.

Greenberg, B. 1990. Nocturnal oviposition behaviour of blow flies (Diptera: Calliphoridae). *J Med. Ent.* 27: 807-810.

Greenberg, B. 1991. Flies as forensic indicators. *J. Med. Ent.* 28(5): 565-577.

Greenberg, B. & J. C. Kunich 2002. *Entomology and the law: flies as forensic indicators*. Cambridge University Press, Cambridge, pp. 306.

Haefner, J. N., J. R. Wallace & R. W. Merritt 2004. Pig decomposition in lotic aquatic systems: the potential use of algal growth in establishing a *post-mortem* submersion interval (PMSI). *J. Forensic Sci.* 49: 1-7.

Harvey, M.L., I.R. Dadour & S. Gaudieri 2003. Mitochondrial DNA cytochrome oxidase I gene: potential for distinction between stages of some forensically important fly species (Diptera) in Western Australia. *Forensic Sci. Int.* 131: 134-139.

Haskell, N. H., R.D. Hall, V. J. Cervenka & M. A. Clark 1997. On the body: insect's life stage presence, their *post-mortem* artifacts. In: *Forensic taphonomy: the post-mortem fate of human remains*. W. D. Haglund and M. H. Sorg (eds), CRC, Boca Raton, Fla., pp. 415-448.

Haskell, N. H., W. D. Lord & J. H. Byrd, 2000. Collection of entomological evidence during death investigations. In: *Forensic entomology: the utility of arthropods in legal investigations*, Byrd, J. H. and J. L. Castner (eds), CRC, Boca Raton, Fla, pp 111-120.

Haskell, N. H., D. G. McShaffrey, D. A. Hawley, R. E. Williams & J. E. Pless, 1989. Use of aquatic insects in determining submersion interval. *J. Forensic Sci.* 34: 622–632.

Haskell, N. H. & R. E.Williams, 1990. Collection of entomological evidence at the death scene. In: *Entomology and death: a procedural guide*. Catts, E. P. and N. H. Haskell (eds), Joyce's Print Shop Inc, Clemson, SC, pp. 1-182.

Henssge, C., B. Madea, B. Knight, L. Nokes & T. Krompecher, 1995. *The Estimation of the Time Since Death in the Early Post-mortem Interval.* Arnold, pp. 262.

Hewadikaram, K.A. & M.L. Goff, 1991. Effect of carcass size on rate of decomposition and arthropod succession patterns. *Am. J. Forensic Medi. Path.* 12(3): 235-240.

Higley, L.G. & N.H. Haskell, 2001. Insect development and forensic entomo-logy. In: *Forensic entomology: the utility of arthropods in legal investigations.* Byrd, J. H. and J. L. Castner (eds) CRC, Boca Raton, Fla, pp. 287-302.

Hobischak, N. R. & G. S. Anderson, 2002. Time of submergence using aquatic invertebrate succession and decompositional changes. *J. Forensic Sci.* 47: 142-151.

Hobischak, N. R., S. L. VanLaerhoven & G. S. Anderson, 2006. Successional patterns of diversity in insect fauna on carrion in sun and shade in the Boreal Forest Region of Canada near Edmonton, Alberta. *Can. Entomol.* 138: 376–383.

Holdaway, F. G. & A. C. Evans, 1930. Parasitism a stimulus to pupation: *Alysia manducator* in relation to the host *Lucilia sericata*. *Nature* 125: 598-599.

Hunt, P., 2005. Third Annual North American Forensic Entomology Association Conference, Orlando, FL.

Introna, F. Jr., C. P. Campobasso & M. L. Goff, 2001. Entomotoxicology. *Forensic Sci. Int.* 120: 42-47.

Ireland, S. & B. Turner, 2006. The effects of larval crowding and food type on the size and development of the *blow fly Calliphora vomitoria*. *Forensic Sci. Int.* 159: 175-181.

Ji, Y.J., D.X. Zhang, G.M. Hewitt, L. Kang & D.M. Li, 2003. Polymorphic microsatellite loci for the cotton bollworm *Helicoverpa armigera* (Lepidoptera: Noctuidae) and some remarks on their isolation. *Mol. Ecol. Notes* 3: 102-104.

Joy, J.E., N.L. Liette & H.L. Harrah, 2006. Carrion fly (Diptera: Calliph-oridae) larval colonization of sunlit and shaded pig carcasses in West Virginia, USA. *Forensic Sci. Int.* 164: 183-192.

Junqueira, A.C.M., A. C. Lessinger, T. T. Torres, F. R. da Silva, A. L. Vettore, P. A. Arruda & M. L. A. Espin, 2004. The mitochondrial genome of the *blow fly Chrysomya chloropyga* (Diptera: Calliphoridae). *Gene* 339: 7-15.

Kashyap, V.K. & V.V. Pillay, 1989. Efficacy of entomological method in estimation of *post-mortem* interval: a comparative analysis. *Forensic Sci. Int.* 40: 245-250.

Keh, B., 1985. Scope and applications of forensic entomology. *Ann. Rev. Entomol.* 30: 137-54.

Keiper, J. B., E. G. Chapman & B. A. Foote, 1997. Midge larvae (Diptera: Chironomidae) as indicators of *post-mortem* submersion interval of carcasses in a woodland stream: a preliminary report. *J. Forensic Sci.* 42: 1074-1079.

Kreike, J. & S. Kampfer, 1999. Isolation and characterization of human DNA from mosquitoes (Culicdae). *Int. J. Legal Med.* 112: 380-382.

LaMotte, L. R. & J. D. Wells, 2000. p-Values for *post-mortem* intervals from arthropod succession data. *J. Agric. Biol. Environ. Stat.* 5: 58–68.

Linville, J. G., J. Hayes & J. D. Wells, 2004. Mitochondrial DNA and STR analyses of maggot crop contents: effect of specimen preservation technique. *J. Forensic Sci.* 49: 1-4.

Linville, J. G. & J. D. Wells, 2002. Surface sterilization of a maggot using bleach does not interfere with mitochondrial DNA analysis of crop contents. *J. Forensic Sci.* 47: 1-5.

Lord, W. D, J. A. DiZinno, M. R. Wilson, B. Budlowle, D. Taplin & T. L. Meinking, 1998. Isolation amplification and sequencing of human mitochondrial DNA obtained from human crab louse *Pthirus pubis* (L) blood meals. *J. Forensic Sci.* 43: 1097-1100.

Malgorn, Y. & R. Coquoz, 1999. DNA typing for identification of some species of Calliphoridae: an interest in forensic entomology. *Forensic Sci. Int.* 102: 111-119.

Marchenko, M. J., 2001. Medicolegal relevance of cadaver entomofauna for the determination of time since death. *Forensic Sci. Int.* 120: 89-109.

McWatters, H. G. & D. S. Saunders, 1998. Maternal temperature has different effects on the photoperiodic response and duration of larval diapause in blow fly (*Calliphora vicina*) strains collected at two latitudes. *Physiol. Entomol.* 23: 369-375.

Merck, M. D., 2007. Veterinary forensics. Animal cruelty investigations. Blackell, Oxford pp. 368.

Merrit, R. W. & J. R. Wallace, 2001. The role of aquatic insects in forensic investigations In: *Forensic entomology – the utility of arthropods in legal investigations*. Byrd, J. H. and J. L. Castner (eds) CRC, Boca Raton, London, pp 177-222.

Mukabana, W., R. W. Takken & B. G. J. Knols, 2002. Analysis of arthropod bloodmeals using molecular genetic markers. *Trends Parasitol.* 18: 505-509.

Mumcuoglu, K., Y. N. Gallili, A. Reshef and P. Brauner, 2004. Use of human lice in forensic entomology. *J. Med. Ent.* 41: 803-806.

Norris, K. R., 1965. The bionomics of blow flies. *Ann. Rev. Ent.* 10: 47-68.

Nuorteva, P., 1959. Studies on the significance of flies in the transmission of poliomyelitis. III. The composition of the blow fly fauna and the activity of the flies in relation to the weather during the epidemic season of poliomyelitis in south Finland. *Ann. Entomol. Fenn.* 25: 137-162.

Nuorteva, P., 1977. Sarcosaprophagous insects as forensic indicators. In: *Forensic medicine: a study in trauma and environmental hazards. Vol 2,* Tedeshi, G. C., W. G. Eckert, and L. G. Tedeshi (eds), Saunders, Philadelphia, pp. 1072-1095.

Nuorteva, P., H. Schumann, M. Isokoski & K. Laiho, 1974. Studies on the possibilities of using *blow flies* (Diptera: Calliphoridae) as medicolegal indicators in Finland. *Ann. Entomol. Fenn.* 40: 70-74.

Payne, J. A., 1965. A summer carrion study of baby pig *Sus scrofa* Linn. *Ecol.,* 46(5): 590-602.

Ratcliffe, S. T., D. W. Webb, R. A. Weinzievr & H. M. Robertson, 2003. PCR-RFLP identification of Diptera (Calliphoridae Muscidae and Sarcophagidae) – a generally applicable method. *J. Forensic Sci.* 48: 1–3.

Reiter, C., 1984. Zum Wachstumsverhalten der Maden der blauen Schmeißfliege *Calliphora vicina*. *Z. Rechtsmed* 91: 295-308.

Replogle, J., W. D. Lord, B. Budlowle, T. L. Meinking & D. Taplin, 1994. Identification of host DNA by amplified fragment length polymorphism analysis: preliminary analysis of human crab louse (Anoplura: Pediculidae) exreta. *J. Med. Ent.* 31: 686-690.

Sadler, D. W., J. Richardson, S. Haigh, G. Bruce & D. J. Pounder, 1997. Amitriptyline accumulation and elimination in *Calliphora vicina* larvae. *Am. J. Forensic Med. Path.*, 18: 397-403.

Saunders, D. S., 1975. Manipulation of the length of the sensitive period, and the induction of pupal diapause in the flesh-fly *Sarcophaga argyrostoma*. *J. Entomol.* 50: 107-118.

Schiro, G. J., 2001. Extraction and quantification of human deoxyribonucleic acid and the amplification of human short tandem repeats and a sex identification marker from fly larvae found on decomposing tissue. MS Thesis University of Central Florida.

Schoenly, K., 1992. A statistical analysis of successional patterns in carrion-arthropod assemblages: implications for forensic entomology and the determination of the *post-mortem* interval. *J. Forensic Sci.* 37: 1489-1513.

Schoenly, K., K. Griest & S. Rhine, 1991. An experimental field protocol for investigating the post-mortem interval using multidisciplinary indicators. *J. Forensic Sci.* 36: 1395-1415.

Schoenly, K. & W. Reid, 1987. Dynamics of heterotrophic succession in carrion arthropod assemblages: discrete series or a continuum of change? *Oecologia,* 73: 192-202.

Singh, D. & M. Bharti, 2001. Further observations on the nocturnal oviposition behaviour of blow flies (Diptera: Calliphoridae). *Forensic Sci. Int.* 120: 124-126.

Singh, D. & M. Bharti, 2008. Some notes on the nocturnal larviposition by two species of *Sarcophaga* (Diptera: Sarcophagidae). *Forensic Sci. Int.* 177(1): e19-e20.

Singh, D. & B. Greenberg, 1994. Survival after submergence in the pupae of five species of blow flies (Diptera: Calliphoridae). *J. Med. Ent.* 31: 757-759.

Smith, K. G. V., 1986. *A manual of forensic entomology*. British Museum, London, pp.1-205.

Smith, K. E & R. Wall, 1997a. Asymmetric competition between larvae of the *blow flies Calliphora vicina* and *Lucilia sericata* in carrion. *Ecol. Entomol.* 22: 467-474.

Smith, K. E. & R. Wall, 1997b. The use of carrion as breeding sites by the *blow fly Lucilia sericata* and other Calliphoridae. *Med. Vet. Ent.* 11: 38-44.

Sorg, M. H., J. H. Dearborn, E. I. Monahan, H. F. Ryan, K. G. Sweeney & E. David, 1997. Forensic taphonomy in marine contexts. In: *Forensic taphonomy: the post-mortem fate of human remains*. Haglund, W. D. and M. H. Sorg (eds), CRC, LLC, Boca Raton, pp. 567–604.

Sperling, F. A. H., G. S. Anderson & D. A. Hickey, 1994. A DNA-bases approach to the identification of insect species used for *post-mortem* interval estimation. *J. Forensic Sci.* 39: 418-427.

Statheropoulos, M., C. Spiliopoulou & A. Agapiou, 2005. A study of volatile organic compounds evolved from the decaying human body. *Forensic Sci. Int.* 153: 147-155.

Stevens, J. and R. Wall, 1996. Species sub-species and hybrid populations of the *blow flies Lucilia cuprina* and *Lucilia sericata* (Diptera: Calliphoridae). *Proc. R. Soc. Lond. B.* 263: 1335-1341.

Stevens, J. & R. Wall, 1997. The evolution of ectoparasitism in the genus *Lucilia* (Diptera: Calliphoridae). *Int. J. Parasitol.* 27: 51-59.

Stevens, J. & R. Wall, 2001. Genetic relationships between *blow flies* (Calliphoridae) of forensic importance. *Forensic Sci. Int.* 120: 116-123.

Stevens, J., R. Wall & J. D. Wells, 2002. Paraphyly in Hawaiian hybrid *blow fly* populations and the evolutionary history of anthropophilic species. *Insect Mol. Biol.* 11: 141-148.

Tabor, K. L., C. C. Brewster & R. D. Fell, 2004. Analysis of the successional patterns of insects on carrion in southwest Virginia. *J. Med. Ent.* 41(4): 785-795.

Tessmer, J. W., C. L. Meek & V. L. Wright, 1995. Circadian patterns of oviposition by necrophilous flies (Diptera: Calliphoridae). *Southwest Entomol.* 24: 439-445.

Thyssen, P. J., A. J. Lessinger, A. M. L. Azeredo-Espin & A. X. Linhares, 2005. The value of PCRRFLP molecular markers for the differentiation of immature stages of two necrophagous flies (Diptera: Calliphoridae) of potential forensic importance. *Neotrop. Entomol.* 34: 777-783.

Torres, T. T. & A. M. L. Azeredo-Espin, 2005. Development of new polymorphic microsatellite markers for the New World screw-worm *Cochliomyia hominivorax* (Diptera: Calliphoridae). *Mol. Ecol. Notes* 5: 815-817.

Torres, T. T., R. P. V. Brondani, J. E. Garcia & A. M. L. Azeredo-Espin, 2004. Isolation and characterization of microsatellite markers in the New World screw-worm *Cochliomyia hominivorax* (Diptera: Calliphoridae). *Mol. Ecol. Notes* 4: 182-184.

Turner, B. & T. Howard, 1992. Metabolic heat generation in dipteran larval aggregations: a consideration for forensic entomology. *Med. Vet. Ent.* 6: 179-181.

Vance, G. M., J. K. VanDyk & W. A. Rowley, 1995. A device for sampling aquatic insects associated with carrion in water. *J. Forensic Sci.* 40: 479–482.

Vinogradova, E. B., 1986. Geographic variation and ecological control of diapause in flies. In: *The evolution of insect life-cycles*. Taylor, F. and R. Karban (eds) Springer, Berlin Heidelberg New York, pp. 35-47.

Vinogradova, E. B., 1991. Diapause in flies and its control (in Russian with English summary). *Proc ZIN RAS* St. Petersburg, 214.

Vinogradova, E. B. & K. B. Zinovjeva, 1972. Maternal induction of larval diapause in the *blow fly Calliphora vicina*. *J. Insect Physiol.* 18: 2401-2409.

Voss, S. C., H. Spafford & I. R. Dadour, 2009. Annual and seasonal patterns of insect succession on decomposing remains at two locations in Western Australia. *Forensic Sci. Int.* 193(1-3): 26-36.

Wallman, J. F., & M. Adams, 2001. The forensic application of allozyme electrophoresis to the identification of *blow fly* larvae (Diptera: Calliphoridae) in Southern Australia. *J. Forensic Sci.* 46: 681-684.

Wallman, J. F. & S. C. Donnellan, 2001. The utility of mitochondrial DNA sequences for the identification of forensically important *blow flies* (Diptera: Calliphoridae) in Southeastern Australia. *Forensic Sci. Int.* 120: 60-67.

Wallman, J. F., Leys, R. & K. Hogendoorn, 2005. Molecular systematics of Australian carrion-breeding *blow flies* (Diptera: Calliphoridae) based on mitochondrial DNA. *Invertebr. Syst.* 19: 1-15.

Wells, J. D., F. Introna, G. Di Vella, C. P. Campobasso, J. Hayes & F. A. H. Sperling, 2001. Human and insect mitochondrial DNA analysis from maggots. *J. Forensic Sci.* 46: 685-687.

Wells, J. D. & L. King, 2001. Incidence of precocious egg development in flies of forensic importance (Calliphoridae). *Pan-Pac. Entomol.* 77: 235-239.

Wells, J. D. & L. R. LaMotte, 2001. Estimating the *post-mortem* interval. In: *Forensic entomology: the utility of arthropods in legal investigations.* Byrd, J. H. and J. L. Castner (eds), CRC, Boca Raton, Fla., pp. 263-286.

Wells, J. D., N. Lunt & M. H. Villet, 2004. Recent African derivation of *Chyrsomya putoria* from *Chrysomya chloropyga* and mitochondrial DNA paraphyly of cytochrome oxidase subunit one in *blow flies* of forensic importance. *Med. Vet. Ent.* 18: 445-448.

Wells, J.D. & F.A.H. Sperling, 1999. Molecular phylogeny of *Chrysomya albic-eps* and *Chrysomya rufifacies* (Diptera: Calliphoridae). *J. Med. Ent.* 36: 222-226.

Wells, J.D. & F.A.H. Sperling, 2001. DNA-based identification of forensically important Chrysomyinae (Diptera: Calliphoridae). *Forensic Sci. Int.* 120: 110-115.

Wells, J.D., R. Wall & J.R. Stevens, 2007. Phylogenetic analysis of forensically important *Lucilia* flies based on cytochrome oxidase I sequence: a cautionary tale for forensic species determination. *Int. J. Legal Med.* 121: 229-233.

Wells, J.D. & D.W. Williams, 2007. Validation of a DNA-based method for identifying Chrysomyinae (Diptera: Calliphoridae) used in a death investigation. *Int. J. Legal Med.* 121: 1-8.

Williams, H. 1984. A model for the aging of fly larvae in forensic entomology. *Forensic Sci. Int.* 25: 191-199.

Zehner, R., J. Amendt & R. Krettek, 2004a. STR typing of human DNA from fly larvae fed on decomposing bodies. *J. Forensic Sci.* 49: 1-4.

Zehner, R., J. Amendt, S. Schütt, J. Sauer, R. Krettek & D. Povolný, 2004b. Genetic identification of forensically important flesh flies (Diptera: Sarcophagidae). *Int. J. Legal Med.* 118: 245-247.

Chapter - 18

MICROBIAL ENTOMOPATHOGENS : A SUSTAINABLE PEST MANAGEMENT STRATEGY

Arti Prasad

Insect Microbial and Herbal Control Laboratory, Department of Zoology, MLSU, Udaipur – 313 001, Rajasthan, India

E-mail: *artimlsu@yahoo.co.in*

INTRODUCTION

"Pest" in a broader sense is any plant and animal species that is in competition with humans for some resources. In agriculture, generally phytophagous insects are labelled as pest when its population is high enough to cause significant damage in yield or quality to any of the economic plants grown by man (Dhaliwal *et al.*, 1998). Insect may cause losses of 25 to 30 percent crop yield in Indian agriculture amounting approximately to Rs 26,000 crores per year. Recent changes in cultural, agricultural and economic patterns throughout the world have a profound effect on insect pest management and their control. The change to continuous cropping pattern over wide adjacent areas has created serious insect pest problems. Simultaneously the demand for agricultural products damaged by insect pest has intensified management and control problems. Further with the increase in food grain production, pesticide consumption has also increased from 20,000 tonnes a year to 80,000 tonnes (Dhaliwal & Singh, 2000). The sale of pesticides has increased 30 percent and are projected to increase from US $ 5.4 million in 2004 to 7.5 million by 2009 (Dhaliwal & Singh, 2001).

Although pesticides have dramatically changed the agricultural scenario but as every bright side has a dark side, similarly the indiscriminate and unscientific use of these pesticides has led to a huge number of threats not only to the environment and food but also to farmers and consumers. We all are now familiar with 3 *Rs, i.e.,* resistance, residue and resurgence. During past two decades the periodical unabated explosion aphids on chillies, white flies, Helicoverpa and Spodoptera on cotton, scales on sugarcane, mealy bug on grapevine, fruit borer and hopper on mango, vector based virus on chillies, leaf on tomatoes, red hairy caterpillar on several crops and diamond back moth on Crucifers are some of the few examples that have made the agriculture less remunerative and highly risk prone (Rao *et al.,* 2003). Therefore, with a motto to reduce the role of these chemicals in plant protection programmes, alternative components, like cultural, biological, mechanical and legal etc. were put forth by various experts of this line. However, individually none of them is as sound as chemical control but the integration of two or more proved profitable economical and sustainable. This combined system is called Integrated Pest Management or IPM. According to FAO, it is a basic step towards shaping a sustainable agro-ecosystem that retains the recyclic phenomenon where beneficial micro and macro life forms are activated and harmful life forms are reasonably reduced.

Use of bio-pesticides of animal and plant origin are one of the important components of IPM. India with a large agrarian society and with agriculture as a prime occupation needs to concentrate on the eco-safe sustainable aspect of pest control and bio pesticides have come true to the expectation.

Entomopathogens or microbial pesticides are a new class of bio-pesticides that work by infecting selected insect populations with bacteria, virus, fungus and protozoans. They are quiet safe since the application of microbial insecticides is specific to the species one is trying to target. Microbial insecticides have no effect on other animal populations besides being free from 3*Rs*.

In the present chapter we are discussing the efficacy of all the three entomopathogens, *i.e.,* bacterial, viral and fungal against two most important pests of agriculture, *Spodoptera litura* (Fab), the tobacco caterpillar and *Helicoverpa armigera* (Hubner), the gram pod borer. The results are based on laboratory as well as field evaluation of all the three microbials against the two selected pests. Although, the detailed studies have been done taking all stages, but we are summarizing the main part of the work that proves the feasibility and compatibility of microbials against chemical pesticides. The present study advocates strongly the inclusion of these microbials in pest control programes for substituting chemical pesticides if not eliminating completely.

MATERIALS AND METHODS

The three selected bio pesticides were procured from different sources and two selected pests were reared in the laboratory. Since *Helicoverpa armigera* has severe cannibalism they were reared individually on artificial diet. After getting favourable results in laboratory these microbials were taken to field for their evaluation in field conditions by comparing them with chemical pesticides on the basis of yield and cost benefit ratio. Following experimental protocols were adopted:

1. Entomopathogen of Bacterial origin

Bacillus thuriengensis var. *kursataki*

Approved name- *Bacillus thuriengeninsis* var. *kursataki*

Trade Name - Dipel, Halt

Target pest - *Spodoptera litura* (Fabricus)

Formulation - Wettable powder

Method of treatment - Leaf dip Method

Stages Taken - V^{th} instar larva of *Spodoptera litura* (Fab)

Doses applied - 0.001, 0.01, 0.1 and 1.0 % of $5x10^7$ spores per mg

2. Entomopathogen of Viral Origin

Nuclear Polyhedrosis Virus

Approved name - *Sl* Nuclear Polyhedrosis Virus

Trade name - Spod-x

Target Pest - *Spodoptera litura* (Fabricus)

Formulation - Liquid

Dose concentrations - 0.1,0.25.0.5 and 0.75 ml of NPV with $2x10^9$ polyhedra /ml

3. Entomopathogen of fungal origin

Beauveria bassiana (Balsamo) Vullemin

Approved name - *Beauveria bassiana* (Balsamo) Vulleimin

Trade Name - Dispel

Target Pest - *Helicoverpa armigera* (Hubner)

Formulation - Wettable powder

Dose Concentrations - 0.1,0.125,0.2 and 0.250 ml of $1x10^8$ conidia/ml

Method of treatment - Topical

The efficacy of the microbials were adjudged on the basis of percent mortality in relation to time period and concentrations and various morphogenetic changes in the treated larvae. ANOVA was applied on the obtained data.

RESULTS

Fifth instar larvae of *Spodoptera litura* (Fabricus) treated with *Sl* NPV and Bt.k.

Experiment with Nuclear Polyhedrosis Virus (*Sl* NPV)

When fifth instar larvae were treated with formulated doses of *Sl* NPV, the mean percent mortality showed an increasing trend during treatment period from 48 hours and further to 96 hours with a mean increase of 6.98 to 28.99 percent respectively, as compared to 1.22 percent in control, but further increase of treatment period upto 144 hours did not significantly increase mean percent mortality which was observed only 30.90 percent. Although, the significant value of 'F' at CD 1 percent for treatment period indicated an overall significant effect on the larval instar. Further, when the different concentrations were compared, no apparent difference could be observed as depicted by mean percent mortality which had slight increase from 26.50 to 32.56 percent at lowest (0.10) and highest applied dose (0.75ml). However, ANOVA depicted significant 'F' value at CD 1 percent (4d.f.) for concentration of doses. 'F' value was further significant at CD 1 percent (8d.f.) when doses and treatment period were considered together.

Experiment with *Bacillus thuringiensis* var *kurstaki* (Bt.k.)

Bacillus thuringiensis var *kurstaki* treatment on fifth instar with various concentrations brought significant effect on larval mortality. Time period was observed to cause significant mortality unto 96 hours after which it did not show further apparent increase from 96 to 144 hours. At highest dose (1 percent) 26.66 percent mortality was observed at 48 hours which significantly increased to 62.22 percent at 96 hours and at 144 hours it increased only upto 64.47 percent. Although as compared to control with 6.06 mean percent mortality the mean percent mortality ranged from 27.86 to 50.92 at lowest and highest doses respectively, and it was well significant and was statistically supported by 'F' value which was significant at CD 1 percent (4d.f.).

When effect of both bio-pesticides on fifth instar larvae were compared the overall treatments of both the bio-pesticides had shown significant results for both the parameters in individual as well as in combination. The significant value of 'F' (297.49) at CD 1 percent (1.d.f.) for effect of both the bio-pesticides for combined analysis proved them to be at par.

Morphological abnormalities: When Vth instar larvae of *Spodoptera litura (Fb)* were treated with different doses of *Sl* NPV, body became swollen as if filled with liquid. Development of yellowish pinkish color was observed Larval-pupal intermediates were very common indicating the effect of virus on chitin synthesis of the insect. When the larvae were treated with different doses of Bacterial preparations, immediate cessation of feeding was observed. Diarrohoea due to gut paralysis was also observed.

Table 1. Effect of Nuclear Polyhedrosis Virus (*Sl* NPV) and *Bacillus thuringiensis* var *kurstaki* (Bt.K.) on fifth instar larvae of *Spodoptera litura* (Fabricus)

(1) Arc sign mean and (retransformed) values of *Sl* NPV treatment

	C1	C2	C3	C4	C5	Mean
B1	0.000	18.418	17.313	19.458	21.404	15.319
	(0.00)	(9.99)	(8.86)	(11.10)	(13.33)	(6.98)
B2	7.007	37.248	38.558	38.559	41.145	32.503
	(1.49)	(36.66)	(38.88)	(38.88)	(43.32)	(28.89)
B3	11.990	37.250	38.558	39.214	41.791	33.760
	(4.32)	(36.66)	(38.88)	(40.00)	(44.44)	(30.90)
Mean	6.332	30.972	31.476	32.410	34.780	27.194
	(1.22)	(26.50)	(27.28)	(28.75)	(32.56)	(20.90)

(2) Arc sign mean and (retransformed) values of Bt.K. treatment

	C1	C2	C3	C4	C5	Mean
B1	10.510	18.423	21.404	24.082	31.076	21.099
	(3.33)	(10.00)	(13.33)	(16.66)	(26.66)	(12.97)
B2	14.947	37.250	39.214	43.069	52.055	37.307
	(6.66)	(36.66)	(40.00)	(46.66)	(62.22)	(36.76)
B3	17.262	39.861	41.791	44.980	53.389	39.456
	(8.81)	(41.11)	(44.44)	(50.00)	(64.47)	(40.41)
Mean	14.240	31.844	34.136	37.377	45.507	32.621
	(6.06)	(27.86)	(31.51)	(36.88)	(50.92)	(29.08)

(3) CRD ANOVA FOR V INSTAR

S.No.	SOURCE	DF	SS	MS	F	SE(m)	CD5%	CD1%
1	Bet A	1	662.622	662.622	297.493**	0.222	0.629	0.837
	Within A1							
	Treat	14	8363.99	597.428	268.223**			
	Bet B	2	3184.93	1592.46	714.959**	0.385	1.090	1.450
	Bet C	4	4973.17	1243.29	558.193**	0.497	1.407	1.872
	B×C	8	205.898	25.7372	11.555**	0.862	2.437	3.243
	Within A2							
	Treat	14	8136.54	581.181	260.929**			
	Bet B	2	3021.56	1510.78	678.285**	0.385	1.090	1.450
	Bet C	4	4764.95	1191.24	534.822**	0.497	1.407	1.872
	B×C	8	350.03	43.7538	19.644**	0.862	2.437	3.243
2.	Error	60	133.641	2.22735				

General mean = 29.91 CV = 4.99

Vth instar larvae of *Helicoverpa armigera* (Fabricus) treated with *Beauveria bassiana* (Balsamo) Vulleimin

Highest dose of 0.25 ml of 1 x 10^8 conidia /ml brought about 73.33 percent mortality against 13.33 percent in control. Results for toxicity further got confirmation by one way ANOVA test showing F value significant at 1 percent level. When cumulative pathogenecity of *Beauveria bassiana* was recorded for Vth larval instars at all the four concentrations, *i.e.*, 0.1, 0.125, 0.2 and 0.25 ml, highest dose of 0.25 ml of 1x10^8 conidia /ml was infectious for the larvae. The fungus brought about characteristic abnormalities indicating clearly its chitin inhibitory nature. During our investigation, the treated larval instars depicted different morphological abnormalities. All the cadavers of *Helicoverpa armigera* gave rise to sporulation of the fungus, indicating that death was caused by infection and that strains were capable of sporulating under existing humidity conditions. Larvae showed blackening of body color (Plate 1, Fig. 1 and 4) and oozing out of fungal conidia from lethargic and ruptured larval integuments.

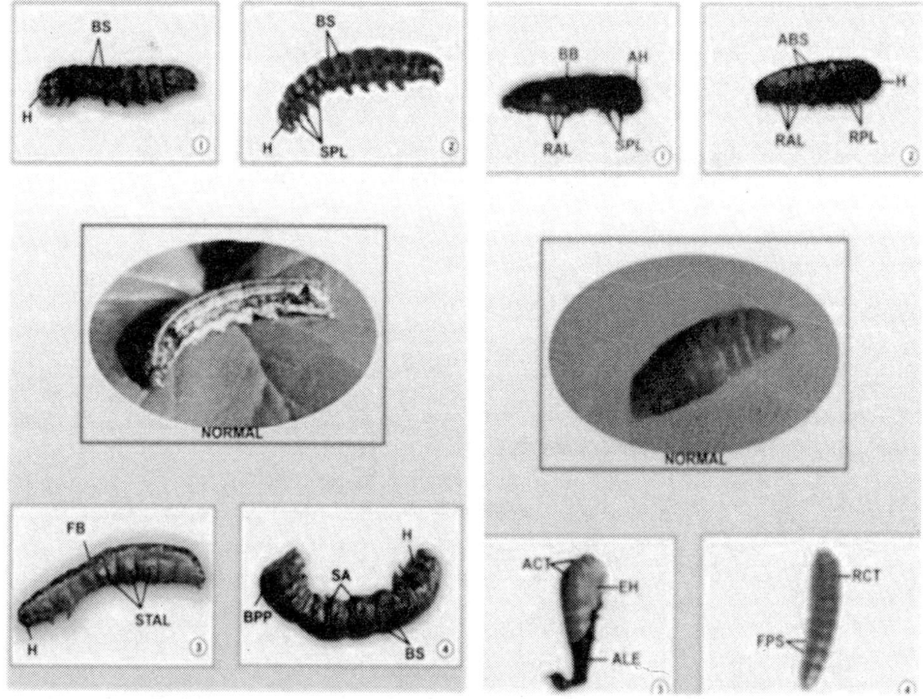

Fig. 1. Morphological abnormalities in treated fifth instar larva with *Sl* NPV and *Bt.k.* (BS – Blackening of skin, H – Head. SPL – Swollen prothoracic legs, FB – Flaccid body. STAL – Stumpy abdominal leg. BPP – Blackening of posterior part, BS – Blackening of the skin, AH – Abnormal head, RAL – reduced abdominal leg, ACT – Abnormal cephalo thorax, EH – Entangled head, ALE – Attached larval exuviae, FPS – Fragile pupal segment)

Table 2. Efficacy of *Beauveria bassiana* (Balsamo) Vuillemin on Vth Instar larve *Helicoverpa armigera* (Hubner)

S No	Mortality	No. of Replica	% Mortality	Standard Deviation	Standard Error
1	Control	3	13.3%	0.58	0.33
2	0.100 ml	3	40.0%	1.00	0.58
3	0.125 ml	3	50.0%	1.00	0.58
4	0.200 ml	3	70.0%	1.00	0.58
5	0.250 ml	3	73.3%	0.58	0.33

ANOVA

Mortality	Sum of Squares	df	Mean Square	F	Result
Between Groups	71.60	4	17.90	24.41	**
Within Groups	7.33	10	0.73		

Note: ** - Significant at 1% (p<0.01); * - Significant at 5% (p<0.05).

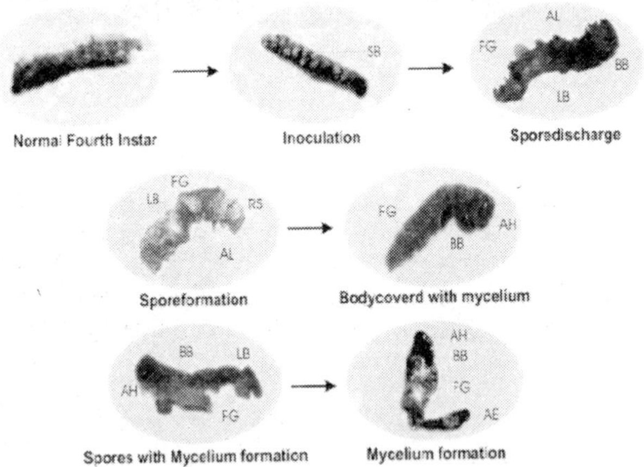

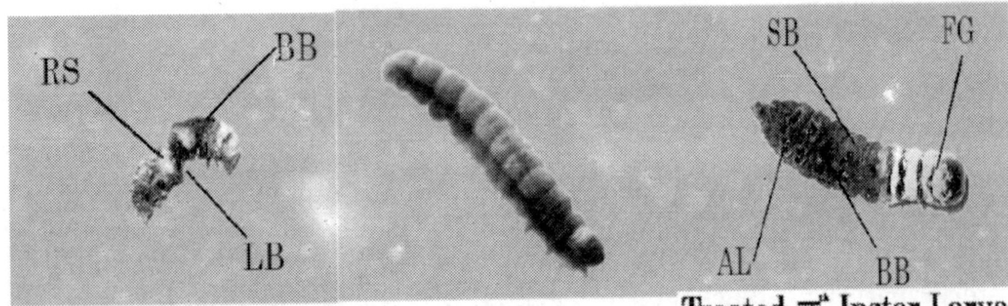

Treated V Instar Larva
Fig-3: 0.2 x10^{8} ml

Normal V Instar Larva

Treated V Instar Larva
Fig-4: 0.25 x10^{8} ml

Insect pathogenic fungal conidia invade insects by penetrating cuticle and producing toxins and draining nutrients from the insects eventually killing them. The body parts were atrophied.mouth parts and legs became short and stumpy. The hyphe grew out through the softer portion of the cuticle, covering the insects with a layer of white mold. Higher doses of 0.2 and 0.25 ml of 1 x 10^8 conidia/ ml treated larvae.

BIO-EFFICACY AND ECONOMICS OF ALL THE THREE BIO-PESTICIDES

Bio-efficacy of *Sl* NPV and Bt.k.1 against *Spodoptera litura* (Fabricus)

The purpose of field trial of both bio-pesticides i.e. Nuclear Polyhedrosis Virus (*Sl* NPV) and *Bacillus thuringiensis kurstaki* (Bt.k.) in the **Randomised Block Design** (RBD) pattern was observed to use them for the effective control of the experimental insect pest *Spodoptera litura* (F). Although cauliflower and cabbage both are the major cole crops grown in and around Udaipur region but the incidence study has shown that cauliflower is more affected by the pest as compared to cabbage. Hence, the efficacy study has been conducted on cauliflower. Same way Fungal entomopathogen *Beauveria bassiana* was used against *Helicoverpa armigera* (Hubner). Assessment of the results were done on the basis of percent plant infestations on number and weight basis as compared to control and by analyzing cost benefit ratio of microbial and synthetic pesticide. Calculations were done by using following formulae:

$$\text{Percent infestation (on weight basis)} = \frac{\text{Weight of infested fruits}}{\text{Weight of Total fruits}} \times 100$$

$$\text{Percent infestation (on number basis)} = \frac{\text{Number of infested fruits}}{\text{Number of Total fruits}} \times 100$$

After getting the total, cost benefit was calculated by comparing botanicals with synthetic recommended pesticide (Endosulfan) by using following formula:

$$\text{Percent gain in yield over control} = \frac{\text{Yield infexted plot} - \text{Yield in controlled plot}}{\text{Yield in treated plot}} \times 100$$

Further the net profit was calculated on the basis of yield gained from different treatments and cost incurred for the production, including cost of insecticide and their application cost etc. (Total yield was calculated for each treatment on the basis of total fruit weight and then converted to quintal/hectare).

Percent infestation on weight basis (Table 3)

Sl NPV treated blocks

When three different recommended doses of *Sl* NPV, *i.e.*, 250, 500 and 750 LE per hectare (*i.e.*, T_1, T_2 and T_3) were sprayed on cauliflower crop, minimum infestation of fruit was obtained with 750 LE per hectare (T_3) where only 8 kgs of curds were infested out of total yield of 63.75 kgs of curds. The percent infestation was only 12.43 as compared to 15.85 (T_2) and 18.20 (T_1) at the dose levels of 500 and 250 LE per hectare respectively. The percent yield was higher in all the *Sl* NPV treated blocks as compared to control where 38.57 percent infestation was 20.75 kgs and total yield obtained was 53.75 kgs.

Bt.k. treated blocks

When average infestation of fruits on weight basis was observed in Bt.k. treated blocks (T_4, T_5 and T_6), although maximum yield was obtained in T_5 and T_6 blocks i.e. 64.75 kgs at the dose levels of 250 gm and 500 gm per litre per hectare respectively but percent infestation was only 6.45 percent in T_6 block as compared to 10.22 percent in T_5 block. At minimum recommended dose (125 grams per litre per hectare) percent infestation was 14.09 with average of 9 kgs of curds infested out of total yield of 63.75 kgs. Here also the yield was far superior when compared with control where 38.57 percent infestation was observed.

Percent infestation on number basis

Percent infestation was also recorded on the basis of number of infected curds obtained out of total number of curds. For this, in each treatment block of (3.17 × 4.2 meters size) 120 plants were planted and number of plants infested were recorded and the percent infestation on number basis was calculated for each treatment.

Sl NPV treated blocks

Out of three NPV treated blocks (T_1, T_2 and T_3), minimum infestation was obtained in T_3, (750 LE per hectare) with only 11 infected curds out of 120 curds giving only 9.16 percent as compared to 37.58 percent (45 out of 120 curds). In T_1, (250 LE per hectare) and T_2 (500 LE per hectare percent infestations on number basis were almost at par having 14.16 and 11.45 percent infestation as compared to 37.58 percent in control.

Bt.k. treated blocks

Bt.k. treated blocks T_4 (125 gms per litre per hectare), T_5 (250 gms per litre per hectare) and T_6 (500 gms per litre per hectare) revealed significant reduction in percent infestation when compared on numbe basis with control. Here also maximum applied dose of 500 gms per litr

(T_6) brought only 6.56 percent infestation (8 curds out of 120 curds) where as 9.37 and 10.10 percent infestations were observed in T_5 and T_4 blocks respectively as compared to 37.58 percent infestations in control where no spray was done.

When both the bio-pesticides were compared and statistically analysed the ANOVA test revealed that 'F' value was significant for both the parameters, *i.e.*, the spray of both the bio-pesticides were effective on weight basis as well as on number basis (CD at 5 percent and 1 percent)

Comparative economics of bio-pesticides and synthetic insecticides

In this part both the bio-pesticides, *i.e., Sl* NPV and Bt.K. were compared with most commonly used synthetic insecticides for cauliflower by the farmers on the basis of their total yield per hectare, net profit and total incurred cost. The calculations have been done on the basis of cost benefit ratio based on the total yield obtained in the different treatments together with the increased yield over control. The sale price of cauliflower was Rs. 500 per quintal, when the two bio-pesticides were compared together Bt.K. was superior to *Sl* NPV in terms of highest net profit. Although, Bt. @ 125 gm per litre per hectare (T_4 block) gave maximum value of cost benefit ratio, *i.e.*, 1:50.75 as compared to Bt. @ 250 gm per litre and 500 gm per litre per hectare (T_5 and T_6 block) with C.B. ratio of 1:36.57 and 1:22.66, respectively, but as far as increase in yield over control was concerned, maximum yield was obtained with highest dose of 500 gm per litre (115 quintals) and with increase in value of Rs. 57500 as compared to Bt.k. @ 250 gms per litre, *i.e.*, (104.83 quintal) with increase in value of Rs. 52415 and 125 gms per litre per hectare with 90.78 quintal increase in yield and increased value of Rs. 45390.

Since 250 gms per litre per hectare produced, 1:36.57 cost benefit (C:B) ratio as compared to 1:50.75 for lesser dose of 125 gms per litre per hectare and only 1:22.66 with highest dose of 500 gms per litre per hectare.

Since lowest dose (125 gm per litre per hectare) produced maximum C:B ratio (1:50.75) as compared to recommended (250 gm per litre per hectare) and highest dose (500 gm per litre per hectare) with C:B ratio 1:36.57 and 1:22.66, respectively. When the cost of bio-pesticide was compared together with yield, the recommended dose proved its worth.

Similarly, lowest dose of *Sl* NPV @ 250 LE per hectare gave maximum C: B ratio followed by recommended @ 500 LE per hectare and highest dose @ 750 LE per hectare with C.B ratio of 1:15.24 and 1:12.56 respectively. But the increase in yield over control was maximum with highest dose (*Sl* NPV @ 750 LE per hectare) i.e. 95.2 quintal per hectare followed by 79.93 and 67.05 for recommended dose of *Sl* NPV of 500 LE per hectare and lowest dose of 250 LE per hectare respectively. Here, also when compared the recommended dose of bio-pesticide *Sl* NPV (500 LE per hectare) proved better than the two, with the increase in yield.

And at last when these tow bio-pesticides were compared with synthetic insecticides i.e. ROCKET (1 litre per hectare) and Methamyl (1/2 Kg per hectare) although increase yield over bio-pesticide was greater (i.e 147.80 and 132.51 quintal per hectare respectively) than bio-pesticides. Even the cost of synthetic pesticides were lesser than those of bio-pesticides when C:B ratios were considered little higher values were obtained with synthetic insecticides as compared to *Sl* NPV treatment and almost at par with Bt.k. treatments.

Our data on economics of bio-pesticides and insecticides revealed that insecticides were although superior than *Sl* NPV but they were at par with Bt.K. (1:42.21 and 1:34.62) for recommended doses of ROCKET and Methomyl as compared to 1:15.24 and 1:36.57 for recommended doses of *Sl* NPV and Bt.K., respectively). But if the residue and pollution of problems of pesticides are considered than instead of ROCKET and Methomyl, Bt.K. and *Sl* NPV can be very well used to control the tobacco caterpillar in cauliflower field.

Fig. 2. Field View of Cauliflower field treated with Sl NPV an Bt.K along with synthetic insecticide ROKET

Table 3. Bio-efficacy of bio-pesticides against *Spodoptera litura* (Fabricius) infesting cauliflower crop

Treatment No.	Treatment	Dosage per hectare	Infestation of fruits on weight basis		Infestation of fruits on number basis	
			Average number	Percent infestation	Average number	Percent infestation
T_1	NPV	250LE	10.92≈11kg of 60 kg	18.20±1.0280	17 of 120	14.16±0.6818
T_2	NPV	500LE	9.83 ≈ 10 of 62kg.	15.85±0.8040	14 of 120	11.45±0.7990
T_3	NPV	750LE	7.92 ≈ 8 of 63.75 kg.	12.43±0.5150	11 of 120	9.16±0.6818
T_4	Bt	125gm/litre	8.98≈ 9 of 63.75 kg	14.09±0.4639	12 of 120	10.10±0.5878
T_5	Bt	250gm/litre	6.61≈ 7 of 64.75 kg	10.22±0.7662	11 of 120	9.37±1.0490
T_6	Bt	500gm/litre	4.17 of 64.75 kg	6.45±0.3637	8 of 120	6.56±0.5239
T_7	Control	--	20.75 of 53.75 kg	38.57±1.5179	45 of 120	37.58±1.5864
CD at5%				1.268		1.339
CD at 1%				1.728		1.825
Sem ±				0.431		0.455
CV				5.21		6.48

Table 4. Economics and cost benefit ratio of the bio-pesticides and synthetic insecticide on cauliflower crop

S.N	Treatment (Per hectare)	No. of Application	Average total yield (quintals per hectare)	Increased yield over control (quintals per hectare)	Approx. sales price Rs. / quintal	Value of increase yield (Rs.)	Approx. cost of insecticide + labor cost	Approx. net price ± in Rs. Per hectare	C.B. ratio
T_1	NPV @ 250 LE	3	204.66	67.05	500	33525	1410	32115	1:22.77
T_2	NPV @ 500 LE	3	217.54	79.93	500	39965	2460	37505	1:15.24

T_3	NPV @ 750 LE	3	232.81	95.2	500	47600	3510	44090	1:12.56
T_4	Bt. @ 125 gms per litre	3	228.39	90.78	500	45390	877	44513.5	1:50.75
T_5	Bt. @ 250 gms per litre	3	242.44	104.83	500	52415	1395	51020	1:36.57
T_6	Bt. @ 500 gms per litre	3	252.61	115.00	500	57500	2430	55070	1:22.66
T_7	Control	--	137.61	--	--	--	--		
General insecticides used by farmers	ROCKET @ one litre	3	285.41	147.80	500	73900	1710	72190	1:42.21
	Methomyl @ ½ kg	3	270.12	132.51	500	66255	1860	64395	1:34.62

First spray was made after one month of transplantation further second and third sprays were done at fifteen days of interval.

ROCKET: profenofos 40+ Cypermethrin 4 percent

40+ Cypermethrin 4 percent EC

Bio Efficacy of *Beauveria bassiana* against *Helicoverpa armigera* (Hubner)

Percent infestation on weight and number basis

As it is clear from the table that percent infestation on weight as well as number basis was minimum in case of endosalfan spray it was 20.83% and 27.17% on weight and number basis respectively. In case of *Beauveria bassiana* fungus spray infestation was higher than in case of endosalfan but lower than in case of control where no spray was done. It is also evident that in case of *Beauveria bassiana* (Balsamo), as the dose concentration is increased the percent infestation decreased on the weight as well as on number basis. Percent infestation was maximum among three treatments (T_2, T_3 & T_4) when *Beauveira bassiana* fungus concentration was 1.00 ml, it was 66.67% on weight basis and 50.36% on number basis. As the dose concentration is increased to 1.5 ml in T_3 the percent infestation on weight basis was 48.33% and 42.39% on number basis which is lower infestation than in case of T_2. In T_4 the percent infestation was minimum among T_2, T_3 & T_4 treatments when the dose concentration was 2 ml/litre. The percent on weight basis was found to be 32.35% and 35.59% on number basis. The infestation was maximum on weight as well as number basis. When no spray was done, *i.e.*, in case of control. The infestation on weight basis was 75% which is very valve was 68.12% on number basis which is also high. Further all treatment groups were significantly different from each other Cost benefit ratio: The data on the effect of different insecticides treatment on yield per hectare is presented in the Table 6. As far as yield is concerned maximum yield was obtained after the application of endosalfan with an average yield of 426.50 quintal/hectare at recommended dose level. After that average yield was calcutated for *Beauveria bassiana* fungus at different concentration level the average yield was 237.50 quiental/hectare 284.50 q/h and 340.00 q/h respectively i.e. yield was increasing with dose concentration. The yield was minimum in case of control (97.00q/h) when no spray was done.

Comparative Economic of *Beauveria bassiana* fungus with chemical pesticide Endosulfan

In this study the cost involved in applying the insecticides on the cost – benefit ratio has been calculated. The calculation is based on total yield obtained in different treatment together with increase yield over control. The sales price of tomato has been taken as Rs. 500 per quintal. The cost benefit ratio was (cost of labour + cost of insecticide) ratio of cost to value of increase yield over control. Among the five treatments T_1-T_5 the maximum profit was earned (Rs, 1,62,180.00) when endosalfan was sprayed the ratio of cost to benefit was 1: 62.37 after that the profit maximum of Rs. 119,233.00 and C.B. ratio 1:52.60 in case of *beauveria bassiana* fungus when dose concentration was 2 ml/litre among the 3 treatment (T_2, T_3 &T_4) as the does concentration was decreased in *Beauveria bassiana* fungus, *i.e.*, 1.5 ml/litre the profit was decreased to Rs. 91800 and C.B. ratio was 1: 47.80 and it was least among three (T_2, T_3 & T_4), *i.e.*, 1: 41.99 and profit was Rs. 68,616. Hence it can be concluded that the profit and Cost benefit ratio was maximum in case of endosalfan in comparision to other extracts. And, also in other extracts the profit and cost benefit ratio increase as the dose concentration is increased.

Table 5. Bioefficacy of *Beauveria bassiana* (Balsamo) against *Helicoverpa armigera* (Hubner) infesting tomato crop

Treat-ment Number	Treatment	Dosage per hectare	Infestation of tomato fruits on weight basis		Infestation of tomato fruits on number basis	
			Average weight (Kg)	Percent Infestation	Average	Percent Infestation
T1	Endosulfan	1000ml	3.75/18	20.83	75 of 276	27.17
T2	*Beauveria bassiana* fungus	1.0ml/litre	9.50/14.25	66.67	139 of 276	50.36
T3	*Beauveria bassiana* fungus	1.5ml/litre	7.25/15	48.33	117 of 276	42.39

T4	*Beauveria bassiana* fungus	2ml/litre	5.50/17	32.35	98 of 276	35.51
T5	control	-	8.25/11	75.00	188 of 276	68.12
CD at 5%			1.28		32.86	
CD at 1%			1.77		45.43	
SEm			0.72		475.40	
CV				46.65%		34.97%

Fig. 3. Field view of Tomato field treated with *Beauveria bassiana* along with synthetic insecticide Endosulfan

Table 6. Economics and cost benefit ratio of the microbial and synthetic pesticide

S.N.	Treatment (Per hectare)	No. of Application	Average total yield (quintals per hectare)	Increased yield over control (quintals per hectare)	Approx. sales price Rs./ quintal	Value of increased yield (Rupees)	Approx. cost of insecticide + labor cost	Approx. net price + in Rs.per hectare	C.B.Ratio
T1	Endoslfan	4	426.50	329.50	500.00	164,750.00	2,600.00	162,150.00	1 : 62.37
T2	*Beauveria bassiana* fungus	4	237.50	140.50	500.00	70,250.00	1,634.00	68,616.00	1 : 41.99

T3	*Beauveria bassiana* fungus	4	284.50	187.50	500.00	93,750.00	1,950.00	91,800.00	1 : 47.08
T4	*Beauveria bassiana* fungus	4	340.00	243.00	500.00	121,500.00	2,267.00	119,233.00	1 : 52.60
T5	Control	4	97.00	0.00	500.00	-	-	-	-

Labour cost is Rs. 250 per day, hence 250 x 4 = 1000/- for four sprays; Cost of Endosulfan Rs. 400/ liter, hence for 1000 ml the cost will be 400 x 4 = 1600, hence total expenditure 1600 + 1000 = 2600.

CONCLUSIONS

Microbial control refers to exploiting disease causing organisms to reduce the population of insect pests below damaging levels. The ideal pest control agents should ensure pest population suppression to significant level it should be economical for mass production it should be available in formulations should disseminate quickly and safe to non-target species. Microbial pesticides fulfil all the above mentioned characteristics, hence they are coming up fastly in the pest control strategies. The major advantage of exploiting microorganisms for pest control strategy their environmentally safe and species specific action. Today a variety of entomopathogenic are used for the control of wide range of pests (Beveges, 1981; Lecey and Kaya, 2000; Lacey *et al.*, 2001).

The present work is a summary of three microbials, *i.e.*, Sl NPV, *Bacillus thuriengensis* var Kursataki and *Beauveria bassiana* gainst two notorious pests *Spodoptera litura* (Fabricus) and *Helicoverpa armigera* (Hubner). Fifth instar larvae of *Spodoptera litura* (Fab) revealed increase in percent mortality with increase IN concentrations and time period for all the three bio-pesticides. The general toxicity of the viral and bacterial biopesticides is through ingestion so as to reach to the alkaline medium of midgut for their activation. Insecticidal properties of NPV and Bt are well documented in literature (Jacques, 1973; Ramakrishnan, 1976; Narayanan, 1980; Douglas *et al.*, 1993; Pingles and Lewis, 1997). Kambarkar *et al.* (2009) reported different isolates of HaNPV from different parts of India and advocated that Gulberga and Koimbtore isolates as an ideal agent for IPM. Similarly Bt.K, a better microbial is well known for its production of delta - endotoxin that paralyses the Lepidopteran larvae (Prasad & Adhwani, 2003). The susceptibility of 525 Species of insects of different orders to *Bacillus thuriengensis* is self explanatory for its efficacy (Jayaraj, 1986). Similarly the third microbial, *i.e.*, *Beauveria bassiana* is a fungl biopesticide and very potent against insects of order Lepidoptera, Coleoptera and Hymnoptera (Puri, 1995; Shimuza, 2004; Reetal, 2008).

Further development of different morphogenetic abnormalities with all the three microbials like larval-pupal intermediates, abnormal development of body parts, death during moulting, abnormal development of pupae, swollen and flaccid body etc. indicate toward the chitin inhibitory nature of the bio pesticides and chitin sysnthesis inhibitor compounds have already proved to be a good pest control agaents (Vogel *et al.*, 1979; Jaymaj, 1986; Arora *et al.*, 2000; Dhaliwal & Arora, 2002; Coates *et al.*, 2002).

The above mentioned characteristic features are one part of the efficacy of microbials. The other better part is their highly competent field applicability. They proved at apr synthetic pesticides in the field for both the pests.The present infestations on weight and basis and number bsis for Slnpv in Rbd for *Spodoptera litura* (Fab) came to be 15.85 and 12.43 at 500 and 750 LE/hectare respectively as compared to 38.57 percent infestation in control whereas it was 10.22 and 6.45 percent with 250 and 500 gms of Bt.K/hectare. Further maximum reduction in number of initial cuds were obtained with Bt.K treated blocks (6.56 percent at 500gms /hectare as compared to 35.78 percent in control. F present yield and cost benefit ratios were analysed, increase in yield ws 115 quintals/ hectare for highest dose of Bt.K (500 gms/hectare) as compared to 147.80 quintals/hectare with synthetic insecticide ROKET over control. Cost : benefit ratio was also at par ie 1:36.56 and 1:42.21 for recommended doses of Bt.K and ROKET. If we analyse the performance of fungus *Beauveria bassiana*, highest dose of 2ml /hectare gave 243.00 percent increase in yield as compared to 329.50 with recommended dose of Endosulfan and Cost benefit ratio were 1:62.37 and 1:52.60 with Endosulfan and *Beauveria bassiana*. Field compatibility of microbial pesticides against synthetic pesticides have been also studied by many workers (Bittondi *et al.*, 1998; Gazzoni *et al.,* 1992; Kamalajayanthi & Padmna-vathamma, 2001; Grazywacz *et al.*, 2008; Kamberkar *et al.*, 2009).

The complete observations show that that these microbials have high potency in field conditions and looking to the adverse effects of synthetic pesticides they should be included in pest control programme at large scale for sustainable development of agriculture in an eco safe non hazardous envornment.

REFERENCES

Arora, Ramesh, Battu, G.S. & Ramakrishnan, 2002. Microbial Pesticides: Current Status and Further Outlook, 344-395.

Bitondi, M.M., I.M. Mora, Z.L.P. Simoes & V.L.C. Figueiredo, 1998. The *Apis mellifera* pupal melanization program is affected by treatment with a juvenile hormone analogue. *J. Insect Science* 3:34-38.

Burges H.D., 1981. (Ed.) Microbial control of pests and plant diseases 1970-1980. Academic Press, London.

Coates, B.S., Hellmich, R.L. & Lewis, L.C., 2002. Allelic variation of a Beauveria bassiana (Ascomycotina: Hyphpocreales) minisatellite is independent of host range and geographic origin. *Genome* 45:1:125-132

Dhaliwal, G.S., 2000. Pesticides and environment. Commonwealth Publishers.

Dhaliwal, G.S., Arora, R. & Dhawan, A.K. (eds), 1999. Emerging trends in sustainable agriculture, Commonwealth Publisher, New Delhi.

Dhaliwal, G.S. & Dhawan, A.K., 2001. Ecological perspectives in pest management- a paradigm shift. In: *Plant Protection: New Horizons* (eds.) Bhardwaj, S.C., Saxena, R.C., Swaminathan, R.S.,pp. 36 – 46.

Dhaliwal, G.S. & Singh, Balvinder, 2000. Pesticides in developing countries: Retrospect and Prospect. In: *Pesticides and environment*; pp. 1-38 Commonwealth Publisher, New Delhi.

Dhaliwal, G.S. and Singh, Balvinder 2000. Pesticides and environment, Commonwealth Publisher, New Delhi. Microbial Pesticides: Current status and Future Outlook (eds.). pp. 344-395.

Douglas, M., Koilodny, Hirsch, Dennis, L., Warkentin, Benito, Avarado-Rodriguez. & Reed, Kirkland, 1993. *Spododptera exigua* Nuclear Polyhedrosis Virus as a candidate viral insecticide for the Beet armyworm (Lepidoptera: Noctuidae). *J. Econ. Entomol.* 86(2): 314-321.

Gazzoni, D. L., Oliveira, E. B., Corsa, I. C., Lorra Ferrera, B. S., Vilasbom, G. L. Embrapa, 1981. Londrina Circ. Tec. 5: 44.

Grazywacz David, Wilfred, L. Mushobozi, Mark Parnell, Flavia Jolliffe & Kennath Wilson, 2008. Evaluation of *Spodoptera exempta* Nuclear Polyhedrosis Virus (SPexNPV) for the field control of African Army Worm Spodoptera exempta in Tanzania.*Crop Protection* 27(1): 17-24.

Gupta, H.C.L., A.U. Siddiqui & Aruna Parihar, 2010. Bio-Pest Management. Agrotech Publishing Company, Udaipur, India.

Jacques, R.P., 1973. Tests on microbial and chemical insecticides for control of *Trichoplusia ni* and *Pieris rapae* on cabbage. *Can. Ent.* 105:21-27.

Jayaraj, S., 1986. Role of Insect Pathogens in Plant Protection Proc. *Indian. Natl. Sci. Acad.* B 52 (1): 91-107.

Jayraj, S., 1996. Bio-pesticides and integrated pest management In: Agrochemicals and Sustainable Agriculture. (ed.) N.K. Roy, Kalyani Publishers, New Delhi, pp. 62-85.

Kamalajaynthi, P.D. & K. Padmnavathamma, 2001. Joint action of microbial and chemical insectticides on *Spodoptera litura* (Fabricus) (Lepidoptera: Noctudiae). *J. Tropical Agriculture* 39:142-144.

Kambrekar, D.N., K.A. Kulkarni, R.S. Giraddi, J.H. Kulkarni, B. Fakurudin, 2009. Management of chikpea pod borer, *Helicoverpa armigera* (Hubner), through Nuclear Polyhedrosis Virus Isolates. *Precision Agriculture* 10.450-457.

Lacey, L.A., Fructos, R., Kaya, H.K., and Vails, P., 2001. Insect Pathogens as Biological Control Agents: Do they have a future? *Biological Control* 21: 230-248.

Narayanan, K., 1980. Field evaluation of a nuclear polyhedrosis virus of *Heliothis armigera* (Hubner) on *Chickpea, Cicer arietinum* L. Prof. Workshop on Biological control of *Heliothis* species, Department of Primary, Industries, Too Womba, Queensland, Australia, 23-25.

Pingel, Randall, L. & Lewis, Leslie, C., 1997. Field application of *Bacillus thuringiensis* and *Anagrapha falcifera* multiple Nuclear Polyhedrosis Virus against the corn earworm, (Lepidoptera: Noctuidae). *J. Econ. Entomol.* 90 (5): 1195-1199.

Ramakrishnan, N., 1976. Development of microbial control agents nuclear polyhedrosis virus of *Spodoptera litura* (F). *Proc. Natn. Acad. Sci. India B* 46: 110-116.

Rao, N. Venugopal, Uma Maheshwari, T., Rajendra Prasad, P., Goverdhan Naidu, V. & Savithri, P., 2003. Integrated Insect Pest Management, Agrobios India, pp. 72-79.

Shimuza, M., 2004. A novel technique to inoculate conidia of entomopathogenic fungi and its application for investigation of susceptibility of the Japanese pine sawyer, *Monochamus alternatus,* to *Beauveria bassiana*. *Appl. Entomol. Zool.* 39:495-490.

Chapter - 19

SIMULATION OF TEMPERATURE DEPENDENT DEVELOPMENT OF INSECTS

Reji Gopalakrishnan[1] and Subhash Chander[2]
[1]Indian Agricultural Research Institute, New Delhi 110012, India
[2]Defence Research Laboratory, Tezpur 784 001, Assam, India

INTRODUCTION

Temperature is an environmental factor having direct influence on development, survival and distribution of insects (Andrewartha & Birch, 1954). Studies on the effects of temperature on duration of development have shown that exposure to high temperature leads to shortening of development time in insects (Williams, 2003). The amount of heat required for completing a particular stage or complete life cycle of a species is fixed and is termed as thermal constant (Uvarov, 1931). Threshold of development or developmental zero is the minimum temperature required for development. Depending upon ambient temperature and threshold of development, the insect accumulates heat units and completes development when thermal requirement is fulfilled.

Tropical insects generally have higher thresholds of development than their counterparts in temperate regions. There exists a negative correlation between the lower development threshold and the sum of effective temperature and the tradeoff between the two parameters enables insects to adapt to their thermal environments (Honek, 1999; Trudgill, 2005). Linear increase in the development rates of insects is observed with increase in temperature up to a certain temperature maximum (Brevault & Quilici, 2000). A species is said to exhibit rate isomorphy if the proportion of total development time spent in a particular development stage does not change with temperature. This occurs if the lower development threshold is the same for all stages (Jarossik, 2002). The relationship between temperature and insect development rate is an ecological variable in population dynamics models (Davidson, 1944; Richards, 1957). Thermal time has been widely used as an approximation of physiological time and linear regression models incorporating the effect of temperature on rate of development have been developed (Forouzan *et al.,* 2008; Jones *et al.,* 2005; Ryoo & Cho, 2003). However, the relationship between temperature and the rates of insect development is generally non linear. Hence, the models employing thermal time approach make use of a linear approximation, which is valid within certain ranges of temperature (Bergant & Trdan, 2006; Liu and Ye, 2009).

Modelling thermal requirements

Thermal requirements for insect development can be modelled using a number of mathematical equations. Assuming a linear relationship between temperature and development rate, the thermal constant $K = D(T-T_0)$, where, D is the period of development, T is the temperature and T_0 is the threshold of development (Davidson, 1944). The rate of development (1/D) can be expressed as $V= aT + b$, where a is the reciprocal of thermal constant (Danks, 2000). However, extrapolation of development rates beyond the range of temperature over which the data was collected is not advised (Liu, 1995). A non-linear model for temperature-dependent development was formulated by Briere *et al.* (1999) which can be mathematically expressed as $r(T) = aT(T-T_0)\sqrt{(T_L-T)}$ where a is an empirical constant and T_L is the lethal temperature (Liu & Ye, 2009).

The study of the relationships between weather parameters and biological processes like insect development is known as phenology. Models of insect phenology are generally based on the concept of accumulation of heat units or degree days. Maximum + minimum, trapezoidal approximation, single sine and double sine methods can be used for calculating degree days in linear models (Wilson & Barnett, 1983). Models in which species specific thresholds of development are subtracted from daily mean temperature to yield a sum of effective temperature expressed in degree days are commonly used for simulating insect population dynamics (Pruess, 1983: Higley, 1986; Oshaghi, 2009).

Mechanistic models have been used to predict the population dynamics of many pests (Graf *et al.,* 1992, Safranyik *et al.,* 1999) and some models contain a mechanism based on cumulative population size (Prajneshu, 1998; Matis *et al.,* 2007). The distribution of insect species is limited by temperature and hence the potential range of invasive species can be predicted by incorporating the influence of temperature on development times in mechanistic models (Abril *et al.,* 2008; Estay *et al.,* 2008). Continuous dynamic simulation models employ the state variable – rate variable approach for schematic representation of biological systems. The current state of the system is characterised by state variables, whereas temperature acts as a driving variable. Simulation is the process of building models and their applications to study the behavior of the system (De wit & Goudriaan, 1974; Rabbinge *et al.,* 1994).

Temperature-dependent development of *Leptocorisa acuta*

Rice bug, *Leptocorisa acuta* (Thunb.) (Hemiptera: Alydidae) is an important pest of rice crop in India. The nymphs feed by sucking plant sap from leaves and stems of rice. The adults feed on the milky juice of the developing grains, which become chaffy, brown and finally dry (Nair, 1995). Studies on the effect of temperature on duration of development of *Leptocorisa acuta* were carried out at the Indian Agricultural Research Institute, New Delhi. *Echnocloa crusgalli*, which is an alternate host of *L. acuta*, was used for rearing the bugs inside cages. The insects were allowed to grow and reproduce under ambient temperature by placing the cages in open field. Three sets of experiments were conducted at different ranges of ambient temperature conditions.

Egg laying was observed on the leaves and leaf sheaths and rarely on panicles. The eggs were boat-shaped and brownish or blackish brown in colour. There were five nymphal instars and the nymphs could be grouped into non-wing padded and wing-padded based on the absence or presence of wing pads. The duration of development of the egg and nymphal stages were recorded and these values were used for determining the threshold of development and thermal constant. The threshold of development (T_0) for the bug was calculated using the formula; $T_0 = (\sum t_1 - \sum t_2) / (D_1 - D_2)$, where $\sum t_1$ and $\sum t_2$ were the sums of daily mean temperatures and D1 and D2 were the durations of development. The thermal constant (K) for the life stages was estimated as $K = (T - T_0) \times D$, where T is the daily mean temperature and D is the corresponding duration of development.

The mean incubation period was 6.1 days while the mean durations of the five nymphal instars were 4.7, 2.4, 4.1, 4.1 and 9.5 days respectively. Therefore, the total duration from egg to adult was found to be 31.1 days under ambient conditions. The threshold of development of eggs was estimated to be 13.6°C while that for the five nymphal instars were 14.7, 13.6, 13.0, 10.0 and 8.5°C, respectively. The thermal constant for the egg stage was estimated to be 79.7 degree days (DD) while that for the nymphal instars were 55.3, 32, 53.7, 63 and 160.9 DD, respectively. Therefore, the total thermal requirement for completion of the nymphal stage was 364.9 DD, whereas that for egg to adult emergence was 444.6 DD (Reji & Chander, 2007).

Simulation of *Leptocorisa acuta* population dynamics

A continous dynamic model was formulated to simulate the population dynamics of *L. acuta* (Figure 1). State and rate variables approach was used for building the model which was written in Fortran Simulation Translator (FST). The number of eggs, nymphs and adults at any point in time were the state variables in the model, while the rates of change in the number of any life stage were the rate variables. In the present model, the duration of development of the life stages were replaced with the corresponding thermal requirements. The threshold of development and thermal constant for the pre-oviposition period of the bug were 9.6 °C and 256 DD, respectively. The thermal requirement for the adult bug was estimated to be 1170.4 DD and the mortality during the egg and nymphal stages were assumed to be 50 and 10%, respectively.

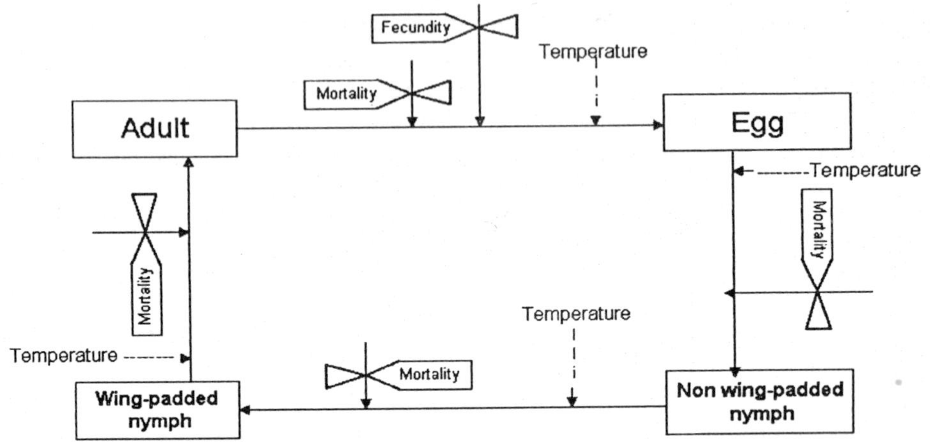

Fig. 1. The variables of *Leptocorisa acuta* population dynamics model

The number of eggs, nymphs and adults at the beginning of the simulation run were described as state variables at the initial part of the model. The number individuals belonging to each stage at any given time was described as a function of the initial number and rate of increase. The rate of formation of nth stage in the life cycle was expressed as $R_n = N_{n-1} \times S_{n-1} \times (ET_{n-1}/K_{n-1})$, where N_{n-1}, S_{n-1}, ET_{n-1} and K_{n-1} represent the number of individuals, survival rate, effective temperature and thermal constant respectively, for the (n-1)th stage. The rate of change in the number of a life stage is dependent on the rates of development of the preceding and succeeding stages as well as the death rate. Provision was made in the model to account for fecundity, pre-oviposition period adult longevity along with mortality due to parasitism and predation. The model was validated using the data on light trap catches of adult bugs (DRR, 2001-2004). Although the model underestimated the population growth initially, the overall performance was found satisfactory (Figure 2). Simulation of rice bug population using New Delhi weather indicated that the population attains its peak during the third week of October (Reji & Chander, 2008). The simulation model developed can be used for predicting the population of *L. acuta* under field conditions and to take necessary management practices.

The simulation model was used for predicting the impact of climate change on *L. acuta* population in New Delhi. Simulation runs with rise in daily mean temperature indicated that there was prolongation of generation time and reduction in bug population. It was predicted that a

3 degree Celsius rise in daily mean temperature would prolong the duration of development by 1-3 days, and lead to 14-29% reduction in bug population. This can be attributed to the fact that the favourable range of temperature for *L. acuta* is 13-35°C within which the development rate is directly proportional to ambient temperature. The development of the bug would be adversely affected if the daily mean temperature rises above 35 °C due to climate change (Chander *et al.*, 2009).

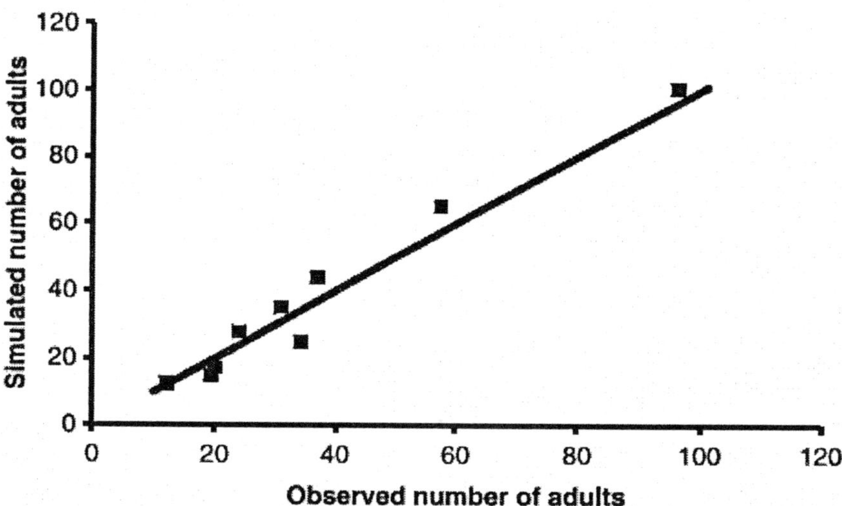

Fig. 2. Observed and simulated population of *Leptocorisa acuta* adults

CONCLUSIONS

Temperature is the most important driving variable in biological systems. Hence, modelling insect development has been largely dependent on functional relationships between temperature and biological processes. Empirical models are liable to change spatially and temporally, whereas mechanistic models based on thermal requirements are universally applicable. Simulation of temperature dependent development in insect of economic importance would be useful for developing decision support systems for pest management as well as for studying the impact of climate change on insect populations. Further, the coupling of population dynamics models with crop growth simulation models would help in forecasting and assessment of crop damage.

REFERENCES

Abril, S., Oliveras, J. & Gomez, C., 2010. Effect of temperature on the development and survival of the Argentine ant, *Linepithema humile*. *J. Insect Science* 10: 97.

Andrewartha, H.G. & Birch, L.C., 1954. *The distribution and abundance of animals*. University of Chicago Press, Chicago, IL, USA, 782 pp.

Bergant, K. & Trdan, S., 2006. How reliable are thermal constants for insect development when estimated from laboratory experiments? *Entomol Exp. Appl*. 120: 251-256.

Brévault, T. & Quilici, Serge, 2000. Relationships between temperature, development and survival of different life stages of the tomato fruit fly, *Neoceratitis cyanescens*. *Entomol. Exp. Appl*. 94: 25-30.

Briere, J.F., Pracros, P., Roux, A.Y.L. & Pierre, J.S., 1999. A novel rate model of temperature-dependent development for arthropods. *Environ. Entomol*. 8:22–29.

Chander, S., Reji, G. & Aggarwal, P.K., 2009. Assessing impact of climate change on rice gundhi big using a population dynamics simulation model. In Aggarwal, PK (ed) *Global climate change and Indian Agriculture – Case studies from the ICAR Network Project*. Indian Council of Agricultural Research, New Delhi, India.

Danks, H.V., 2000. Measuring and reporting lifecycle duration in insects and arachnids. *Eur. J. Entomol*. 97: 285-303.

Davidson, J., 1944. On the relationship between temperature and rate of development of insects at constant temperature. *J. Animal Ecol*. 13: 26-38.

DeWit, C.T., Goudriaan, J., 1974. *Simulation of ecological processes*. Simulation monographs. Centre for Agricultural Publishing and Documentation (Pudoc), Wageningen.

DRR (2001–2004). Annual Progress Reports. All-India Coordinated Rice Improvement Project (AICRIP). Directorate of Rice Research, Hyderabad.

Estay, S.A., Lima, M. & Labra, F.A., 2009. Predicting insect pest status under climate change scenarios: combining experimental data and population dynamics modelling. *J. Appl. Entomol*. 133: 491-499.

Forouzan, M., Amirmaafi, M. & Sahragard, A., 2008. Temperature-dependent development of *Habrobracon hebetor* (Hym.: Braconidae) reared on larvae of *Galleria mellonella* (Lep.: Pyralidae). *J Ent. Soc. Iran*. 28:67-78.

Graf, B., Lamb, R., Heong, K.L., Fabellar, L., 1992. A simulation model for the population dynamics of rice leaf folders (Lepidoptera: Pyralidae) and their interactions with rice. *J. Appl. Ecol*. 29: 558–570.

Higley, L.G., Pedigo, L.P., Ostile, K.R., 1986. DEGDAY: a program for calculating degree-days, and assumptions behind the degree-day approach. *Environ. Entomol*. 15: 999-1016.

Honek, A., 1999. Constraints on thermal requirements for insect development. *Entomological Sciences* 2: 615-621.

Jarosik, V., Honek, A. & Dixon, A. F. G., 2002. Developmental rate isomorphy in insects and mites. *Am. Naturalist* 160: 497-510.

Jones, V.P., Doerr , M.D., Brunner, J.F., Baker, C.C., Wilburn, T.D., Wiman, N.G., 2005. A synthesis of the temperature dependent development rate for the oblique banded leafroller, *Choristoneura rosaceana*. *J. Insect Science*. 5: 24.

Liu, S.S., Zhang, G.M. & Zhu, J., 1995. Influence of temperature variations on rate of development in insects: analysis of case studies from entomological literature. *Ann. Entomol. Soc. Am.* 88:107-119.

Liu, X. & Ye, H., 2009. Effect of temperature on development and survival of *Bactrocera correcta* (Diptera: Tephritidae). *Scientific Research and Essay*. 4: 467-472.

Matis, J.H., Zhou, L., Kiffe, T.R., Matis, T.I., 2007. Fitting cumulative size mecha-nistic models to insect population data: a nonlinear mixed effects model analysis. *J. Ind. Soc. Agril. Stat.* 61: 147-155.

Nair, M.R.G.K., 1995. *Insects and mites of crops in India*. ICAR, New Delhi.

Oshaghi, M.A., Ravasan, N.M., Javadian, Rassi, Y., Sadraei, J., Enayati, A.A., Vatandoost, H., Zare, Z., Emami, S.N., 2009. Application of predictive degree day model for field development of sandfly vectors of visceral leishmaniasis in northwest of Iran. *J. Vector Borne Dis.* 46: 247-254.

Prajneshu, 1998. A nonlinear statistical model for aphid population growth. *J. Ind. Soc. Agri. Stat.* 51: 73-80.

Pruess, K.P., 1983. Day-degree methods for pest management. *Environ. Entomol.* 12: 613-619.

Rabbinge, R., Rossing, W.A.H. & vander Werf ,W., 1994. Systems approaches in pest management: the role of production ecology. In: *Proceedings of the Fourth International Conference on Plant Protection in the Tropics*. (Ed. Rajan, A., Ibrahim, Y.) Malaysian Plant Protection Society, Kualalumpur, pp. 25-46.

Reji, G. & Chander, S., 2007. Thermal requirements for development of rice bug, *Leptocorisa acuta* (Thumb.) under variable temperature conditions. *J. Entomol. Res.* 31(3): 229-232.

Reji, G. & Chander, S., 2008. A degree day simulation model for the population dynamics of rice bug *Leptocorisa acuta* (Thunberg). *J. Appl. Entomol.* 132(8): 646-653.

Richards, A.G., 1957. Cumulative effects of optimum and suboptimum temperatures on insect development. pp. 57-65. In: Johnson, F. H. (Ed.) *Influence of temperatures on biological systems*. American Physiological Society.

Ryooa, M.I. & Choa, K.J., 1988. A model for the temperature-dependent developmental rate of *Sitophilus oryzae* L. (Coleoptera: Curculionidae) on rice. *J. Stored Product Res.* 24: 79-82.

Safranyik, L., Barclay, H., Thomson, A., Riel, W.G., 1999. A population dynamics model for the mountain pine beetle, *Dendroctonus ponderosae* Hopk. (Coleoptera: Scolytidae). *Information Report Pacific Forestry Centre*, Canadian Forest Service No. BC-X-386,35.

Trudgil, D.L., Honek, A. & Van Straalen, N.M., 2005. Thermal time – concepts and utility. *Ann. Appl. Biol.* 146:1-14.

Uvarov, B. P. (1931). Insects and climate. *Transactions of Royal Entomological Society, London* 79: 1-247.

Williams, R.S., Lincoln, D.E. & Norby, R.J., 2003. Development of gypsy moth larvae feeding on red maple saplings at elevated CO_2 and temperature. *Oecologia* 137: 114-122.

Wilson, L.T. & Barnett, W. W., 1983. Degree-Days: An aid in crop and pest management. *California Agriculture*. 37: 4-7.

Chapter - 20

ASSESSMENT AND MAPPING OF VECTORS' DISTRIBUTION AND DIVERSITY USING REMOTE SENSING AND GIS

M. Govindaraju and P. Rani Lalitha

Department of Environmental Biotechnology, Bharathidasan University, Tiruchirappalli – 620 024, TN, India
E-mail: *mgrasu@gmail.com; mgrasu@bdu.ac.in*

ABSTRACT

Vector-borne diseases are one of the major causes of human suffering. There has been a worldwide resurgence of vector-borne diseases since 1970s including malaria, dengue, yellow fever, louse-borne typhus, plague, leishmaniasis, sleeping sickness, West Nile encephalitis, Japanese encephalitis etc. After nearly three decades chikungunya's resurgence in India in 2005 is a revelation to the disease's changing epidemiological features. As per the World Health Organisation (WHO), about 260 million people worldwide are infected with malaria and 1.5 to 2.7 million patients die annually due to this most significant infectious parasite. If larval habitats could successfully be identified, necessary mosquito control measures could be used to reduce the risk of vector-borne diseases. Remote sensing (RS) is increasingly being used as a tool for mapping the vector habitats for the control of vector-borne diseases which is important for the eradication of vectors. Scientists are now adopting remote sensing technologies to study the effect of vector-borne diseases and to plan the control strategies. IKONOS and LANDSAT images are usually used to identify mosquito larval habitat. Remote sensing data becomes an important input for Geographic Information System (GIS) in mapping the vector breeding sites. GIS allows the overlaying of thematic layers to analyse databases and maps that combine spatial and temporal information from various sources. Remote sensing covers vast area, including unsurveyed and inaccessible areas and thus helps in understanding the breeding and distribution patterns of vectors spatially. Remotely sensed data is now being used to predict the risk of vector-borne diseases worldwide.

REMOTE SENSING AND GEOGRAPHIC INFORMATION SYSTEM

Remote sensing and Geographic information system (GIS) is gaining importance as vital tools in the analysis and integration of spatio-temporal data. GIS and remote sensing from earth-observing satellites are sophisticated and powerful technologies that are finding applications far beyond those originally intended. Remote sensing is the stand-off collection through the use of a variety of devices for gathering information on a given object or area. GIS is a computer-aided database management and mapping technology that is becoming more and more popular among decision makers as it enables them to quickly refer the GIS outputs which help them to solve problems and to make right decisions. Visualization of features, conversion of data into thematic maps and the capability of providing solutions by taking into account the overall scenario of an area are some of the virtues of GIS which it is being implemented across a number of sectors and departments including health. Remote sensing and GIS allow near real-time access to data on temperature, soil, elevation, patterns of land use, and phases of vegetation in addition to the precise geographic location of water bodies, population centres, buildings, roads, and other infrastructure.

REMOTE SENSING AND GIS IN EPIDEMIOLOGY

The use of remote sensing techniques for vector mapping and disease risk assessment has evolved significantly over the past 25 years. Recently, Geographical Information System (GIS) has emerged as an innovative and important component of many projects in public health and epidemiology. GIS and remote sensing have capabilities that are ideally suited for infectious disease surveillance and control, particularly for many vector-borne diseases in remote rural areas. One of the most useful functions of GIS in epidemiology continues to be its utility in basic mapping. GIS has been used in the surveillance and monitoring of vector-borne diseases, in environmental health, analysis of disease policy and planning, health situation in an area, generation and analysis of research hypotheses, identification of high-risk health groups, planning and programming of activities, and monitoring and evaluation of interventions. GIS helps to show regional variation in health problems, environmental risks and the use of health services. GIS applications related to health have been introduced and used in the surveillance and monitoring of vector-borne diseases and its influence on public health. Application of remote sensing data in epidemiology involves retrieving environmental variables that characterize the vector ecosystem such as land cover, temperature, humidity or vapor pressure and precipitation. GIS is particularly well suited for studying these associations because of its spatial analysis and display capabilities. WHO has been using mapping techniques coupled with surveillance to monitor the global health situation and present it through GIS which has been widely used in the surveillance and monitoring of vector-borne diseases.

USES OF GIS IN PUBLIC HEALTH

GIS is being used by public health administrators and professionals, including policy makers, statisticians, epidemiologists, regional and district medical officers (Weekly Epidemiological Record, 1999). Some of its applications in public health are mentioned below:

- Find out geographical distribution and variation of diseases
- Analyse spatial and temporal trends
- Identify gaps in immunizations
- Map populations at risk and stratify risk factors
- Document health care needs of a community and assess resource allocations
- Forecast epidemics
- Plan and target interventions
- Monitor diseases and interventions over time
- Manage patient care environments, materials, supplies and human resources
- Monitor the utilisation of health centres
- Route health workers, equipments and supplies to service locations
- Publish health information using maps on the Internet
- Locate the nearest health facility.

VECTOR HABITAT MAPPING

In recent years, vector-borne diseases (VBD) have emerged as a serious public health problem in countries of the South-East Asia Region, including India. Approximately 2.1 billion people, in 100 countries, are at risk of malaria alone (Oaks *et al.*, 1991). In order to ensure better planning and implementation of vector control strategies improved methods are required for forecasting, early detection, and prevention of vector-borne diseases due to the increasing trend of large scale epidemics such as malaria (Teklehaimanot *et al.*, 2005). People living in tropical and temperate regions are at higher risk to get exposed to one or more vector borne diseases. This current global situation increases the need of new technologies for the development of a surveillance system to accurately estimate the abundance of vector in aquatic habitats and predicting adult vector population outbreaks. Endemic areas should be monitored continuously for active surveillance of vector-borne diseases. Vector surveillance includes monitoring the presence and density of vector species their breeding sites the effectiveness and efficiency of vector control.

Epidemiologists now adopt spatial analysis and Remote Sensing (RS) technologies as an effective tool for mapping the vector habitats for the eradication of vectors by developing effective disease control strategies. Remote sensing data becomes an important input for GIS in mapping the vector breeding sites. If larval habitats could successfully be identified, necessary mosquito control measures could be used to reduce the risk of vector borne diseases. Environmental factors such as land and sea surface temperature and the type of vegetation its potential linkages with various diseases can also be mapped through remote sensing (Beck *et al.,* 2006). RS data provides early prediction which is beneficial to mosquito control operations by determining the area which needs focus for efficient mosquito control and provides information on the distribution of vector habitats.

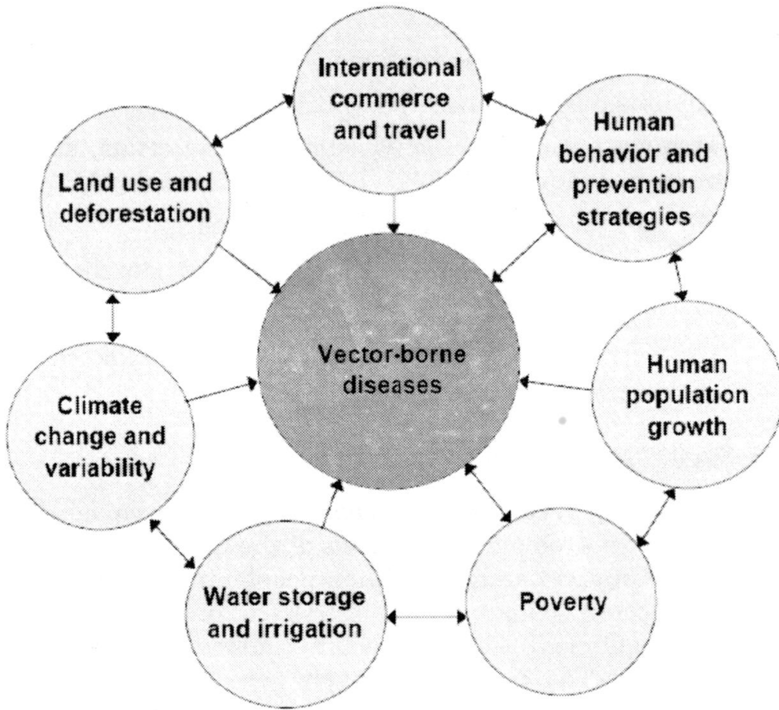

Fig. 1. Impact of Vector-borne diseases

Vector Mapping

In current day scenario, GIS is finding application in diverse fields including health. In any disease control programme, there are several factors, involved, namely estimation of disease burden, monitoring of disease trend, identification of risk factors, planning, allocation of resources, implementation etc. GIS owing to its inherent ability to manage both spatial and non-spatial information provides an excellent framework for disease monitoring and control.

As the first step the data on different types of vectors causing infectious diseases, their survival conditions and other data relevant to the vectors are collected from various sources. The base map of the particular area to be studied is obtained and all the features in the map are digitized into their respective themes. The vector breeding habitats in that particular area is identified using suitable interpretation and image processing techniques. The association between satellite-derived environmental variables such as temperature, humidity and land cover type and vector density is used to identify and characterize the vector habitats.

Buffers are created for the given population categorizing them into commercial area, institutes, industries and residential areas. Depending upon the flight range of different vectors, suitable buffers are created from the vector breeding sites considering all the environmental and seasonal factors. All the buffers are analysed using suitable operations such as union, intersection, overlay, network, etc., the vulnerable areas and the vulnerable groups of populations are assessed. It enables researchers to locate high prevalence areas and populations at risk, identify areas in need of resources, and make decisions on resource allocation.

Remote Sensing data for Vector Mapping

Remotely sensed data is now being used to predict the risk of vector-borne diseases worldwide. Landsat TM data is the best choice for the base imagery of vector habitat assessment. Landsat spectral data, particularly band 4 (infrared) and band 5 (mid-infrared) are well suited for vegetation and water content analysis. Geospatial mapping by using remote sensing offers the potential to identify larval habitats on a large area basis to a degree that is difficult or impossible using conventional ground survey. Eventhough, a variety of numerical techniques for vector map creation of vector distribution in time and space from satellite data are available, only those techniques that aid our understanding of biological processes provide meaningful information in epidemiology and vector control.

MALARIA SURVEILLANCE SYSTEM

Remote sensing applications in epidemiology have been most widely used to study mosquito borne diseases. Globally malaria clinical cases are reported as 300 to 500 million and 1.5 to 2.7 million deaths annually. Malaria situation is worsening with large-scale epidemics and increasing mortality. If larval habitats could successfully be identified, necessary mosquito control measures could be used to reduce the risk of vector-borne diseases. RS studies largely focused on identifying mosquito breeding habitats such as marshes and wetlands through land-use and land-cover mapping using remote sensing data (Barnes & Cibula,1979; Wagner *et al.*, 1979; Pope *et al.*, 1992).

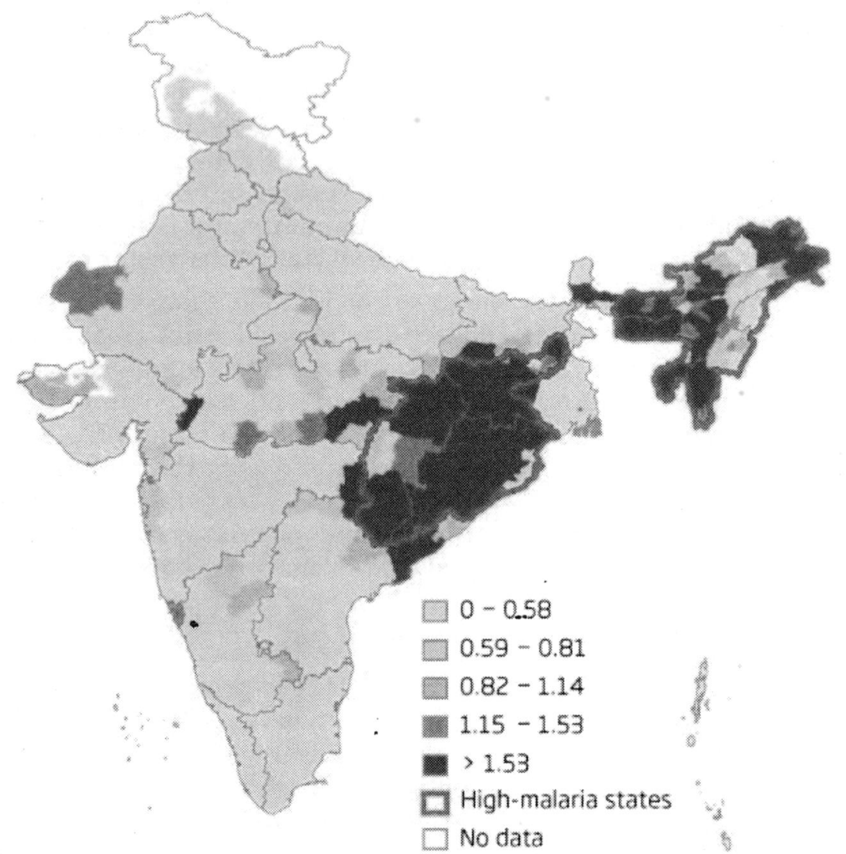

Fig. 2. Distribution of breeding habitats of malaria vectors in India using RS/GIS (Source: www.newscientist.com)

An efficient information management system is highly required for surveillance of disease and control programme. A GIS based surveillance system for Dindigul municipality, has been developed. The district initially composed of 44 wards (now 48), the map was digitized and streetwise information on 33 parameters of each ward has been attached. Information of any ward can be retrieved at the click of the mouse. Any desired area can be selected and zoomed into reach from ward to street and street to house, and house-wise information can be attached. One can study malaria dynamics both in space and time and relate the increase in malaria incidence to specific breeding sites. The major advantages of this system are: (i) web hosting completely eliminates the traditional method of flow of information and the information is instantly available, (ii) once the basic structure is ready, it is easy to convert it to surveillance system; for any other diseases viz. dengue, filarial, DHF one needs to replace malaria data with the data of respective disease, (iii) Buffer zone creation capability of GIS can map the impact zone of any major breeding site, where control activity needs to be strengthened, and (iv) network analysis can identify catchment areas for some health facility and also appropriate sites for location of a new health facility.

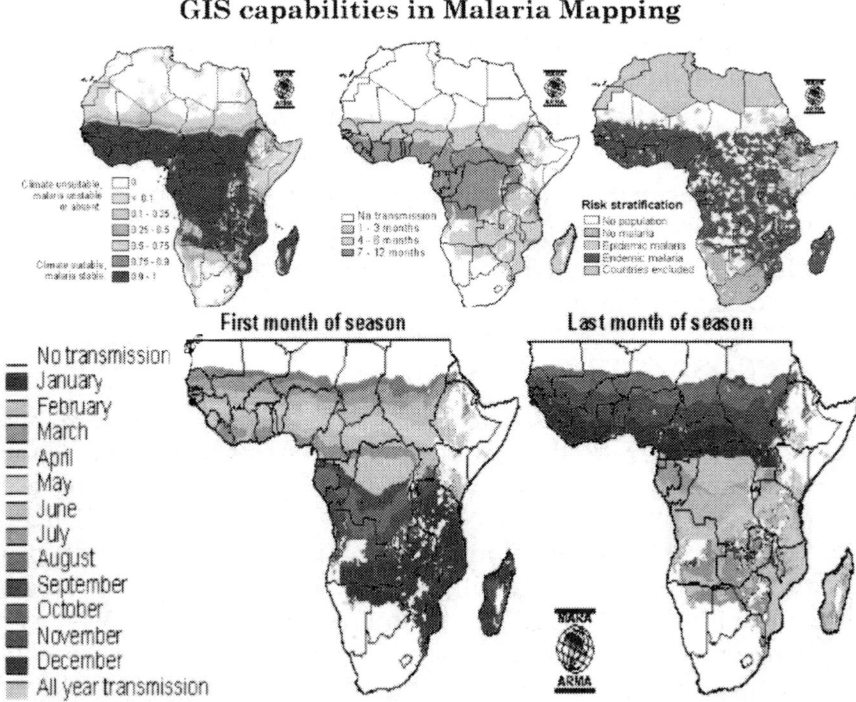

Fig. 3. GIS capabilities in Malaria Mapping (Source: www.gisdevelopment.net)

MOSQUITO MAP

Mosquito map is a geospatially referenced clearinghouse for mosquito species collection records and distribution models. Users can pan and zoom to anywhere in the world to view the locations of past mosquito collections and the results of modeling the predicts the geographic extent of individual species. Collection records are searchable and downloadable, users can map and upload their own georeferenced collection data or distribution models, and all contributions have full attribution. Currently mosquito map has 1,26,031 records, for over 3,07,543 individual specimens/observations, and over 1891 unique scientific names. Mosquito map is designed to preserve and make available the results of past collecting and distribution modeling activity. The utility of mosquito map will increase as more records and models are added.

Uses of Mosquito Map: It is useful for the following causes -

- Informing decisions about where mosquito collection efforts should be directed.
- Identifying areas relevant to the study of mosquito biogeography, evolution and biodiversity.
- Allow predictions about the potential spread of exotic mosquito introductions.
- Allow predictions about the potential effects of global warming on mosquito distributions.
- Allow insights into mosquito community structure, and environmental and climatic correlates to species occurrence.
- Allow continent-wide rather than just local studies of vector-borne disease.
- Identifying cryptic evolutionary lineages that differ in geographic or ecological space.

FUTURE PROSPECTS

The availability of moderate resolution imagery from new NASA satellites such as Terra and Aqua at up to 250-m spatial resolution offers improved capabilities in modeling epidemiology from space at a finer spatial resolution than AVHRR (Advanced Very High Resolution Radiometer). Technologies such as global positioning systems and GIS provide geo-location and mapping capabilities on the ground at unprecedented accuracies, making it possible to merge more accurately ground observations of vector demographics and disease incidence with satellite data. GIS and statistical software packages are available making it feasible for epidemiologists and biologists to experiment with new spatial analysis techniques. Maps showing seasonal risks of vector borne diseases will be necessary to monitor the impacts of global change on vector ecologies. The use of remote sensing techniques for modeling and forecasting vector-borne diseases in an emerging field, and a sustained use of these applications could be ensured by collaboration between remote sensing scientists and epidemiologists from the onset of research projects (Herbreteau *et al.,* 2005). Remote sensing of vector ecosystems and interpretation of these patterns is likely to provide both challenges and opportunities in epidemiology.

REFERENCES

Barnes, C.M., Cibula, C.G., 1979. Some implications of remote sensing technology in insect control programs including mosquitoes. *Mosquito News* 39: 271-282.

Beck, L.R., Lobitz, B.M. & Wood, B.L., 2000. Remote sensing and human health: New sensors and new opportunities. *Emerg. Infect. Dis.* 6(3):217-27.

Herbreteau, V., Salem, G., Souris, M., Hugot, J.P. & Gonzalez, J.P., 2005. Sizing up human health through remote sensing: uses and misuses. *Parassitologia* 47:63-79.

Oaks, S.C., Mitchell, V.S., Pearson, G.W. & Carpenter, C.J. (Eds), 1991. Malaria: Obstacles and Opportunities. Institute of Medicine, National Academy Press, Washington, D. C.

Pope, K.O., Sheffner, E.J., Linthicum, K.J., Bailey, C.L., Logan, T.M., Kasischke, E.S., Birney, K., Njogu, A.R. & Roberts, C.R., 1992. Identification of central Kenyan Rift Valley Fever virus vector habitats with Landsat TM and evaluation of their flooding status with airborne imaging radar. *Remote Sens Environ.* 40:185.

Teklehaimanot, A., Singer, B., Spielman, A., Tozan, Y. & Schapira, A., 2005. Coming to grips with malaria in the new millennium. UN Millennium Project. London: Earthscan.

Wagner, V.E., Hill-Rowley, R. & Newson, H.D., 1979. Remote sensing: A rapid and accurate method of data acquisition for a newly formed mosquito control district. *Mosquito News* 39: 271.

Wood, B.L., Beck, L.R., Washino, R.K., Palchick, S.M. & Sebesta D., 1991. Spectral and spatial characterization of rice field mosquito habitat. *Int. J. Remote Sens.* 12: 621.

Chapter - 21

VICIOUS CYCLE OF VECTOR, VIRUS AND VICTIM IN DENGUE/DENGUE HAEMORRHAGIC FEVER

Sajal Bhattacharya

Post Graduate Department of Zoology, Asutosh College
(University of Calcutta), Kolkata – 700 026, India

INTRODUCTION

Annals of history of medical sciences reveal that during the course of human evolution, man has always encountered the prowess of a tiny insect called mosquito. Malaria, filariasis, to name a few, have plagued mankind for centuries and have added to the embarrassingly long list of sufferings that the best creation of nature has had at the "hands" of mosquitoes. A few mosquito-borne viral diseases have shown signs of resurgence. To make things worse for human being, several diseases which never existed in some parts of the world, have revealed high prospects of a possible invasion in near future. There are considerable evidences for the resurgence and progressive increase of mosquito-borne diseases. For example, dengue and chikungunya have flared up in many countries including India that formerly had them under control. Sudden outbreak of West Nile (WN) virus has been reported from USA and some parts of Europe in recent years. Although yellow fever has never been reported in India, yet the region is believed to be a "Yellow fever receptive area" (Park & Park, 1977). Compared to nine reporting countries in the 1950s, today the geographic distribution of dengue includes more than 100 countries worldwide (Guha-Sapir & Schimmer, 2005). The WHO estimates that almost 40% (2.5 billion) of the world population are at risk of dengue infection. Current estimates of dengue fever per annum are reported between 50 and 100 million cases worldwide. Of these cases 5,00,000 develop into severe forms of the disease, *i.e.*, dengue hemorrhagic fever (DHF) and dengue shock syndrome (DSS) (Guzman & Kouri, 2002).

Dengue Serotype

The dengue virus belongs to the family flaviviridae occurs as four antigenically distinct serotypes (DEN 1,2,3 and 4), that have emerged or re-emerged throughout the world since the 17th Century (Gubler, 1998). All four dengue viruses are circulating in Asia including India. There is genetic variation within each serotype in the form of phylogenetically distinct subtypes. Three subtypes can be identified for DEN-1, six for DEN-2, four each for DEN-3 and DEN-4. Four distinct serotypes exhibit upto 30% divergence across their polyproteins. Southeast Asia harbours the greatest amount of dengue virus genetic diversity, suggesting it acts as a viral 'source' population (Zhang *et al*, 2005). Little is known about the evolution of virulence among dengue viruses and the selection pressure of many factors including passive serotype immunity, likely contribute to disease severity (Twiddy *et al.*, 2003). Immunity developed against a given virus serotype is limited to this only. It does not provide cross-protection against any one of the other three serotypes, and upon contracting a second infection subsequently with a different serotype, a patient with a strain-specific immunity may stand a greater risk of developing dengue haemorrhagic fever (Tyagi, 2010). Antibody dependent cellular cytotoxicity (ADCC) has been associated with severe forms of dengue (Garcia *et al.*, 2006).

Table 1. District-wise distribution of cases and deaths due to Dengue/ DHF in West Bengal in 2005

District /Area	Number	Cases (%)	Deaths
Kolkata	4013	63.76	13
24 Parganas(N)	524	8.33	7
Howrah	264	4.19	4
24 Parganas(S)	253	4.02	3
Hooghly	171	2.71	0
Maldah	155	2.46	0
Salt Lake City	135	2.15	1
Burdwan	104	1.65	1
Nadia	85	1.35	1
Midnapur (W)	50	0.75	0
Birbhoom	47	0.7	1
Midnapur (E)	27	0.43	0
Dinajpur (S)	13	0.21	0
Bankura	9	0.14	1
Darjeeling	4	0.06	1
Purulia	2	0.03	1
Dinajpur (N)	2	0.03	0

Coochbehar	2	0.03	0
Jalpaiguri	1	0.02	0
Outside State	16	0.24	0
Not known	416	6.61	0
Total	6293	100	34

Epidemiology

During dengue epidemic in 2005, out of 19 districts of the State of West Bengal, Kolkata (erstwhile Calcutta) was worst affected. Kolkata city alone shared 63.76% of the total dengue cases (6293) (Hati, 2009). All the remaining districts were affected (Table 1). Attempt had been made to know the man biting *Aedes* mosquitoes in Kolkata Municipal Corporation area during this outbreak. Only *Aedes aegypti* was encountered in the urban Kolkata while both *Ae. albopictus* and *Ae. aegypti* were recorded in suburban Kolkata (Bhattacharya & Roy, 2006). Dengue was essentially an urban disease earlier, confined to Kolkata, but in 1994 dengue was recorded for the first time in rural areas too in the state and dengue virus was isolated from locally caught *Aedes aegypti* (Bhattacharya *et al.* 1997). These observations indicate the adaptability of *Ae. aegypti* in different ecological situations. In Thailand, dengue is now more common in rural areas than urban areas (Chareonsook *et al.*, 1999).

Table 2. Age-wise case incidence of primary and secondary Dengue infection (2005-2007)

Age (Yr.)	Patients suffering from primary dengue	Patients suffering from secondary dengue	Total cases
0-5	27(26.73)	15(7.46)	42
6-10	21(20.79)	17(8.45)	38
11-20	32(31.68)	59(29.35)	91
21-30	12(11.88)	49(24.37)	61
31-40	7(6.93)	25(12.43)	32
41-50	1(0.99)	15(7.46)	16
51-60	1(0.99)	10(4.97)	11
61-70	0	5(2.48)	5
71-80	0	3(1.49)	3
>80	0	3(1.49)	3
Total	101	201	302

Source: Hati (2009), Figures in parentheses are percentage.

The age group affected by dengue has varied across regions in India. At present it is predominantly a disease of young adults (21-30 years, median 30-35 years) and an increasing proportion of infection are secondary in nature (Hati, 2006; Vijaykumar *et al.*, 2005). Hati (2009) found during a dengue serosurveillance study in Kolkata that 79.2% of primary dengue cases occurred in age group 0–20 year and no primary case was found in persons beyond 60 year of age (Table 2). Secondary dengue cases were distributed in all age groups; 0-20 year age group contributed 45.26% cases. Since secondary dengue cases were distributed in all the age groups and were much more than that of primary dengue cases, the chances of developing DHF/DSS would be there in any age group. Since dengue is endemic in Kolkata, naturally a large proportion of people in Kolkata harbour IgG antibodies. Hence, this population could be vulnerable as they can be attacked with a different dengue virus and DHF may occur.

Table 3. Infecting dengue serotypes versus disease severity

Disease severity	Infecting virus				Total
	DEN – 1	DEN – 2	DEN – 3	DEN – 4	
DF (%)[a]	0	3(30)	7(70)	0	10
DF + HM (%)	1(17)	0	4(67)	1(17)	6
DHF grade I (%)[b]	1(10)	1(10)	5(50)	3(30)	10
DHF grade II (%)	0	0	2(100)	0	2
Total(%)[c]	2(7.1)	4(14.3)	18(64.3)	4(14.3)	

DF : dengue fever; DF+HM : DF with haemorrhagic manifestations; DHF: dengue haemorrhagic fever; a : Percentage based upon specific disease category; b : A single DHF grade I patient was co-infected with both DEN-3 and DEN-4; c : Percentage based upon specific dengue serotype (Source: Suwandono *et al.*, 2006).

In 2004 the outbreak of dengue in Jakarta, Indonesia was characterised by the circulation of all four dengue viruses and resulted in a relatively high percentage of the representative population of hospitalised patient developing DHF. DEN-3 virus was the predominant serotype and was prominently associated with DF+HM and DHF (Table 3). The emergence of a new dengue serotype in a dengue-endemic area often results in an epidemic. Although dengue infections are often found to be asymptomatic, the incidence of DHF increases with heterotypic dengue infections (Suwandono *et al.*, 2006)

Table 4. State-wise dengue cases (C) and deaths (D) in India (2001-2008)

State	2001		2002		2003		2004		2005		2006		2007		2008	
	C	D	C	D	C	D	C	D	C	D	C	D	C	D	C	D
Andhra Pradesh	1	0	61	3	95	5	230	1	99	2	197	17	587	2	313	2
Bihar	0	0	1	0	0	0	0	0	0	0	4	0	0	0	1	0
Chandigarh	0	0	15	0	0	0	0	0	2	0	182	0	99	0	167	0
Delhi	322	3	45	2	2882	35	606	3	1023	9	3366	65	548	1	1312	2

Goa	1	0	0	0	12	2	3	0	1	0	1	0	36	0	43	0
Gujarat	69	0	40	0	249	9	117	4	454	11	545	5	570	2	1065	2
Haryana	260	5	3	0	95	4	25	0	183	1	838	4	365	11	1137	9
J&K											24	1	0	0	0	0
Karnataka	220	0	428	1	1226	7	291	2	587	17	109	7	230	0	339	3
Kerala	41	0	219	2	3546	68	686	8	1028	8	981	4	603	11	733	3
Maharashtra	54	2	370	18	772	45	856	22	349	56	736	25	614	21	743	22
Orissa	0	0	0	0	0	0	0	0	0	0	1	0	4	0	0	0
Sikkim	0	0	0	0	0	0	12	0	0	0	0	0	0	0	0	0
Punjab	49	0	27	2	848	13	52	0	251	2	1166	6	28	0	4349	21
Rajasthan	1452	35	325	5	685	11	207	5	370	5	1805	26	540	10	682	4
Tamil Nadu	816	8	392	0	1600	8	1027	0	1142	8	477	2	707	2	530	3
Pondicherry	0	0	0	0	6	0	0	0	0	0	0	0	274	2	35	0
Uttar Pradesh	21	0	0	0	738	8	8	0	121	4	639	14	132	2	51	2
D&N Haveli	0	0	0	0	0	0	1	0	0	0	0	0	0	0	0	0
West Bengal	0	0	0	0	0	0	32	0	6293	34	1230	8	95	4	1038	7
Madhya Pradesh											16	0	51	2	3	0
Manipur													51	1	0	0
Uttarakhand															20	0
Total	3306	53	1926	33	12754	215	4153	45	11903	157	12317	184	5534	71	12561	80

Table 5. Case, Death and Case Fatality Rate (CFR) of Dengue in India

Year	Case	Death	CFR (%)
2001	3306	53	1.60
2002	1926	33	1.71
2003	12754	215	1.68
2004	4153	45	1.08
2005	11903	157	1.30
2006	12317	184	1.49
2007	5534	71	1.28
2008	12561	80	0.63
Total	64,454	838	1.30

In India total number of dengue cases per year varies from 1926 to 12754 between 2001 and 2008 and case fatality rate varies from 0.63 to 1.71 (Table 5). Total number of cases was 64,454 encompassing 22 states during that period (Table 4). The CFR varies from 1.5% to 10.9% across different regions in India (NVBDCP, 2007; Garg *et al.*, 2008).

Concurrent Chikungunya and Dengue Virus Infections

An outbreak of febrile illness occurred in Gabon in 2007. Chikungunya or dengue-2 virus infections were identified in 321 patients; 8 patients had documented co-infections. *Aedes albopictus* was identified as the principal vector for the transmission of both viruses (www.cdc.gov, Leroy *et al.*, EID, Vol.15 No. 4, April, 2009). *Ae. aegypti* mosquitoes are common vectors for dengue virus and chikungunya virus. In areas where both viruses co-circulate, they can be transmitted together. During a dengue outbreak in Delhi in 2006, six serum samples were positive for both viruses.

Table 6. Per Man Hour landing (PMHL) of *Aedes aegypti* and *Ae. albopictus* off human bait

Season (2006-2007)	Dumdum		Bangaon	
	Ae. aegypti	*Ae. albopictus*	*Ae. albopictus*	*Ae. aegypti*
Summer (March - May)	0.16	0.08	0.29	0.08
Rainy (June-August)	0.33	0.16	0.45	0.16
Autumn (September-November)	0.45	0.25	0.83	0.37
Winter (December – February)	0.20	0.12	0.54	0.16
Total	**0.29**	**0.15**	**0.53**	**0.19**

Vector ecology

Aedes aegypti and *Ae. albopictus* are found in different parts of the world including Indian sub-continent and can be referred to as global vectors. *Aedes aegypti* is thought to have been introduced to Asia from Africa. *Aedes albopictus* was first described from Bengal. It has spread from South East Asia to the USA, Europe and Africa. These two species are potential vectors of several arboviruses of public health importance viz., Yellow fever, dengue and chikungunya. *Ae. aegypti* is confined to the tropics and subtropics, while the latter species also occurs extensively in temperate regions thereby extending the geographical range over which chikungunya and dengue transmission may occur. Both the species have drought-resistant eggs that remains viable over few months. *Ae. albopictus* exploits a wider range of larval habitats, both natural and artificial, than does *Ae. aegypti* explaining its abundance in rural areas as well as parks and suburbs. *Ae. aegypti* is more closely associated with human habitation, breeding indoors and outdoors in artificial habitats (Townson and Nathan, 2008). *Aedes aegypti* has been considered to be the principal vector of chikungunya and dengue virus in the urban transmission cycle, with *Aedes albopictus* and other anthropophilic *Aedes* sp. serving as secondary vectors/maintenance vectors (Moncayo *et al.*, 2004; Gratz, 2004). The outbreak of dengue and chikungunya in India was generally associated with *Ae. aegypti* but in the kerala state *Ae. albopictus* was recently identified as the primary vector (Townson & Nathan, 2008, Tyagi, 2010). Some arboviruses, *viz.*, West Nile, Potosi, Tensaw, Keystone, Eastern equine encephalitis and Cache valley virus have been isolated from *Ae. albopictus* in the USA (Mitchell, 1998). Amongst the above mentioned viruses, DEN, chikungunya and West Nile are endemic in India. Both these mosquitoes are capable to perpetuate dengue viruses in nature through trans-ovarial transmission (Gokhale *et al.,* 2001, Ahmed *et al.*, 1997). Once infected the vector mosquito remains infected for its life time (Hati, 2001). In tropical countries, transmission of dengue infection can be perennial albeit it is increased in the monsoon. The overall distribution of *Aedes* mosquitoes is rapidly changing in the changing climatic and anthropogenic impacts, specially, *Ae. albopictus* (the Asian tiger mosquito) has dispersed globally into new territories previously occupied by *Ae. aegypti*. As a consequence, the characteristics of DEN virus and CHIK virus circulation and their outbreak dynamics are likely to be modified. *Ae. aegypti* and *Ae. albopictus* are diurnal mosquitoes biting principally in the morning and late afternoon. Man biting activity of *Ae. aegypti* has also been observed in the late evening in urban Kolkata in the months of August and September, 2005 (Bhattacharya & Roy, 2006). This species will bite humans in the night when hungry (Chandler & Read, 1981). *Aedes albopictus* preferred to feed on humans but would feed on birds if no other host was available (Hawley, 1988; Hess *et al.*, 1968). An attempt had been made (2006-07) to find out the prevalence of man biting mosquitoes in the fringes of Bongaon (close to Indo-Bangladesh international border) and near Netaji Subhash Chandra Bose International Airport, Dumdum (close to Kolkata), West Bengal.

Both the above mentioned *Aedes* species were encountered on human bait in these two strategically important areas. *Aedes aegypti* predominates over *Ae. albopictus* in Dumdum while the situation was found to be reverse in Bongaon. Per man hour landing (PMHL) of *Aedes aegypti* and *Aedes albopictus* mosquitoes was 0.19 and 0.53 respectively near the international border while PMHL of these two species was found to be 0.29 and 0.15 respectively near the international airport (Table 6). Sustained vector, virus and epidemiological surveillance would help to forecast the future outbreak of above mentioned arboviral diseases, if any. Molecular and biochemical taxonomy of the captured mosquitoes from the above mentioned sites are required to ascertain whether the species are indigenous or of exotic strain (Bhattacharya, 2008). As there is no political boundary to which virus and mosquito vectors restrict their activities, especially in this globalised world, the possibility of the entry of YF virus and other named arboviruses in India by aircrafts and ships having infected mosquitoes, man and animal reservoirs is a matter of concern. Surveillance on suspected migratory birds in this context is also needed. This is more so because some of the above mentioned viruses have bioterrorism potential. Naturally, this issue demands our serious attention (Bhattacharya, 2009).

Variation in the vector competence

Wide variation exists within and among species in their ability or inability to transmit important arboviral diseases. A particular geographic strain is often associated with different form of oral susceptibility and horizontal transmission rates. Boromisa *et al.,* 1987 examined vector competence of eight geographic strains of *Aedes albopictus* from N. America and Asia and one North American strain of *Aedes aegypti* to Fiji strain of Dengue-1 serotype virus. All strains tested except the HOUSTON strain, showed high susceptibility to midgut infection. However, there were marked differences in percent disseminated infection and transmission rates among the various strains tested. In a presumptive genetic investigation, strain differences in basal lamina thickness, assessed by measurement of transmission electron micrographs, and disseminated infection rates of DEN-1 virus were compared among three laboratory strains of *Aedes albopictus* (Thomas *et al.*, 1993) Mean basal lamina thickness for the NEW ORLEANS and the HOUSTON strains were significantly greater than those for the OAHU strain, which demonstrated a higher disseminated infection rate than the former two (Rai, 2010).

Serratia odorifera modulates vector susceptibility to dengue virus

Deobagkar *et al.* (2009) have recently identified among the diversity of micro flora in the midgut of mosquitoes (*Anopheles* as well as *Aedes* species), a Gram negative bacterium, *Serratia odorifera* which is maintained in the life cycle of vector mosquitoes. They observed that presence of this bacterium in the midgut significantly increased susceptibility (from 30% to about 65%) of the mosquito to chikugunya (CHIK) as well as Dengue (DEN-2) viruses. A 40kDa (P40) extracellular glycoprotein synthesized by *Serratia odorifera* was identified to be interacting with the midgut brush border. Expression of this protein was increased by enhanced external temperature. The total proteomic analysis also revealed that P40 selectively binds to the mosquito protein, identified as Prohibitin. Both CHIK and DEN-2 could also interact with P40, possibly through their E protein. P40, however, did not interact with other viral receptors in the midgut. Prohibitin on the other hand is a non receptor protein interacting with these viruses. Prohibitin has been thus identified as an important intrinsic refractory molecule in the vectors. Interestingly, even the plasmodium would interact with prohibitin suggesting prohibitin as a universal refractory molecule in mosquitoes. Prohibitin is also a part of innate immunity controlling pathway. Thus prohibitin would be visualized to play a crucial role in confining resistance to pathogenic infection in mosquitoes.

Identification and characterization of DEN – 2 binding proteins from vector mosquitoes

The function of cell surface protein and other cellular proteins in DEN-2 attachment, entry and assembly are not yet completely understood. Deobagkar and his group (2009) have recently identified DEN-2 binding polypeptides from midgut brush border membrane fraction of *Aedes aetypti* adult and larvae and also from the *in vitro* cultured cell lines, viz., C6/36 (*Aedes albopictus)* and A7 (*Aedes aegypti)* by employing virus Overlay Protein Binding Assay (VOPBA). Identities of Vav-1, hsc70, orisis, actin, tubulin and prohibitin were confirmed by MALDI analysis and also using monospecific antibodies.

Dengue Zoonosis

There are a number of animal viruses that multiply in the haematophagus arthropods and are transmitted by their bites. Mosquitoes and ticks harbour large varieties of these viruses. Most of these vector-borne viruses exist primarily or primitively in wild birds or mammals and are disseminated by sylvan or "wild" vectors. Like yellow fever, dengue is probably a sylvan disease of monkeys, transmitted among them, by forest dwelling mosquitoes such as *Ae. albopictus* and some other members of the *Ae. scutellaris* group (Chandler & Read, 1981). The preferred vertebrate hosts for the mosquito vectors of dengue are humans, but in Africa, Southeast Asia and perhaps India there is evidence that the virus can be maintained in enzootic and epizootic sylvan cycles in non-human primates (Gubler, 1997). Monkeys probably serve as interepidemic reservoirs of the disease (Chandler & Read, 1981). It has been demonstrated that the dengue is a zoonotic disease of monkeys in Malaysia, maintained by *Aedes pseudoniveus / subniveus* (Rudnick, 1978).

Impact of climate change on vector ecology

Global warming is likely to reshape the ecology of many vector mosquitoes and will have wide-ranging consequences on our health and well-being (Bannoth & Ramachari, 2009). Dengue is an important climate sensitive disease. Global warming may contribute to wider spread of dengue infections (Patz *et al.*, 1996). It was estimated that, in the 2080s,

5-6 billion people would be at risk of dengue as a result of climate change and population increase, compared with 3.5 billion people if the climate remain unchanged (Hales *et al.*, 2002).Vector competence of *Ae. aegypti* increases with increasing temperature for dengue virus. The temperature required for such efficiency is at least 32°C to 35°C (Watts *et al.*, 1987, Kuno, 1995). Global warming can also have a bearing on disease transmission by shortening the pathogen's incubation period within the vector. Warmer temperatures have been shown to directly increase mosquito vector breeding, biting, pathogen transmission despite shortening mean daily survival. The greater effect of climate change on transmission is likely to be observed at the extremes of temperatures at which transmission occurs. For many diseases, these lie in the range of 14°C-18°C at the lower end and about 35°C - 40°C at the upper end (Githeko et al. 2000). Related changes in rainfall coupled with El Nino/ Southern Oscillation (ENSO) may alter the quality and availability of mosquito (vector) breeding sites.

Climate change as selection agent

Change of climate and the effect it causes may bring genetic changes. Climate change can here serve as the selection agent. In North America, there is evidence of recent micro-evolutionary (genetic) responses of the mosquito species *Wyeomyia smithii* to increased average land surface temperatures and earlier arrival of spring in the past two decades (Bradshaw & Holzapfel, 2001). This species is closely related to important vector species that may be going through similar evolutionary changes.

Climate change and mosquito-borne diseases

Climate change has been implicated in changes in the migratory and reproductive behaviour, abundance, population dynamics and geographic range of several bird species (Brommer, 2004). Several species of wild birds can act as reservoirs/ amplifiers of infectious diseases, especially in case of mosquito-borne viral diseases viz., JE, WN, MVE, SLE. The possible consequences of these changes can be the shifts in the geographical distribution of the vectors and pathogen and the changes in the life cycle of bird-associated pathogens due to the mistiming between bird breeding and the breeding of vectors.

Discussion

It is reasonable to suppose, to some extent, that neither viruses nor diseases they cause are new, but our awareness, understanding and recognition of them are rather new (Schaeffer *et al.,* 1950). Naturally, more viruses and their subtypes will be found and more voids in our understanding of ecology, distribution and dynamics of infectious viral diseases will be filled up in future. Mutation rates in viruses with RNA genomes (*viz.,* dengue, Japanese encephalitis etc.) can be six orders of magnitude greater than the viruses with DNA genomes, as no proof reading enzymes are available for RNA dependent RNA polymerases. Consequently, anti-genic variants arise rapidly and inter-virus genetic recombination occur (Holmes & Twiddy, 2003). Phylogenetic studies have yielded important insights into the selection and evolution of arboviruses. Genetic change in the virus either leads to an increase or decrease in viral fitness albeit it is not very clear why some viruses have greater replicative fitness than others.

The finding of the virus/pathogen in a particular species of mosquitoes is not by itself a proof that the species is a vector to man/ animal. Additional evidences such as that the species is common enough and has reasonable contact with humans or both animals and humans (bridge vector in case of a zoonotic disease), may therefore, be helpful. The very brief period during which various reservoir hosts are usually infectious to mosquitoes, before developing immunity (Hale *et al.,* 1957), mean that the source of virus at any particular moment from which mosquitoes can acquire the infection, will usually be few and widely scattered. This would seem to require a vector both numerous and long lived for the persisting prevalence of the infection in the population (Reid, 1961). *Ae. albopictus* may take blood from diverse animals but it rarely takes blood from more than one host during a single gonotropic cycle, thus decreasing the probability of transmission (Hayes *et al.*, 2006). *Aedes albopictus* may not be an efficient vector for dengue epidemic but it may play a crucial role in inter-epidemic periods as a maintenance vector until the population immunity is substantially lowered and the *Ae. aegypti* population is significantly dense to support epidemic transmission (Gubler, 1987, 2004). Yellow fever (YF) belongs to group-B arbovirus (flavivirus), now confined in Africa and some parts of America, may find its way in India just as Dengue, Japanese Encephalitis etc. did. India is a 'yellow fever receptive area', *i.e.,* an area in which YF does not exists but where conditions would permit its development, if introduced. The non-occurrence of YF in India is still inexplicable. A plausible explanation can be that group-B (flavivirus) dengue antibody is present here and there are chances of antigenic sharing, as YF virus too, is a group-B (flavivirus) arbovirus. Widespread presence of group-B (flavivirus) antibodies of dengue, Japanese encephalitis in India might deter outbreaks of YF.

Ecological changes associated with population growth and developmental activities meant for the human welfare are often responsible for the increase of some mosquito-borne diseases. Dengue is an enigmatic disease owing to its sudden appearance and not having a specific antidote. There is also no vaccine available against the pathogen (Tyagi, 2010). When outbreaks of arbovirus disease occur in human communities, vector control programs generally aim at reducing the size and/or survival of the adult mosquito population in order to reduce or obstruct transmission of the virus; relatively small reductions in daily survival rates will result in much greater reductions in proportions reaching the infective age. However, there is inadequate information on the natural history of the Indian arboviruses, particularly their methods of survival between periods of activity. If the pathogen (*viz.*, dengue/ chikungunuya viruse) is able to reappear in subsequent generations by virtue of vertical transmission mechanisms such as transovarial transmission, then vector control measures will need to be directed at the breeding sites and the juvenile stages, since adults may be infective at emergence. Studies towards a greater understanding of the natural history of dengue virus, vector competence and immune response of the victim in heterotypic infections are required in order to contain the disease more effectively.

Data Collection

Data on Dengue cases and deaths were obtained from National Vector-Borne Disease Control Programme (NVBDCP), Ministry of Health and Family Welfare, Government of India, New Delhi and Health Department, Government of West Bengal.

ACKNOWLEDGMENTS

I am grateful to Dr. D.K. Kar, Principal, Asutosh College, Kolkata for providing encouragement and facilities for the work. I am indebted to Prof. A.K. Hati, former director, Calcutta School of Tropical Medicine and to Prof. S.C. Santra, Department of Environmental Science, Kalyani University, West Bengal for their kind support and valuable suggestions during the work and while writing this article.

REFERENCES

Ahmed, R., Ismail, A., Saat, Z. & Lim, L.H., 1997. Detection of dengue virus from field *Aedes aegypti* and *Aedes albopictus* adults and larvae. *Southeast Asian J. Trop. Med. Public Health.* 28: 38.

Bannoth, Reddya Naik & Ramachari, A., 2009. Ronald Ross: Global warming-malaria warming. *J. Asiatic Soc.* LI (2): 53-61.

Bhattacharyya, N., Neogi, D.K., Hati, A.K., Pramanik, N., Banerjee, A., Mukherjee, K.K.,1997. An outbreak of Dengue fever in a rural area of West Bengal. *Indian J. med. Microbiol*; 15: 139-141.

Bhattacharya, Sajal, 2009. Mosquito-borne diseases in India with special reference to malaria vectors and their control. *Journal of the Asiatic Society*, Li.(2): 15-34.

Bhattacharya, S., 2008. Studies on *Aedes* mosquitoes near an international border and an international Airport in India. Paper presented in the 9th

International Symposium on Vectors and Vector-borne diseases, held at Puri, India. Pp. 84.

Bhattacharya, S. & Ray, S., 2006. Man-biting *Aedes* mosquitoes in Kolkata during out break of dengue, abstract book, National congress of Entomology, Punjabi University Patiala, India. pp. 70.

Boromisa, R.D., Rai, K.S. & Grimstad, P.R., 1987. Variation in the vector competence of geographic strains of *Aedes albopictus* for dengue-1 virus. *J. Am. Mosq. Cont. Assoc.* 3: 378-386

Bradshaw, W.E. & Holzapfel, C.M., 2001. Genetic shift in photoperiodic response correlated with global warming. *Proc. Natl. Acad. Sci., USA* 98:14509-14511.

Brommer, J.E., 2004. The range margins of northern birds shift polewards. *Ann. Zool Fenn.* 41: 391-397

Chandler, A.C. & Read, C.P., 1981. Introduction to Parasitology. John Wiley and sons, Inc., New York, Pp.822.

Chareonsook, O., Fog, H.M., Teeraratkul & Silarug, N., 1999. Changing epidemiology of dengue haemorrhagic fever in Thailand.Epidemiol. *Infect.* 122: 161-166.

Deobagkar, D.N., 2009. Vector competence in mosquitoes: Interaction between intrinsic and extrinsic factors. Presidential address, Xth International Symposium on vector and vector-borne diseases, 4-6 November, 2009, Goa, India.

Garcia, G *et al.*, 2006. Antibodies from patients with dengue viral infection mediate cellular cytotoxicity. *J. of Clinical Virology* 37: 53-57.

Garg, P., Nagpal, J., Khairnar, P. & Seneviratne, S.L., 2008. Economic burden of dengue infections in India. *Tran. Roy. Soc. Trop. Med. Hyg.* 102: 570-577.

Githeko, A.K., Lindsay, S.W., Confalonieri, U.E. & Patz, J.A., 2000. Climate change and vector-borne disease: a regional analysis. *Bull. World Hlth. Org* 78(9): 1136-1147

Gokhale, M.D., Barde, P.V., Satkal, G.N., Gore, M.M. & Mourya, D.T., 2001. Vertical transmission of dengue-2 virus through *Aedes albopictus* mosquito *J. Commun. Dis.* 33: 212,.

Gratz, N.G., 2004. Critical review of the vector status of *Aedes albopictus. Med Vet Entomol.* 18:215-27.

Gubler, D.J., 1987. Current research on dengue In. Harris K.F. (editor). Current topics in vector Research. New York, Springer verlog, Inc. pp. 37-56.

Gubler, D.J., 1997. Dengue and dengue haemorhagic fever : its history and resurgence as a global public health problem. In: Gubler, D.J., Kuno, G. (Eds), Dengue and Dengue Haemorrhagic Fever. CAB International, New York, pp. 1-22.

Gubler, D.J., 1998. Dengue and dengue haemorrhagic fever, *Clin. Microbiol. Rev.* 11:480-296.

Gubler, D.J., 2004. Cities spawn epidemic dengue viruses. *Nature Medicine* 10: 129-130.

Guha-Sapir, D. & Schimmer, B., 2005. Dengue fever: new paradigms for a changing epidemiology. *Emerg. Themes Epidemiol.* 2: 1.

Guzman. M.G. & Kouri, G., 2002. Dengue: an update. *Lancet infect.Dis.* 2:33-42.

Hati, A.K., 2006. Studies on dengue and dengue hemorrhagic fever (DHF) in West Bengal state. *India. J. Commun. Dis.* 38,124-129.

Hati, A.K., 2009. Dengue serosurveillance in Kolkata, facing an epidemic in West Bengal, India. *J. Vector Borne Dis.* 46: 197-204.

Hati, A.K., 2001. Medical Entomology Allied Book Agency, Kolkata-700073, India. Pp.201.

Hales, S., de Wet, N., Maindonald, J. & Woodward, A., 2002. Potential effect of population and climate changes on global distribution of Dengue fever: An empirical model. *Lancet* 360: 830-834.

Hale, J.H., Lim, K.A. & Coloss, D.H., 1957. Investigation of domestic pigs as a potential reservoir of Japanese encephalitis virus on Singapore island. *Ann. Trop. Med. & Parasitol*: 51: 374-379.

Hayes, J.M., RigauPerez, J.G., Reiter.P., Effler, P.V.,Pang, L.,Vorndam,V., Hinten, S.R., Mark, K.E., Myers, M.F., K., Bergau, L. Meyer, C., Amador, M., Napier, M., Clark, G.G., Biggerstaff, B.J., and Gubler, D.J., 2006. Risk factors for infection during a Dengue-1 outbreak in Maui, Hawaii, 2001. *Trans. Roy. Soc. Trop. Med. & Hyg.* 100:559-566

Hawley, W.A., 1988. The biology of Ae. albopictus. *J. Am. Mosq. Control Association* 4 (Suppl. 1): 1-40

Hess, D.S., Hayes, R.O. & Tempelis, C.H., 1968. The use of the forage ratio technique in mosquito host preference studies. *Mosq. News* 28:386-389.

Holmes, E.C. & Twidy, S.S., 2003. The origin, emergence and evolutionary genetics in dengue virus, Infect. *Genet. Evol.* 3: 19-28.

Kosasih H., Nurhayati, Kusriastuti R., Harun S., Roef M. C., Wuryadi S., Herianto B., Yuwono D., Porter R.K., Beckett G.C., Blair J. P., 2006. Four dengue virus serotypes found circulating in an outbreak of dengue fever and dengue haemorrhatic fever in Jakarta, Indonesia, during 2004. *Trans. Roy. Soc. Trop. Med. & Hyg.* 100: 855-862.

Kuno, G., 1995. Review of the factors modulating dengue transmission. *Epidemiol. Rev.* 17: 321-335.

Moncayo, A.C., Fernandez, Z., Ortiz, D., Diallo, M., Sall, A., Hartman, S., 2004. Dengue emergence and adaptation to peridomestic mosquitoes. *Emerg. Infect. Dis.* 10: 1790-1796.

Mitchell, C.J., 1998. Public health implications of the presence of Aedes albopictus (skuse) (Diptera:Culicidae) in the USA. *Acta Parasitologica portuguesa* 5 (1): 66.

Park, J.E. & Park, K., 1977. Text book of preventive and social medicine, Banarashidas Bhanot publishers, Jabalpur, India. Pp. 758.

Patz, J.A., Epstein, P.R. & Burke T.A., 1996. Global climate change and emerging infections disease. *JAMA* 275: 217-223.

Rai, K.S., 2010. Insights from mosquito evolution: Patterns, tempo and speciation. In: Nature at work- ongoing saga of evolution, V.P. Sharma (Editor) published by National Academy of Sciences, India and Springer, pp. 197-217.

Reid, J.A., 1961. The attraction of mosquitoes by human or animal baits in relation to the transmission of disease. *Bull. Ent. Res* 52: 43-62.

Rudnick, A., 1978. Ecology of dengue virus. *Asian. J. Infect. Dis.* 2: 156

Schaeffer, M., Robinson, R.Q. & Fodor, A.R., 1959. New viruses and viral disease of man. *Ann. Rev. Microbiol.* 13:335-348.

Thomas, R.E. Wu, W.K., Verleye, D. & Rai, K.S., 1993. Midgut basal lamina thickness and dengue -1 virus dissemination rates in laboratory strains of *Aedes triseriatus* (Diptera culicidae). *J.med. Entomol.* 30: 326-331.

Townson, H. and Nathan, M.B., 2008. Resurgence of chikungunya. *Trans. Roy. Soc. Trop. Med. Hyg.* 102: 308-309.

Tyagi B.K., 2010. Impact of climate change on *Aedes albopictus* skuse, the Asian tiger mosquito, and Emergence of Dengue fever in Kerala state (South India), lecture delivered in the National Seminar on Climate change and Vector-borne Diseases, held at the Asiatic Society, Kolkata, 6th February, 2010, pp. 11-16.

Twiddy, S.S., Holmes, E.C. & Rambaut, A., 2003. Inferring the rate and timescale of dengue virus evolution. *Mol. Biol. Evol.* 20, 122-129.

Vijaykumar, T.S., Chandy, S., Narayanan, S., Abraham, M., Abraham, P. & Gopalan, S. (2005). Is dengue emerging as a major public health problem? *Indian J. Med. Res.* 121, 100-107.

Watts, D.M. Burke, D.S., Harrison, B.A., Whitmire, R.E. & Nisalak, A., 1987. Effect of temperature on the vector efficiency of *Aedes aegypti* for dengue-2 virus. *Am. J. Trop. Med. Hyg.* 36: 143-152.

Zhang, C *et al.,* 2005. Clade replacements in dengue virus serotypes 1 and 3 are associated with changing serotype prevalence. *Journal of virology* 79: 15123-15130.

Chapter - 22

ILLUMINATING THE FIELD OF ENTOMOLOGY: AN OVERVIEW OF APPLICATION OF MARKER BASED TECHNOLOGY

Neelima Arora[1], Amit Kumar Banerjee[2] and U.S.N. Murty[2]

[1] Centre for Cellular and Molecular Biology, Hyderabad, A.P., India
[2] Bioinformatics Group, Biology Division,
Indian Institute of Chemical Technology, Hyderabad, A.P., India
E-mail: *aroraneelima123@yahoo.co.in* ; *neelimaiict@gmail.com*

ABSTRACT

Insects represent the most diverse group of animals that dwell on earth and represent about 90% of metazoa. The diversity shown by insects in shape, size, forms, habit and habitat is enormous. Correct identification of insects and their characterization is of prime importance keeping in mind their economical and medical importance. Genetic diversity displayed by insects often perplexes entomologists. Earlier practices of identification merely based on morphological characters which are turning to be futile in crucial cases and can no longer be relied upon. Advances in molecular biology have led to the development of biochemical and molecular markers that can be exploited to explore the vast spectrum of genetic diversity. As molecular markers directly reflect the diversity or variation existing at DNA level and are not influenced by environment or developmental stage or time, they have becomes potent tools for entomologists. Molecular markers are being applied to a wide range of applications in entomology such as forensics, species identification, phylogenetics etc. Molecular markers also provide an opportunity to study insect behavior, plant-insect interaction, insecticide resistance and insect-pathogen interactions in great detail. Availability of such molecular marker technique has illuminated the field of entomology and has given much needed impetus to the genetic mapping efforts. This review focuses on application of molecular markers in entomology and will also provide an insight on specific genomic regions used as molecular markers.

INTRODUCTION

Nature follows the prime rule of "Unity in Diversity". All organisms are unified by a common theme of evolution as beads coupled by an elongated string. Though various organisms seem to be different in appearance and forms from each other yet they are connected with a strong link sharing some genetic traits with a common ancestor. Understanding this common link of conservation while appreciating the variations that emerged as consequence of evolution has become foremost task for the modern biologists. Solving this complex riddle posed by nature both excites and overwhelms the researchers across the world.

In spite of various efforts to reach to the "grail number" (total number of species on earth), we are far from even identifying all the life forms that inhabit the planet earth. Species that have been catalogued till date do not even represent the complete spectrum of the vast diversity and are just a fraction. Considering their tremendous variability, varied natural habitats, genotypic or phenotypic variability, it is almost impossible to know each species in great detail. Gone is the era when taxonomists relied on morphological characters for identification and classification of a species. Reliance on such characters had led to erroneous identification and positioning of species in classification systems in past thus, forcing the researchers to look for other plausible and unswerving alternatives. This led to subsequent upsurge in development of biochemical and molecular marker systems for accurate identification of organisms and also for assessing the variability existing within and among different species.

Man acquires inspiration for demystifying complex problems of nature from the splendid school of nature. Here, we are reminded of an analogy. In an ultra modern mall, for streamlining the storage and management of a wide array of products, unique bar codes are assigned to products. Similarly, at the time of origin of species, nature tagged each species with a novel and unique natural bar. This idea led to a revolution of a sort where researchers were tempted to employ biochemical and molecular markers, and morphological markers were used just as substitutes to obtain superficial information for preliminary investigation. Though we are discussing about looking for the unique fingerprint for each species yet it is not so simple technically as millions of species originated from a fabulous magical combination of only four nucleotide bases, *i.e.*, "A","T","G" and "C" at the genomic level and only 20 amino acid combination at the protein level. Unveiling the whole diversity arising from permutation and combination of these nucleotides by merely relying on phenotypic characters is similar to imagining the intricacy of a sculpture by looking at its shadow. The advent of molecular techniques has provided us with a potent way of examining genetic variations in a more direct manner.

In the animal kingdom, the maximum diversity is found in Class Insecta (Price, 1997). It is estimated that 900,000 species of insects dwell on the earth in all imaginable habitats. It is due to the enormous ecological, economic, scientific and aesthetic value of insects that drive efforts to know and understand them in a better way. Who can imagine the splendid beauty of nature and its bounty, had it not been for pollinating insects. One can never deny the importance of insects. On one hand, the beneficial insects bless us with wonderful products like honey, silk, lac while, on the other hand, we face the biggest danger from them when we come across insect pests, which rob us from our share hard earned fruits of toil. While some insects ward off many deadly pests by acting as their natural enemies, others act as vectors and/or transmitter of many deadly diseases. The adage "Love them or hate them, you cannot ignore them" suits to the "T" to these six legged creatures. By far, the class insect has emerged as one of most successful group in animal kingdom and its diversity is not presented only in the forms, shapes but also in habits and habitats. Their success is attributed to their immense adaptability of interacting with environment and clever strategies in thriving in unfathomable conditions (Dempster & McLean, 1999).

This chapter will provide a flavor of various biochemical and molecular techniques and genome regions of present or probable importance in unveiling the diversity of insects.

MARKER BASED METHODS

Study of genetics would have remained incomplete and lackluster in absence of markers. For many decades, entomologists have used morphological characters as means of identification and classification, and also in elucidating phylogeny. Though phenotypic characters are easily identified, available during various stages of life cycle and do not require sophisticated techniques and often provide firsthand information in genetics, their use is becoming obscure owing to difficulty in scoring and less number of distinct traits. Clearly, morphological markers have been instrumental in enhancing our knowledge of insect genetics but fall short in providing dense genetic maps. Due to strong conservatism towards the traditional dichotomous keys based on morphological characters, molecular marker techniques have not been exploited fully in entomology. But the scenario has now changed due to technological advances and increasing knowledge base generated due to advances in field of genetics. Several advanced and state-of-art techniques are now available that can be applied for exploring, identifying and quantifying diversity existing in various insect species. This century has witnessed the development of biochemical and molecular markers, which can be exploited in entomological studies. The stupendous ability of these markers in resolving the genetic differences has made them popular yardsticks for gauging the variability among insects. Progress in development of such marker based has given much needed impetus to the field of molecular entomology. Progressive technologies harnessed in modern biological research have generated a wealth of information that can be applied effectively in various arenas of entomology (Hoy, 2003; Crampton *et al.,* 1996). Marker based methods have illuminated the field of phylogeny, population dynamics and genome mapping in insects (Loxdale & Lushai, 1998; Severson *et al*., 2001; Heckel, 2003).

Biochemical markers

Owing to difficulties faced in resolution of genetic traits by use of morphological markers, protein markers became popular and flourished during early phase (Loxdale & Lushai, 1998).

Protein based methods

Generic term used to describe any enzyme variant revealed by electrophoresis is "isoenzyme/isozyme" (Bush & Kitto, 1978) while the term allozyme was proposed by Prakash (1969) for those enzymes, which are produced by different alleles at a single locus. Allozymes directly reflect upon subtle or major variations in DNA sequences through amino acid change observed as different electrophoretic mobilities (Rosendahl & Sen, 1992). Allozymes have been used in revealing genetic diversity of various organisms. These show Mendelian inheritance and absence of pleiotropic and epistatic interaction. The alleles of most isozyme loci show codominant expression, which allows identification of homozygotes and heterozygotes (Weeden & Wendel, 1989). Being relatively rapid and inexpensive, this technique gained instant popularity among researchers. Their use is limited by lack of polymorphism, a comparatively low number of loci and alleles for analysis and the need to use a variety of methods for marker detection (Bernatsky & Tanksley, 1989). However, the problem with allozymes is their inability to separate various races because allozymes only quantify variation at restricted number of sites and hence, these may not reflect overall patterns of genetic variation throughout the genome (Whitt, 1981) and hence, are not preferred. Another potential shortcoming of isozyme markers is that they are considered to be phenotypic markers since assay is for gene product and not gene itself (Richardson *et al*., 1986). Furthermore, for many enzymes, the expression of the gene product is specific to a particular tissue or development stage. As a result, this assay can be futile due to spatio-temporal expression or dependence on external stimuli for expression. Another shortcoming of allozyme based markers is the problem of storage as the specimen should be either alive or in frozen state until usage which becomes a liability in field condition. This method offers limited resolution in species that have not been reproductively isolated for long owing to its ability for identifying little fraction of variation where net charge is not affected. Allozymes have been applied to many insect genera yet the application of this technology in medically important insects is in its infancy.

Direct analysis of proteins

Though 2DGel electrophoresis (O'Farrell, 1975) has been the mainstay in the analysis of complex protein mixtures yet such highly sensitive techniques have not yet been widely applied to study insects. Two-dimensional (2-D) gel electrophoresis separates a single sample separated initially by charge (isoelectric focusing) and then by molecular weight. This technique can prove to be an asset as one can study a number of proteins in one instance but spots cannot be identified easily and comparison of individuals becomes difficult in such case, as it requires examination of more gels, which is not the case with one dimensional slab electrophoresis. One such study focused on comparing protein differences in *Drosophila melanogaster* and *D. simulans* (Ohnishi *et al.*, 1982). 2D gel electrophoresis has also been performed for analyzing hemolymph proteins in *Anopheles gambiae* (Paskewitz & Shi, 2005) and salivary proteins in *Triatoma infestans*, a vector of Chagas' disease (Assumpção *et al.*, 2008).

MOLECULAR MARKERS

Molecular marker systems to study the genetics of insects are nowadays common in entomological research (Crampton *et al.,* 1996; Osborne *et al.,* 2002). As DNA is quite stable and is easily preserved by desiccation or immersion in alcohol and offers virtually unlimited polymorphic markers in coding and non-coding regions, past few decades have witnessed a shift towards DNA-based markers (Weising *et al.*, 1995; Fajardo, 1998). The advent of PCR technology led to the development of many variants in technique (Innis *et al.,* 1990), which made DNA polymorphism studies a lot easier (Saiki *et al.*, 1988; Weising *et al.,* 1995). A major breakthrough occurred with the development of markers based on DNA sequence (Bolstein *et al.,* 1980). DNA based genetic markers provide a genetic diagnostic tool that permits direct identification of a genotype in any tissue and development stage in environment- independent manner. Molecular markers used to study DNA polymorphism falls in 2 categories, *viz.*, hybridization and PCR based markers.

The hybridization assays simply make use of sequence complementarity between labeled nucleic acid probes and samples for identifying specific nucleic acid sequence (Crampton *et al.,* 1997). But this technique is not considered quite unswerving as it displays sensitivity to unequal quantity of target DNA loaded on a membrane and variation in copy number across the species considered (Cooper *et al.*, 2002; Beebe *et al.*, 1996).

A major advantage of PCR based technology is the requirement of small quantity of DNA sample, thus making it quite non- invasive process for analysis, which eliminates the need of sacrificing the specimen (Paskewitz & Collins, 1997). Properties desirable for ideal DNA marker (Joshi *et al.*, 1999) are:

1. Highly polymorphic nature
2. Co-dominant inheritance so as to facilitate identification of homozygotes and heterozygotes
3. Frequent occurrence in genome
4. Selective neutral behavior
5. Easy access
6. Easy and fast assay
7. High reproducibility
8. Easy exchange of data between laboratories

The choice of molecular markers depends entirely on the kind of study and one molecular marker cannot satisfy all conditions.

Restriction Fragment Length polymorphism (RFLP)

Major strength of this technique is their capability to reveal an almost unlimited number of polymorphism (Kan & Dozy, 1978). This technique relies on detection of genetic polymorphism in DNA fragments and provides substantially more polymorphism than isozyme markers. RFLPs have their origin in DNA arrangements that occur due to evolutionary processes, point mutations within restriction enzyme recognition site sequences, insertions or deletion within fragments. RFLP markers were used for the first time for construction of genetic map by Bolstein (Bolstein *et al.*, 1980). They are inherited as Mendelian characters. RFLPs depend on use of probes to identify single or low copy sequences in DNA. However, the assay is time-consuming and labour-intensive and also the necessity of radioactive probes makes the process costly. Moreover, their inability to detect base changes restricts their use in detection of polymorphism. This method also requires large amount of DNA.

Amplified Fragment Length polymorphism (AFLP)

This technology developed by KEYGENE provides a powerful tool for detection and assessment of genetic variation (Vos *et al.*, 1995). Originally coined selective restriction fragment amplification (SRFA) (Zabeau & Vos, 1993), the method is now popularly known as Amplified Fragment Length polymorphism (AFLP) (Vos *et al.*, 1995) apparently to reveal its similarity to restriction fragment length polymorphism (RFLP). This technique which was devised during 1990s proffers a comparatively novel method for determining genetic diversity and has been increasingly applied in studies of arthropods (Vos *et al.,* 1995; Zabeau & Vos, 1993) is a technique based on detection of genomic restriction fragments by PCR amplification and can be used for depicting distinctive fingerprints in DNAs of any origin or complexity. AFLP technique is based on selective PCR amplification of restriction fragment from a total restriction enzyme digest of genomic DNA. After digesting oligonucleotide adapters are ligated to restriction fragment. Selective amplification is achieved using amplimers designed from both restriction enzymes recognition sequence and adapter sequence extending a short way (about 3 bases) into the restriction fragment. Only those primers that match the restriction fragment at their 3' end are amplified. AFLP screens many loci quickly and reveals large numbers of polymorphism. No previous genetic information on the study organism is necessary before use of AFLP. The AFLP technique is consistent and does not alter with minor variations in experimental conditions since stringent reaction conditions are used for primer annealing. This technique shows resourceful combination of PCR and RFLP techniques and is extremely useful in detection of polymorphism among closely related genotypes. Most AFLP fragments correspond to unique positions on genomes and hence, can be exploited in genetic and physical mapping. AFLPs are found to be less labour-intensive and more informative when compared with RFLPs (Jones *et al.,* 1997). AFLP markers show a high marker index as compared to other marker technologies (Pejic *et al.,* 1998). Major shortcoming of this technique is its inability to distinguish homozygotes from heterozygotes and this limits the use of these markers for population genetic studies.

Microsatellites

Litt and Lutty coined the term "microsatellite". Microsatellite also called Short Tandem Repeats (STRs) or Simple Sequence repeats (SSRs) (Creg, 1992) consist of tandemly recurring units such as $(TG)_n$ or $(AAT)_n$ where n usually diverges from 11-60 (Brudford & Wayne, 1993). Microsatellites occur frequently across the eukaryotic genome. SSR reflect polymorphism based on number of repeat units in a defined region of genome being investigated (Jacob *et al.,* 1991). High information content of data obtained by microsatellite loci makes these markers one of the most preferred molecular tools for many population and biodiversity studies. This technique involves the construction of DNA genomic libraries which are screened with microsatellite-type repeated sequence probes (Estoup *et al.,* 1993). On identification of locus, sequencing is done and primers are designed for the unique microsatellite flanking regions. This step can be omitted if primers applicable in one species amplify microsatellite regions in related species. Polymorphism of these loci can be detected in a straightforward way by determining the size of the PCR-amplified fragments (Navajas and Fenton, 2000). Microsatellite loci are now known in many organisms and the analysis is facilitated by advances and availability of robust statistical and computational methods (Luikart & England, 1999). High marker abundance, vast polymorphism, simple identification are some of the factors that have contributed to the popularity of microsatellites in genetic studies (Milligan *et al.,* 1994; Paszek *et al.,* 1998; Powell *et al.,* 1996). Main drawbacks of this technique are the homoplasy of MS and extensive stutter bands of PCR products (Zhang *et al.,* 2002).

Sequence-tagged microsatellite site markers

This method includes DNA polymorphism using specific primers designed from sequence data of specific primers from sequence data of a specific locus. Primers complementary to flanking regions of simple sequence repeat loci yield highly polymorphic amplification products. Polymorphism is due to variation in number of Tandem repeats (VNTR loci) in a given repeat motif (Nakamura *et al.,* 1987). Tri and tetra-nucleotide microsatellites are more popular for STMS analysis owing to clear banding pattern after PCR and gel electrophoresis.

Direct amplification of minisatellite DNA markers (DAMD-PCR)

This technique has been investigated as a way of producing DNA probes useful for detecting polymorphism. DAMD-PCR clones can yield individual specific DNA fingerprinting pattern and thus, have the potential as markers for species differentiation.

Inter Simple Sequence repeats (ISSR)

This technique was devised by Zietkiewicz *et al.* (1994). This involves utilization of primers based on microsatellites for amplifying inter-SSR DNA sequences. In this technique, various microsatellites anchored at 3' end are used for amplifying genomic DNA, thus enhancing their specificity. These are known to be most dominant markers, but occasional reports have shown co-dominance. Infinite primers can be synthesized for various combinations of di-, tri-, tetra - and penta-nucleotides etc with an anchor made up of few bases and can be used for diverse applications. SSR regions consist of 2-6 bp sequences (Weising *et al.,* 1991) and are major source of genomic variation (Tautz *et al.,* 1986).

Randomly Amplified Polymorphic DNA (RAPD)

Williams *et al.* (1990) established a DNA polymorphism assay based on PCR amplification of random DNA fragments with single primers of arbitrary nucleotide sequence (Williams *et al.,* 1990; Welsh & McClelland, 1990). Amplification is carried out using a single short primer (8–10 base pairs) under low temperature conditions) increasing multiple binding at sites spread throughout the genome. PCR products are then separated on an agarose gel matrix and visualized (Navajas & Fenton, 2000). Polymorphism can be detected as DNA segment which amplify from one parent but not other and are inherited in a Mendelian fashion. They suggested that these polymorphisms should be called Randomly Amplified polymorphic DNAs (RAPDs). They also stated that nucleotide sequences of primer should be selected with restriction of 9-10 nucleotides. RAPDs reveal high level of polymorphism within and among species. Martin *et al.,* (1991) emphasized on their applicability in constructing genetic map by comparing fingerprints resulting from same primer but DNA isolated from closely related species. RAPD polymorphism is due to presence of band in one parent of a mapping cross and absence of that band in another parent (Reiter *et al.,* 1992). RAPD markers are relatively short and hence, a mismatch of even a single nucleotide can often prevent the primer from annealing, giving rise to RAPD polymorphism. RAPD markers have enabled creation of high-density genetic maps and saturation of already existing genetic maps.

The advantages of RAPDS over RFLPs and isozymes are as follows:

1. Non- radioactive assay.
2. It requires small quantities of genomic DNA.
3. Simple and fast technique.
4. Non- labour intensive, inexpensive technique.
5. Prior knowledge of nucleotide sequence is not required.
6. Species independent markers.
7. Construction and maintenance of genomic library is not required.
8. It provides unlimited number of markers throughout the genome.

The RAPD profiles are sensitive to small changes in reaction conditions such as the source of Taq polymerase, template DNA quantity and quality, type of PCR machine, ramp rates, and application of enhancers and inhibitors (Williams *et al.,* 1990; Ellsworth *et al.,* 1993; Gallego & Martinez, 1997). Thus, main difficulty encountered in interpreting data remains with the poor repeatability of the results (Black, 1993). Being dominant markers, RAPD markers are unable to differentiate between homozygous or heterozygous loci (Lynch & Milligan, 1994). Owing to its rapid nature, RAPD technique is often used as the first source of information.

Direct amplification of length polymorphism (DALP)

This method employs random PCR primers for generating multi-band genome DNA profiles and facilitates the sequencing of polymorphic bands (Navajas & Fenton, 2000). These bands are then defined and make the study of distinct loci possible (Desmarais *et al.,* 1998). Use of specific primers aid in achieving the specificity that is reflected in reproducibility of results, often a prime issue in such studies and also offer the advantage of *a priori* definition of polymorphism.

Sequencing of DNA fragments

Recent proliferation in sequencing technologies along with cut down in the cost of sequencing has brought this technique within the reach of entomologists. The nucleotide sequence thus obtained mirrors the variability at the molecular level in the most precise manner. Such data is stored, managed and catalogued in public repositories such as EMBL, DDBJ and GENBANK. Access to such data enables researchers across the world to explore phylogenetic relationships as well as variability among various insect species.

Specific genomic regions as markers

The question that arises now is whether there is any ideal marker for a species available in the nature, especially for insects? The answer is not yet known to the scientific community. The astonishing diversity of insects always cautions the molecular taxonomists that it is difficult to differentiate two species. Researchers are still in search of that magical marker which may have all the ideal properties of a marker. There are few specific regions in the genome, which are being used for past several years as standard markers. In the following section, we would like to discuss such regions in detail.

Non-coding regions

It is a well evident fact that majority of the genome is not used for coding. This has led to speculation of junk DNA concept. But it has to be kept in mind that nature that believes in economy of genome will never retain anything useless and pass the so-called junk DNA from generation to generation. Many mysteries unfolded with time regarding the non coding regions.

As far as specific regions of interest in genomes of insects are concerned, there are a few regions that have attracted a lot of attention during past few years. Non-coding regions consist of introns of nuclear genes (Freisen, 2000), mitochondrial control region (Zhang & Hewitt, 1997) and the internal transcribed spacer regions of nuclear ribosomal genes. These regions are solely useful within the species or closely related species as they remain under minute selection pressure and evolve quite fast. The observed substitution rate is high in these regions.

Among the most used genetic and phylogenetic markers, nuclear ribosomal gene clusters are quite popular.

Internal Transcribed Spacer 2 (ITS2)

Most widely used marker in insect and other organisms is the second internaltranscribed spacer region (ITS2) of the nuclear ribosomal gene cluster. The organization of this particular cluster is provided in Fig. 1. The cluster consists of three genes viz. 18S rDNA, 5.8S rDNA and 28S rDNA. These are transcribed into RNA but fail to make the ultimate product, protein molecules. These RNA molecules are further used for forming ribosome. These important genes are separated from each other by two spacer regions, namely, Internal Transcribed Spacer 1 and Internal transcribed Spacer 2 respectively. They are spliced out later on during transcription and destined for no use similar to lariats. This spacer region shows fast substitution due to less selection pressure as it is quite unimportant and useless region of the genome. Distinguishing very closely species also becomes easy when one rapidly evolving spacer region is flanked by two slowly evolving gene or vice versa. Designing and synthesis of effective primers have yielded tremendous useful results in this regard and wide variety of taxa has been identified and classified using this molecular tool. Evidences of such studies are available in the literature for plants (Buckler and Holtsford, 1996), fungi (Huang *et al.*, 2010), pathogens (Erensoy *et al.*, 2009; Areekit *et al.*, 2009) and insects (Schliitterer *et al.*, 1994; Fritz *et al.*, 1994; Kambhampati, 1995; Schulenburg *et al.*, 2001).

Apart from the traditional gel based marker analysis, latest bioinformatics analysis is providing more insight with minute details on these marker regions. Construction of secondary structure of the spacer regions or ribosomal genes and their analysis aids in distinguishing closest species and even sibling species. Even throughout the eukaryotes, conservation and variation has been determined at the secondary structure level (Coleman, 2007). Specialized databases of ITS2 (Auria *et al.*, 2006; Schultz *et al.*, 2006; Selig *et al.*, 2008) and other regions have been developed in this regard. This supports interactive direct analysis of a new species. More advanced *in silico* studies revealed the interrelation and influence of secondary structural parameters on each other (Banerjee *et al.*, 2007), application of robust and advanced mathematical models developed by artificial neural network architecture and self organizing maps on ITS2 sequential and structural parameters were also found to assist in identification, classification and clustering of mosquito species with greater accuracy and precision (Banerjee *et al.*, 2008; Banerjee *et al.*, 2009).

Internal Transcribed Spacer 1 (ITS1)

Though less popular and useful than ITS2, ITS1 (having similar properties to ITS2) has been explored as suitable marker region by many groups. It is reported that ITS1 display more variability than ITS2, making the alignment more difficult. Nominal polymorphism was seen even within the species in case of ITS1. ITS1 has been used for phylogenetic analysis in *Culicoides* biting midges (Diptera: Ceratopogonidae) (Perrin *et al.,* 2006) and for evolutionary analysis in triatomines (Tartarotti & Ceron, 2005).

The other regions such as the genes 5.8s, 18s and 28.5s are also used as a marker. Application of 28.5s is comparatively less in this regard and there are lots of arguments on the using a single specific marker. The cluster arrangement is provided in Fig. 1.

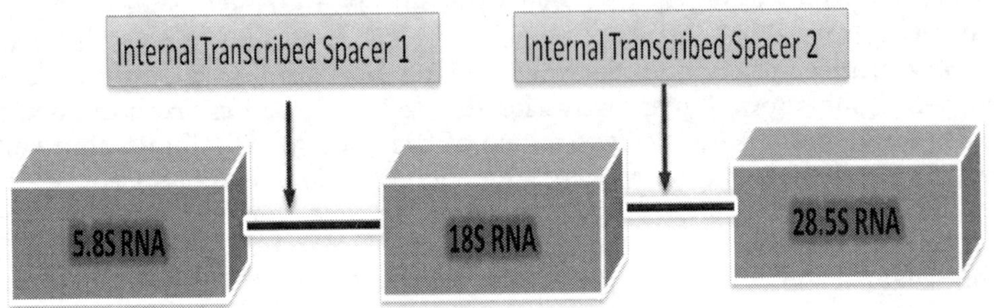

Fig.1. Arrangement of ribosomal gene cluster with their specific spacer regions

Mitochondrial genes

Mitochondrial genes hold a special place as candidate region for molecular systematics. Maternal inheritance and high copy number make it top priority region as high copy number provide technical edge over single copy number nuclear genes and maternal inheritance allows conducting studies at intra-species level.

Two categories of mitochondrial genes are protein coding genes and ribosomal genes. Further, similar to nuclear ribosomal genes, mitochondrial ribosomal genes are of two types, 12S and 16S rDNA except that they are not isolated by spacer regions. Studies on these regions and 18S region forwarded our understanding about these regions and their application as a molecular tool for resolving the interrelationship of different species and determining their exact taxonomic position (Norris *et al.,* 1999; Murrell *et al.,* 2000). It has been reported that 18S is better than 12S and 16S in application to molecular phylogenetic analysis (Norris *et al.,* 1999; Klompen *et al.,* 2000).

Cytochrome Oxidase Sub Unit I (COI), II and III (COIII)

Cytochrome oxidase is also extensively used as phylogenetic marker. The genetic organization of COI, COII and COIII is provided in figure 2. The arrangement of the mitochondrial genes in white flies was experimentally proved by Thao and coworkers in 2004. The CO-I gene is followed by Leu tRNA gene (L2), CO-II, tRNA for Lys (K),tRNA gene for Asp (D), ATP Synthase subunit 8, 6 and CO-III respectively (Figure 2).

They differ from ITS2 in having faster evolution rate. Extensive studies with this region in insect and other species were performed by Navajas et al. in 1998, and they witnessed noticeable amount of polymorphism in *Tetranychus urticae*. Valuable inference was reported from similar studies, which justified the application of this specific portion of mitochondrial genome (Anderson and Trueman, 2000; Toda, 2000). A number of experiments reports about this region as a quality marker (Funk, 1999; Morlais and Severson, 2002; Agustí *et al.*, 2003, Zhang and Hewitt 1996; Mahendran *et al.*, 2006; Kranthi *et al.*, 2006). In similar fashion, COII and III are extensively used as a phylogenetic determinant (Kourti, 2006).

Fig 2. Arrangement of Cytochrome Oxidase subunit I, II and III genes in mitochondrial genome.

APPLICATION OF MOLECULAR MARKERS IN ENTOMOLOGY

Molecular markers have provided important clues and have illuminated the field of entomology. Some of the areas where molecular markers have been employed extensively are described below:

Insecticide resistance

Changing practices in intensive agriculture and over-reliance on synthetic insecticides has given rise to many problems; resistance to insecticides is one of them. Resistance to insecticides has been a growing concern among research communities. It is speculated that hundreds of species have developed resistance to many chemical control agents (Georghiou, 1990). Common mechanisms of resistance to the insecticides involve P450 microsomal oxidase enzymes, excessive production of esterases for degrading the insecticides, insensitivity of acetyl cholineesterases to organophosphates and insensitivity of the sodium gated channels to synthetic pyrethroids (Marshall, 1995; Devonshire *et al.*, 1998; Denholm *et al.*, 1999; Elbert & Nauen, 2000; Foster *et al.*, 2000).

Study of mechanism of insecticide resistance, its extent and the response of different species towards insecticides is important not only for medically important insects but also in crop pests. This also holds significance as to determine the quantity and type of insecticides so as they harm only pests and exclude beneficial insects. Molecular markers have been frequently used for identification of resistance genes in such insects. Spirits of Public health specialists were marred by failure in controlling the surmounting problem of malaria after golden era of DDT. This was accounted to growing resistance of mosquito vectors to DDT. Ranson and coworkers determined DDT resistance markers in *Anopheles gambiae*, major malaria vector in Africa by using microsatellites (Ranson et al. 2000). RAPD markers have been used for mapping genetic loci in Rhyzopertha *dominica* Fabricius that render resistance to phosphines (Schlipalius *et al.*, 2002). AFLP markers were also used in mapping resistance loci in *Leptinotarsa decemlineata* to pyrethroid (Hawthorne, 2001) and in diamondback moth to *Bacillus thuringiensis* toxins (Heckel *et al.,* 1999). Biochemical markers have also been applied in screening resistance to methyl-parathion and neonicotinoid cross-resistance in western corn rootworms (Zhou et al. 2002) and in *Bemisia tabaci* (Rauch & Nauen 2003).

Insect Plant Interaction

Though modern approaches used in agriculture today have enhanced the production manifold yet a big proportion of our population is starved. Insect pests cause huge losses of crop and result in substantial economic loss every year. One of the interesting applications of molecular markers is characterization of plant-insect interaction. These markers can be used in tagging and mapping genes conferring resistance to harmful insect pests and also in identification of avirulence genes in insects interacting with host plant. Information generated from marker based strategies can also enable us in elucidating the relationship between adaptability of these insects and/or specificity to host with genomic constitution (Behura, 2006) as observed in distinct selectivity of host in winged and wingless forms in aphids (Lushai et al. 2002) and Sitobion *avenae* (Lushai *et al.*, 2002). RAPD markers have been found useful in measuring the level of virulence in pea aphids in response to natural resistance posed by alfalfa (Bournoville *et al., 2000*) and also in identifying distinct loci specific to individual strains of Asian rice gall midge (Behura *et al.*, 1999).

Insect behaviour

Insects epitomize highly advanced social behavior and man who consider himself sitting at the top of intelligence pyramid can learn a lot from these lowly creatures when it come to behavioral pattern. This has attracted the attention from various spheres of researchers including molecular biologists. Colony behavior of ants that leaves one mesmerized by its sheer complexity has been studied by microsatellites and mtDNA markers (Crozier *et al.,* 1997). The basis of kinship selection in *Iridomyrmex purpureus*, a highly evolved social trait, was established by tracking movement of females between colonies using mitochondrial DNA markers (Crozier *et al.,* 1997). In *Leptothorax acervorum* (F.), mitochondrial haplotypes were studied to explore intra- and inter-nest variation (Stille & Stille, 1992). AFLP and microsatellite markers have been utilized recently to reveal guarding and stinging behavior in honeybee (Arechavaleta-Velasco *et al.,* 2003) and molecular basis of ecological implications of foraging range and nest density behaviors in bumble bee (Knight *et al.,* 2005). Various groups have attempted in dissecting social behavior of insects at molecular level using honeybee as model organism (Robinson *et al.,* 2005). Wide range of markers have successfully unearth the genetic basis of nursing and foraging behavior (Hunt & Page 1995; Whitfield *et al.,* 2003) as well as colony level behaviours such as stinging behaviour, body size, pheromone alarm level, traits for reversal learning and hygienic behavior (Breed *et al.,* 2004). Molecular markers also provide an excellent mean for ascertaining parentage and kinship relation in insects and also for studying breeding pattern (Behura, 2006).

Phylogenetics

Because of the dawn of era of molecular markers, we are now able to use refined methods for phylogenetic reconstruction than the age-old parsimony based methods, which are available for the analysis of morphological data (Lewis, 2001). This has boosted the field of phylogenrtics research in insects. Mitochondrial and ribosomal DNA gene sequences are used often for constructing evolutionary history of insect species and these markers shed ample light on evolutionary trajectories of closely related species (Luque *et al.,* 2002; Chatterjee & Mohandas, 2003; Chatterjee & Tanushree, 2004; Prasad *et al.,* 2005). Considerable success has been achieved in resolving ancestral relationships in various insects like mosquitoes (Besansky *et al.,* 1997), vespid wasps (Schmitz & Moritz, 1998), braconid wasps (Belshaw & Quicke, 1997; Dowton & Austin, 1994), tiger beetles (Vogler & Pearson, 1996) and apocritan wasps (Dowton & Austin, 2001).

Insect pathogen interaction

Progress in marker based technology has opened up a window of opportunity where one can study and figure out mode of interaction between dreaded pathogens and insect vectors responsible for their spread at the genetic level (Crampton *et al.,* 1997). Using RAPD, information was obtained on the vectorial ability of *Triatoma infestans* (Klug), a reduviid vector that transmits *Trypanosoma cruzi,* the causative agent of Chagas' disease (Pinto *et al.,* 1998). While assessing the malarial transmission, knowledge of vectorial capacity of *Anopheles* is a must before devising an effective strategy for intervention for malaria control. RFLPs have been employed for determining vectorial capacity of mosquitoes by mapping, quantitative trait loci (QTL) to determine the vectorial ability of mosquitoes (Severson *et al.,* 2001; Severson *et al.,*1995;

Bosio *et al.,* 2000). Various studies aimed at identifying the potential target genes involved in immunity to specific pathogens causing malaria in mosquito vector have exploited Expressed sequence tag (EST) markers (Bartholomay *et al.,* 2004).

Studies on insect-pathogen interaction have found wide applications in agriculture and horticulture. Many insect-borne viruses wreak havoc on agriculture and forestry. Power of ESTs were unleashed to study the gene expression pattern in life cycle of the host aphid for gaining insight of genetic basis of its vectorial ability to transmit the citrus tristeza virus (Hunter *et al.,* 2003). Marker based technology is also used to identify better cell lines for production of viable pathogens as biopesticides (Sudeep *et al.,* 2005). Understanding the relation between entomopathogens and insect host can aid in developing effective bio-control agents and can be achieved by utilizing molecular markers.

Study of fossils

Many times, treasure of information is hidden well beneath relics of past while we are scratching the surface of present. Scientists worldwide have accepted the significance of fossil samples as important tools in conservation biology, forensic science and entomology and these can serve as source for ancient DNA, which can contribute to the interpretation of history of extinct population or species (Brown & Brown 1994). But the success of this approach depends o the quality of DNA obtained from such samples. Goldstein & Desalle (2003) discovered a fixation event of anthropogenic character linked with the extinction and fragmentation of populations in tiger beetle *Cicindela d. dorsalis* (Coleoptera: Carabidae).

Species identification

For any given purpose, entomologists need to identify the species correctly. Usually morphological features are enough proof for identifying any species according to classical concept. But now this boundary seems to be blurred and fail to distinguish between closely related species and more importantly, cryptic species. Cryptic species perplex researchers and add to confusion in taxonomical research. Genetic markers have been intensively used as a tool for species identification when morphological characters are unknown or of difficult interpretation (Brunner *et al.,* 2002; Naeole & Haymer, 2003, Schama *et al.*, 2005). Mitochondrial DNA was used to identify sibling species in red wood ants (*Formica rufa* group) (BernasconI *et al.,* 2010). Allozymes and restriction patterns of the mitochondrial regions 16S and Cyt b have been applied to identification of *Euglossa* spp. Latreille (Reference). Identification of species identification is of paramount importance in devising efficient vector control program and cryptic species often compound the problem. For resolving such issue in cryptic species identification in mosquitoes, RFLPs were used (Collins *et al.* 1988). AFLP clearly separates isomorphic members of the *A. dirus* complex, an important malaria vector in Southeast Asia (Yasothornsrikul *et al.,* 1988), as well as the *An. quadrimaculatus* (Mitchell *et al.,* 1992) complex in North America. RAPDS have been employed to resolve the status of five closely related species in the *Aedes (Stegomyia) scutellaris* group (Kambhampati *et al.,* 1992). RAPDS were found to be of immense use in *Aedes aegypti* for comparing the geographically isolated species.

Forensic entomology

"Corpses tell the tales of crime". Forensic experts cannot agree more to this statement. One of the major areas of forensic sciences is medico-legal entomology. The role of insects in crime investigation is undeniable (Erzinclioglu, 1983, Catts & Goff, 1992). Necrophagous Insects collected from dead bodies can provide clues for estimating the post-mortem interval (PMI) which is calculated by considering time taken by insects to reach the stage found on body (Ames *et al.,* 2006; Anderson & Cervenk, 2002; Amendt *et al.,* 2004). Even the species present on a dead body can be indicative of PMI as insects colonization takes place in a series of succes-sion events (Bornemisszza, 1957). Some insects have defined geographical distributions and their presence outside their confined habitat can indicate post-mortem movement. For gathering such critical information in criminal investigations, rapid and unequivocal identification of insects collected from the crime scene is needed. Reliance on only morphological features in carrion–feeding insects can be misleading and require expertise in such crucial cases. Sometimes, early life stages of such insects are hard to distinguish if samples are not conserved with caution (Otranto *et al.,* 2002). Another problem often encountered in such cases is loss of time in rearing of immature stages. Best option in such cases seems to embrace the DNA based marker technology. Mitochondrial cytochrome oxidase has often been used as a marker of choice in identification of blowflies, flesh flies (Diptera: Sarcophagidae) and other forensically important insects (Vincent *et al.,* 2000; Wallman *et al.,* 2001; Harvey *et al.,* 2003; Caine *et al.,* 2006; Sperling *et al.,* 1994; Loxdale & Lushai, 1998; Malgorn & Coquoz, 1999; Caterino *et al.,* 2000, Wells & Sperling, 2001; Stevens & Wall, 2001; Zehner *et al.,* 2004). Molecular markers have been used to differentiate between *Calliphora vicina* and *Calliphora vomitoria* (Ames *et al.,* 2006). Similarly Cytochrome b (Cyt-b) gene was used successfully for identification of the necrophagous fly species Stearibia nigriceps (Pablo *et al.,* 2006). In case of highly deteriorated bodies, it becomes impossible to identify toxic compounds, drugs or poison involved as cause of death. In such cases, analysis of insects feeding on body can come as rescue (Goff & Lord, 2001; Nolte *et al.,* 1992; Goff & Lord, 1994, 2001; Introna *et al.,* 2001; Campobasso *et al.,* 2001).

Genetic diversity and genetic mapping

Genetic mapping has been made possible due to progress in molecular markers as it was not feasible to trace the genotype of loci across the whole genome using the available technology (Tanksley, 1993; Via & Hawthorne, 1998). Genetic mapping can be used to study many interesting traits such as resistance to toxins, biomass accumulation, silk production, fecundity, sex ratio, longevity, emergence rate etc. Among the invertebrates, most of the genome maps have been constructed in insects (Solignac et al. 2004). A landmark was the construction of mapping of model organism Drosophila which initiated a cascade of events, which resulted in new genome, maps as well as saturation of already existing dense genome maps (O'Brien, 1993). Genetic maps have been constructed in silkworm using RFLP (Shi *et al.*, 1995), RAPD (Promboon et al. 1995), RAPD with double primer pairs (Yasukochi, 1998), and AFLP markers (Tan et al. 2001). 309 markers that segregated among F2 interspecific hybrid individuals of Hawaiian Cricket Laupala were identified using AFLP markers (Parsons and Shaw, 2002). Genetic maps are available in medically important species such as mosquitoes. For instance, genetic map of *Anopheles gambiae,* major vector of malaria in Africa, were constructed using microsatellites (Zheng *et al.*, 1996), RAPD (Dimopoulos *et al.,* 1996), and RFLP (Mori *et al.,* 1999) and also in *Aedes aegypti,* vector of Chikungunya & Dengue, using RFLP (Severson *et al.*, 1993, 1995), cDNA/single-strand conformation polymorphism (SSCP) (Fulton *et al.*, 2001) and RAPD/SSCP (Antolin *et al.*, 1996). Similarly, genetic maps have been generated in *Culex pipens* using RFLP (Mori *et al.,* 1999); and *A. albopictus* using RFLP (Severson *et al.,* 1995) and RAPD- SSCP (Mutebi *et al.,* 1997). RAPD markers have been frequently employed in genetic mapping in various insects such as *Tribolium castaneum* (Beeman and Brown 1999), *Apis mellifera* (Hunt & Page, 1995; Solignac et *al.* 2004), Tri*chogramma brassicae* (Laurent *et al.*, 1998) and Nasonia (Gadau *et al.,* 1999). AFLP markers have been used to construct linkage maps for understanding the resistance to Bacillus thuringiensis toxins in *Plutella xylostella,* diamondback moth (Heckel *et al.*, 1999), color pattern evolution in Mimetic butterfly (Jiggins *et al.,* 2005), resistance to pyrethroid insecticides in Colorado Potato Beetle (Hawthorne, 2001), Hawaiian Cricket Laupala (Parsons & Shaw, 2002) and red flour beetle Tribolium (Zhong *et al.,* 2004). Molecular markers have also aided in exploring genetic diversity in Red Palm Weevil (Gadelhak & Enan, 2005), bollworm (Subramanian & Mohan Kumar, 2005) and muga silkworm (Neog *et al.,* 2005) and butterflies (Takami *et al.,* 2004).

CONCLUSION

Though numerous markers are available, yet recent extensive analyses suggest that individually they are not capable to shed light on controversial cases where correct taxonomic position is yet to be determined. Questions have been raised in specific cases (Banerjee *et al.*, 2007). Application of more than one specific marker is suggested for confirmation of the obtained results. In the present scenario, most of the studies are performed combining different markers. Still, there is always a hunt for a novel marker, which can match up with most or all the unique properties of marker. It is hoped that with the prosperity of sequencing technology and comparative genomics in near future, we will be able to find the magical marker from the haystack of various genomes.

REFERENCES

Agustí, N., S.P. Shayler, J.D. Harwood, I.P. Vaughan, K.D. Sunderland & W.O.C. Symondson, 2003. Collembola as alternative prey sustaining spiders in arable ecosystems: prey detection within predators using molecular markers. *Molecular Ecology* 12: 3467–3475.

Alessandrini, F., M. Mazzanti, V. Onofri, C. Turchi & A. Tagliabracci, 2008. MtDNA analysis for genetic identification of forensically important insects. Forensic Science International. *Genetics Supplement Series* 1(1):584-585.

Amendt, J., R. Zehner and R. Krettek, 2004. Forensic Entomology. Naturwisse-nschaften 91:51–65.

Ames, C., B. Turner & B. Daniel, 2006. Estimating the post-mortem interval (I): The use of genetic markers to aid in identification of Dipteran species and subpopulations. *International Congress Series* 1288: 795-797.

Assumpção, T.C., I.M. Francischetti, J.F. Andersen, A. Schwarz, J.M. Santana & J.M. Ribeiro, 2008. An insight into the sialome of the blood-sucking bug Triatoma infestans, a vector of Chagas' disease. *Insect Biochemistry & Molecular Biology* 38(2): 213-232.

Anderson, D.L. & J.W.H. Trueman, 2000. Varroa jacobsoni (Acari: Varroidae) is more than one species. *Experimental & Applied Acarology* 24:165-189.

Anderson, G.S. & V.J. Cervenka, 2002. Insects associated with the body: their use and analyses. In: Haglund WD, Sorg MH (eds) Advances in forensic taphonomy – method, theory and archaeological perspectives. CRC Press, Boca Raton, pp 173–200.

Antolin, M.F., C.F. Bosio, J. Cotton, W. Sweeney & M.R. Strand, *et al.* 1996. Intensive linkage mapping in a wasp (*Bracon hebetor*) and a mosquito (*Aedes aegypti*) with single-strand conformation polymorphism analysis of random amplified polymorphic DNA markers. *Genetics* 143: 1727–1738.

Arechavaleta-Velasco, M.E., G.J. Hunt & C. Emore, 2003. Quantitative trait loci that influence the expression of guarding and stinging behaviors of individual honeybees. *Behavioral Genetics* 33:357–364.

Areekit, S., P. Singhaphan, S. Khuchareontaworn, P. Kanjanavas, T. Sriyaphai, A. Pakpitchareon, P. Khawsak and K. Chansiri, 2009. Intraspecies variation of Brugia spp. in cat reservoirs using complete ITS sequences. *Parasitol. Res.* 104: 1465–1469.

Auria, G. D., R. Pushker & F. Rodriguez-Valera, 2006. IWoCS: analyzing ribosomal intergenic transcribed spacers configuration and taxonomic relationships. *Bioinformatics* 22(5): 527–531.

Banerjee, A.K., K. Kiran, U.S.N. Murty & Ch. Venkateswarlu, 2008. Classification and identification of mosquito species using artificial neural networks. *Comp. Biol. Chem.* 32:442-447.

Banerjee, A.K., N. Arora & U.S.N Murty, 2007. Stability of ITS 2 structure in *Anopheles*: What Lies Beneath? International Journal of Integrative Biology. 1(3): 232-238.

Banerjee, A.K., N. Arora & U.S.N. Murty, 2009.Clustering and Classification of Anopheline Spacer Sequences using Self Organizing Maps. *The Internet J. of Genomics & Proteomics* 4(1).

Banerjee, A.K., N. Arora & U.S.N. Murty, 2007. How Far is ITS2 Reliable as a Phylogenetic Marker for the Mosquito genera? *Electronic Journal of Biology* 3 (3): 61-68.

Bartholomay, L.C., W.L. Cho, T.A. Rocheleau *et al.,* 2004. Description of the transcriptomes of immune response-activated hemocytes from the mosquito vectors *Aedes aegypti* and *Armigeres subalbatus*. *Infection Immunity* 72: 4114–4126.

Beebe, N.W., D.H. Foley, R.D. Cooper, J.H. Bryan & A. Saul, 1996. DNA probes for the Anopheles punctulatus complex. *Am. J. Trop. Med. Hyg.* 54:395–98.

Beeman, R.W. & S.J. Brown, 1999. RAPD-based genetic linkage map of *Tribolium castaneum. Genetics* 153: 333–338.

Behura, S.K., 2006. Molecular marker systems in insects: current trends and future avenues. *Molecular Ecology* 15: 3087–3113.

Behura, S.K., S.C. Sahu, S. Rajamani, A. Devi, R. Mago, S. Nair & M. Mohan, 1999. Differentiation of Asian rice gall midge, Orseolia oryzae (Wood-Mason), biotypes by sequence characterized amplified regions (SCARs). *Insect Mol. Biol.* 8(3):391–397.

Belshaw, R. & D.L.J. Quicke, 1997. A molecular phylogeny of the Aphidiinae (Hymenoptera: Braconidae). Mol. Phylogenet. Evol. 7:281–93.

BerNasconi, C., P. Ilo & D. Cher, 2010. Molecular markers allow sibling species identification in red wood ants (*Formica rufa* group). *Systematic Entomology* 35:243–249.

Besansky, N.J. & G.T. Fahey, 1997. Utility of the white gene in estimating phylogenetic relationships among mosquitoes (Diptera: Culicidae). *Mol. Biol. Evol.* 14(4): 442–54.

Black, W.C., 1993. PCR with arbitrary primers: approach with care. *Insect Mol. Biol.* 2:1-6.

Bolstein, D., R.L. White, M. Skolnick & R.W. Davis, 1980. Construction of genetic linkage map in man using restriction fragment length polymorphisms. *Am. J. Hum. Genetics* 32: 314-331.

Bornemissza, G.F., 1957. An analysis of arthropod succession in carrion and the effect of its decomposition on thesoil fauna. *Aust. J. Zool.* 5: 1-122.

Bosio, C.F., R.E. Fulton, M.L. Salasek *et al.,* 2000. Quantitative trait loci that control vector competence for dengue-2 virus in the mosquito *Aedes aegypti. Genetics* 156:687–698.

Bournoville, R., J.C. Simon, I. Badenhauser, C. Girousse, T. Guillox & S. Andre, 2000. Clones of pea aphid, Acyrthosiphon pisum (Hemiptera: Aphididae) distinguished using genetic markers, differ in their damaging effect on a resistant alfalfa cultivar. *Bull. of Entomol. Res.* 90: 33–39.

Breed, M.D., E. Guzman-Novoa & G.J. Hunt, 2004.Defensive behavior of honeybees: organization, genetics, and comparisons with other bees. *Annual Review of Entomology* 49: 271–298.

Breuker, C.J., K. Victoir, P.W. De Jong, E. Van der Meijden, P.M. Brakefield & K. Vrieling, 2005. AFLP markers for the R-gene in the flea beetle, Phyllotreta nemorum, conferring resistance to defenses in Barbarea vulgaris. *Journal of Insect Science* 5:38.

Brisse, S., J. Dujardin & M. Tibayrenc, 2000. Identification of six Trypanosoma cruzi lineages by sequence-characterized amplified region markers. *Mol. Biochem. Parasitol.* 111:95-105.

Brown, T.A. & K.A. Brown, 1994. Ancient DNA: using molecular biology to explore the past. *Bioessays* 16:719–726.

Brudford, M.W. & R.K. Wayne, 1993. Microsatellites and their application to population genetics studies. *Current opinion in Genetics & Development* 3: 939-943.

Brunner, P.C., C. Fleming & J.E. Frey, 2002. A molecular identification key for economically important thrips species (Thysanoptera: Thripidae) using direct sequencing and a PCR-RFLP-based approach. *Agr. & For. Entomol.* 4:127–136.

Buckler, E.S. IV. & T. P. Holtsford, 1996. Zea Systematics: Ribosomal ITS Evidence. Mol. Biol. Evol. 13(4):612-622.

Bush, G.L., G.B. Kitto, 1978. Applications of genetics to insect systematics and analysis of species differences. In Biosystematics in Agriculture. (ed.: J. A. Rhomberger, R. H. Foote, L. Knutson, P. L. Lentz), New York, Wiley, pp. 89-118.

Byrd, J. & J. Castner (Eds.), 2001. Forensic Entomology. The Utility of Arthropodes in Legal Investigation, CRC Press.

Cainé, L.M., F.C. Real, M.I. Saloña-Bordas, M.M. Pancorbo, G. Lima, T. Magalhães & F. Pinheiro, 2009. Typing of Diptera collected from human corpses in Portugal. *Forensic Science International* 184(1-3): e21-e23.

Caine´, L., F.C. Real, G. Lima, L. Pontes, D. Abrantes *and* M.F. Pinheiro, 2006. Genetic identification of forensically important Calliphoridae in Portugal. *International Congress Series* 1288 846–848.

Campobasso, C.P. & F. Introna, 2001. The forensic entomologist in the context of the forensic pathologist's role. *Forensic Sci. Int.* 120: 132-139.

Caterino, M.S., S. Cho & F.A. Sperling, 2000. The current state of insect molecular systematics: a thriving Tower of Babel. *Annu Rev Entomol.* 45:1-54.

Catts, E.P. & M.L. Goff, 1992. Forensic entomology in criminal investigations. *Annu. Rev. Entomol.* 37: 253-272.

Chatterjee, S.N. & T. Tanushree, 2004.Molecular profiling of silkworm biodiversity in India. *Genetika* 40: 1618–1627.

Chatterjee, S.N. & T.P. Mohandas, 2003. Identification of ISSR markers associated with productivity traits in silkworm, *Bombyx mori L. Genome* 46: 438–447.

Clark, M. J. (ed.), 1995. Molecular Action of Insecticides on Ion Channels. 207 National Meeting of the American Chemical Society, San Diego, California, 13 – 17th March, 1994, (Washington, DC, USA: American Chemical Society).

Coleman, A.W., 2007.Pan-eukaryote ITS2 homologies revealed by RNA secondary structure.Nucleic Acids Research. 35 (10): 3322–3329.

Collins, F.H., V. Petrarca, S. Mpofu, A.D. Brandling-Bennett, J.B.O. Were, M.O. Rasmussen & V. Finnerty, 1988. Comparison of DNA probe and cytogenic methods for identifying field collected *Anopheles gambiae* complex mosquitoes. *Am. J. Trop. Hyg.* 39: 545-550.

Cooper, R.D., D.G.E. Waterson, S.P. Frances, N.W. Beebe & A.W. Sweeney, 2002. Speciation and distribution of the members of the *Anopheles punctulatus* (Diptera: Culicidae) group in Papua New Guinea. *J. Med. Entomol.* 39: 16–27.

Crampton, J.M. & S.M. Hill, 1997. Generation and use of species-specific DNA probes for insect vector identification. (In: The Molecular Biology of Insect Disease Vectors: A Methods Manual, ed. J.M. Crampton, C.B. Beard & C. Louis). Chapman & Hall, London. pp. 384– 98.

Crampton, J.M., B. Beard & C. Loui (eds), 1996. The Molecular Biology of Insect Disease Vectors. Kluwer Academic Publishers, The Netherlands.

Crampton, J.M., C.B. Beard & C. Louis, 1997. The Molecular Biology of Insect Disease Vectors: A Methods Manual. Kluwer Academic Publishers, Boston, Massachusetts.

Creg, P.B. 1992. Simply sequence repeat DNA length polymorphism. *Probe* 2:18-22.

Crozier, R.H., B.P. Oldroyd, W.T. Tay *et al.* 1997. Molecular advances in understanding social insect population structure. *Electrophoresis* 18:1672–1675.

Dempster, J. & I. Mclean, 1999. Insect populations: In theory and in Practice. Kluwer Academic Publishers, Boston, Massachsetts.

Denholm, I., J.A. Pickett & A.L. Devonshire (eds.), 1999. Insecticide Resistance: From Mechanisms to Management. CABI Publishing, UK.

Desmarais, E., I. Lanneluc & J. Lagnel, 1998. Direct amplification of length polymorphisms (DALP) or how to get and characterize new genetic markers in many species. *Nucleic Acids Res.* 26:1458–1465.

Devonshire, A.L., L.M. Field, S.P. Foster, G.D. Moores, M.S. Williamson & R.L. Blackman, 1998. The evolution of insecticide resistance in the peach-potato aphid, Myzus persicae. *Philosophical Transactions of the Royal Society of London Series B Biological Sciences* 353: 1677-1684.

Dimopoulos, G., L. Zheng, V. Kumar, A. Della Torre & F.C. Kafatos, 1996. Integrated genetic map of *Anopheles gambiae:* Use of RAPD polymorphisms for genetic, cytogenetic and STS land marks. *Genetics* 143: 953–960.

Dowton, M. & A.D. Austin, 1994. Molecular phylogeny of the insect order Hymenop tera: Apocritan relationships. *Proc. Natl. Acad. Sci. USA.* 91(21): 9911–9915.

Dowton, M. & A.D. Austin, 2001. Simultaneous analysis of 16S, 28S, COI and morphology in the Hymenoptera: Apocrita - evolutionary transitions among parasitic wasps. *Biol. J. of the Linn. Soc.* 74: 87–111.

Elbert, A. & R. Nauen, 2000. Resistance of Bemisia tabaci (Homoptera: Aleyrodidae) to insecticides in southern Spain with special reference to neonicotinoids. *Pest Management Science* 56: 60-64.

Ellsworth, D.L., K.D. Rittenhouse & R.L. Honeycutt, 1993. Artifactual variation in randomly amplified polymorphic DNA banding patterns. *BioTechniques* 14: 214–217.

Erensoy, A., S. Kuk & M. Ozden, 2009. Genetic identification of Fasciola hepatica by ITS-2 sequence of nuclear ribosomal DNA in Turkey. *Parasitol Res.* 105: 407–412.

Erzinclioglu Y.Z., 1983. The application of entomology to forensic medicine. *Med. Sci. Law* 23: 57-63.

Estoup, A., M. Solignac, M. Harry & J.M. Cornuet, 1993. Characterization of (GT)n and (CT)n microsatellites in two insect species: *Apis mellifera* and *Bombus terrestris*. *Nucleic Acids Res.* 21: 1427–1431.

Fajardo, D., F. Angel, M. Grum, J. Tohme, M. Lobo, W.M. Roca & I. Sanchez, 1998.Genetic variation analysis of the genus *Passiflora L.* using RAPD markers. *Euphytica* 101: 341–347.

Fernandez, S., A.M. Katsuyama, A.Y. Kashiwabara, A.M.B.N. Madeira, A.M. Durham & A. Gruber, 2004. Characterization of SCAR markers of Eimeria spp. of domestic fowl and construction of a public relational database (The Eimeria SCARdb). *FEMS Microbiol. Lett.* 238:183-188.

Foster, S.P., I. Denholm & A.L. Devonshire, 2000. The ups and downs of insecticide resistance in Peach-potato aphids (Myzus persicae) in the UK. Crop Protection.19:873-879.

Freisen, V., 2000. Introns. (In: Baker, A.J. (ed.). Molecular Methods in Ecology. Blackwell, pp. 274-294.

Fritz, G.N., J. Conn, A. Cockburn & J. Seawright, 1994. Sequence Analysis of the Ribosomal DNA Internal Transcribed Spacer 2 from Populations of Anopheles nuneztovari (Diptera: Culicidae). *Mol. Biol. Evol.* 11(3): 406-416.

Fulton, R.E., M.L. Salasek, N.M. DuTeau & W.C.IV. Black, 2001. SSCP analysis of cDNA markers provides a dense linkage map of the *Aedes aegypti* genome. *Genetics* 158: 715–726.

Funk, D.J., 1999. Molecular systematics of cytochrome oxidase I and 16S from neochlamisus leaf beetles and the importance of sampling. *Mol. Biol. Evol.* 16(1): 67–82.

Gadau, J., R.E. Page Jr. & J.H. Werren, 1999. Mapping of hybrid incompatibility in *Nasonia. Genetics* 153: 1731–1741.

Gadelhak, G.G. & M.R. Enan, 2005. Genetic Diversity Among Populations of Red Palm Weevil, *Rhynchophorus ferrugineus* Olivier (Coleoptera: Curculionidae), Determined by Random Amplified polymorphic DNA Polymerase Chain Reaction (RAPD-PCR). *Int. J. Agri. Biol.* 7(3): –395–399.

Gallego, F.J. & I. Martinez, 1997. Methods to improve reliability of random-amplified polymorphic DNA markers. *Bio-Techniques* 23:663–664.

Georghiou, G.P., 1990. Overview of Insecticide Resistance. (In: Managing Resistance to Agrochemicals. M. B. Green, H. M. Le Baron & W. K. Moberg (eds). American Chemical Society, Washington, DC. pp. 18 –41.

Goff, M.L. & W.D. Lord, 1994. Entomotoxicology: A new area for forensic investigation. *Am. J. Forensic Med. Pathol.* 15: 51-57.

Goff, M.L. & W.D. Lord, 2001. Entomotoxicology: Insects as toxicological indicators and the impact of drugs and toxins on insect development, p.331-340. (In: .), Forensic entomology: The utility of arthropods in legal investigations. J.H. Byrd & J.L. Castner (eds). CRC, Boca Raton, pp. 418.

Goldstein, P.Z. & R. Desalle, 2003. Calibrating phylogenetic species formation in a threatened insect using DNA from historical specimens. *Molecular Ecology* 12:1993–1998.

Harvey, M.L., I.R. Dadour and S. Gaudieri, 2003. Mitochondrial DNA cytochrome oxidase I gene: potential for distinction between immature stages of some forensically important fly species (Diptera) in Western Australia. *Forensic Sci. Int.* 131:34-139.

Hawthorne, D.J., 2001. AFLP-based genetic linkage map of the Colorado potato beetle Leptinotarsa decemlineata: sex chromosomes and a pyrethroid-resistance candidate gene. *Genetics* 158:695-700.

Hawthorne, D.J., 2001. AFLP-Based Genetic Linkage Map of the Colorado Potato Beetle Leptinotarsa decemlineata: Sex Chromosomes and a Pyrethroid-Resistance Candidate Gene. *Genetics* 158: 695–700.

Heckel, D.G., 2003.Genomics in pure and applied entomology. *Annual Review of Entomology* 48:235–260.

Heckel, D.G., L.J. Gahan, Y.B. Liu and B.E. Tabashnik, 1999. Genetic mapping of resistance to *Bacillus thuringiensis* toxins in diamondback moth using biphasic linkage analysis. *Proc. Natl. Acad. Sci. USA.* 96: 8373–8377.

Heckel, D.G., L.J. Gahan, Y.B. Liu *et al.,* 1999.Genetic mapping of resistance to Bacillus thuringiensis toxins in diamondback moth using biphasic linkage analysis. *Proceedings of National Academy of Sciences, USA* 96: 8373–8377.

Hoy, M.A., 2003. Insect Molecular Genetics, 2nd edn, Academic Press/Elsevier, San Diego, California.

Huang, T.P., Y. Yeh & D.D.S. Tzeng, 2010. Heteroduplex Mobility Assay for Identification and Phylogenetic Analysis of Anthracnose Fungi. *J Phytopathol.* 158:46–55.

Hunt, G.J. & R.E. Page, 1995. Linkage map of the honey bee *Apis mellifera*, based on RAPD markers. *Genetics* 139: 1371–1382.

Hunt, G.J. & R.E. Page, 1995. Linkage map of the honeybee, *Apis mellifera*, based on RAPD markers. *Genetics* 139: 1371–1382.

Hunter, W.B., P.M. Dang, M.G. Bausher *et al.*, 2003. Aphid biology: expressed genes from alate *Toxoptera citricida*, the brown citrus aphid. *Journal of Insect Science* 3:23.

Innis, M.A., D.H. Gelfand, J.J. Sninsky & T.J. White, 1990. PCR protocols: A guide to methods and applications. Academic Press, San Diego.

Insuan, S., S. Deowanish, S. Klinbunga, S. Sittipraneed, H.A. Sylvester & S. Wongsiri, 2007. Genetic Differentiation of the Giant Honey Bee (*Apis dorsata*) in Thailand Analyzed by Mitochondrial Genes and Microsatellites. *Biochemical Genetics* 45(3-4):345-361.

Introna, F., C.P. Campobasso & M.L. Goff, 2001.Entomotoxicology. *Forensic Sci. Int.* 120: 42-47.

Jacob, H.J., K. Lindpaintner, S.E. Lincoln, K. Kusumi, R.K. Bunkler, T.P. Mao, D. Ganten., V.J. Dzau & E.S. Landner, 1991. Genetic mapping of a gene causing hypertension in stroke prone spontaneously hypertensive rat. *Cell* 67: 213-224.

Jiggins, C.D., J. Mavarez, M. Beltra´n, W. O.McMillan, S. Johnston & E. Bermingham, 2005. A Genetic Linkage Map of the Mimetic Butterfly Heliconius melpomene. *Genetics* 171: 557–570.

Jones, C.J., K.J. Edwards, S. Castaglione, M.O. Winfield, F. Sala, C. van de Wiel, G. Bredemeijer, B. Vosman, M. Matthes & A. Daly *et al.,* 1997. Reproducibility testing of RAPD, AFLP, and SSR markers in plants by a network of European laboratories. *Mol. Breed.* 3:381–390.

Joshi, S.P., P.K. Ranjekar & V.S. Gupta, 1999. Molecular markers in plant genome analysis. *Curr Sci.* 77: 230–240.

Kambhampati, S., 1995. A phylogeny of cockroaches and related insects based on DNA sequence of mitochondrial ribosomal RNA genes. *Proc. Natl. Acad. Sci. USA* 92: 2017-2020.

Kambhampati, S., W.C. Black & K.S. Rai, 1992. Random amplified polymo-rphic DNA of mosquito species and populations (Diptera: Culicidae): T echniques, statistical analysis and applications. *J.Med. Entomol.* 29:939–45.

Kan, Y. & A. Dozy, 1978. Antenatal diagnosis of sickle cell anaemia by DNA analysis of amniotic fluid cells. *Lancet* 2: 910-912.

Kartik, N., H. R. Singh, B. Unni & A. K. Sahu, 2010. Analysis of genetic diversity of muga silkworm (Antheraea assamensis, Helfer; Lepidoptera: Saturniidae) using RAPD-based molecular markers. *African Journal of Biotechnology* 9(12): 1746-1752.

Kengne, P., H.D. Trung, V. Baimai, M. Coosemans & S. Manguin, 2001. A multiplex PCR-basedmethod derived from random amplified polymorphic DNA (RAPD) markers for the identification. *Insect Mol. Biol.* 10(5): 427-435.

Keyghobadi, N., L.A. Crawford & S.A. Maxwell, 2009. Successful analysis of AFLPs from non-lethally sampled wing tissues in butterflies. *Conserv Genet.*10: 2021–2024.

Klompen, J.S.H., W.C. Black, J.E. Keirans & D.E. Norris, 2000. Systematics and biogeography of hard ticks, a total evidence approach. *Cladisitcs* 16: 70-102.

Knight, M.E., A.P. Martin, S. Bishop *et al.,* 2005. An interspecific comparison of foraging range and nest density of four bumblebee (Bombus) species. *Molecular Ecology* 14: 1811–1820.

Kourti, A., 2006. Mitochondrial DNA Restriction Map and Cytochrome c Oxidase Subunits I and II Sequence Divergence of Corn Stalk Borer Sesamia nonagrioides (Lepidoptera: Noctuidae). *Biochemical Genetics* 44:321-332.

Kranthi, S., K.R. Kranthi, A.A. Bharose, S. N. Syed, C.S. Dhawad, R.M. Wadaskar, G.T. Behere & E.K. Patil, 2006. Cytochrome oxidase I sequence of *Helicoverpa* (Noctuidae: Lepidoptera) species in India- Its utility as a molecular tool. *Indian Journal of Biotechnology* 5: 195-199.

Laurent, V., E. Wajnberg, B. Mangin, T. Schiex, C. Gaspin *et al.,* 1998. A composite genetic map of the parasitoid wasp *Tricho gramma brassicae* based on RAPD markers. *Genetics* 150: 275–282.

Lewis, P.O., 2001. A likelihood approach to estimating phylogeny from discrete morphological character data. *Systematic Biology* 50:913-925.

Loxdale, H.D. & G. Lushai, 1998. Molecular markers in entomology. *Bull. Entomol. Res.* 88: 577–600.

Luikart, G. & P.R. England, 1999. Statistical analysis of microsatellite DNA data. *Trends Ecol. Evol.* 14: 253–255.

Luque, C., L. Legal, H. Staudter *et al.,* 2002. ISSR (inter simple sequence repeats) as genetic markers in *noctuids* (Lepidoptera). *Hereditas*136: 251–253.

Lushai, G., H.D. Loxdale, C.P. Brookes *et al.,* 1997.Genotypic variation among different phenotypes within aphid clones. *Proceedings of Royal Society of London. Series B, Biological Sciences* 264: 725–730.

Lushai, G., O. Markovitch & H.D. Loxdale, 2002. Host-based genotype variation in insects revisited. *Bull. Entomol. Res.* 92:159–164.

Lynch, M. & B.G. Milligan, 1994. Analysis of population genetic structure with RAPD markers. *Mol. Ecology* 3:91-99.

Mahendran, B., S. K. Ghosh & S.C. Kundu, 2006. Molecular phylogeny of silk-producing insects based on 16S ribosomal RNA and cytochrome oxidase subunit I genes. *J. Genet.* 85:31-38.

Makert, G.R., R.J. Paxton & K. Hartfelder, 2006. An optimized method for the generation of AFLP markers in a stingless bee (*Melipona quadrifasciata*) reveals a high degree of intracolonial genetic polymorphism. *Apidologie* 37:687–698.

Malgorn, Y. & R. Coquoz, 1999. DNA typing for identification of some species of Calliphoridae. An interest in forensic entomology. *Forensic Sci. Int.* 102: 111–119.

Manguin, S., P. Kengne, L. Sonnier, R.E. Harbach, V. Baimai, H.D. Trung & M. Cossemans, 2002. SCAR markers and multiplex PCR-based identification of isomorphic species in the *Anopheles dirus* complex in Southeast Asia. *Med. Vet. Entomol.* 16:46–54.

Martin, G.B., J.G.K. Williams & S.D. Taksley, 1991. Rapid identification of markers linked to a Pseudomonas resistance gene in tomato by using random primers and near isogenic lines. *Proc. Natl. Acad. Sci. USA* 88: 2336-2340.

Milligan, B.G., J. Leebens-Mack & A.E. Strand, 1994. Conservation genetics: Beyond the maintenance of marker diversity. *Mol. Ecol*. 3:423–435.

Mitchell, S.E., S.K. Narang, A.F. Cockburn, J.A. Seawright & M. Goldenthal, 1992. Mitochondrial and ribosomal DNA variation among members of the Anopheles quadrimaculatus species complex. *Genome* 35:939–50.

Mori, A., D.W. Severson & B.M. Christensen, 1999.Comparative genetic maps for the mosquitoes (*Culex pipiens* and *Aedes aegypti*) based on common RFLP loci. *J. Hered.* 90: 160–164.

Morlais, I. & D.W. Severson, 2002.Complete Mitochondrial DNA Sequence and Amino Acid Analysis of the Cytochrome C Oxidase Subunit I (COI) from *Aedes aegypti. Mitochondrial DNA* 13 (2): 123-127.

Murrell, A., N.J.H. Campbell & S.C. Barker, 2000. Mitochondrial 12S rDNA indicates that the Rhipicephalinae (Acari: Ixodida: Ixodidae) is paraphyletics. *Molecular Phlogenetics & Evolution* 16:1-7.

Mutebi, J.P., W. C. IV. Black, C.F. Bosio, W.P.Jr. Sweeney & C.B. Craig Jr., 1997. Linkage map for the Asian tiger mosquito [*Aedes* (*Stegomyia*) *albopictus*] based on SSCP analysis and RAPD markers. *J. Hered*. 88: 489–494.

Naeole, C.K.M. & D. Haymer, 2003. Use of oligonucleotide arrays for molecular taxonomic studies of closely related species in the oriental fruit fly (*Bactrocera dorsalis*) complex. *Mol. Ecol. Notes*. 3: 662-665.

Nakamura, Y., M. Leppert, P. O' Connell, R. Wolff, T. Holm, M. Culver, C. Martin, E. Fujimoto, M. Hoff, E. Kumlin & R. White, 1987. Variable number of tandem repeat (VNTR) markers for human genetic mapping. *Science* 235:1616-1622.

Navajas, M. & B. Fenton, 2000.The application of molecular markers in the study of diversity in acarology: a review. *Experimental & Applied Acarology* 24: 751–774.

Navajas, M., 1998. Host plant associations in the spider mite *Tetranychus urticae* (Acari: Tetranychidae): Insights from molecular phylogeography. *Experimental & Applied Acarology* 22:201-214.

Nolte, K.B., R.D. Pinder & W.D. Lord, 1992. Insect larvae used to detect cocaine poisoning in a decomposed body. *J. Forensic Sci.* 37: 1179-1185.

Norris, D.E., J.S.H. Klompen & W.C. Black, 1999. Comparison of the mitochondrial 12S and 16S ribosomal DNA genes in resolving phylogenetic relationships among hard ticks (Acari: Ixodidae). *Annals of the Entomological Society of America* 92:117-129.

O'Brien, S.J., 1993. *Genetic Maps*, Ed. 6. Cold Spring Harbor Laboratory Press, Cold Spring Harbor, NY.

O'farrell, P.H. 1975. High-resolution two dimensional gel electrophoresis of proteins. *J. Biol.Chem.* 250:4007-4021.

Ohnishi, S., A. J.L. Brown, R.A. Voelker, C.H. Langley, 1982. Estimation of genetic variability in natural populations of Drosphila simulans by two dimensional and starch gel electrophoresis. *Genetics* 100:127-36.

Osborne, J.L., H.D. Loxdale & I.P. Woiwod, 2002. Monitoring insect dispersal: methods and approaches. In J. M. Bullock, R. E. Kenward and R. S. Hails (eds) Dispersal Ecology, British Ecological Society Symposium, 3 – 5 April, 2001, Reading University, UK. (Oxford, UK: Blackwell Publishing), pp. 24 – 49.

Otranto, D. & J. Stevens, 2002. Molecular approaches to the study of Myiasis causing larvae. *Int. J. Parasitol.* 32 (11): 1345–1360.

Pablo, R.R., M. Salon, E. Sarasola, S. Cardoso & M.M. Pancorbo, 2006. Molecular identification of Stearibia nigriceps: An example of the usefulness of Cytochrome b gene for the identification of entomofauna species. *International Congress Series* 1288:864–866.

Paran, I. & R.W. Michelmore, 1993. Development of reliable PCR-based markers linked to downy mildew resistance genes in lettuce. *Theor. Appl. Genet.* 85:985–993.

Parsons, Y. M. & K. L. Shaw, 2002. Mapping Unexplored Genomes: A Genetic Linkage Map of the Hawaiian Cricket Laupala. *Genetics* 162:1275–1282.

Parsons, Y.M. & K.L. Shaw, 2002. Mapping Unexplored Genomes: A Genetic Linkage Map of the Hawaiian Cricket Laupala. *Genetics* 162: 1275–1282.

Paskewitz, S.M. & L. Shi, 2005. The hemolymph proteome of Anopheles gambiae. *Insect Biochemistry & Molecular Biology* 35 (8): 815-824.

Paskewitz, S.M. & F.H. Collins, 1997. PCR amplification of insect ribosomal DNA. (In: The Molecular Biology of Disease Vectors: A Methods Manual; ed. J.C. Crampton, C.B. Beard, C. Louis). Chapman & Hall, London, pp. 374–83.

Paszek, A. A., G. H. Flickinger, L. Fontanesi, C.W. Beattie, G.A. Rohrer, L. Alexander & L.B. Schook, 1998. Evolutionary divergence with microsatellites. *J. Mol. Evol.* 46:121–126.

Pejic, I., P. Ajmone-Marsan & M. Morgante, 1998. Comparative analysis of genetic similarity among maize inbred lines detected by RFLPs, RAPDs, SSRs and AFLPs. *Theor Appl Genet.* 97:1248–1255.

Perrin, A., C. Cetre-Sossah, B.Mathieu, T. Baldet, J.C. Delecolle & E. Albina, 2006. Phylogenetic analysis of Culicoides species from France based on nuclear ITS1-rDNA sequences. *Med. & Vet. Entomol.* 20: 219–228.

Pinto, A.S., M. de Lana & B. Bastrenta *et al.,* 1998. Compared vectorial transmissibility of pure and mixed clonal genotypes of *Trypanosoma cruzi* in *Triatoma Infestans. Parasitology Research* 84: 348–353.

Powell, W., M. Morgante, C. Andre, M. Hanafey, J. Vogel, S. Tingey & A. Rafalski, 1996. The comparison of RFLP, RAPD, AFLP and SSR (microsatellite) markers for germplasm analysis. *Mol. Breeding* 2: 225–238.

Prakash, S., 1969. Genic variation in a natural population of Drosophila persimilis. *Proc. Nat. AC/ld. Sci. USA* 62: 778-84.

Prakash, S., R.C. Lewontin & J.L. Hubby, 1969. A molecular approach to the study of genic heterozygosity in natural populations. IV. Patterns of genic variation in central, marginal and isolated populations of Drosophila pseudoobscura. *Genetics* 61: 841-858.

Prasad, M.D., M. Muthulakshmi & M. Madhu *et al.,* 2005. Survey and analysis of microsatellites in the silkworm, *Bombyx mori*frequency, distribution, mutations, marker potential and their conservation in heterologous species. *Genetics* 69:197–214.

Price, P.W. 1997. Insect ecology. Wiley, New York, pp. 874.

Promboon, A., T. Shimada, H. Fujiwara & M. Kobayashi, 1995. Linkage map of random amplified polymorphic DNAs (RAPDs) in the silkworm, *Bombyx mori. Genet. Res*. 66:1-7.

Ranson, H., B. Jensen and X. Wang *et al.* 2000.Genetic mapping of two loci affecting DDT resistance in the malaria vector Anopheles gambiae. *Insect Molecular Biology* 9: 499–507.

Rauch, N. & R. Nauen, 2003.Identification of biochemical markers linked to neonicotinoid cross resistance in Bemisia tabaci (Hemiptera: Aleyrodidae). *Archives of Insect Biochemistry and Physiology* 54:165–176.

Reiter, R.S., J.G.K. Williams, K.A. Feldmann, J.A. Rafalski, S.V. Tingley & P.A. Scolnik, 1992. Global and local genome mapping in Arabidopsis thaliana by using recombinant inbred lines and random amplified polymorphic DNAS. *Proc. Natl. Acad. USA* 89:1477-1481.

Robinson, G.E., C.M. Grozinger & C.W. Whitfield, 2005. Sociogenomics: social life in molecular terms. *Nature Reviews Genetics*. 6: 257–270.

Rosendahl, S. and & R. Sen, 1992. Isozyme analysis of mycorrhizal fungal and their mycorrhizas". (In: Methods in Microbiology: Techniques for the Study of Mycorrhiza; ed. A.K. Varma, D.J. Read & J.R. Norris). Academic Press, London. pp. 169-194

Saiki, R.K., D.H. Gelfand, S. Stoffel, S.J. Scharf, R. Higuchi, G.T. Horn, K.B. Mullis & H.A. Erlich, 1988. Primer-directed enzymatic amplification of DNA with a thermostable DNA polymerase. *Science* 239: 487–497.

Schama, R., A.M. Solé-Cava & J.P. Thorpe, 2005. Genetic divergence between east and west Atlantic populations of *Actinia* spp. sea anemones (Cnidaria: Actiniidae). *Mar. Biol. 146*: 435-443.

Schliitterer, C., M.T. Hauser, A.V. Haeseler & D. Tautz, 1994. Comparative Evolutionary Analysis of rDNA ITS Regions in *Drosophila. Mol. Biol. Evol.* 11(3):5 13-522.

Schmitz, J. & R.F.A. Moritz, 1998. Molecular phylogeny of Vespidae (Hymenoptera) and the evolution of sociality in wasps. Mol.Phylogenet. Evol. 9:183–91.

Schulenburg, J.H.G.V.D., J.M. Hancock, A. Pagnamenta, J.J. Loggett, M.E.N. Majerus & G.D.D. Hurst, 2001. Extreme Length and Length Variation in the First Ribosomal Internal Transcribed Spacer of Ladybird Beetles (Coleoptera: Coccinellidae). *Mol. Biol. Evol.* 18(4): 648–660.

Schultz, J., T. Muller, M. Achtziger, P.N. Seibel, T. Dandekar & M.Wolf, 2006. The internal transcribed spacer 2 database-a web server for (not only) low level phylogenetic analyses. *Nucleic Acids Research* 34:W704–W707.

Selig, C., M. Wolf, T. Muller, T. Dandekar & J. Schultz, 2008. The ITS2 Database II: homology modelling RNA structure for molecular systematic. *Nucleic Acids Research* 36:D377–D380.

Severson, D.W., A. Mori, V.A. Kassner & B.M. Christensen, 1995. Comparative genetic map for the mosquitoes *Aedes albopic tus* and *Aedes aegypti*, based on common RFLP loci. *Insect Mol. Biol.* 4: 41–45.

Severson, D.W., J.K. Meece, D.D. Lovin *et al.*, 2002. Linkage map organization of expressed sequence tags and sequence tagged sites in the mosquito, *Aedes aegypti*. *Insect Molecular Biology* 11:371–378.

Severson, D.W., S.E. Brown & D.L. Knudson, 2001.Genetic and physical mapping in mosquitoes: molecular approaches. *Annual Review of Entomology* 46:183–219.

Severson, D.W., V. Thathy, A. Mori *et al.*, 1995.Restriction fragment length polymorphism mapping of quantitative trait loci for malaria parasite susceptibility in the mosquito *Aedes aegypti*. *Genetics* 139: 1711–1717.

Shi, J., D.J. Heckel and M.R. Goldsmith, 1995. A genetic linkage map for the domesticated silkworm, *Bombyx mori*, based on restriction fragment length polymorphism. *Genet. Res.* 66:109–126.

Solignac, M., D. Vautrin & E.A. Baudry, 2004. Microsatellite-Based Linkage Map of the Honeybee, *Apis mellifera* L. *Genetics* 167: 253–262.

Solignac, M., D. Vautrin, A. Loiseau, F. Mougel, E. Baudry *et al.*, 2003. Five hundred and fifty microsatellites markers for the study of the honeybee genome (*Apis mellifera* L). *Mol. Ecol. Notes* 3: 307–311.

Sperling, F.A., G.S. Anderson & D.A. Hickey, 1994. A DNA-based approach to the insect species used for postmortem interval estimation. *J Forensic Sci.* 39:418–427.

Stevens, J. & R. Wall, 2001.Genetic relationships between blowflies (Calliphoridae) of forensic importance. *Forensic Sci Int.* 120: 116–123.

Stille, M. & B. Stille, 1992. Intra- and inter-nest variation in mitochondrial DNA in the polygynous ant *Leptothorax acervorum* (Hymenoptera; Formicidae). *Insect Soc.* 39:335–340.

Subramanian, S. & S. Mohankumar, 2006. Genetic *variability of the bollworm, Helicoverpa armigera,* occurring on different host plants. Journal of Insect *Science* 6:26.

Sudeep, A.B., R. Khushiramani & S.S. Athawale *et al.,* 2005. Characterization of a newly established potato tuber moth (*Phthorimaea operculella* Zeller) cell line. *Indian Journal of Medical Research* 121:159–163.

Takami, Y., C. Koshio, M. Ishii, H. Fujii, T. Hidaka & I. Shimizu, 2004. Genetic diversity and structure of urban populations of Pieris butterflies assessed using amplified fragment length polymorphism. *Mol Ecol.* 13: 245–58.

Tanksley, S.D., 1993. Mapping polygenes. *Annu. Rev. Genet.* 27: 205–233.

Tartarotti, E., & C.R. Ceron, 2005. Ribosomal DNA ITS-1 Intergenic Spacer Polymorphism in Triatomines. *Biochemical Genetics* 43(7/8): 365-373.

Tautz, D.M., M. Trick & G.A. Dover, 1996. Cryptic simplicity in DNA is a major source of genetic variation. *Nature* 322:652-656.

Thao, M.L., L. Baumann & P. Baumann, 2004. Organization of the mitochondrial genomes of whiteflies, aphids, and psyllids (Hemiptera, Sternorrhyncha). *BMC Evolutionary Biology* 4:25.

Toda, S., M.H. Osakabe & S. Komazaki, 2000. Interspecific diversity of mitochondrial COI sequences in Japanese Panonychus species (Acari: Tranychidae). *Experimental & Applied Acarology.* 24:821-829.

Via, S. & D.J. Hawthorne, 1998. The genetics of speciation: promises and prospects of quantitative trait locus mapping, pp.352–364. (In: Endless Forms: Species and Speciation; ed. D.J.Howard & S. H. Berlocher. Oxford University Press, NewYork.

Vincent, S., Vian, J.M. & M.P. Carlotti, 2000. Partial sequencing of the cytochrome oxydase b subunit gene I: A tool for the identification of European species of blow flies for postmortem interval estimation. *J. For. Sci.* 45 (4): 820–823.

Vogler, A.P. & D.L. Pearson, 1996. A molecular phylogeny of the tiger beetles (Cicindelidae): congruence of mitochondrial and nuclear rDNA data sets. *Mol. Phylogenet.Evol.* 6:321–38.

Vos, P., R. Hogers, M. Bleeker, M. Reijans, D.L.T. Van, M. Hornes, A. Frijters, J. Pot, J. Peleman, M. Kuiper & M. Zabeau, 1995. AFLP: a new technique for DNA fingerprinting. *Nucleic Acids Res.* 23:4407–4414.

Wallman, J.F. & S.C. Donnellan, 2001. The utility of mitochondrial DNA sequences for the identification of forensically important blowflies (Diptera: Calliphoridae) in southeastern Australia. *Forensic Sci. Int.* 120 (1–2):60-67.

Weising, K., H. Nybom, K. Wolff & W. Meyer, 1995. DNA Fingerprinting in Plants and Fungi. CRC Press, Inc., pp. 1-332.

Weising, K., J. Ramser, D. Kaemmer, G. Kahl & J.T. Epplen, 1991. Oligonucleotide fingerprinting in plants and fungi, In: T. Burke G. Dolf., A.L. Jeffreys and R. Wolff(eds) DNA fingerprinting approaches and application. Birkhauser Verlag, Basel/Switzerland,pp. 312-319.

Wells J.D. & Sperling F.A.H.. 2001. DNA-based identification of forensically impo-rtant Chrysomyinae (Diptera: Calliphoridae). *Forensic Sci. Int.* 120:110–15

Welsh, J. & M. McClelland, 1990. Fingerprinting genomes using PCR with arbitrary primers. *Nucleic acid Research* 18:7213-7218.

Whitfield, C.W., A.M. Cziko & G.E. Robinson, 2003.Gene expression profiles in the brain predict behavior in individual honeybees. *Science* 302:296–299.

Whitte, G.S., 1981. Evolution of isozyme loci and their differential regulation. In: Scudder, G.G.E., Reveal, J.L. (Eds.), Evolution Today: Proc. 2nd Int. Congress of Systematic and Evolutionary Biology. Hunt Inst. Bot. Documentation, Carnegie-Mellon Univ., Pittsburgh, pp. 271-289.

Wilkerson, R.C., T.J. Parsons, D.G. Albright, T.A. Klein & M.J. Braun, 1993. Random amplified polymorphic DNA (RAPD) markers readily distinguish cryptic mosquito species (Diptera: Culicidae: Anopheles). Insect Mol. Biol. 1:205–11.

Wilkerson, R.C., T.J. Parsons, T.A. Klein, T.V. Gaffigan, E. Bergo & J. Consolim, 1995. Diagnosis by random amplified polymorphic DNA polym-erase chain reaction of four cryptic species related to Anopheles (Nyssor-hynchus) albitarsis (Diptera: Culicidae) from Paraguay, Argentina, and Brazil (Diptera: Culicidae). *J. Med. Entomol.* 32:697–704.

Williams, J.G.K., A.R. Kubelik, K.J. Livak, J.A. Rafalski & S.V. Tingey, 1990. DNA polymorphisms amplified by arbitrary primers are useful as genetic markers. *Nucleic Acids Res.* 18:6531–6535.

Yasothornsrikul, S., S. Panyim & R.Rosenberg, 1988. Diagnostic restriction fragment patterns of DNA from the isomorphic species of Anopheles dirus. *Southeast Asian J. Trop. Med. Public Health*19: 703–8.

Yasukochi, Y., 1998. A dense genetic map of the silkworm, *Bombyx mori*, covering all chromosomes based on 1018 molecular markers. *Genetics* 150:1513–1525.

Zabeau, M. & P. Vos, 1993. Selective restriction fragment amplification: a general method for DNA fingerprinting. European Patent Application of 92402629 (Publ No. 0534859A1)

Zehner, R., J. Amendt, S. Schutt, J. Sauer, R. Krettek & D. Povolny, 2004. Genetic identification of forensically important flesh flies (Diptera: Sarcophagidae). *Int J Legal Med.* 118: 245–247.

Zhang, D.X. & G.M. Hewitt, 1996. Assessment of the universality and utility of a set of conserved mitochondrial COI primers in insects. Insect Molecular *Biology* 6 (2): 143-150.

Zhang, D.X. G.M.& Hewitt, 1997. Insect mitochondrial control region: a review of its structure, evolution and usefulness in evolutionary studies. Bioche-mical *Systematics and Ecology* 25: 99-120.

Zhang, X, F. C. Leung, D. K.O. Chan, Y. Chen & C. Wu, 2002. Comparative Analysis of Allozyme, Random Amplified Polymorphic DNA, and Microsatellite Polymorphism on Chinese Native Chickens. *Poultry Science* 81:1093–1098.

Zheng, L., M.Q. Benedict, A.J. Cornel, F.H. Collins & F.C. Kafatos, 1996. An integrated genetic map of the African human malaria vector mosquito, *Anopheles gambiae*. *Genetics* 143: 941–952.

Zhong, D., A. Pai & G. Yan, 2004. AFLP-Based Genetic Linkage Map for the Red Flour Beetle (Tribolium castaneum). *Journal of Heredity* 95 (1):53–61.

Zhou, X., M.E. Scharf, S. Parimi *et al.*, 2002. Diagnostic assays based on esterase-mediated resistance mechanisms in western corn rootworms (Coleoptera: Chrysomelidae). *Journal of Economic Entomology* 95:1261–1266.

Zietkiewcz, E., A. Rafalski & D. Labuda, 1994. Genome fingerprinting by simple sequence repeat (SSR) anchored PCR amplification. *Genomics* 20:176-183.

Chapter - 23

NEUROPTERA AS BIOCONTROL AGENTS IN INDIA: PRESENT STATUS AND FUTURE THRUSTS

N. Bakthavatsalam

National Bureau of Agriculturally Important Insects, Post Bag No. 2491, H.A.Farm Post, Bangalore – 560024, India

INTRODUCTION

Linnaeus (1758) erected the order Neuroptera which was represented by a large number of heterometabolic and holometabolic insects. These were later separated into several orders of their own, besides, true Neuroptera. The adult neuropterans are characterized by the two pairs of membranous wings with primitive type of venation comprising definitive accessory branches, many accessory marginal branches, and numerous cross veinlets. The larvae are distinguished by the suctorial jaws formed of mandibles and maxillae, generally very active and always carnivorous.

Popularly the neuropterans are called Dobson flies, Alder flies, Fish flies, Snake flies, Lacewings, Ant lions and Owl flies. The order Neuroptera is economically important as both the larvae and in many species, the adults, are predacious on pests of agricultural crops such as Aphidae, Coccidae and mites Some authors divided Neuroptera into three orders namely Megaloptera, Raphidioptera and Neuroptera under the super order Neuropterida while others split neuropterans into two suborders, Megaloptera and Planipennia. These two sub orders contain 7 superfamilies which include 20 families. Families such as Sialidae, Ithonidae, Polystoechotidae, Psychopsidae, Neurorthidae, Nymphidae and Stilbopterygidae were not represented in India (Ghosh, 1998). Worldwide around 6000 species have been recorded (New, 2001).

The most important families from the point of view of agricultural importance are the Chrysopidae and Hemerobiidae and to some extent the Coniopterygidae, Mantispidae and Myrmeleontidae. McEwen *et al.*, (2001), in their book on *Lacewings in the Crop Environment,* have given detailed account of the use of neuropterans in agricultural ecosystems, especially species belonging to the families Coniopterygidae,

Hemerobiidae and Chrysopidae. In India, the recent past has witnessed an intensified interest in the utilization of the Neuropterans, especially chrysopids and hemerobiids, in the suppression of the several pest species. In this review, present status of research on neuropterans and future strategies in India has been discussed in detail.

BIOSYSTEMATICS

Ghosh (1976a,b, 1977a-d, 1978, 1980, 1981a-c, 1982, 1983a,b, 1984a,b, 1985a,b, 1985a, 1987, 1988, 1990a-c,1990d, 1991a,b, 1997, 1998, 2000) has made significant contribution to the faunistic studies of Neuropterans in India.

The order Neuroptera is represented by 22 families in two subdivisions namely, Megaloptera and Planipennia (Ghosh, 2000). Megaloptera includes alder flies and snake flies and Planipennia includes the ant lions and the lace wings. The sub order Megaloptera has families like Sialidae, Corydalidae, Inocellidae and Raphidiidae of which only Corydalidae and Inocellidae were reported from India (Ghosh, 1998). The family Sialidae comprises insects commonly called alder flies. Species like *Sialis lutaria* Linnaeus were commonly reported from several regions of the World. However, none of the species were reported from India.

The species belonging to the Corydalidae are recorded as predators of dipterans. They are referred as dobson flies. *Protohermes grandis* Thurnberg is common species from the World. In India, 24 species under 7 genera were reported (Ghosh, 1998). Taxonomic studies were conducted on the species belonging to the genera *Neochauliodes* by Ghosh (1981d). Species such as *Neochauliodes khasianus* (Weele), *N. sinensis* (Walker), *N. indicus* (Weele), *N. simplex* (Walker), *N. obscurus* Weele, *Protohermes albipennis* (Walker), *P. anticus* (Walker), *P. arunachalensis* Ghosh, *P. montanus* (MacLachlan), *Hermes costastriata* Weele, *H. maculipennis* Gray, *H. selysi* Weele, *Acanthacorydalis asiatica* (Wood-Mason), *A. orientalis* (MacLachlan), *A. horrenda* Navas, *Corydalus territans* Needham, *Neoneuromus fenestralis* (MacLachlan), *N. fenestralis fenestralis* (MacLachlan), *N. latratus* (MacLachlan), *N. latratus latratus* (MacLachlan), *N. sikkimenis* (Weele) and *Neuromus decemmaculatus* (Walker) were the species recorded from different parts of India (Ghosh, 2000).

Family Inocelidae was represented by only one species *Inocellia (Inocellia) crassinervis* (Schummel) (Ghosh, 1998).

In India, the suborder Planipennia is represented by families like Coniopterygidae, Sisyridae, Hemerobiidae, Mantispidae, Chrysopidae, Nemopteridae, Myrmeleontidae, Dilaridae, Berothidae, Osmylidae and Ascalaphidae. Among these Chrysopidae and Hemerobiidae have been sufficiently represented in India.

Insects belonging to the family Coniopterygidae were referred as wax flies or dusty wings. Species like *Conwentzia psociformis* (Curtis) were recorded from several parts of the world. They usually predate on the armored scales and hoppers. In India 11 species belonging to 7 genera were reported (Ghosh, 1998). Ghosh and Maulik (1985) studied the taxonomy of Coniopterygidae. *Coniopteryx* (*Coniopteryx*) *ambigua* Withycombe, *C. exigua* Withycombe, *Conwentzia inverta* Withycombe, *Heteroconis terminalis* (Banks) and *Coniocompsa indica* Withycombe are some of the commonest species recorded from India (Ghosh, 2000). *Nimboa basipunctata* Withycombe was reported from India as predator of *Pyrilla perpusilla* Walker on sugarcane (Sharma *et al.,* 1985).

The mantispids look like mantids of Dictyoptera. The characteristics anterior legs of this family distinguishes them from all other families. The larvae are parasites of hymenopterans and spiders and have a complicated development known as hypermetamorphosis (Ghosh, 2000). Species like *Mantispa minuta* Fab., *M. viridis* Walter, *M. styriaca* (Poda) and *M. interrupta* Say were often recorded from several parts of the world. In India, 4 genera comprising 12 species were recorded (Ghosh 1998). Redborg (1998) gave detailed description of Mantispidae from the World.

In India other families like Nemopteridae (6 species in 4 genera), Berothidae (3 species in 3 genera), Osmylidae (14 species in 4 genera), Dilaridae (3 species in 3 genera) and Sisyridae (4 species in 2 genera) were also reported. However, no recent studies were available on insects belonging to the families in India.

Myrmeleontidae comprise the ant lions. Forty genera comprising 126 species were reported from India (Ghosh, 1998). Species belonging to the genera *Indophanes, Macroneumurus*, *Stenares*, and *Palpares* were described by Ghosh (1980b, 1982, 1987, 1990c, 1990d, 1991). *Myrmeleon mobilis* Hagen, *Euroleon nostras* Geoffrey and *Macroleon quinquem-aculatus* (Hagen) are some of the species recorded from different parts of the World. Adults of some species of Myrmeleontidae were found to be flower visitors on *Eriobotrya haponica* Thun. (Mann & Sagar, 1987). Bhatti & Bhatti (1988) studied the predatory habits of *Myrmeleon formicarius* Linnaeus. A species of Myrmeleontid was recorded as predator of tea aphid, *Toxoptera aurantii* Boyer (Das & Kakoty, 1992).

Hemerobiidae is represented by 22 species belonging to 10 genera in India. The adult Hemerobiids are generally dull brown in colour with lacy brownish wings. They have prominent eyes and in some genera the hind wings may be rudimentary. Hemerobiids have been recorded as predators on several crops and pests such as *Pentalonia nigronervosa* Coquerel (Rajan, 1981), other aphids (Johnson, 1983*), Aphis crassivora* Koch (Joshi *et al.,* 1997), and mites (Gupta, 2001).

The predominant species among the Hemerobiids are *Micromus timidus* Hagen and *Micromus igorotus* Banks. These were recorded from several pests (Patnaik *et al*, 1977; Agarwala *et al.,* 1983).

Chrysopidae comprise adults distinguishable by their lace green wings, prominent green eyes, and large left mandible and in some species the left mandible has a more prominent tooth like projection on the inner margin. The chrysopids have been extensively used in the suppression of several pest species world wide because of their bio diversity, amenability to laboratory rearing, and easy availability due to commercial production. Extensive studies have been made on chrysopids by several workers on the biology, rearing techniques, field release techniques, utilization in biorationale integrated pest management (BIPM) and conservation technologists. Canard *et al.* (1984) in their book on "*Biology of Chrysopidae*" have given the detailed account on chrysopids.

The first chrysopid to be known from Indian subcontinent was *Ankylopteryx octopunctata* Fabricius, 1798. Later, several workers have contributed to the faunistic studies of Chrysopidae. Species *Ankylopteryx tesselatus* Needham was recorded in 1909 by Needham (1909). Brauer (1864) recorded species like *Ankylopteryx anomala* Brauer.

Of the 67 species of reported from India (Singh & Narasimham, 1992), a few species are considered agriculturally important in India.

The common species recorded are *Brinckochrysa scelestes* (Banks) and *Chrysoperla carnea* (Stephens). *C. carnea* is the common species encountered in several agro ecosystems. The identification of this species in question as recent identification based on the vibrational song revealed this as *Chrysoperla zastrowi arabica* Henry *et al.* (Venkatesan *et al.,* 2008). Similarly the other species *Mallada boninensis* (Okamoto) has been identified as *M. desjardinsi* (Navás) and *M. astur* (Banks) as *Dichochrysa astur* (Banks) based on taxonomic characters and vibrational songs. However, for the practical purpose the old names have been retained in this article.

CRYPTIC SONGS FOR THE IDENTIFICATION OF CHRYSOPIDS SPECIES

Chrysoperla carnea species of green lacewings is globally distributed in North hemisphere but with a southern range extension through African continent to Cape region (Henry *et al.,* 2006). Its morphological homogeneity caused earlier entomologists to treat the complex as a single species *Chrysoperla carnea* (Stephens) (Tjeder, 1960). However, recent work has shown that *C. carnea* comprises many distinct species, each identifiable by its own substrate-borne, vibrational mating signal produced by courting males and females in a precise duetting format (Henry 1980, 2006). Venkatesan *et al.* (2008) proved that through the cryptic song analysis, the so called *C. carnea* is in fact *Chrysoperla zastrowi arabica* and not *C. carnea* as claimed earlier. Such cryptic song based taxonomic analysis will become a boon for identification of several other species of chrysopids.

MOLECULAR TOOLS FOR THE IDENTIFICATION OF NEUROPTERAN

DNA sequence data have been used recently to explore the genetic integrity of song defined species and inter phylogenetic relationships among them (Wells & Henry, 1998). Henry *et al.* (2001) described the importance of molecular systematics for the chrysopids. Molecular phylogenetic survey of chrysopids was performed with the nuclear genes, wingless (546 bp) phosphoenolpyruvate carboxylases (483bp) and sodium/ potassium ATPs alpha subunit (410bp) and based on these genes phylogenetics was worked out (Haruyama *et al.,* 2008). Some authors used a single nuclear and 2 mitochondrial genes: carbomoyl phosphate synthase (CPS), large subunit ribosomal gene (16S) and cytochrome oxidase genes were used for phylogenetic studies (Winterton *et al.,* 2006).

In India Venkatesan *et al.* (2008) studied the internal transcribed spacer (ITS1 and ITS2) region of the *C. zastrowi arabica* and identified 538bp fragment. Gautam *et al.* (2008) identified differences between *C. z. arabica* (*C. carnea*) and *M boninensis* in the ITS region with *C.z. arabica* having 1166bps and *M. boninensis* having 1142 bps.

HOST RECORDS OF COMMON SPECIES

The majority of the neuropterans have been recorded from the crop agro ecosystems, probably, because of their economic importance. The neuropterans collected from other ecosystems like forest, agro forestry are meager. Several species of chrysopids have been reported from both annual crops and plantation crops.

Species like *Apertochrysa* was recorded from coconut, mulberry, *Michelia champaka*, *Nerium* sp. (Anonymous, 1998a, 1999, 2000b). The most common chrysopid which was reported from different crop ecosystems is *B. scelestes* often quoted as *Chrysopa scelestes* Banks and *Chrysoperla carnea.* The latter species was often recorded on the crops like cotton, sunflower etc. Species like *M. boninensis, M. astur* and *Apertochrysa* were often recorded on several crop ecosystems. During the surveys the authors have come across one species of *Plesiochrysa* recorded from coconut ecosystem.

Species like *Ankylopteryx octopunctata* and *A. octopunctata candida* (Fabricius) were recorded from coconut feeding on *Stephanitis typicus* Distant and *Opisina arenosella* Walker respectively (Sathiamma *et al.,* 1998; Sathiamma *et al.,* 1985, Pillai & Nair, 1993). Four species of

Ankylopteryx were collected from *Nerium* sp., *Michaela champaka*, brinjal and coconut (Anonymous, 1998a,1999, 2000b). *Odontochrysopa lacciperda* (Kimmins) (*Chrysopa lacciperda* (Kimmins) is of common occurrence in some of the horticultural crops. Some of the species like *Apertochrysa* sp. were often recorded on coconut plantations while some like *Plesiochrysa* sp. was recorded from mango trees.

Mallada boninensis (Okamoto) is a predominant species identified from several crops such as cotton, mango, grapevine, tobacco, citrus, grapevine, mustard, *Michelia champaka* and mulberry. This species was recorded to feed on several species of insects such as *Aphis gossypii* Glover, *Amrasca biguttula biguttula* Ishida, *Empoasca* sp., *Myzus persicae* Sulzer, *Spodoptera litura* Fabricius, *Lipaphis erysimi* Kaltenbach, *Aleurocantus woglumi* Ashby, *Idioscopus* spp. *Planococcus citri* (Risso), *Amritodus atkinsoni* (Lethierry), *Ferisia virgata* Cockerell and *Liriomyza trifolii* Burgess. *Mallada boninensis* was found in abundance on citrus feeding on citrus black fly, Psylla, leaf miner, citrus aphids and mealy bugs (Shivankar and Singh, 2001). It preyed more on the larvae of citrus leaf miner reared on acid lime and Nagpur mandarin than those reared on rough lemon and Rangpu lime (Rao *et al.,* 2001). *M. astur* was recorded from *M. champaka*, guava, coconut, and palm feeding on aphids, scales, spiraling whitefly and mealy bugs.

Species such as *Chrysopa dasyptera* (Mc Lachlen), *C. himalayana* Ghosh, *C. murreensis* Tjeder, *Chrysopidia garhwalensis* Ghosh were also reported from *Cedrus deodara* feeding on *Lachnus* sp. and *Chromaphis hirutustibis* Kumur and Lavigne (Debnath *et al.,* 1988). *Brinckochrysa scelestes* was reported from *Cassia occidentalis*, apple, mesta, rice bean and chickpea feeding mostly on mealy bugs, *Pyrilla perpusilla* Walker, *Maconellicoccus hirsutus* Green and *Catopsilia pyranthe* (Linnaeus). *Mallada basalis* (Walker) *(Anisochrysa basalis* Walker*)* was reported to feed on *P. citri* on citrus plants. *O. lacciperda* (Kimmins) (*Plesiochrysa lacciperda* (Kimmins)) was reported to feed on hoppers on mealy bugs on mango.

Several species of unidentified chrysopids were recorded to feed on different species of aphids, mealy bugs and mites on different crops. The preference was observed even for host insect. Though the chrysopids are usually referred to as aphid lion, all the species of aphids were not preferred by the chrysopids. Generally, aphids which are larger in size and move fast or with more tubercles or with deterring secretion were not preferred by chrysopids. The host range of the important chrysopids is appended in Table 1.

Table 1.

Species	Prey	Host plant	Reference
Mallada boninensis (Okamoto)	*Lipaleyrodes euphorbiae* David and Subramaniam	Star goose berry	Mani and Krishnamoorthy, 1995
	Kanakarajiella cardamomi (David and Subramaniam)	Cardamom	Selvakumaran *et al.*, 1996
	Aphis gossypii Glover	Cotton	Jalali and Singh, 1994
	Amrasca biguttula biguttula Ishida	Several plants	Singh *et al.*, 1993
	Empoasca sp.		
	Peregrinus maidis Ashmead		
	Kolla ceylonica (Meliker)		
	Exitianus indicus (Distant)		
	Idioscopus clypealis Lethierry	Mango	Fasih and Srivastava, 1990
	Maconellicoccus hirsutus Green	Grape vine	Krishnamoorthy and Mani, 1989
	Bemisia tabaci Gennadius	Cotton	Joshi and Yadav, 1990
	Lipaphis erysimi (Kaltenbach)	Mustard	Kalra, 1988
	Aleurocanthus woglumi Ashby		Satpute *et al.*, 1986
	Myzus persicae Sulzer	Tobacco	Rao and Chandra, 1985
	Spodoptera litura Fabricius	Tobacco	Rao and Satyanarayana, 1984
	Greenidea (Trichosiphum) formosana heeri Raychaudhuri, Ghosh, Banerjee and Ghosh	*Psidium guajava*	Debnath *et al.*, 1988
	Idioscopus clypealis Lethierry; *I. nitidulus* (Walker); *Amritodus atkinsoni* (Lethierry)		

	Planoccus citri Risso; *Ferrisa virgata* Cockerell; *P. lilacinus* (Cockerell)	Vineyards	Mani and Krishnamoorthy, 1990
M. astur	*A. dispersus*	Guava	Mani and Krishnamoorthy, 2007.
Mallada sp.	*Ferrisia virgata* Cockerell	Coffee	Balakrishnan *et al.*, 1991
Anisochrysa obvia Hotzel	*Rhopalosiphum* sp.	Grass	Debnath *et al.*, 1988
Chrysopa dasyphlebia (Mclachlan)	*Betacallis sikkimensis* Basu, Ghosh and Raychaudhuri	*Betula alnoides*	Debnath *et al.*, 1988
Chrysopa himalayana Ghosh	*Chromaphis hirssiibis*	*Juglans regia*	Debnath *et al.*, 1988
Chrysopa murreensis Tjeder	*Lachnus* sp.	*Cedrus deodara*	Debnath *et al.*, 1988
Chrysopidia garhwalesnsis Ghosh	*Lachnus* sp.	*Cedrus deodara*	Debnath *et al.*, 1988
Cunctochrysa jubingensis Holzel	*Aphis kurosawai* *Brevicoryne brassicae* *Chaitophorus kapuri* *Greenidea* sp.	*Artemisa vulgaris* *Brassia campestris* *Poulus ciliata* *Psidium guajava*	Debnath *et al.*, 1988
Ankylopteryx octopunctata	*Opisina arenosella*	Coconut	Pillai and Nair, 1993
Unidentified Chrysopid	*Macrosiphum rosaeiformis* Das	Rose	Kakar and Sood, 1989
	Sitobion avenae Fab.	Wheat	Bhagat *et al.*, 1990
	H. armigera		Sinivas and Jayaraj, 1898
	Heteropsylla cubana Crawford	*Leucaenia leucocephala*	Gopalan *et al.*, 1988
	Thrips	Pearl millet	Bhagat and Makhmoor, 1997
	Brevicoryne brassicae	Cauliflower	Devjani *et al.*, 1997
	Lipaphis erysimi (Kaltenbach)	Mustard	Singh and Singh, 1994
Chrysopa sp.	*Aphis craccivora* Koch	Chick pea	Mehto *et al.*, 1986

	Thrips tabaci Lind	Onion	Saxena, 1981
Chrysoperla carnea	*Lipaphis erysimi*	Mustard	Kalra 1988; Akhtar *et al.*, 2006
	Hyadahis coriandri Das	Coriander	Sagar and Kumar, 1996
	F. virgata	Coriander	
	P. citri	Citrus	Gautam, 1994 Krishnamoorthy and Mani 1989
	B. tabaci	Cotton	Rao *et al.*, 1989
	Lipaphis erysimi (Kaltenbach)	Mustard	Rao *et al.*, 1988
	A. craccivora	Groundnut	Patel and Vyas (1984) Patel *et al.*, 1976
	H. peltigera (Schiff)	*Acanthospermum hispidum*	Manjunath *et al.*, 1976
	Bemisia tabaci	Cotton	Rao *et al.*, 1989
	Aphids	Pigeon pea	Sharma *et al.*, 2007
	Aleurodicus disperses		Aishwarya *et al.*, 2007
	Sucking insects	Linseed	Patel and Thakur, 2005
	Thrips tabaci Lind	Onion	Kadri and Goud, 2005
	Nausinoe geometralis	Jasmine	Rajabaskar, 2007
	Trialeurodes vaporarioum		Kumar and Gupta, 2006).
	Unidentified	Linseed	Patel and Thakur 2005
Brinckochrysa scelestis (Banks)	*Catopsilia pyranthe*	*Cassia occidentalis*	Patel and Yadav, 1991
	Eriosoma lanigerum	Apple	Thakur and Pawar, 1986; Thakur *et al.*, 1988
	Mealy bug	Mesta	Rao *et al.*, 1984
		Rice bean	Stayanarayana *et al.*, 1995
	H. armigera	Chick pea	Thakur *et al.*, 1995
	Macconellicoccus hirsutus	Mesta	Rao *et al.*, 1984.
	Pyrilla perpusilla	Sugarcane	Asre *et al.*, 1983.

Mallada basalis (*Anisochrysa basalis*)	*P. citri*	Citrus	Krishnamoorthy and Mani, 1989
Apertochrysa	*M. hirsutus*	Vineyard	Krishnamoorthy and Mani, 1987
Plesiochrysa lacciperda	*Idioscopus clyeaplis; I. nitidulus; Adisora atkinsoni; P. citri*	Mango	Fasih and Sriva-stava, 1990; Krishna-moorthy and Mani, 1989; Mani and Kris-hnamoorthy, 1990a
Chrysopa sp.	*Heteropsylla cubana* Crawford	Eucaena	Sen and Verma, 1998
	Ceratovacuna lanigera Zehnt	Sugarcane	Tripathi (1995)
	Aphis craccivora	Chick pea	Mehto *et al.*, 1989
Chrysopa sp.	Mango gall midge	Mango	Grover, 1986.
Chrysopa sp.	*Ureleucon carthamimi*	Safflower	Rathore, 1983
Chrysopa sp.	Aphids and eggs of borers	Maize	Pawar *et al.*, 1985.
Chrysopa sp.	*Spodopotera litura*	Tobacco	Thontadarya and Nangia, 1983
Chrysopa sp.	*Maconellicoccus hirsutus*	Grapevine	Manjunath, 1985.
Chrysopa sp.	*Oligonychus indicus* Hirst; *Raoiella indica* Hirst; *Tetranychus fijiensis* Hirst	Arecanut	Daniel, 1979
	Myzus persicae	Tobacco	Rao *et al.*, 1981
Chrysopa orestes	*Aphis gossypii*	Brinjal	Bhagat and Masood, 1986
	Pyrilla perpusilla	Sugarcane	Dhaliwal and Bains, 1983.
	Aphis craccivora *Lipaphis erysimi*	Leguminous and Crucifer plants	Ghosh *et al.*, 1981.
Micromus igoratus	*Ceratovacuna lanigera*	Sugarcane	Chkravarthy and Thiagarajan, 2005, Patil and Bhosale, 2007)
Micromus australis Hagen	*Melanaspis sacchari* (Zehntner)	Sorghum	Vidya *et al.*, 2008 c:2008 d

BIOLOGY OF NEUROPTERANS

The biology of several species of chrysopids and a few species of hemerobiid was studied in detail. However, species belonging to other families were seldom studied.

The eggs of chrysopids were laid on the peduncle of the leaf, buds, stem and even fruiting bodies. There are 3 larval instars and one pupal instar, usually in a cocoon. Males and females can be easily distinguished by their genitalia. Females are generally bigger than males. There is no parthenogenesis reported from the chrysopids. Mating takes place on the 2 day of adult emergence. The oviposition starts usually on the 5 th day of emergence or 3 day of mating, which however, varies with the species. Striking differences were noticed on the egg laying patterns of different species of chrysopids. Generally most of the species lay the eggs singly. Species like *Apertochrysa* laid the eggs in groups that varied in numbers from 9-23 eggs (Bakthavatsalam *et al.,* 1996). Stalks of the eggs bundled together was observed in an unidentified species. The length of egg stalk was more in species where the eggs are laid in groups compared to eggs laid singly. The egg stalk of *Apertochrysa* sp. was very long compared to *C. carnea.* There are differences in the length of the egg stalk and posture of the eggs. Even though the egg stalk of *C. carnea* and *M. astur* looks alike, the eggs of *M. astur* remains erect while that of the *C. carnea* bends.

The larva of *C. carnea* is longer than broader while that of *M. astur* broader than longer. The trash carrying capacity also appears to be species specific. The larvae of *C. carnea* seldom carries trash while *M. astur, M. boninensis* and *Apertochrysa* always carries the trash on their back.

The variation between the species in terms of larval and pupal duration is drastic. Generally, species which has shorter life duration and low mortally rate are considered amenable to laboratory rearing. Accordingly *C. carnea* and *M. boninensis* were considered to be highly amenable for laboratory rearing and field releases (Bakthavatsalam *et al.,* 1994,1996). Most of the species could complete their life cycle in the laboratory conditions. However species like *Plesiochrysa, Ankylopteryx* could not be reared in the laboratory on the laboratory hosts.

A. octopunctata candida larvae consumed the eggs and larvae of *O. arenosella*. The larval period was from 9-11 days and pupal period from 10-12 days (Sathiamma *et al.,* 1985). However, attempts to rear this predator in the laboratory using the eggs of *C. cephalonica* were not successful. Though the adults could be obtained, no egg laying was noticed (*author's unpublished data*).

In the case of *Odontochrysa lacciperda* (*Chrysopa laccicperda)* the developmental period was 22.52 days with 10-15 days old nymphs and

21.27 days on the eggs, and the larvae consumed a total of 747 nymphs when fed on *P. citri* (Krishnamoorthy, 1988a).

The biology of *B. scelestis* (often referred to as *Chrysopa scelestes*) was studied extensively by several workers (Rahman, 1940, Nasir, 1947, Brooks & Barnard, 1990; Patel & Vyas, 1985a). The adults preferred to lay the eggs on the lower surface of the leaves followed by branches (Patel & Vyas, 1985a). A single larva consumed 665.4 eggs or 410 neonate larvae of *Helicovperpa armigera* (Huebner) and the larval duration was 8.60 day when fed on eggs and 11.70 days when fed on larvae of *H. armigera* (Krishnamoorthy & Mani, 1982). On the eggs of *C. cephalonica* the larval, pupal, adult duration were 6.77. 613 and 23.42 days respectively (Patel & Vyas, 1985b). Krishnamoorthy (1988b) studied the biology of *C. scelestes* on hosts like *H. armigera*, *Earias* spp., *S. litura*, *O. arenosella*, *Crocidolomia binotalis*, *Plutella xylostella*, *Phthorimaea operculellla*, *M. persicae* and *Aphis gossypii*. The fecundity was 59.30 ± 14.72 eggs per female and larvae consumed around 169.9 nymphs of *P. perpusilla* (Mishra *et al.*, 1996a).

Chrysoperla carnea is another species whose biology has been studied in detail by several workers. The egg, larval and pupal periods are generally 4-5. 8-10, 8-10 days, respectively. However, variation was observed between different strains/ecotypes from different geographical locations, ecosystems and the prey (Varma & Shenhmar, 1983; Krishnaswamy & Yaseen, 1972; Bakthavatsalam *et al.*, 1994, 1996, Mani & Krishnamoorthy, 1989, 1990b; Balakrishnan *et al.*, 2005; Patil & Hallolli, 2007).

The life cycle was completed in 28.2 days on eggs and 25.6 days on nymphs of *Heteropsylla cubana* Crawford with a fecundity of 52 eggs (Jalali & Singh, 1992). Development was faster on aphids than aleyrodids (Kapadia & Puri, 1992). Balasubrmani & Swamiappan (1994) and Khulbe *et al.* (2005) observed that the development was rapid on the eggs of *C. cephalonica*, while prolonged on larvae of *H. armigera*. Each larvae consumed 662.53 eggs of *H. armigera,* 419.18 *Aphis gossypii*, 409.55 larvae of *H. armigera*, 329.7 pupae of *Bemisia tabaci* Gennadius and 288.45 nymphs of *A. biguttulla*. The net reproductive rate was greatest in *C. carnea* (558.78) followed by *M. boninensis* (191.71) (Bakthavatsalam, *et al.,* 1994; Geethalakshmi et al, 2000). The duration was shortest in the case of *C. carnea* and longest in the case of *Apertochrysa* sp. The aphids were consumed lesser by *C. carnea*. Based on these studies, *C. carnea* and *M. boninensis* were recommended for mass rearing and field release. (Bakthavatsalam *et al.,* 1996). Around 4.8, 20.6 and 59.6 nymphs of *A. biguttula biguttula* were consumed by 1.,2 and 3 instars respectively (Balasubramani & Swamiappan, 1998a). On *Tetranychus neocaledonicus* Andre, a single larvae of *C. carnea* consumed 10968, 3793 and 2889 eggs, nymphs and adults respectively (Sharana-basava *et al.,* 1998; Sharanabasava & Manjunath, 1998). The feeding potential of *C. carnea* was studied on *A. gossypii* (Thite & Shivpuje, 1999). Larvae of *C. carnea* consumed 3783.73 eggs or 728.52 nymphs or 96.39 adults of *P. citri* respectively (Malleshaiah *et al.,* 2000). The larval duration was shorter on the eggs of *C. cephalonica*, followed by nymphs of *Uroleucon compositae* Theobald, *A. gossypii* and eggs and neonates of *H. armigera* (Bansod and Sarode, 2000). The total life cycle was completed in 19.36 ± 1.18 days, the per day consumption was 442 ± 52 aphids during its development (Patro & Behera, 2002). No variation in the biology of *C. carnea* fed on aphids reared on Bt cotton (Manju *et al.,* 2007). Higher larval period and longer developmental period of *C. carnea* was noticed when fed on *Phenacoccus solenopsis* (Kaur *et al.,* 2008). *C. carnea* consumed 540.17, 463.29 and 449.30 individuals of *L. erysimi, Myzus persicae and Brevicoryne brassicae*, respectively on cabbage (Mandal & Patnaik, 2006).

Mehra (1965, 1966) and Madan *et al.* (1988) studied the biology of *Mallada madestes* (Banks) and *Plesiochrysa lacciperda* (Kimmins). Madan *et al.* (1988) studied the biology of *Mallada* spp. *Apertochrysa parla* was recorded as a predator on *Eurema blanda siletana* (Wallace) which is a pest on *Lathyrus sativus* (Pophaly & Gupta, 1996) and on *Maconellicoccus hirsutus* on grape vine (Krishnamoorthy & Mani, 1989).

Though the adults of *M. boninensis* resemble the adults of *C. carnea*, the larvae are short and broader than long and are essentially trash carriers. The biology and feeding potential was studied on *Aphis gossypii*, eggs of *H. armigera* and *C. cephalonica* (Bakthavatsalam *et al.*, 1996). The net reproductive rate was 191.71 (Bakthavatsalam *et al.,* 1994). Mani & Krishnamoorthy (1990) observed that *M. boninensis* preyed upon *F. virgata*, *P. citri* and *P. lilacinus* and the type of prey did not influence the biology. The larval duration was 12 days and prepupal and pupal periods 5-7 and 13.79 days, respectively, when fed on *B. tabaci* and larva consumed 453 nymphs of *B. tabaci* (Joshi & Yadav, 1990). Mani & Krishnamoorthy (1989) observed that the development of *M. boninensis* was completed in 25.80 days and 237.90 nymphs were consumed by a single larvae on *Maconellicoccus hirsutus*. The biology of *M. boninensis* was studied on *Aleurocanthus woglumi* Ashby (Lavhe, 2000; Lavhe & Nimbalkar 2000; Ramkumar *et al.,* 2005a; 2005b). The fecundity and net reproductive rate of *M. boninensis* was more in larvae of *Tribolium castenum* Herbst than the eggs of *C. cephalonica* (Elsiddiq *et al.,* 2006).

Mallada astur is quite distinct from *M. boninensis* in taxonomic characters. This species was recorded from several crops like guava, mango, coffee (Singh & Narasimham, 1994). The prey included leaf-hoppers, spiralling whitefly, (*Aleurodicus dispersus* Russel) (Mani & Krishnamoorthy, 1999), *Chlorpulivanaria polygonata* (Ckll.) (Mani and Krishnamoorthy, 1998). *M. astur* could complete development faster at densities 25 to 100 nymphs per day (Jalali & Singh, 1994).

Patnaik & Bhagat (1989b) studied the biology of *Chrysopa orestes* Banks on *Melanispis sacchari* (Zehntner) ((*Longiunguis sacchari* (Zehntner)), *Ferrisia virgata*, *Thrips tabaci* Lindemann and *Tetranychus urticae* Koch. Around 86.33 eggs were laid by a single female and larvae consumed 25.5 aphids/day.

The biology of some unidentified chrysopids was also studied by several workers (Dhaliwal & Bains, 1983). Among the hemerobiids, the biology of *Micromus igorotus* was studied in detail (Magar *et al.*, 2007; Chakravarthy & Thyagaraj, 2005; Chakrabarthy *et al.*, 2006). Vidya *et al.* (2009a) studied the biology of *M. igorotus* on *Ceratovacuna lanigera* Zehntner and the net reproductive rate of *M. igorotus* was 856.05 with a fecundity of 1712 eggs per female on *C. lanigera* (Singh *et al.*, 2009b). No competitive interaction was noticed between *M. igorotus*, *Dipha aphidivora* (Meyrick) and *Macrosyrphus confrator* Wiedmann in both no choice and choice condition (Vidya *et al.*, 2009b). *M. igorotus* consumed 1.95 4.55. 6.80, 10,65 and 9.40 aphids at unstarved and 2.35, 4.93, 7.15, 11.50 and 10.23 first, 2nd, 3rd, 4th instars and adults of *C. lanigera* respectively (Singh, 2008). The cumulative predatory potential of single larvae of *M. igorotus* during life (approximately 25-33 days) is 260-400 (Rajendran, 2005). The predatory efficiency of *M. timidus* was studied on aphid *Cervaphis quercus* Takahashi (Shantibala *et al.*, 1994). *M. timidus* was recorded as a predator on *Myzus persicae* Sulzer (Singh *et al.*, 1994) and *Cydia leucostoma* (Meyrick) (Selvasundaram & Muraleedharan, 1987). Raychaudhari *et al.* (1981) studied the biology of *M. timidus* species on *Lipaphis erysimi* (Kalt.) and *A. craccivora*. The larval and pupal duration ranged from 7-8 and 6-7 days respectively. The 1, 2, and 3 instars consumed 10-16, 31-40 and 99-123 respectively and adults fed 27-30 aphids. Radhakrishnan & Muraleedharan (1989) observed the total duration was 19 ± 0,54 days and each predator consumed 309 ± 49.14 *Toxoptera aurantii* Boyer de Fonsolome. Biology of another species *M australis* Hagen was studied on the sorghum aphid *(Melanaspis sacchari* (Vidya *et al.*, 2008a*)*. Biology of an unidentified species of Hemerobiid on *Hysteraoneara setariae* (Thomas) was studied by Jagadish *et al.* (1996).

STRAIN IMPROVEMENT

Emphasis was laid on selecting and utilizing best strains of neuropterans for biological control programs. Biotypes of *C. carnea* from Bangalore, Anand, Rajamundry and New Delhi did not show marked difference in their fecundity. Similarly the hybridization of these strains did not improve their fecundity (Singh *et al.*, 1994a). Variation in oviposition, longevity, predatory potential and other biological parameters was noticed in the different strains of *C. carnea* (Kumar *et al.*, 2007a,b, 2008).

Efforts were made to develop strains with definite traits and increased levels of insecticide resistance. The developed strain showed manifold levels of resistance to monocrotophos than other strains (Patel & Yadav, 1996). These strains also showed resistance to other organophophorus compounds like dimethoate, acephate, phosphomidon and methyl demeton (Patel & Yadav, 1997). The resistant strain showed more esteryl esters and free fatty acids, and the activity of the acetylcholine esterase was double in the resistant strains than the susceptible strain (Patel *et al.,* 1996).

All the strains collected from different regions developed resistance to insecticides in 3-4 generations after exposure to pesticides, however, withdrawal of insecticides brought back the resistance level drastically (Bakthavatsalam *et al.,* 1997b; Singh *et al.,* 1994a). *C. carnea* exposed to insecticides developed resistance to endosulfan, chrlorpyrifos to many folds at 11-12 generation of exposure (Maurya *et al.,* 2009). The strains which showed resistance to organophosphorous compounds also developed cross resistance to synthetic pyrethroids on continuous exposure to the pesticides (Anonymous, 2000a). Populations of *C. zastrowi arabica* collected from geographical areas with higher sprays of insecticides, showed more LC_{50} than other populations and populations of *C.z. arabica* from cotton recorded more resistance than other crops probably due to large quantity of pesticide sprays (Venkatasan *et al.,* 2009). However, the inheritance to pesticide resistance was through the recessive multiple genes as reported in other species of chrysopids. Therefore, there are greater chances of losing the resistance in mating with the susceptible strain. The successive generations in the field will eventually have least resistance due to inter breeding with the field populations.

REARING OF NEUROPTERANS

Emphasis has been laid in rearing the chrysopids and to some extend on hemerobiids because of their economic importance. Rarely attempts were made to rear the other groups of neuropterans. Singh *et al.* (1994b) and Jalali *et al.* (2003). provided detailed rearing procedures for the chrysopids which are widely practiced even in the insectaries. Depending on the facilities, demand and requirements, commercial insectaries have been successfully rearing *C. carnea* with slight modifications. The adults were reared on the adult diet and the larvae on the eggs of the laboratory hosts such as *Corcyra cephalonica*. Species like *C. carnea* and *M. boninensis* are highly amenable to laboratory rearing. However some species of chrysopids like *Ankylopteryx*, *Plesiochrysa* could not be reared in the present rearing methods. In group rearing, at higher feeding rates of 1300 eggs per larva, 150-200 larvae produced around 80% pupation (Patel *et al.,* 1988), Krishnamoorthy and Nagarkatti (1983) refined the rearing procedures by using honey comb cells made of paper. The larvae were reared individually in each cells and were fed with the frozen eggs of *C. cephalonica*. Group rearing of with wood shaving recorded higher pupation (91.33%) (Hegde & Kulkarni, 2000a). However, isolation rearing is more economical and cost effective than group rearing for the product-ion of *C. carnea* (Kumar *et al.,* 2007c). Singh *et al.* (1994b) gave a detailed procedure for rearing the chrysopids. The infrastructure includes a room of 6 x 6' with temperature and RH maintained at 27 ± 1°C and 70% respectively.

Adults numbering 200 pairs were kept in an oviposition cage, measuring 75 x 30 x 15 cm. The sides of the cage are lined with smooth nylon wiremesh (not preferred for oviposition) and the sliding top cover is fitted with black cloth for obtaining eggs. To prevent damage to the eggs the top is slid over a comb fitted on both the sides of the cage. The sliding cover is replaced on alternate days starting from 4 th day onwards. The oviposition cage was maintained for 30 days and the dead adults were removed every alternate day. The adults in oviposition cage were fed on alternate days on swabs (kept in plastic plates) of drinking water, 50% honey, protinex mixture (equal quantity in volume of protinex@ + fructose + honey + powdered yeast dissolved in small quantity of water) and castor pollen.

One day old eggs were dislodged from the black cloth top cover of oviposition cage by gently working with a piece of sponge. Eggs collected can be used for further multiplication or for field release. In the first step of larval rearing, three day old chrysopid eggs numbering 120 are mixed with 0.75 ml of UV radiated *Corcyra* eggs and kept in a flat container with paper fringes. The larvae on emergence start feeding the eggs and the larvae, on the third day they were transferred to cubical cells of 2.5^3 cm in louvers and reared individually. Another 3.5 ml of *C. cephalonica* eggs may be needed for feeding 100 larvae in the louvers which may be given either as a single feed or as two feed in divided doses. Louver is secured on one side by brown paper sheet and after transfer the louver was covered with acrylic sheet and clamped. Brown paper is used for clear visibility *of C. cephalonica* eggs and for pupal formation. Cocoons were collected after 24 hours of formation by removing the brown paper sheets. The cocoons are used for further multiplication. Around 4.25 ml of *C. cephalonica* eggs were required for feeding 100 larvae of *C. carnea* during their whole life time (Bakthavatsalam *et al.,* 1995). Higher quantity of the eggs (around 6 ml) is required for culturing *M. boninensis*. The eggs may be stored conveniently at 10°C for 15 days without any deterioration in the quality of hatching (Bakthavatsalam *et al.,* 1997a) or for 7 days at 4-5°C (Kaur *et al.,* 2001).

Rearing of the adult chrysopids was attempted using several diet combinations. Patnaik and Bhagat (1984) reared the adults of *C. orestes* on a diet consisting of lactalbumin, yeast extract and cane sugar in the ratio of 5:6:10. The diet with 40% honey and the castor pollen was found to record highest fecundity followed by protinex mixture with fructose and castor pollen. The protinex diet is still followed for culturing the adults (Krishnamoorthy, 1984). Kulkarni (1998) described the mass rearing procedures for chrysopids and Mohana *et al.* (1998) suggested improved larval rearing methods for *C. carnea.*

Geetha *et al.* (1998) recorded highest fecundity of *C. carnea* when fed on Baker's yeast hydrolysate diet (389.80 eggs) followed by caesin hydrolysate and protinex. However, similar fecundity was observed even with the protinex. The advantage of protinex is that it is readily available and does not require any fermentation process or hydrolyzing process. Irrespective of the diets used, the pollen is considered to be main factor in increasing the fecundity of the chrysopids.

Several attempts were made to substitute the protinex with powders from the pulses. The powdered blackgram when mixed with other ingredients like honey, sugar and yeast produced same fecundity such as that of the protinex (Singh *et al.,* 1994a). However, among all the ingredients tested in the adult diet, honey is the costliest. It remains impossible to substitute honey as it also acts as phagostimulant apart from being best nutritional source. Deotale *et al.* (1998) observed that among the pollen grains from 5 species of plants, castor pollen recorded highest fecundity by *C. carnea*. Isolated rearing resulted in the production higher cocoons compared to group rearing of *C. carnea* (Kumar *et al.,* 2007). Kumar and Gautam (2007) recorded that feeding the adults with *Hibiscus rosasinensis* increase the biotic potential of *C. carnea C. carnea* could be successfully reared under fluctuating temperature (11.3-42.1°C) for 12 and 9 generations during the 2 years at out door conditions, respectively (Devi & Yadav, 2005). Viji and Gautam (2005) attempted to rear the larvae of *C. carnea* on the larvae of *Tribolium castaneum* and larvae of *Drosophila melanogaster* in comparison with the eggs of *C. cephalonica* and observed that the cost of production of 1000 eggs was Rs. 3.68, 7.06 and 7.16 respectively on *T. castaneum*, *C. cephalonica* and *D. melanogaster*.

Gautam (1994 & 1996) suggested that group rearing of 5-10 numbers of neonate larvae of *C. carnea* can be reared with 1.5 to 2.50 cc of eggs of *C. cephalonica* as prey in 1.5 x 6.5 cm tubes produced around 64-76% pupation. Axenic rearing of larvae on the frozen eggs of *C. cephalonica* over 20 generations recorded reduction in fecundity and acceptability of aphid prey. Saminathan *et al.* (1999) suggested frequent rearing of larvae of *C. carnea* on *A. gossypii* to rejuvenate the fecundity *A. gossypii* is the suitable host for rearing *M. igorotus* and cotton and thread were good oviposition substrate (Vidya *et al.,* 2007c).

REARING IN SYNTHETIC DIETS

The larvae of chrysopids are generally reared on the eggs of *C. cephalonica*. The production cost of 1 cc of eggs of *C. cephalonica* varies between Rs 45 and Rs. 55. Considering that around 4.25 cc of eggs are required for producing 80 adults, the cost of producing 1 adult will be around Rs 2.35. Besides, the production of *C. cephalonica* is seriously affected by factors like rearing conditions, demand, parasitization by the parasitoids (especially *Bracon brevicornis*) etc. Therefore, the need of developing suitable alternate food instead of depending on the laboratory host, *C. cephalonica* or *Sitotroga cerelella* (Oliv.) had been felt by researchers over several years. The suitable larval diets were successfully developed and tested in other parts of the world by several workers (Hassen and Hagen, 1978; Vanderzant, 1969; Cohen & Smith, 1998).

Gautam & Paul (1987) reared the larvae of *B. sceletes* for 3 generations on synthetic diet and found that they are comparable to the larvae reared on the *Corcyra* eggs. Singh & Varma (1989) observed that the spent and dead adults did not provide adequate nourishment for the larvae of *C. carnea*.

Initial attempts were made in India on the development of suitable using the *Corcyra* eggs as a paste along with some ingredients. Though some success could be achieved, it still partially depended on the host culture. Attempts were made to develop synthetic diets at Project Directorate of Biological Control since early 1990s. Diets were developed earlier using protein hydrolysates, yeast flakes, yeast hydrolysate, and honey encapsulated with paraffin wax (of low boiling point) vaseline mixture. However, the number of larvae surviving and the percentage of adult emergence was low. An improved diet consisting of hydrolyzed soybean powder (1.3%) + egg yolk (32.3%)+ honey 16.1%) + yeast extract (1.3%) + water (38.7%) + petroleum jelly (0.7%) + paraffin wax (9.6%) was the best with higher pupation (84.5%) and adult emergence (71%) and fecundity and the fitness of larvae reared from synthetic dieted was on par with the larvae reared on natural host (*C. cephalonica*) (Venkatesan *et al.*, 2000). This diet with slight modifications was well suited for allied species *Mallada astur* (Venkatesan *et al.*, 2002). The shelf life of the diet was 200 days at 5°C (Anonymous, 2002a). Pushpalatha *et al.* (1994) observed that spent adults of *S. litura* mixed with diet ingredients were found to result in higher adult emergence.

Large scale production of chrysopids using mechanized methods will definitely improve the production using only synthetic diet will definitely improve the production capacity of the rearing units.

BIOINTENSIVE IPM

The role of neuropterans as efficient predators in agricultural systems is well known. Several studies were conducted to demonstrate the successful use of chrysopids and hemerobiids for the management of several pests. Earlier, it was reported that the chrysopids were killed for mistaken identity (Yadav *et al.*, 1983). Chrysopids are being released as larvae in the field, for obvious reasons to avoid egg parasitization. Usually larvae were released manually mixed in saw dust or with paper fringes. Srivastava *et al.* (1996) devised an applicator for the release of chrysopids. The dosage of release, however varies. Patel *et al.* (1976) noticed control of *A. craccivora* on ground nut by predators *C. carnea* and *Cheilomenes sexmaculata* (Fabricius). Asre *et al.* (1983) concluded that the release of *B. scelestes* did not reduce the population of *Pyrilla perpusilla* on sugarcane. *B. scelestes* was found to effectively control pests such as *Myzus persicae*, *Bemisia tabaci*, apple wooly aphid (Rao & Chandra, 1985; Thankur *et al.,* 1988; Dhandapani *et al.,* 1992). Periodic release of *B. scelestes* along with other parasitoids such as *Trichogramma chilonis* Ishida, *Tetrastichus howardii* (Olliff) and *Telenomus remus* Nixon decreased the population of *Spodoptera litura* on potato (Ansari *et al.,* 1992). Mani & Krishnamoorthy (1990a) observed that the popula-tion of *Planococcus citri* and *Ferrisia virgata* was suppressed by *Plesioc-hrysa lacciperda* in combination with *Spalgis epeus* (Westwood) and *Cryptolaemus montrouzieri* Mulsant.

Release of *C. carnea* at varying doses was found to effectively reduce the population of the pests such as *H. armigera* on cotton; chick pea, cardamom, safflower, sunflower for various pests (Patel *et al.,* 1976; Manjunath, 1993; Bharpoda *et al.,* 2000; Rabindra *et al.,* 1994; Hegde & Lingappa, 1998; Mathew *et al.,* 1998; Venkatesan *et al.,* 1997; Badgujar *et al.,* 2000, Rajaram *et al.,* 2006). Santharam *et al.,* (1994) concluded that release of *C. carnea* along with sprays of *Bacillus thuringiensis* and NPV was as effective as endosulfan alone in reducing populations of *H. armigera* on pigeon pea. The *C. carnea* released at 1:10 ratio achieved good control over the population of *Pentalonia nigronervosa* (Coquerel) on cardamom (Mathew *et al.,* 1998). Pod borer was reduced on the pigeon pea when the larvae of *C. carnea* was released @ 1,25,000 per hectare (Hegde & Lingappa, 1998).

Release of *C. carnea* followed by spray of monocrotophos (0.04%) or endosulphan (0.07%) greatly reduced the incidence of castor mite *Tetranychus dellaris* on castor (Rajasekhar *et al.,* 1999). Release of *C. carnea* along with *T. chilonis* effectively controlled *H. armigera* and aphids on Hybrid-6 cotton in Gujarat (Bharpoda *et al.,* 2000). Biorationale pest management, including release of *C. carnea* @ 2500 eggs -ha was found to reduce incidence of *P. xylostella* on cabbage (Reddy and Guerrero, 2000). Release of 2 or 3 larvae of *C. carnea* per plant resulted in the reduction population of *U. carthami* on safflower (Badgujar *et al.,* 2000). IPM module involving release of *T. chilonis*, *C. carnea*, Spray of NPV, nimbecidine, dipel ad synthetic chemicals imposed at different intervals depending on the pheromone catches, could effectively control the pest economically (Reddy & Manjunatha, 2000). Use of *C. carnea* along with the application of neem seed kernel extract (NSKE) reduced the population of

Lipaphis erysimi on oilseed crops in terms of socio economic and environmental values (Pandey and Singh, 2008; Sanjeev Rai *et al.*, 2007). Farkade *et al.* (1999) made studies on the adoption of the various biological control practices and found that use of chrysopa was adopted by around 7% of the agriculturist and ranks 4 th in the adoption.

Release of *C. carnea* at 25 days after sowing along with the spray of neem oil recorded maximum rupee equivalent on *Vigna radiata* (Singh *et al.*, 2009a). Two releases of *Chrysopa scelestes* @ 10000 eggs per ha effective when the population of *Saccaricoccus sacchari* (Chelvi & Kandasamy, 2009). Releases of *Mallada boninensis* @50 and 60 eggs per tree effectively controlled *Aleurocanthus woglumi* on citrus (Zade *et al.*, 2007; Bhagat *et al.*, 2007) and in combination with the release of parasitoids effectively controlled incidence of citrus leafminer (Rao *et al.*, 2005; Ingole *et al.*, 2005). Release of *C. carnea* eggs @ 20000 per acre on 60 ad 70 days after sowing along with other techniques like pheromone, decreased the incidence of *L. orbonalis* Guenee on brinjal (Elanhezhyan and Baskaran, 2008). *C. carnea* @ 50000/ha at 40 and 70 DAT in combination with *Beauveria* @1.5 kg/ha reduced the incidence of *S. litura* on cabbage (Das & Mandal, 2005). Combined releases of *T. chilonis* (1.5 lakh parasitized eggs/ha) with different rates of *C. carnea* were comparable with endosulfan (0.07%) and reduced the infestation in squares, flowers and green bolls on cotton (Panchbhai *et al.*, 2005).

C. carnea is not effective for effective for controlling *H. armigera* on chickpea (Mandal & Mishra, 2004) similarly *B. scelestes* is not effective against *Pyrilla perpusilla* (Asri *et al.*, 1983).

With the sudden outbreak of sugarcane wooly aphid, *Ceratovacuna lanigera* (Hemiptera: Aphididae) during 2002, the potential of *Micromus igorotus* along with *Dipha aphidivora* as an effective predator was proved at field conditions (Patil *et al.*, 2007). An IPM module consisting of soil application of neem cake, application of NPK at 250 : 75 : 185 kg/ha and releases of *M. igorotus* at 2600 grubs/ha and release of *D. aphidivora* at 1000 larvae/ha recorded highest reduction in sugarcane wooly aphid population (Mallapur *et al.*, 2006).

Intercrop with lucerne (1:1) population, NKSKE (5) two sprays and release of *C. carnea* 0.75 and 1.0 lakhs /ha showed effective control of sucking pests and *H. armigera* on cotton (Hanumantharaya *et al.*, 2008). IPM module-consisting of release of *T. chilonis*, *C. carnea* @ 10000/ha and need based application of insecticides can prove very useful in curtailing the excessive and indiscriminate use of insecticides and ensuring the sustainable cotton yield (Brar *et al.*, 2001). BIPM practices including release of larvae of *C. carnea* on Bt cotton recorded efficient management of pests (Prasad *et al.*, 2005 and Shanmugam *et al.*, 2006). *C. carnea* release along with NSKE was effective against *Lipaphis erysimi* (Sanjeev Rai *et al.*, 2007).

ECOLOGY

The seasonal incidence and population dynamics of the neuropterans is rarely studied except for a few species of the chrysopids. Population dynamics of the *C. carnea* was correlated with the prey *Myzus persicae* on sunflower while no such relationship was observed with prey *Amrasca biguttula biguttua* and *Aphis gossypii* (Mannan *et al.*, 1995). The period of activity was from September on sorghum for chrysopids (Duffield, 1995). Rao *et al.* (1989) observed that the population of predators decreased after rainfall. The abundance of *C. scelestes* was generally negatively correlated with maximum daily temperature, wind speed and sunshine and positively correlated with minimum daily temperature, RH and rainfall (Faleiro *et al.*, 1990). Population of *C. carnea* peaked during third week of June coinciding with the aphid population (Patil *et al.*, 2008). Population of chrysopids had a negative correlation with minimum and maximum temperature (Dhaka & Pareek, 2007).

Type II functional response in the predatory response of *C. carnea* on *Aphis fabae* (Khan & Zaki, 2008). Population of aphids positively correlated with the population of chrysopids on Lucerne (Ingawale and Tambe, 2007. Activity of *Micromus* larvae was observed during November and thereafter decreased (Sharanabasappa *et al.*, 2008).

NATURAL ENEMIES OF CHRYSOPIDS

The eggs, larvae and pupae of the neuropterans are often parasitized by a variety of parasitoids. Approximately 61 species of parasites have been recorded from chrysopidae, the majority of which belong to Chalcidoidea and Ichneumonidea, Serphoidea contains 12 chrysopid parasites, Cleridae (Coleoptera), 2 species and Chironomidae 1 species (Clancy, 1936). Singh & Narasimham (1992) enumerated the different species of parasitoids recorded from India. However, the natural enemies of other neuropterans are scantily available. In India several species of parasitoids were recorded on chrysopids. *Telenomus* near *perplexus* (Nixon) parasitised eggs of *C. carnea* in Gujarat (Patel *et al.*, 1975). *Telenomus* sp. were commonly recorded as parasitoids on the eggs of *C. carnea* (Balasubramani & Swamiappan, 1997; Krishnammoorthy & Mani, 1989). Krishnamoorthy & Mani (1989) recorded *Tetrastichus* sp. and *Isodromus axillaris* from the pupae of *C. carnea* and *C. lacciperda*. The level of parasitism was less than 4% and might not influence the predatory potential in field conditions. These parasitoids however did not negate the production of these natural enemies in commercial insectaries and laboratories. *Telenomus ampullaceus* and *Trichogramma* sp. were found to parasitize the eggs of *C. carnea* while *Brachycyrtus* sp. and *Tetrastichus* sp. nr. *sempronius* parasitized the pupae (Kapadia & Puri, 1991a). *Inkaka keraliensis* Sureshan and Narendran parasitized on *M. igorotus* (Ramegowda *et al.*, 2006).

CONSERVATION OF NEUROPTERANS

Strategies for the conservation of neuropterans were suggested by several researchers. Methods include use of safe pesticides, using of kairomones and attractants, provision of shelter and food for the larvae or adults, spraying of protein hydrolysates as supplementary food, selection of cultivars or varieties which are not detrimental to the neuropterans and suitable cropping systems like inter crops/ border crops and multiple crops. IPM fields with treatments including biopesticides, NSKE, *Trichogramma brasilensis* in cotton and organic amendments and biotrationale in crops such as cotton and soyabean helped in the conservation of chrysopids population (Sardana *et al.,* 2008; Abhilash & Patil, 2008; Ameta & Bunker, 2009).

TRITROPHIC INTERACTIONS

Plants often influence the colonisation of neuropterans. Some of the species of plants or cultivars of a species were preferred, probably because of their preferential morphological or chemical characteristics. Males of *C. carnea* had higher preference to sunflower and females had preference to sunflower and cotton, while pigeon pea was least preferred. *M. boninensis* males did not show any preference for any of the host plants (Ballal and Singh, 1999). *M. boninensis* preyed more on the larvae of citrus leafminer reared on acid lime and Nagpur mandarin than those reared on rough lemon and Rangpu lime (Rao *et al.,* 2001). *C. carnea* was 4 times abundant on hybrids than open pollinated cultivars of cotton (Vennila, 1998). Mohite and Uthamaswamy (1998) observed that predation of *H. armigera* by *C. carnea* was more on *Gossypium armourianum*, MCU-9 and *G. harkeressi*.

Some of the sunflower cultivars like GKSFH 2002 recorded lower population of phytophagous insects and greater population of *C. carnea* and early sown sunflower crop recorded greater numbers of predators than late sown crop (Arora *et al.,* 1998). *C. carnea* laid more eggs on cotton and sunflower but lesser on aster and coriander (Geetha and Swamiappan, 1998). The female *C. carnea,* showed comparatively better response than males. Among the cotton varieties tested Savitha, LRK-516, C256.4 evoked better response than others. Least response was observed in Sahana variety. Least response was noticed in HLS 72.Among the sunflower varieties, females of *H. armigera* showed more response than the males, KBSH-1 recorded more response followed by 6D -1R. Least response was noted in the variety PCSH -243.

Cotton cultivars like MCU-10, MCU-9, NSPHH-5, Gujarat Hybrid 10, DHB 290, MCU-7 recorded more oviposition and among the sunflower cultivars, PCSH 243, KBSH-1, SFL 103 recorded highest oviposition under multiple choice conditions. The volatile profile of the varieties suggested the positive role of compounds such as pinene, caryophyllen and other terpenoids (Anonymous, 2002b; Bakthavatsalam, 2001b, 2002). Tobacco types such as Burley, Beedi and cigar wrapper tobacco was favourbale to *C. carnea* in tritrophic interaction studies (Rao *et al.,* 2008). Electrophysiological studies indicated that the cultivars of safflower such as DSH 129, A-1, GMU 2518, NARI–H15 elicted higher response than other cultivars (Hanumantharaya *et al.,* 2008b). Higher oviposition was recorded on LRA 5166 variety and higher predation by larvae on RCH-2 and trichome density showed negative relation with predation of *C. carnea* (Balakrishnan *et al.,* 2006).

Singh *et al.* (1993b) observed higher incidence of *C. carnea* and *H. armigera* on okra and sunflower and suggested the possibility of exploiting these crops in cultural control of *H. armigera* when inter cropped with cotton. Raising two rows of trap crop of tobacco or chick pea around flue cured tobacco increased the population of chrysopids and decreased the incidence of *H. armigera* (Rao *et al.,* 1994). Intercropping rice bean with redgram increased the population of chrysopids on rice bean (Satyanrayana *et al.,* 1998). Duffield & Reddy (1997) found that in sorghum- short duration pigeon pea inter crop system, egg laying by chrysopids was recorded only on sorghum and not on adjacent pigeon pea plants and chrysopids did not migrate to pigeon pea despite the presence of potential prey. Spraying alternate rows (skip row coverage) with monocrotophos or endosulfan against *A. biguttula* and *Aphis gossypii* resulted in larger populations of *C. carnea* on cotton (Surulivelu & Kumaraswami, 1989). Kundu *et al.* (1998) suggested methods for the conservation of chrysopids in cotton ecosystem.

C. carnea and *Coccinella septempunctata* were more on safflower intercropped with coriander and sorghum (Hanumantharaya *et al.,* 2008a,b; Rao *et al.,* 2007). BIPM modules and Insecticide resistance management modules developed by several authors conserved the chrysopids quite effectively (Ameta & Bunker 2009, Suryawanshi *et al.,* 2007). IPM modules encouraged *C. carnea* population on cotton and brinjal (Ameta *et al.,* 2006; Sardana *et al.,* 2006; Bhosle *et al.,* 2007; Shanmugam *et al.,* 2006; Tanwar *et al.,* 2007). However, intercropping soyabean with cotton did not encourage the populations of Chrysopids on cotton (Vennila *et al.,* 2007). Cotton interspersed with *Cassia occidentalis* at 6:1 ratio + sowing of maize and *Binnia elegans* with 10% cotton plants and release of *T. chilonis* and *C. carnea* @5000 larva / ha recorded higher abundance of natural enemies (Godhani *et al.,* 2009).

B t and organic amendments like vermicompost, NSKE, neem cakes, cow urine, vermivash were safer to chrysopids in brinjal and okra (Jyoti & Goud, 2008; Jayakumar *et al.,* 2008; Sharma *et al.,* 2007; Deepika *et al.,* 2009) and transgenic cotton conserved the chrysopids probably due to the reduced usage of pesticides than conventional cotton. (Wadhwa & Gill,

2007). Bt cotton hybrid MECH 64 conserved more chrysopids than conventional hybrids (Channakeshava & Patil, 2006).

ALLELOCHEMICALS TO INCREASE THE EFFICIENCY OF THE NEUROPTERANS

Kairomones are defined as chemicals released by an organism in the natural context which evokes a behavioural response adaptively favorable to the receiver. Chrysopids have been found to respond to different kairomones. The adult chrysopids were found to respond to the honey dew secretions from the aphids. On chemical analysis it was found that L-tryptophan, the break down product of the honey dew is the primary component attracting the adult chrysopids.

The adult chrysopids showed behavioural changes like reduced walking speed, more resting period and less speed on contact with the kairomones like tryptophan (Bakthavatsalam *et al.,* 1997b). Electroantennogram studies showed that the fertile female *C. carnea* responded to the acid hydrolyzed L- tryptophan (Bakthavatsalam *et al.,* 2000a). Under laboratory conditions, the adults of *C. carnea* were attracted to the acid hydrolyzed L-tryptophan in wind tunnel studies. In net house conditions, the L-tryptophan treated plants recorded more ovipositional compared to untreated plants (Bakthavatsalam & Singh, 1996). Further studies were conducted at field conditions and it was found that oviposition was more in the cotton plants treated with L-tryptophan than untreated plants (Bakthavatsalam *et al.,* 1999, Anonymous, 2000b). Tryptophan was also attempted on the tomato, the least preferred host of chrysopids, however, the L-tryptophan was ineffective in increasing the oviposition on tomato (Bakthavatsalam *et al.,* 2001a). Bakthavatsalam & Tandon (2007) recorded that indole acetaldehyde and acid hydrolysed L-Tryptoph- an elicited better response in *C. carnea* through electrophysiological studies.

The larvae of *C. carnea* responded to the scale extracts of *H. armigera* by spending more time on the wax droplets treated with kairomone. However, in the absence of the three dimensional structures they did not show any response to the kairomones (Bakthavatsalam *et al.,* 1997: Hegde & Kulkarni, 2000b). The larvae of chrysopids showed response the scales of the hosts like *H. armigera*. The scale extract of *H. armigera* was found to attract the larval chrysopids in olfactometer studies (Bakthavatsalam & Singh, 1999). On chemical analysis it was found that several compounds like tricosane, pentacosane were observed to be important in the attraction of chrysopids. Attempts were made to utilize the scale extract of *H. armigera/C. cephalonica* (laboratory waste) to increase the predatory potential of larval chrysopids. The predatory potential was more in the plants treated with the kairomonal formulation cotton (Anonymous, 2001; 2002a). These formulations however did not increase the predatory potential of chrysopids on tomato, the least preferred host plants for chrysopids (Bakthavatsalam *et al.*, 2001a).

Apart from the kairomones, synomones released form the cotton bolls, infested with *H. armigera* also evoked behavioural responses in *C. carnea* (Bakthavatsalam *et al.*, 2000b). Kairomone septa impregnated with n-tricosane along with the spraying of 3 day old acid hydrolysed L-tryptophan increased the population of chrysopids on cotton (Bakthavatsalam *et al.*, 2007). Acetone extracts of onion leaves and cluster bean enhanced the predatory activity of *C. carnea* (Elanchezhyan *et al.*, 2009).

SAFER PESTICIDES

Several studies conducted by researchers have identified various insecticides as toxic for the neuropterans. Singh (1996) gave a detailed account on the safer pesticides for the chrysopids. Endosulfan, abemectin, phosalone, dichlorovos, malathion, methyl oxy demeton, emamectin benzoate and spinosad were generally considered safe for the chrysopids (Anonymous, 1992, 1993, 1998, Singh, 1989, Surulivelu & Kumaraswamy, 1989, Mani, 1993, Reddy & Divakar, 1998, Singh & Varma, 1986, Srinivasan & Babu, 2000; Raghuraman & Uthamasamy, 2005; Sontakke *et al.*, 2007; Dhanalakshmi and Mallapur 2008). Spiromesifen 240 SC caused 0.55% egg mortality of *C. carnea* (Kavitha *et al.*, 2006). Insecticides such as malathion, endosulfan, azadirachtin, monocrotophos, acephate were safer to eggs of *M. igorotus*. Azadirachtin is safer larvae and pupae are susceptible to different insecticides' (Vidya *et al.*, 2008b).

Insecticides such as quinalphos, fenvalerate, monocrotophos, alphamethrin, diflubenzuron, triazophos, cypermethrin, deltamethrin, dicofol were generally considered toxic to a number of chrysopids (Balasubramani & Swamiappan, 1997; Patel & Yadav, 1995, Bhuvaneswari & Uthamasamy, 1994, Arora *et al.*, 1993, Thirumurthy 1995; Kapadia & Puri, 1991; Mani & Krishnamoorthy, 1991; Muraleedharan & Chari, 1990; Rao *et al.*, 1990; Natarajan 1990; Krishnamoorthy, 1995; Naik *et al.*, 2006; Aggarwal & Brar, 2006; Sharma *et al.*, 2008). Fipronis @ 11.25G was safer to *Chrysoperla* sp. (Gosalwad & Wadnerkar, 2009). Emamectin benzoate recorded mortality up to 75% when the insecti-cides were treated in different forms (Kanna & Chandrasekaran, 2006).

Seed treatment of imidacloprid and thiamethoxam attracted the population of *C. carnea* on cotton (Satpute *et al.*, 2002; Ameta and Sharma 2005) and other crops (Preetha *et al.*, 2009).

Neem products like neemark, Wellgro, Neem seed kernel suspension, nicotine sulphate, nimbicidine, neem azal and neemrich were ovipositional repellant to *B. sceletes* (Yadav & Patil, 1993; Kumar & Babu, 1996: Deole *et al.*, 2000; Reddy & Manjunatha, 2000; Rao & Raghuraman, 2005). Sarade & Sonalkar (1999) screened the ovicidal effect of nine insecticides against *C. carnea* and found neem seed extract to be comparatively safer and chlorpyrifos, deltamethrin and cypermethrin to be toxic to *C. carnea*. Other botanicals such as Palmarosa oil 50 EC was safe to *C. carnea* (Thilgavathy *et al.,* 1997). Among the botanicals tested, *Alpini galaganga* and *Acorus calamus* were safer to *M. boninensis* and *Cathranthus roseus* for *C. carnea* (Shivankar *et al.,* 2001; Alagar and Sivasubramanian, 2007a; 2007b; Alagar *et al.,* 2007). *Eucalyptus tereticornis* extracts (50) is safer to the third instar larvae of *C. carnea* (Alagar *et al.,* 2007).

Laboratory studies showed that NPV of *H. armigera* and *S. litura* caused least mortality (0.94-1.24%) in the larvae of *C. carnea* (Balasubramani & Swamiappan, 1998b; Jeyamani *et al.,* 2008). NPV of *S. litura* was not cross infective to *C. carnea* (Jayanthi and Padmavathamma, 1996). Spicturin (*B.t* formulation) was least toxic to the larvae of *C. carnea* (Loganathan *et al.,* 2000). *Verticillium lecanii* (Zimmerman) is not pathogenic to *C. carnea* (Derakhshan *et al.,* 2007). Imdacloprid @ 0.28 ml/l and diafenthiuron are harmless to *C. carnea* (Preetha *et al.,* 2009). Fipronis @ 11.25G ai/ha is safer to *Chrysoperla* sp. (Gosalwad *and* Wadnerkar, 2009). Plant growth regulators such as triacontanol did not have any adverse effect on chrysopids on potato I (Samui *et al.,* 2007)

ANTAGONISTIC EFFECTS

Often the role of chrysopids as an beneficial insect is confronted by several workers. It is a known fact the *Chrysopa madestis* has been recorded as predator on Indian lac insect, *Kerria lacca* (Kerr.). In fact studies were initiated to screen the insecticides or the use of *Bacillus thuringiensis* for controlling Chrysopids (Anonymous, 1976; Mishra *et al.,* 1996). The parasitized eggs larva/pupae of pests were devoured by the chrysopids (Krishnamoorthy & Mani, 1985). Kapadia & Puri (1990) recorded that *C. carnea* attacked the pupae of *B. tabaci* parasitized by *Encarta transvena.* Jadhav & Varma (1988) observed that the larvae of *Chrysopa* sp. were found to feed on the cocoons of *melanoleuca* (Fletcher) on sugarcane and thus affecting the efficiency of this parasitoid. *T. chilonis* parasitized eggs of *H. armigera*, *T. achaeae* parasitized eggs of *E. vitella* and *Telenomus remus* parasitized eggs of *S. litura* were predated by *B. scelestes* and there are antagonistic effects between parasites and predators.

CONCLUSION AND FUTURE STUDIES

There are different families belonging to the order neuropterans. Unfortunately only a selected group is often studied and the other groups are seldom studied. In fact for groups like Megaloptera even the faunistic studies are not available. It is utmost important to collect data on the fauna of various families of Neuroptera in India. A systematic approach is needed to record the various species along with their host plant and the pest species which will enable us to study consider to utilise them as predator. Virtually no systematist is engaged in this kind of studies in India. It is therefore essential to encourage researchers on the studies like bio systematics and biodiversity of this group.

The biology of only selected chrysopids was studied. It is very essential to study the biology, feeding potential and ecology of all the species available. This study will be an effective guide for the selection of various species for different pests in different agro ecosystem. Though the mass rearing technologies are available for the species like *C. carnea*, there are other species like *M. boninensis*, *Apertochrysa crassinervis* etc. for which mass rearing techniques need to be developed. There is a need to develop for developing rearing methods for the other economically important neuropterans, especially hemerobiids. Automation of the production of neuropterans on synthetic diet is essential to increase the production capacity as well to meet the demand. Technology is necessary to develop long term storage of neuropterans using methods like diapause induction in the adults and storing them at lower temperatures for prolong periods to achieve optimal production.

Production of economically important neuropterans may be encouraged at village level units with the help of trained personnel to meet the demand. Financial assistance in terms of soft loans may also be provided to ensure availability of the bioagents as per the demand. Presently release of only the larvae were recommended for the control of the pests for obvious reasons that the eggs are parasitized by several Hymenopteran parasitoids. The release technique presently followed is tedious and labour intensive. A simple mechanized device is necessary for the release of chrysopids.

The chrysopids were recommended for field release only on a few crops such as cotton and sunflower. However, there are other crops where there are ample opportunities to utilize chrysopids especially plantations like coffee/ tea which needs to be looked into. The dosage and periodicity of releases correlated with the pest intensity needs to be standardized to various pests in different crop ecosystems. The cost of production and availability also needs to taken into account while the dosage is decided. The identification of species through the cryptic songs and molecular tools need to be undertaken for most of the species so that the identification becomes easier.

Studies on the conservation technologies using habitat manipulation techniques, safer pesticides, behavioural manipulation techniques etc should be intensified. The role of border crops/inter-crops and collateral hosts in increasing the activity of the neuropterans has to be studied. Technology for usage of kairomones and supplementary food spraying has to be perfected to conserve the neuropterans. Influence of cultivars on the neuropterans and selection of suitable cultivars for cultivation and development of entomophage parks are other conservation technologies worth attempting. Collection of cess from pesticide sale for giving subsidies to those who adopt conservation methods may also be considered.

Research grant needed for basic studies like biosystematics, bioecology, standardization of rearing techniques and conservation technologies on neuropterans should be given priority. Training and teaching of farmers about the importance of neuropterans as predators and community programs for the propagation of the usage of neuropterans are needed for the realization of the role of the neuropterans as efficient predators.

ACKNOWLEDGEMENT

The author is grafeful to the Director, National Bureau of Agriculturally Important Insects, Bangalore, for the facilities provided.

REFERENCES

Abhilash, C. & Patil, R. H., 2008. Effect of organic amendments and biorationals on natural enemies in soybean ecosystem. *Karnataka J. Agril. Sci.* 21:396-398

Agarwala, B.K., Das, S. & Saha, S., 1983. New host and prey records of some aphidophagous insects from India. *Akitu* 49:5-6.

Aggarwal, N. & Brar, D.S., 2006. Effects of different neem preparations in comparison to synthetic insecticides on the whitefly parasitoid *Encarsia sophia* (Hymenoptera: Aphelinidae) and the predator *Chrysoperla carnea* (Neuroptera: Chrysopidae) on cotton under laboratory conditions. *J. Pest Sci.* 79: 201-207.

Akhtar, M.S., Choudhury, R.A. & Shujauddin, 2006. Natural enemies of mustard aphid, *Lipaphis erysimi* (Kaltenbach) in Aligarh (India). *Env. Ecol.* 24S: 781-784.

Alagar, M. & Sivasubramanian, P., 2007a. Influence of botanicals and insecticides on predatory potential and biology of *Chrysoperla carnea* Stephens. *Indian J Ent.* 69: 117-121.

Alagar, M. & Sivasubramanian, P., 2007b. Safety of botanicals and insecticides to *Chrysoperla carnea* Stephens and egg parasitoid, *Trichogramma chilonis* Ishii. *Indian J. Ent.* 69: 71-77.

Alagar, M., Sivasubramanian, P., Saravanan, P A & Amutha, M., 2007. Safety of botanicals and insecticides to *Chrysoperla carnea* Stephens and *Trichogramma chilonis* Ishii. *J Pl. Prot. Env.* 4: 7-14.

Ameta, O. P. & Bunker, G. K. 2009. Validation and popularization of integrated pest management technology in cotton (*Gossypium hirsutum*). *Indian J. agric. Sci.* 79: 527-530.

Ameta, O. P. & Sharma, K. C., 2005. Bioefficacy of imidacloprid (Confidor) against sucking insect pests of cotton. *Insect Environment* 11: 9-13.

Ameta, O. P., Sharma, K C., Rana, B. S. & Bambawale, O. M., 2006. Validation and popularization of IPM technology in cotton through farmers participatory approach in tribal area of Southern Rajasthan. *Ann. Agric. Res.* 27: 162-166.

Anonymous, 2000c. Final Technical Report. National Repository of Natural Enemies of Crop Pests and Weeds. 23 pp.

Anonymous, 1976. *Annual report 1974-76.* Indian Lac Research Institute. 136 pp.

Anonymous, 1988. *Annual Report 1987-88.* AICRP on Biological Control of Crop Pests and Weeds. Biological Control Centre, Bangalore technical Document No 39. 339pp.

Anonymous, 1992. Annual Report 1991-92. AICRP on Biological Control of Crop Pests and Weeds. Biological Control Centre, Bangalore technical Document No 40. 313pp.

Anonymous, 1993. Annual Report 1992-93. AICRP on Biological Control of Crop Pests and Weeds. Biological Control Centre, Bangalore technical Document No 41. 260 pp.

Anonymous, 1998a Annual Report 1-4-1997 to 31-3-1998. National Repository of Natural Enemies of Crop Pests and Weeds. 23 pp.

Anonymous, 1998b. *Annual Report 1997-1998.* Project Directorate of Biological Control, Bangalore 217pp; Pp 37-38.

Anonymous, 1999. Annual Report 1-4-1998 to 31-3-1999. National Repository of Natural Enemies of Crop Pests and Weeds. 21 pp.

Anonymous, 2000a. Annual Report 1-4-1999 to 31-3-2000. National Repository of Natural Enemies of Crop Pests and Weeds. 23 pp.

Anonymous, 2000b. Final technical Report Use of Semiochemicals to increase the biocontrol potential of some important predators and parasitoids. Project Directorate of Biological Control. Bangalore.

Anonymous, 2001. *Annual Report 2000-2001.* Project Directorate of Biological Control, Bangalore 217pp; Pp 37-38.

Anonymous, 2002a. *Annual Report 2001-2002.* Project Directorate of Biological Control, Bangalore 221 pp; Pp 36-37.

Anonymous, 2002b. Final Technical Report. Implications of tritrophic interaction in the Integrated Pest Management of some important crop pests. Project Directorate of Biological Control, Bangalore 221pp; Pp 36-37.

Ansari, M.A., Pawar, A.D. & Kumar, D.A. 1992. Possibility for Biocontrol of tropical armyworm, *Spodoptera litura* (Lepidoptera: Noctuidae) on potato. *Pl. Prot. Bull. Faridabad* 44(4): 27-31.

Arora, R. Kaur, S., & Dhillon, A.S. 1998. Manipulation of date of sowing for natural control of insect pests in sunflower agro ecosystem. In Dhaliwal, D.S. and Arora, R. (eds). *Ecological Agriculture and Sustainable develop-*

ment Volume 1. Proceedings of an International Conference on Ecological Agriculture: towards sustainable development. Chandigarh

Arora, R., Gill, J.S. & Varma, G.C. 1993. Adverse effects of a chitin synthesis inhibitor, diflubenzuron on *Chrysoperla carnea* (Stephens) and *Trichogramma chilonis* Ishi. *J. Insect Sci.* 6: 279-280.

Asre, R., Gupta, P.K. & Pawar, A.D., 1983. Control of sugarcane pyrilla by its natural enemies. *Indian Fmg.* 33: 37-38.

Badgujar, M.P., Deotale, V.Y., Sharnagat, B.K. & Nandanwar, V.N. 2000. Performance of *Chrysoperla carnea* against safflower aphid, *Dactynotus carthami* (HRL). *J. Soils and Crops* 10: 125-127.

Bakthavatsalam, N & Tandon, P L., 2007. Behavioural and electrophysiological responses of *Chrysoperla carnea* (Stephens) to kairomones - acid hydrolyzed/ oxidized L-tryptophan and its breakdown products. *J. Biol. Control.* 21 (Special): 85-87.

Bakthavatsalam, N. & Singh, S.P., 1996. L-tryptophan as an ovipositional attractant for *Chrysoperla carnea* (Stephens) (Neuroptera: Chrysopidae). *J. Biol. Control*, 10: 21-27.

Bakthavatsalam, N. & Singh, S.P., 1999. Behavioural responses of larvae of *Chrysoperla carnea* to kairomones. *I. Insect Sci.*12: 34-36.

Bakthavatsalam, N. Singh, S.P., Tandon, P.L., Chaudhary, M. & Preethi, S. 2000a. Electrophysiological responses of *Chrysoperla carnea* (Stephens) (Neuroptera: Chrysopidae) to potential kairomones. *J. Ent. Res.* 24:109-114.

Bakthavatsalam, N. Singh, S.P., Tandon, P.L., Chaudhary, M. & Preethi, S. 2000b. Synomone mediated behavioural responses of *Chrysoperla carnea* (Stephens) (Neuroptera: Chrysopidae) to cotton infested by *Helicoverpa armigera* (Hubner) (Lepidoptera: Noctuidae) *J. Biol. Control.* 14:1-6.

Bakthavatsalam, N. Singh, S. P. & Tandon, P.L. 2001a. Attempts to increase the efficiency of *Chrysoperla carnea* (Stephens) (Neuroptera: Chrysopidae) on tomato (The Non Preferred host plant) through use of Kairomonal formulations. (In: *IPM in Horticultural Crops; Emerging Trends in the New Millennium*. Held on October 17-19, 2001 at Bangalore), pp.188-189.

Bakthavatsalam, N. Singh, S.P., Tandon, P.L., Hanumantharaya, L. & Vellaikumar, S. 2001b. Influence of different cultivars of sunflower on the electroantennogram response and behaviour of *Chrysoperla carnea* (Stephens) (Neuroptera : Chrysopidae). In Symposium on *Biocontrol based pest management for Quality crop Protection in the Current Millennium.* July 13-19,2001, Punjab Agricultural University, Ludhiana. Pp. 52-53.

Bakthavatsalam, N. Singh, S.P., Tandon, P.L., Hanumantharaya, L. & Vellaikumar, S. 2002. Electroantennogram and ovipositional responses of *Helicoverpa armigera* (Hubner) and *Chrysoperla carnea* (Stephens) to volatiles of different cultivars of cotton in a tritrophic context. Pap. Abstr. Symp. *Biological control of Lepidopteran Pests.* 17-18 July 2002, Bangalore.

Bakthavatsalam, N., Singh, S.P., Pushpalatha, N.A. & Bhummannavar, B.S. 1994. Life tables of four species of chrysopids (Neuroptera: Chrysopidae). *J. Ent. Res.* 18: 357-360.

Bakthavatsalam, N., Singh, S.P., Chaudhary, M. & Preethi, S. 1999. Kairom-ones for *Chrysoperla carnea* (Stephens) (Neuroptera: Chrysopidae). Pap. Abstr. *Natl. Symp. Emerging Trends in Biotechnological Approaches in the Integrated Pest Management,* Chennai; 26-27 February 1999.

Bakthavatsalam, N., Tandon, P L., Patil, S B., Bhemanna Hugar & Hosamani, A. 2007. Kairomone formulations as reinforcing agents for increasing abundance of *Chrysoperla carnea* (Stephens) in cotton ecosystem. *J. Biol. Control.* 21: 1-8.

Bakthavatsalam, N., Singh, S.P. and Tandon, P.L. 1997c. Behavioural responses of *Chrysoperla carnea* (Stephens) (Neuroptera: Chrysopidae) and *Cheilomenes sexmaculata* Fabricius (Coleoptera: Coccinellidae) to the kairomonal substances- Analysis using Digital Tracking System. *National Symposium of Pest Management in Horticultural Ecosystems;* Bangalore, 25-27 October 1997.

Bakthavatsalam, N., Singh, S.P., Pushpalatha, N.A. & Bhummannavar, B.S. 1995. Optimum dosage of eggs of *Corcyra cephalonica* Stainton for mass rearing *Chrysoperla carnea* (Stephens) (Neuroptera :Chrysopidae). *J. Insect Sci.* 8: 90-92.

Bakthavatsalam, N., Singh, S.P., Pushpalatha, N. A. & Bhummannavar, B.S. 1996. Comparative Biology and feeding potential of four species of chrysopids (Neuroptera: Chrysopidae). *Hexapopoda,* 8 : 45-53.

Bakthavatsalam, N., Singh, S.P., Pushpalatha, N.A. & Bhummannavar, B.S. 1997a. Optimum temperature for short term storage of *Chrysoperla carnea* (Stephens). *J. Biol. Conrol.* 9 : 45-46.

Bakthavatsalam, N., Singh, S.P., Pushpalatha, N.A. & Bhummannavar, B.S. 1997b. Evaluation of strains *of Chrysoperla carnea* (Stephens) (Neuroptera: Chrysopidae) resistant to monocrotophos. *National Symposium of Pest Management in Horticultural Ecosystems* held at Bangalore 25-27th October 1997.

Balakrishnan, M.M., Sreedharan, K., Venketesha, Krishnamoorthy, P. & Bhat, P.K. 1991. Observations on *Ferrisia virgata* (Ckll.) (Homoptera: Pseudococcidae) and its natural enemies of coffee, with new records of predators and host plants. *J. Coffee Res.* 21: 11-19.

Balakrishnan, N., Baskaran, R. K. M. & Mahadevan, N. R. 2005. Development and predatory potential of green lacewing *Chrysoperla carnea* (Stephens) (Neuroptera : Chrysophidae) on different prey insects. *Agri. Sci. Digest* 25: 194-197.

Balakrishnan, N., Baskaran, R. K. M. and Mahadevan, N. R. 2006. Influence of different cotton varieties/hybrid on the ovipositional preference and predation of *Chrysoperla carnea* Stephens and parasitization of *Trichogramma chilonis* Ishii. *J. Exptl Zool.* 9: 93-96.

Balasubramani, V. & Swamiappan, M., 1994. Development and feeding potential of the green lacewing, *Chrysoperla carnea* Stephens (Neuroptera: Chrysopidae) on different insect pests of cotton. *Anz. Sch. Pflanz.* 67: 165-167.

Balasubramani, V. & Swamiappan, M. 1997. Record of chrysopid *Apertochrysa* and its egg parasitoid *Telenomus* sp. (Scelionidae: Hymenoptera) from Tamil Nadu. *Insect Environment* 3: 71.

Balasubramani, V. & Swamiappan, M.,1998a. Feeding potential of *Chrysoperla carnea* Stephens on cotton leafhopper *Amrasca biguttula biguttula* Ishida. *Insect Environment* 4: 69-70.

Balasubramani, V. & Swamiappan, M., 1998b. Safety of Ha NPV and Sl NPV to green lacewing *Chrysoperla carnea* Stephens (Chrysopidae: Neuroptera). *Insect Environment* 4:70-71.

Balasubramani, V. & Swamiappan, M., 1998c. Suitability of groundnut aphid *Aphis craccivora* Koch. for rearing green lacewing *Chrysoperla carnea* Stephens. *Intl Arachis Newsletter* 18: 30-31.

Ballal, C.R. & Singh, S.P., 1999. Host plant mediated orientational and ovipositional behaviour of thee species of chrysopids (Neuroptera: Chrysopidae). *Biological Contriol* 16:47-53.

Bansod, R.S. & Sarode, S.V., 2000. Influence of different prey species on biology of *Chrysoperla carnea* (Stephens). *Shashpa,* 7: 21-24.

Bhagat, K.C. & Makhmoor, H.D., 1997. Pests of pearl millet (bajra) and their natural enemies in Jammu (Jammu and Kashmir). *J. Adv. Zool.* 18: 42-45.

Bhagat, K.C. and Masood, M.A. 1986. Record of green lacewing fly *Chrysopa orestes* Banks (Chrysopidae: Neuroptera), as a predator of the brinjal aid, *Aphis gossypii* Glover (Aphididae: Homoptera). *Indian J. Pl. Prot.* 13: 132.

Bhagat, S.M., Deotale, V.Y., Zade, S.M., Makanwar, P.S. & Hiwase, A.B. 2007. Bioefficacy of *Mallada boninensis* (Okamoto) against citrus blackfly *Aleurocanthus woglumi* Ashby. *J. Soils crops.* 17: 125-127.

Bharpoda, T.M., Patel, H.P., Patel, H.P., Patel, G.P., Patel, J.J. & Patel, J.R., 2000. Integrated Pest Management (IPM) in cotton H-6 cultivated in middle Gujarat. *Indian J. Ent.* 62: 327-331.

Bhattacharya, A., Jaiswal, A. K. & Singh, J P., 2008. Management of lac insect predators through IPM based biorational approaches. (In: Goel, S.C. (Ed); *Emerging trends of researches in insect pest management and environmental safety* No. 1: 221-226.

Bhatti, U. S. & Bhatti, S. K., 1988. Observation on the predatory habits of an ant lion larva *Myrmeleon formicarius* (Myrmeleontidae: Neuroptera). *Bull. Ent.* 29: 45-49.

Bhosle, B B., More, D.G., Bambawale, O.M., Sharma, O.P. & Patange, N.R. 2007 Effectiveness of cotton IPM module in rainfed Marathwada Region. *Annals Pl. Prot. Sci.* 15: 21 25.

Bhuvaneswari, K. & Uthamawamy, S., 1994. Toxicity of alphamethrin to beneficial insects in cotton ecosystem. *J. Insect Sci.* 7:228-229.

Brar, K.S., Sekhon, B.S., Singh., J., Shenhmar, M. & Bakhetia, D.R.C., 2001. Evaluation of biocontrol based IPM modules for the control of bollworm complex on cotton. *J. Insect Sci.* 14: 19-22.

Brauer, F., 1864. *Verh. Zool. Bot. Gurs in Wien* 14: 896-902.

Brooks, S.J. & Barnard, P.C., 1990. The green lacewings of the World: a generic Review (Neuroptera: Chrysopidae). *Bull. Br. Museum (Nat. Hist.). Ent. Series* 59: 117-286.

Canard, M., Semeria, Y. & New, T.R., 1984. *Biology of Chrysopidae*. Dr. W. Junk Publishers. The Hague, The Netherlands. 294 pp

Chakravarthy, A.K. & Thyagaraj, N.E. 2005. Role of predators in suppressing sugarcane woolly aphid, *Ceratovacuna lanigera* Zehntner (Homoptera: Aphididae) infestation in South Karnataka. *Insect Environment* 11: 72-74.

Chakravarthy, A K., Gowda, K.N.M. & Thyagaraj, N.E., 2006. Field activity and feeding potential of two insect predators on woolly aphid in sugarcane cultivated ecosystem. *Advances in Indian Entomology: Productivity and Health a Silve - Jubilee, Supplement No 3, Volume II - Insec and Environment.* 111-114 Muzaffarnagar, India: Uttar Pradesh Zoological Society.

Channakeshava, R & Patil, B V., 2006. Performance and economics of Bt cotton hybrid compared to conventional cotton hybrid against major insect pests under irrigated ecosystem. *Karnataka J. Agric. Sci..* 19: 949-951.

Chelvi, C.T. & Kandasamy, R. 2009. Field efficacy of *Chrysopa scelestis* Bank against sugarcane mealy bug *Saccaricoccus sacchari* Cockerell. *Cooperative Sugar* 41: 37-39.

Clancy, D.W., 1936. *The insect parasites of the Chrysopidae (Neuroptera). Univ. California Pub. Entomol.* Pp. 403-496.

Cohen, A.C. & Smith, L.K. 1998. A new concept in artificial diet for *Chrysoperla rufilbris:* The efficacy of solid diets. *Biol. Control.* 13: 49-54.

Daniel, M., 1979. Survey of the indigenous predators of arecanut phytophagous mites. In Venkataram C.S. (Ed.) *Proceeding of the Second Annual Symposium on Plantation Crops,* pp. 227-236.

Das, S.C. & Kakoty, N.N., 1992. Biological studies on tea aphid *Toxoptera aurantii* Boyer and its natural enemy complex. *Two & a Bud* 39: 29-33.

Das, T.K. & Mandal, S.M.A., 2005. Biological control of major lepidopteran pests in cabbage. *J. Pl. Prot. Env.* 2: 68-71.

Debnath, N., Ghosh, D. & Chakrabarti, S., 1988. Predators and parasites of aphids from north west and western Himalaya. II. Records of eight aphidophaogus neuropterans (Insecta) from India. *Entomon* 13: 137-139.

Deepika K., Dahiya, K.K. & Lal, R., 2009. Impact analysis of genetically modified (Bt) cotton cultivars on natural enemies. *J. Cotton Res. Dev.* 23: 111-116.

Deole, S.A., Bodhade, S.N., Mahajan, L.B., Deotale, V.Y. & Sharnagat, B.K. 2000. Residual toxicity of some pesticides used in cotton pest management against a chrysopid (*C. carnea*). *J. Soils Crops* 10: 279-281.

Deotale, V.Y. Nimbalkar, S.A., Shamagat, B.K. & Nandanwar, V.N., 1998. Effect of feeding different pollengrains on fecundity of *Chrysoperla carnea* (Stephens). *J. Soils Crops* 8:223.

Derakhshan, S A., Rabindra, R J & Ramanujam, B. 2007. Impact of entomopathogenic fungus, *Verticillium lecanii* (Zimmerman) Viegas on natural enemies of cabbage aphid, *Brevicoryne brassicae* (Linnaeus) and other beneficial insects.: *J. Biol. Control*. 21(Special): 133-140.

Devi, P.K. & Yadav, D.N. 2005. Performance of *Chrysoperla carnea* (Stephens) (Neuroptera: Chrysopidae) in an outdoor insect rearing facility. *Pest Mgmt. Econ. Zool.* 13: 257-263.

Devjani, P., Singh, T.K. & Singh, M.S., 1997. Prey predator interactions with reference to *Lipaphis erysimi* (Kalt.) on cauliflower. *Annals Pl. Prot. Sci.* **5**: 149-153.

Dhaka, S. R. & Pareek, B. L., 2007. Seasonal incidence of natural enemies of key insect pests of cotton and their relationship with weather parameters. *J. Pl. Prot. Res.* 47: 417-423.

Dhaliwal, Z.S. & Bains, S.S., 1983. Relative role of various parasitoids in limiting the population of sugarcane pyrilla in the monsoon and post monsoon seasons in Punjab. *Indian J. Ecol.* 10: 294-302.

Dhanalakshmi, D.N. & Mallapur, C.P., 2008. Evaluation of promising molecules against sucking pests of okra. *Annals Pl. Prot. Sci.* 16: 29-32.

Dhandapani, N., Kalyanasundaran, M., Swamiappan, M., Babu, P.C.S. & Jayaraj, S. 1992. Experiments on management of major pests of cotton with biocontrol agents in India. *J. Appl. Ent.* 114: 52-56.

Duffield, S.J., 1995. Crop specific differences in the seasonal abundance of four major predatory groups on sorghum and short duration pigeon pea. *International Chickpea and Pigeon pea News letter* 2: 74-76.

Duffield, S.J. and Reddy, Y.V.R., 1997. Distribution and movement of predators of *Helicoverpa armigera* in intercropped sorghum and short duration pigeon pea. *Crop Res. Hissar* 14: 315-335.

Elanchezhyan, K., Baskaran, R. K. M. & Rajavel, D. S., 2009. Kairomone effect of acetone extracts of intercrops on *Chrysoperla carnea* (Stephens) and *Trichogramma chilonis* Ishii. *J. ent. Res.* 33: 51-54.

Elsiddiq, S.I.Y., Gautam, R.D. & Subhash Chander 2006. Life-tables of the predator, *Mallada boninensis* (Okamoto) (Chrysopidae: Neuroptera) on the eggs of *Corcyra cephalonica* Stainton and larvae of *Tribolium castaneum* Herbst. *J. Ent. Res.* 30: 301-307.

Fabricius, 1798. *Ent. Syst. Suppl.* 202.

Faleiro, J.R., Singh, K.M. & Singh, R.N., 1990. Influence of abiotic factors on the population build up of important insect pests of cowpea *Vigna unguiculata* (L) Walp and their biotic agents recorded at Delhi. *Indian J. Ent.* 52: 675-680.

Farkade, B.C., Ahire, R.D., Patange, N.R. & Ahir, P.S., 1999. Extent of adoption of biological pest control in cotton. *J. Soils Crops* 9: 108-110.

Fasih, M. & Srivastava, R.P., 1990. Parasites and predators of insect pest of mango. *Intl. Pest Control* 32: 39-41.

Gautam, R.D., 1994. Present status of rearing of chrysopids in India. *Bull. Ent.* 35: 31-39.

Gautam, R.D., 1996. Effect of space and population density on the development of green lace wing *Chrysoperla carnea* (Stephens). *Annals Pl. Prot. Sci.* 4:134-137.

Gautam, R. D. & Paul, A. V. N., 1987. An artifical diet for the larvae of green lacewing, *Chrysopa scelestes* Banks (Neuroptera: Chrysopidae). *J. Ent. Res.* 11: 69-72.

Gautam, R.D. & Paul, A.V.N., 1988. Influence of adult food supplements of *Chrysopa scelestes* Banks (Chrysopidae: Neuroptera) *J. Ent. Res.* 12(1):25-27.

Gautam, S. Singh, A K., Thomas, A., Malathi, V. G. & Gautam, R. D., 2008. Morphological and molecular differentiations among the green lacewings collected from cotton field in Delhi. *J. Ent. Res.* 32: 291-294.

Geetha, B. & Swamiappan, M., 1998. Influence of crop plants on adult biology of the predator *Chrysoperla carnea* (Chrysopidae:Neuroptera). *Insect Environment* 3: 110.

Geetha, B., Swamiappan, M. & Mohana, D., 1998. Cost effective improved adult diets for the predator, *Chrysoperla carnea* Stephens (Chrysopidae: Neuroptera). In Reddy, P.P., Kumar, K.K, and Verghese, A. (Eds). *Advances in IPM for Horticultural Crops. Proceedings of the First National Symposium on Pest Management in Horticultural Crops: Environmental Implications and Thrusts,* Bangalore India 15-17 October 1997. Pp. 218-221.

Geethalakshmi, L., Muthukrishnan, N., Chandrasekharan, M. & Raghuraman, M., 2000. Chrysopids biology on *Corcyra cephalonica* and feeding potential on different host insects. *Ann. Pl. Port. Sci.* 8: 132-135.

Ghosh, S.K., 1976. A new species of *Chrysopa* (Neuroptera: Chrysopidae) from India. *Entomon* 1: 189-191.

Ghosh, S.K., 1976. *Chrysopa* (*Chrysoperla*) *gujaratensis* n. sp. from India (Neuroptera: Chrysopidae). *Entomologia Scandinavica* 7:74-75.

Ghosh, S.K., 1976. *Hemerobius solanensis* sp. from India (Neuroptera: Hemerobiidae). *Entomologia Scandinavica* 7: 75.

Ghosh, S.K., 1977. A new genus and a new species of Neuroptera (Fam. Hemerobiidae) from India. *Proc. Indian Acad. Sci. B.* 86-235-237.

Ghosh, S.K., 1977. A new species of *Chrysopa* (Neuroptera: Chrysopidae) from India. *Entomon* 2: 103-104.

Ghosh, S.K., 1977. Fauna of Rajasthan, India. Neuroptera. *Rec. Zool. Sur. India* 72: 309-311.

Ghosh, S.K., 1977. New record of *Chrysopa* (*Chrysoperla*) *congrua* Walker (Neuroptera: Chryspidae) from India. *Newsletter Zool. Surv. India* 3:144.

Ghosh, S.K., 1977. On a collection from Himachal Pradesh. *Newsletter Zool. Surv. India* 3:36-38.

Ghosh, S.K., 1977. Two new species of *Chrysopa* (Neuroptera: Chrysopidae) from India. *Entomon* 2:244-245.

Ghosh, S.K., 1978. A new record of a species of *Spilosmylus* (Neuroptera: Osmylidae) from India. *J. Bombay Nat. Hist. Soc.* 75: 506.

Ghosh, S.K., 1980. New record of the species *Indophanes barbarus* (Walker) (Myrmeleontidae: Neuroptera) from India. *Bull. Zool. Surv. India* 3: 119.

Ghosh, S.K., 1980. On a small collection of Neuroptera from Andaman and Nicobar islands. *Rec. Zool. Surv. India* 77: 247-254.

Ghosh, S.K., 1981. A new species of the genus *Italochrysa* Principi (Neurotpera: Chrysopidae) from India. *Bull. Zool. Surv. India* 3: 145-146.

Ghosh, S.K., 1981. New records of Neuroptera : Plannipennia from Laccadive islands. India, coupled with a short note on the taxonomic variation of *Centroclisis eustalacta* (Gerst.) *Bull. Zool. Surv. India* 4: 199-202.

Ghosh, S.K., 1981. On new and little known species of Planipennia (Order. Neuroptera) from India. *Bull Zool. Surv. India* 4: 131-139.

Ghosh, S.K., 1982. A new species of *Macronemurus* Costa (Myrmeleontidae: Planipennia, Neuroptera) from India. *Entomon* 7:111-113.

Ghosh, S.K., 1983. Neuroptera from Western peninsular India with new record. *Rec. Zool. Surv. India* 81:77-87.

Ghosh, S.K., 1983. Notes on the biogeography of Neuroptera: Planipennia from certain area of the north west Himalayan and northern peninsular sector of India. *Rec. Zoo. Surv. India* 80:291-300.

Ghosh, S.K., 1984. Contribution to the taxonomical studies of Neuroptera (sub order Planipennia) from eastern India. 1. Family Myrmeleontidae. *Rec. Zool. Surv. India. Misc. Publ.* 52: vi+ 63pp

Ghosh, S.K., 1984. Contribution to the taxonomical studies of the family Ascalphidae (Neuroptera) from eastern India. *International Congress of Entomology* 17:898.

Ghosh, S. K., 1985. Insecta: neuroptera. *Rec. Zool. Surv. India* 82: 69-71.

Ghosh, S.K ,1985. On a few interesting species of the family Chrysopidae (Neuroptera) from norh -west Himalayas India. *Bull. Zool. Surv. India* 7: 215-218.

Ghosh, S. K., 1987. Insecta: Neuroptera: Planipennia: Myrmeleontidae. A check list. (In: *Fauna of Orissa Part I.* Ed. Lamba, B.S.), Zoological Surv. of India, Calcutta. Pp. 340.

Ghosh, S. K., 1988. Contribution to the taxonomical studies of Neuroptera (suborder Planipennia) from eastern India. II. Family Ascalaphidae. *Rec. Zool. Surv. India* 87: 61-63.

Ghosh, S. K., 1990. Contribution to the taxonomical studies of Neuroptera (Sub order Planippennia) from eastern India. III. Family Chrysopidae. *Rec. Zool. Surv. India* 86:329-354.

Ghosh, S. K., 1990. Description of an unknown female along with two new locality of the family Myrmeleontidae (Neuroptera) from Lakshadweep islands, Inda. *Rec. Zool. Surv. India* 87: 259-261.

Ghosh, S. K., 1990. Interesting species of Neuroptera from Himachal Pradesh. *Rec. Zool. Surv. India* 89: 89-94.

Ghosh, S. K., 1990. On little and known species of Stenrares Hagen (Myrmeleontidae: Neuroptera) from India. *Rec. Zool. Surv. India* 86: 453- 459.

Ghosh, S. K., 1991. On a few intersting species of the fmaily Corydalidae (suborder Megaloptera: order Neuroptera) from India. *Rec. Zool. Surv. India* 88: 47-51.

Ghosh, S. K., 1991. On a new species of the genus *Palpares* (Neuroptera: Myrmeleontidae) from India. *Rec. Zool. Surv. India* 88:71-73.

Ghosh, S. K., 1997. Neuroptera. (In: Alfred, J.R.B. ed.) *Fauna of Nanda Devi Biosphere Reserve. A world heritage site*). Zool. Surv. India, Calcutta. Fauna of Conservation area 9: pp. 166.

Ghosh, S. K., 1998. Faunal Diversity in india: Neuroptera. Pp. 251-257. In Alfred, J.R.B., Das, A.K., Sanyal, A.K., (Eds) *Faunal Diversity in India: a commemorative volume in the 50 th year of India's Independence.* Zool. Surv. India, Calcutta viii + 497 pp.

Ghosh, S. K., 2000. Records of the Zoological Survey of India: Neuroptera fauna of Noth East India. Zool. Survey of Idia. *Occ. Paper* 184: pp.179.

Ghosh, D., Poddar, S. & Raychaudhuri, D. N., 1981. Natural enemy complex of *Aphis craccivora* Koch and *Lipaphis erysimi* (Kalt.) in and around Calcutta, West Bengal. *Sci. Cult.* 47: 58-60.

Ghosh, S.K. & Maulik, D. R., 1985. On a small collection of Conipterygidae (Neuroptera) from West Bengal, India. *Bull. Zool. Surv. India* 7 : 237-239.

Ghosh, S.K. & Sen, S., 1977. Checklist of Indian Plannipennia (Order Neuroptera). *Rec. Zool. Surv. India* 73: 277-326.

Godhani, P. H., Patel, R M., Jani, J J., Yadav, D N., Korat, D M & Patel, B H. 2009. Impact of habitat manipulation on insect pests and their natural enemies in hybrid cotton. *Karnataka J. Agric. Sci.* 22(1): 104-107.

Gopalan, M., Jayaraj, S., Arivanam, M. Pillai, K. & P. V.S. 1988. New record of *Heteropsylla cubana* Crawford (Psyllidae: Homoptera) on subabul, *Leucaena leucocephala* (Lam) de Wit in India. *Curr. Sci.* 51(20): 1124-1125.

Gosalwad, S. S. & Wadnerkar, D. W., 2009. Effect of insecticides on green lace wing and ladybird beetle in okra (Abelmoschus esculentus L.) (Moench). *J Pl. Prot. Env.* 6: 11-16.

Grover, P., 1986. Integrated control of midge pests. *Cecidol. Intl.* 7: 1-28.

Gupta, S.K., 2001. A conspectus of natural enemies of phytophagous mites and mites as potential biocontrol agents. In Halliday, R.B., Walter, D.E., Procher, H.C., Norton, R.A. and Colloff, M.J. (Eds). *Acarology Proceedings of the 10 th International Congress,* pp. 484-497.

Hanumantharaya, L., Mallapur, C P., Venkateshalu, Naik, V. R. & Kumar, C. J. 2008a. Pest status of safflower, *Carthamus tinctorius* L. in northern parts of Karnataka. In Knights, S E and Potter, T D (eds) *Safflower: unexploited potential and world adaptability* 7th International Safflower Conference, Wagga Wagga, New South Wales, Australia, 3-6-November, 2008, pp. 1-3.

Hanumantharaya, L., Parameshwarappa, K. G., Suma, S. B. & Kumar, C. J. 2008b Electroantennogram and ovipositional responses of *Helicoverpa armigera* (Hubner) and *Chrysoperla carnea* (Stephens) to volatiles of different cultivars of safflower, *Carthamus tinctorius* In: Knights, S.E. and Potter, T D (eds) *Safflower: unexploited potential and world adaptability* 7th International Safflower Conference, Wagga Wagga, New South Wales, Australia, 3-6-November, 2008, pp.1-6.

Hanumantharaya, L., Venkateshalu., Kubasad, V. S. & Raju, S. G. 2008c. Role of cropping pattern for the management of insect pests of safflower, *Carthamus tinctorius* In: Knights, S.E. and Potter, T.D. (eds) *Safflcwer: unexploited potential and world adaptability* 7th International Safflower Conference, Wagga Wagga, New South Wales, Australia, 3-6-November, 2008, pp. 1-4.

Hanumantharaya, L., Goud, K.B. & Naik, L.K. 2008d. Use of green lacewing, *Chrysoperla carnea* (Stephens) and neem seed kernel extract for management of insect pests on cotton. *Karnataka J. Agric. Sci.* 21: 41-44.

Haruyama, N, Mochizukli, A., Duelli, P., Naka, H. & Nomura M., 2008. Green lacewing phylogeny based on three nuclear genes (Chrysopidae: Neuroptera). *Syst. Entomol.* 33: 275-288.

Hassen, S.A & Hagen, K.S., 1978. A new artificial diet for rearing *Chrysopa carnea* larvae (Neuroptera: Chrysopidae). *Z. Zng. Ent.* 86: 315-320.

Hegde, M. & Kulkarni, K.A., 2000a. Mass rearing of *Chrysoperla carnea* (Stephens) using wood shavings as filler material. *Karnataka J. Agric. Sci.* 13: 445-447.

Hegde, M. & Kulkarni, K.A., 2000b. Response of *Chrysoperla carnea* (Stephens) to kairomonal substances. *Karnataka J. Agric. Sci.* 13: 184-186.

Hegde, R. & Lingappa, S., 1998. Performance of *Chrysoperla carnea* (Stephens) against pigeon pea pod borer, *Helicoverpa armigera*. *Karnataka J. Agric. Sci.* 11: 658-661.

Henry, C.S., 1980. The courtship call of *Chrysopa downesi* Banks (sic) (Neuroptera: Chrysopidae) in evolutionary significance. *Psyche* 86: 291-297.

Henry, C.S., Brooks, S.J., Thierry, D., Duelli, P. & Johnson, J. B., 2001. The common lace wings (*Chrysoperla carnea* S. lat) and the sibling species problem. (In: *Lace wing in the crop eniivronment*), McEwen, P., New, T.R. and Whittington, A. E. (Eds), pp. 29-42.

Henry, C. S., Brooks S.J., Duelli, P. & Johnson, J. B., 2006. Courtship song of the South African lacewing *Chrysoperla zastowi* (Esben-Peterson) (Neuroptera: Chrysopidae): evidence for a trans equatorial geographic range. *J. Natural History* 40: 2173-2195.

Ingawale, U. V. & Tambe, A.B., 2007. Seasonal abundance of aphids and their natural enemies on lucerne. *Asian J. Bio Sci.* 2: 8-10.

Ingole, D A; Deotale, V. Y; Deotale, R. O. & Ramkumar, J., 2005. Preliminary study on the performance of *Mallada boninensis* (Okamoto) against *Aleurocanthus woglumi* Ashby on citrus. *J. Biol. Control*. 19: 77-79.

Jadhav, R.B. & Varma, A., 1988. Record of a hyperparasitoid and two predators on *Epicarina melanoleuca* in Maharastra. *J. Biol Control* 2: 133-134.

Jagadish, P.S., Prabhuaraj, A., Manjunatha, B. N. & Seetharam. A., 1996. Biology of *Coccinella septempunctata* L., *C. transversalis* Fab. (Coccinellidae: Coleoptera) and an unidentified species of hemerobiidae in *Hysteraoneara setariae* (Thomas) (Aphididae: Homoptera). *Mysore J. Agric. Sci.* 30 : 52-55.

Jalali, S.K., Rabindra, R. J., Rao, N. S. & Dasan, C. B., 2003. *Mass production of Trichogrammatids and Chrysopids*. Tech Bull. 33, Project Directorate of Biological Control, Bangalore.

Jalali, S.K. & Singh, S.P., 1992. Biology and feeding potential of *Curinus coeruleus* (Mulsant) and *Chrysoperla carnea* (Stpehens) on subabul psyllid *Heteropsylla cubana*. *J. Insect Sci.* 5: 89-90.

Jalali, S.K. & Singh, S.P. 1994. Effect of *Aphis gossypii* Glover number on *Mallada astur* (Banks) and *Cheilomenes sexmaculata* (Fabricius). *Biological Control* 4: 45-47.

Jayakumar, R., Naganagoud, A. & Rajashekarappa, K. 2008. Influence of indigenous components on natural enemies in okra ecosystem. *Annals Pl. Prot. Sci.* 16(1): 102-106.

Jayanthi, P.D.K. & Padmavathamma, K., 1996. Cross infectivity of nuclear polyhedrosis virus (NPV), *Bacillus thuringiensis* and *Beauveria bassiana* (Balsamo) Vuille to pests of ground nut (*Arachis hypogaea* Linn.) and their natural enemies. *J. Ent Research.* 20: 211-215.

Jeyarani, S., Karuppuchamy, P., Sathiah, N. & Manimegalai, S. 2008. Safety of UV selected *Helicoverpa armigera* nucleopolyhedrovirus to non target beneficial organisms. *J. Biol. Control*. 22: 107-112.

Johnson, J., 1983. A note on some common aphidivorus insects of Kerala. *Pranikee* 4: 415-418.

Joshi, B. C. & Yadav, D. N., 1990. Biology and feeding potential of *Mallada boninensis* (Okamoto), a chrysopid predator of whitefly *Bemisia tabaci* Gennadius. *J. Biol. Control* 4: 18-21.

Joshi, S., Venkatesan, T. & Rao, N.S. 1997. Host range and predatory fauna of *Aphis craccivora* Koch (Homoptera: Aphididae) in Bangalore, Karnataka. *J. Biol. Control* 11: :59-63.

Jyoti, D. P. & Goud, K. B., 2008. Safety of organic amendments and microbial pesticides to natural enemies in brinjal ecosystem. *Annals Pl. Prot. Sci.,* 16: 111-114.

Kadri, S. & Goud, K. B., 2005. Survey and seasonal incidence of onion thrips, *Thrips tabaci* (Lindeman). *Karnataka J. Agric. Sci.* 18: 510-512.

Kakar, K. L. & Sood, A. K. 1989. Bioecological studies and control of rose aphid, *Microsiphum rosaeiformis* Das. *J. Aphidology* 3: 113-118.

Kalra, V. K., 1988. Population dynamics of various predators associated with mustard aphid, *Lipaphis erysimi* (Kalt). *J. Biol. Control* 2:77-79.

Kanna, S. S. & Chandrasekaran, S., 2006. Safety of emamectin 5 SG on *Trichogramma chilonis* and *Chrysoperla carnea* under laboratory conditions. *J. Ecotoxicol. Env. Monitoring* 16: 509-514.

Kapadia, M.N. & Puri, S.N., 1990. Feeding behaviour of *Chrysoperla carnea* (Stephens) on the parasitized pupae of *Bemisia tabaci* (Gennadius). *Entomon* 15: 283-284.

Kapadia, M.N. & Puri, S.N., 1991a. Parasitzation of *Chrysoperla carnea* (Stephens) in cotton field under Parbahani conditions. *J. Maharastra Agric Univ.* 16: 453-454.

Kapadia, M.N. & Puri, S.N., 1991b. Persistance of different insecticides on cotton leaves against the larvae of *Chrysoperla carnea* (Stephens). *International J. Tropical Agric.* 9: 85-87.

Kapadia, M. N. & Puri, S. N., 1992. Development of *Chrysoperla carnea* reared on aphids and whitefly. *J. Maharastra Agric. Univ.* 17:163-164.

Kaur, R., Kaur, R. & Brar, K.S., 2008. Development and predation efficacy of *Chrysoperla carnea* (Stephens) on mealy bug, *Phenacoccus solenopsis* (Tinsley) under laboratory conditions. *J. Insect Sci.* 21: 93-95.

Kaur, S., Shenhmar, M. & Brar, K. S., 2001. Effect of cold storage on eggs and pupae of *Chrysoperla carnea* (Stephens). *J. Insect Sci.* 14: 58-59.

Kavitha, J., Kuttalam, S., Chandrasekaran, S & Ramaraju, K., 2006. Effect of spiromesifen 240 SC on beneficial insects. *Annals Pl. Prot. Sci.* 14: 343-345.

Khan, A.A. and Zaki, F.A., 2008. Predatory response of *Chrysoperla carnea* (Stephens) (Neuroptera: Chrysopidae) feeding on the Euonymus aphid, *Aphis fabae solanella* Theobald (Homoptera: Aphididae) in Kashmir. *J. Biol. Control* 22: 149-154.

Khulbe, P., Maurya, R P. & Khan, M. A., 2005. Biology of *Chrysoperla carnea* (Stephens) on different host insects. *Annals Pl. Prot. Sci.* 13: 351-354.

Krishnamoorthy, A.,1985. Effect of several pesticides on eggs, larvae and adults of the green lacewing *Chrysopa scelestes* Banks. *Entomon* 10: 21-28.

Krishnamoorthy, A., 1988a. Development and voracity of *Chrysopa lacciperda* Kimmins on *Plancoccus citri* (Risso). *J. Biol. Control* 2: 97-98.

Krishnamoorthy, A., 1988b. Biology of *Chrysopa scelestes* Banks (Neuroptera: Chrysopidae). *Bull. Entomol.* 29: 69-72.

Krishnamoorthy, A. & Mani, M., 1982. Feeding potential and development *of Chrysopa scelestes* Banks on *Heliothis armigera* (Hubn.) under laboratory conditions. *Entomon* 7: 385-388.

Krishnamoorthy, A. & Mani, M., 1985. Feeding behaviour of *Chrysopa scelestes* Banks on the parasitized eggs of some lepidopterous pests. *Entomon* 10:17-19.

Krishnamoorthy, A. & Mani, M., 1989. Records of lacewings preying on mealy bugs in India. *Curr. Sci.* 58: 155-157.

Krishnamoorthy, A. & Mani, M., 1997. Biological control of Oriental mealybug, *Planococcus lilacinus* (Ckll) on chow-chow. In Reddy, P.P., Kumar, K.K, and

Verghese, A. (Eds). *Advances in IPM for Horticultural Crops. Proceedings of the First National Symposium on Pest Management in Horticultural Crops: Environmental Implications and Thrusts*, Bangalore India 15-17 October 1997. Pp. 207-209.

Krishnamoorthy, A. & Nagarkatti, S., 1981. A mass rearing technique for *Chrysopa scelestes* Banks (Neuroptera: Chrysopidae). *J. Ent. Res*. 5: 93-98.

Krishnanoorthy, A., 1984. Influence of adult diet on the fecundity and survival of the predator *Chrysopa scelestes* (Neur.: Chrysopidae). *Entomophaga* 29: 445-450.

Krishnaswamy, S. & Yaseen, M., 1972. Neuroptera: Chrysopidae. (In: Rao, V.P. and Ghani, M. (Eds). *Studies on predators of Adelges spp. in the Himalayas*). Misc. Publ. No.3. CIBC, pp. 116.

Kulkarni, K.A., 1998. Culturing and mass multiplication of natural enemies. In. Radder, G.D., Gidnavar, V.C., Chittapur, B.M., Itnal, C.J. and Patil, B.N. (Eds). *Organics on sustaining soil fertility and productivity*, pp. 257-262.

Kumar, H. L. C., Chauhan, S. V., Ramkumar, J. & Irulandi, S., 2007a.Biology of *Chrysoperla carnea* (Stephens) strains collected from different agro climatic zones of Tamil Nadu. *Insect Environment* 13: 42-44.

Kumar, H.L.C., Chauhan, S.V., Ramkumar, J. & Irulandi, S., 2007b. Predatory potential of different strains of *Chrysoperla carnea* larvae. *Insect Environment* 13: 41-42.

Kumar, H. L. C., Swamiappan, M., Chauhan, S. V., Rajendran, R. & Thirumurugan, V., 2008. Ovipositional period and longevity of different strains of *Chrysoperla carnea* in Tamil Nadu. *Insect Environment* 13: 147-148.

Kumar, H. L. C., Swamiappan, M., Ramkumar, J. & Chauhan, S., 2007. Comparison of group and isolated rearing of *Chrysoperla carnea* (Stephens) on eggs of *Corcyra cephalonica* (Staint.) and assessing the cost of production of cocoons. *Environ. Ecol.* 25: 678-681.

Kumar, M. S & Gautam, R. D., 2007. Influence of different floral feeds on the biological attributes of *Chrysoperla carnea* (Stephens). *Annals Pl. Prot. Sci.,* 15: 4-7.

Kumar, S. P. & Babu, P.C.S., 1996. Evaluation of some neem formulations for safety to certain beneficial insects. *Pest Mgmt. Econ. Zool.* 6: 19-24.

Kundu, S.K., Jeyakumar, P., Meenakshi & Gupta, G.P., 1998. Conservation of *Chrysoperla carnea* in cotton ecosystem for sustainable IPM in cotton echosystem for sustainable IPM programmes. *Indian J. Ent.* 60 : 297-300.

Lavhe, N.V., 2000. Studies on the life cycle on chrysopid predator *Mallada boninensis* (Okamoto*). Insect Environment* 5:176.

Lavhe, N.V. & Nimbalkar, S. A., 2000. A record of predation of *Mallada boninensis* (Okamoto) on citrus blackfly. *Insect Environment* 5: 172.

Loganathan, M., Babu, P. C. S. & Balasubramaniam, G., 2000. Testing of indigenous *Bacillus thuringiensis* var *galleriae* against the predatory green lace wing *Chrysoperla carnea*. *Indian J. Ent.* 62: 286-288.

Madan, Y. P., Chaudhary, J. P., Yadav, S. R. & Sharma, S.K., 1988. Biology of *Chrysopa (Anisochrysa)* sp. a predator of *Pyrilla prepusialla* Walker eggs, under laboratory condtions. *Agri. Sci. Digest* **8**: 170-172.

Magar, S. B., Shinde, V. D. & Patil, A. S., 2007. Biology of *Micromus igorotus* Bank (Neuroptera: Hemerobidae), a potential predator of *Ceratovacuna lanigera* Zehnt. (Hemiptera: Aphididae) in sugarcane. XXVI Congress, International Society of Sugar Cane Technologists, ICC, Durban, South Africa, 29 July 2 August, 2007, pp. 854-857.

Mallapur, C P., Sharanabasappa., Kambrekar, D N., Gundannavar, K P., Kulkarni, K A Integrated management of sugarcane woolly aphid, *Ceratovacuna lanigera* (Zehntner). *Annals Pl. Prot. Sci.* 14: 337-339.

Malleshiah, Rajagopal, B. K. & Gowda, K. N. M., 2000. Feeding potential of *Chrysoperla carena* (Neuroptera: Chrysopidae) of different stages of citrus mealy bug, *Planococcus citri* (Hemiptera: Pseudococcidae) *Crop Res. Hissar* 20: 126-127.

Mandal, S. M. A. & Mishra, B. K., 2004. Evaluation of some biocontrol agents against *Helicoverpa armigera* (Hubner) in chickpea. *J. Pl. Prot. Env.* 1: 27-29.

Mandal, S. M. A & Patnaik, N. C., 2006. Predatory potential of aphidophagous predators associated with cabbage crop. *J. Pl. Prot. Env.* 3: 81-86.

Mani, M. & Krishnamoorthy, A., 1989. Feeding potential and development of *Mallada boninensis* (Okamoto) on the grape mealy bug, *Maconellicoccus hirsutus* (Green). *Entomon* 14: 19-20.

Mani, M. & Krishnamoorthy, A., 1990a. Natural suppression of mealybugs in guava orchards. *Entomon* 15: 245-247.

Mani, M. & Krishnamoorthy, A., 1990b. Predation of Mallada boninensis on Ferrisia virgata, *Planococcus citri* and *P. lilacinus*. *J. Biol. Control* 4: 122-123.

Mani, M. & Krishnamoorthy, A., 1991. Contact toxicity of synthetic pyrethroids to some parasitoids and predators of mealybugs. *Indian J. Pl. Prot.* 19: 93-95.

Mani, M. & Krishnamoorthy, A., 1995. Natural enemies of whitefly, *Lipaleyro-des euphorbiae* David and Subramaniam (Homoptera: Aleyrodidae). *J. Biol. Control* 9: 102-104.

Mani, M. & Krishnamoorthy, A., 1998. Biological control studies on the mango green scale *Chloropulvinaraia polygonata* (Ckll.) (Homoptera: Coccidae) in India. *Entomon* 23: 105-110.

Mani, M. & Krishnamoorthy, A., 1999. Natural enemies of host plants of spiraling whitefly *Aleurodicus dispersus* Russel (Homoptera: Aleyrodidae) in Bangalore, Karnataka. *Entomon* 24: 75-80.

Mani, M., Thontadarya, T. S. & Singh, S. P., 1987. Reconrd of natural enemies on the grape mealy bug, *Maconellicoccus hirsutus* (Green). *Curr. Sci.* 56: 624-625.

Manju, A. C., Patil, B. V., Bheemanna, M., Hosamani, A. C., Hanchinal, S. G. & Bansi, A. B., 2007. Influence of *Bt* cotton fed aphids on the feeding potential and development of *Chrysoperla carnea* (Stephens). *J. Cotton Res. Dev.* 21: 118-119.

Manjunath, T.M., 1985. *Maconellicoccus hirsutus* on grapevine. *FAO Pl. Prot.* 33:74.

Manjunath, T.M., 1993. Biocontrol of *Helicoverpa armigera* and other cotton bollworms. *Indian Fmg.* 42: 18-23.

Mann, G. S. & Sagar, P., 1987. Activity and abundance of flower visiting insects of loquat, *Eriobtrya japonica* (Thumb.) Linsw. *Indian J. Hort.* 44: 123-125.

Mannan, V. D., Varma, G. C. & Brar, K. S., 1995. Seasonal fluctuations and host predator releatiohsip of *Chrysoperla carnea* (Stephens) (Chrysopidae: Neuroptera). *Indian J. Ecol.* 22: 21-26.

Mathew, M. J., Saju, K. A. & Venugopal, M. N., 1998. *Chrysoperla carnea* Stephens (Chrysopidae: Neuroptera) - a potential biocontrol agent of cardamom aphid (*Pentalonia nigeronervosa* van der Goot). *J. Spices Aromatic Crops* 7: 155-156.

Maurya, R P., Khan, M A., Kumar, S & Kumar, A., 2009. Artificial selection of green lacewing, *Chrysoperla carnea* Stephens for tolerance to insecticides. *Annals Pl. Prot. Sci.* 17: 1-4.

McEwen, P. K., New, T. R & Whittington, A. E., 2001. (Eds). *Lacewings in the Crop Environment.* Cambridge University Press, Cambridge, pp.546.

Mehra, B. P., 1965. Biology of *Chrysopa lacciperda* Kimmins. *J. Bombay Nat. Hist. Soc.* 63: 215-218.

Mehra, B. P., 1966. Biology of *Chrysopa madestes* Banks (Neuroptera: Chrysopidae). *Indian J. Ent.* 27: 398-407.

Mehto, D. N., Singh, K. M. & Singh, R. N., 1986. Natural enemy complex on insect pest complex in chick pea *Cicer arietinum* Linn. *Bull. Entomol.* 27:1-12.

Mishra, B.K., Jena, B.C. & Mishra, P.R., 1996a. Biology and feeding potential of *Chrysopa scelestes* Banks feeding on the eggs of *Pyrilla perpusilla* Walker. *Indian Sugar* 45:757-759.

Mishra, Y.D., Sushi, S.N., Sharma, K. K., Bhattacharya, A. & Jaiswal, A. K., 1996b. Efficacy of selected organophosphorus insecticides for control *of Chrysopa madestes* (Neuroptera: Chrysopidae)- a serious sporadic predator of Indian lac insect, *Kerria lacca* (Kerr.). *New Agriculturist* 7:17-20.

Mohana, D., Swamiappan, M. & Geetha, B., 1998. Improved larval rearing methods for green lacewing, *Chrysoperla carnea* (Stephens) (Chrysopidae: Neuroptera). (In: Reddy, P.P., Kumar, K.K, & Verghese, A. (Eds). *Advances in IPM for Horticultural Crops*). *Proceedings of the First National Symposium on Pest Management in Horticultural Crops: Environmental Implications and Thrusts*, Bangalore, Pp. 225-229.

Mohite, P. B. & Uthamasamy, 1998. Host plant resistance and natural enemies interaction in the management of *Helicoverpa armigera* (Hubner) on cotton. *Indian J. Agric Sci.* 32: 28-30.

Muralidharan, C.M. & Chari, M.S., 1990. Effect of electrostatic and conventional spraying system on biocontrol agents in hybrid 6 cotton. *Pl. Prot. Bull.* 42: 21-24.

Naik, R.L., Lolage, G.R., Kale, V.D. & Dethe, M.D., 2006. Biosafety of flufenzin and fenpyroximate to certain beneficial arthropods. *J. Biol. Control*. 20: 147-152

Nasir, M.M., 1947. Biology of *Chrysopa scelestes* Banks. *Indian J. Entomol.* 9: 177-192.

Natarajan, K., 1990. Natural enemies of *Bemisia tabaci* Gennadius and effect of insecticides on their activity. *J. Biol. Control* 4: 86-88.

Needham, J.G., 1909. Notes on the Neuroptera in the collection of Indian Museum, *Rec. Ind. Mus.* 3: 185-210.

New, T.R. 2001. Introduction to the Neuroptera: what are they and how do they operate? (In: McEwen, P.K., New, T.R & Whittington, A.E. (Eds). *Lacewings in the Crop Environment*). Cambridge University Press, Cambridge, pp. 3-5.

Panchbhai, P.R., Sharnagat, B.K., Dhawad, C.S., Wadaskar, R.M. & Bagade, I.B., 2005. Effect of augmentation of *Trichogramma chilonis* Ishii and *Chrysoperla carnea* (Stephens); two natural enemies of *Helicoverpa armigera* (Hubner) on larval infestation and yield in cotton. *Pest Mgmt. Econ. Zool.* 13: 77-81.

Pandey, R. & Singh, N.N., 2008. Effectiveness of biocontrol based IPM modules against *Lipaphis erysimi* Kaltenbach (Hemiptera: Aphididae). *J. Pl. Prot. Res*. 48: 1-7.

Patel, A.G., Yadav, D.N. & Patel, R.C., 1988. Improvement in mass rearing technique of green lacewing *Chrysopa scelestes* Banks (Neuroptera: Chrysopidae). *GAU Res. J.* 14: 1-4.

Patel, H.K. Patel, J.R. & Patel, S.N., 1975. Records of predators and their parasites from Gujarat. *Entomolgists Newsletter* 5: 40.

Patel, H. M. & Yadav, D. N., 1991. Occurrence of mottled emigrant *Catopsila pyranthe* (Lepidoptera: Pieridae) and its natural enemies at Anand. *Indian J. agric. Sci.* 61: 789-791.

Patel, I.S. & Yadav, D.N., 1995. Susceptibility of *Amrasca biguttula biguttula* and *Chrysopa scelestes* in cotton (*Gossypium* species) to three systemic insecticides. *Indian J. Agric. Sci.* 65: 308-309.

Patel, I.S. & Yadav, D.N., 1996. Studies on genetic improvement of green lacewing, *Chrysopa scelestes* Banks (Chrysopidae: Neuroptera) selection for monocrotophos resistance. *GAU Res. J.* 21:155.

Patel, I.S. & Yadav, D.N., 1997. Cross resistance of monocrotophos -resistant strain of green lace wing. *Indian J. agric. Sci.* 67:83-84.

Patel, I.S., Yadav, D.N. & Shukla, Y.M. 1996. Biochemical basis of mechanisms in a monocrotophos resistant strain of the predator, green lace wing, *Chrysopa scelestes* Banks. *Annals of Plant Prot. Sci.* 4: 151-153.

Patel, K.G. & Vyas, H.N., 1985. Ovipositional site preference by green lacewing, *Chrysopa (Chrysoperla) scelestes* cotton and green gram. *GAU Res. J.* 10: 79-80.

Patel, K.G. & Vyas, H.N., 1985a. Biology of green lacewing *Chrysosa (Chrysoperla) scelestes* Banks (Neuroptera: Chrysopidae) an important predator in Gujarat. *GAU Res. J.* 11:18-25.

Patel, K.G. & Vyas, H.N., 1985b. Ovicidal evaluation of certain insecticides against the eggs of green lacewing, *Chrysopa scelestes* Banks - an important predator under laboratory condition. *Indian J. Ent.* 47: 32-35.

Patel, R K & Thakur, B. S., 2005. Insect pest complex and seasonal incidence in linseed with particular reference to bud fly (Dasineura lini Barnes). *J. Pl. Prot. Env.* 2: 102-107.

Patel, R. C., Yadav, D. N. & Patel, J. R., 1976. Natural control of groundnut aphid, *Aphis craccivora* Koch in Central Gujarat. *Curr. Sci.* 45: 34-35.

Patil, A. S., Magar, S B. & Shinde, V. D., 2007. Biological control of the sugarcane Woolly Aphid (*Ceratovacuna lanigera*) in Indian sugarcane through the release of predators. XXVI-Congress, International Society of Sugarcane Technologists, ICC, Durban, South Africa, pp. 797-804.

Patil, N. G. & Bhosale, P. V., 2007. Impact of sugarcane woolly aphid, *Ceratovacuna lanigera* Zehntner (Aphididae: Homoptera), on production and recovery of sugarcane and their natural enemies distribution in Western Maharashtra. *Indian Sugar* 57: 31-38.

Patil, R H. & Hallolli, S. P., 2007. Biocontrol potential of *Chrysoperla carnea* (Stephens) on *Melanaphis sacchari* (Zehntner) and *Uroleucon compositae* (Theobald). *J. Biol. Control.* 21(Special): 163-166.

Patel, Y. R., Yadav, D. N. & Pathak, H. C., 1997. Heterosis in green lacewing, *Chrysoperla carnea* Stephens (Neuroptera: Chrysopidae). *Annals Pl. Prot. Sci.* 5(1): 48-52.

Patnaik, N. C. & Bhagat, K. C., 1984. Studies on the life history of *Chrysopa orestes* Banks (Neuroptera: Chrysoipdae) with notes on its predatory habits. *Pranikee* 5:1-5.

Patnaik, N. C. & Bhagat, K. C. 1989a. Note on the occurrence of aphidophagous insect predators in Puri district (Orissa) and their predation on the sorghum aphid, *Longiunguis sacchari* (Zhnt). *Indian J. Agric. Sci.* 47 : 585-586.

Patnaik, N. C. & Bhagat, K. C., 1989b. Life history of predator, *Chrysopa orestes* Banks and effect of an artificial diet on adult longevity and fecundity. *Bull. Ent.* 30: 143-147.

Patnaik, N. C., Satpathy, J. M. & Bhagat, K. C. 1977. Notes on the occurrence of aphidophagous insect predators in Puri district (Orissa) and their predation on the sorghum aphid, *Longiunguis sacchari* (Zhnt). *Indian J. Agric. Sci.* 47: 585-586.

Patro, B. & Behera, M. K., 2002. Biology and feeding potential of *Chrysoperla carnea* (Stephens) (Neuroptera: Chrysopidae) on the bean aphid. *J. Biol. Control* 16: 77-79.

Pawar, C. S., Bhatnagar, V. S. & Jadhav, D. R., 1985. Some predatory insects and their parasites. *Sci. Cul.* 51: 101-102.

Pillai, G. B. & Nair, K. R., 1993. A check list of parasitoids and predators of *Opisina arenosella* Wlk on coconut. *Indian coconut J. Cochin* 23 : 2-9.

Pophaly, D. J. & Gupta, A. K., 1996. Preliminary studies on insect pests and associated parasitoids on *Lathyrus sativus. Curr. Res. UAS Bangalore* 25: 51-53.

Prasad, T. M., Ghorpade, S. A., Pokharkar, D. S., Bade, B. A & Chandele, A. G., 2005. Effect of biointensive pest management on sucking pests, bollworms and natural enemies in MECH 162 Bt cotton in Maharashtra. *Insect Environment* 11(2): 64-66.

Preetha, P., Stanley, J., Manoharan, T., Chandrasekaran, S. & Kuttalam, S., 2009. Toxicity of imidacloprid and diafenthiuron to *Chrysoperla carnea* (Stephens) (Neuroptera: Chrysopidae) in the laboratory conditions. *J. Plant Prot. Res.* 49: 290-296.

Pushpalatha, N.A., Singh, S. P., Bakthavatsalam, N. & Bhummannavar, B. S., 1994. Semisynthetic larval diet for *Chrysoperla carnea* (Stephens) (Neuroptera: Chrysopidae). *J. Biol. control* 8 : 54-56.

Rabindra, R. J., Muthuswami, M. & Jayaraj, S. 1994. Feasibility of nuclear polyhedrosis virus and *Chrysoperla carnea* (Stephens) use in *Helicoverpa armigera* management on chick pea. *Pest Mgmt. Eco. Zool.* 2:31-34.

Radhakrishnan, B. & Muraleedharan, N., 1989. Life history and population dynamics of *Micromus timidus* Hagen a predator of the tea aphid, *Toxoptera aurantii* (Boyer de Fons Colombe). *J. Plantn Crops* 16: 189-194.

Raguraman, S. & Uthamasamy, S., 2005. Laboratory and field evaluations of spinosad, an actinomycete derivative, on cotton bollworms and their natural enemies. *Biopesticides International* 1: 174-184.

Rahman, R.A., 1940. Important insect predators of India. *Proc. Indian Acad. Sci.* 12:67-74.

Rajabaskar, D., 2007. Stage preference and predatory potential of the bug *Chrysoperla carnea* against jasmine leaf webworm Nausinoe geometralis. *J. Ecobiol.* 19: 97-99.

Rajan, P., 1981. Biology of *Pentalonia nigronervosa* f. Caladii van der Goot, vector of Katte disease of cardamom. *J. Plantn. Crops* 9: 34-41.

Rajaram, V., Mathirajan, V. G & Krishnasamy, S., 2006. IPM in cotton under dry farming condition. *Interntl J. Agric. Sci.* 2: 557-558

Rajasekhar, D. W., Rachappa, V. H. & Awaknavar, J. S., 1999. Role of *Chrysoperla carnea* Stephens and insecticides in suppression of castor mite. *Insect Environment* 4: 151.

Rajendran, B., 2005. Life cycle of sugarcane woolly aphid and the potentiality of its predators at Cuddalore Region, Tamil Nadu. *Indian Sugar* 55:75-78.

Ramegowda, G. K, Vidya Mulimani., Patil, R. K. & Lingappa, S., 2006. Record of pteromalid parasitoid, *Inkaka keraliensis* Sureshan and Narendran on *Micromus igorotus* Banks. *Insect Environment* 12: 12-14.

Ramkumar, J., More, G.D; Rajkumar, M.B & Thiruvarasan, S., 2005a. Predation potential of *Mallada boninensis* (Okamoto) on different stages of *Aleurocanthus woglumi* Ashby. *J. Applied Zool. Res.* 16: 19-20.

Ramkumar, J. and More, G.D; Rajkumar, M.B. & Thiruvarasan, S. 2005b Biology of *Mallada boninensis* (Okamoto) on different stages of *Aleurocanthus woglumi* Ashby. *J. Applied Zool. Res.* 16: 17-18.

Rao, C.N., Shivankar, V.J. & Singh, S., 2003. Predatory potential and development of the green lacewing (*Mallada boninensis*) on citrus leaf miner (*Phyllocnistis citrella*). *Indian J. agric. Sci.* 73: 60-61.

Rao, C.N., Shivankar, V.J. and Shyam Singh, 2005. Evaluation of bio agents against citrus leaf miner, *Phyllocnistis citrella* Staint. *J. Appl. Zool. Res.* 16: 23-25.

Rao, G.M. V.P., Venkateswarlu, P., Raju, K. & Sivaramakrishna, M., 2007. Role of tall growing millet crops in the biological suppression of aphid, *Myzus nicotianae* Blackman in bidi tobacco. *J. Biol. Control* 21(Special): 55-58.

Rao, N.S. & Raguraman, S., 2005. Influence of neem based insecticides on egg parasitoid, *Trichogramma chilonis* and green lace wing predator, *Chrysoperla carnea*. *J. Ecobiol.* 17: 437-443.

Rao, N.V., Reddy, A.S. & Reddy, D.D.R., 1989. Natural enemies of cotton whtiefly, *Bemisia tabaci* Gennadius in relation to hsot population and weather factors. *J. Biol. Control* 3: 10-12.

Rao, N.V., Reddy, A.S. & Reddy, D.D.R., 1990. Effect of some insecticides on the parasitoids and predator of the cotton whitefly, *Bemisia tabaci* Genn. *J Biol. Control* 4: 4-7.

Rao, P.R.M., Kanakaraju, A., Apparao, R.V. & Azam, K.M., 1984. Predators on mealy bug on mesta. *FAO Asia Pacific Pl.Prot. Commission* 27: 4.

Rao, R.S.N. & Chandra, I.J., 1985. Biological control of *Myzus persicae* Sulz. *Indian J. agric. Sci.* 55: 654-656.

Rao, R.S.N. & Satayanaraya, S.V.V., 1984. Note on more additions to the natural enemy complex of *Spodoptera litura* F. and *Myzus persicae* Sulz. in tobacco. *Curr. Sci.* 53: 201.

Rao, R.S.N., Satayanaraya, S.V.V. & Soundararajan, V., 1981. Note on new additions to the natural enemies of *Spodoptera litura* and *Myzus persicae* Sulz. on flue cured tobacco in Andhra Pradesh. *Sci. Cult.* 47: 98-99.

Rao, R.S.N., Sreedhar, U., Rao, S.G. & Satyavani, J.V.R., 1994. Bio- environmental control of *Heliothis armigera* in FCV tobacco. *Tobacco Res.* 20: 36-39.

Rao, S.G., Venkateswarlu, P., Prasad, J.V., Raju, K.S., Singh, S.P & Naik, P. K., 2008. Tobacco type mediated effects on the development of pink aphid, *Myzus nicotianae* Blackman and its predator, *Chrysoperla carnea* (Stephen) (Neuroptera: Chrysopidae). *J. Biol. Control* 22: 155-161.

Rathore, V.S., 1983. Bionomics and management of safflower aphid. *Indian Fmg.* 33: 32-33.

Raychaudhuri, D.N., Ghosh., Poddar, S.C. & Ghosh, S.K., 1981. Notes on phidophagous insect *Micromus timidus* Hagen (Neuroptera: Hemerobiidae). *Sci. Cul.* 47: 223-224.

Redborg, K.E., 1998. Biology of Mantispidae. *Ann. Rev. Ent.* 43:175-194.

Reddy, G.V.P. & Manjunatha, M., 2000. Laboratory and field studies on the integrated pest management of *Helicoverpa armigera* (Hubner) in cotton, based on pheromone trap catch threshold level. *J. Appl. Ent.* 124 : 213-221.

Reddy, G.V.P. & Guerro, A., 2000. Pheromone based integrated pest management to control the diamond back moth *Plutella xylostella* in cabbage fields. *Pest Mgmt. Sci.* 56: 882-888.

Reddy G. R. S. & Divakar, B.J., 1998. Relative toxicity of nine insecticides *to Bracon kikpatricki* and *Chrysoperla carnea. J.Appl. Ent.* 124: 213-221.

Sagar, P. & Kumar, N., 1996. Natural enemies of *Hyadaphis corandri* and feeding rate of *Coccinella septempunctata* on it in the Punjab. *Intl. Pest Control* 38:26-27.

Saminathan, V.R., Baskaran R.K.M & Mahadevan, N.R., 1999. Biology and predatory potential of green lace wing (*Chrysoperla carnea*) (Neuroptera: Chrysopidae) on different host insects. *Indian J. Agric. Sci.* 69: 502-505.

Samui, R.C. & Roy, A., 2007. Effect of growth regulators on growth, yield and natural enemies of potato. *J. Crop Weed* 3: 35-36.

Sanjeev Rai., Singh, N.N & Monobrullah, M., 2007. Efficacy of management modules against *Lipaphis erysimi* on mustard. *Annals Pl. Prot. Sci.* 15: 251-253.

Santharam, G., Victoria, D.R., Rabindra, R.J. & Jayaraj, S., 1994. Studies on biological control of pigeon pea pod borers in India. *Anz. Sch. Pflanz. Umw.* 67: 103-106.

Sarade, S.V. & Sonalkar, V.U., 1999. Ovicidal effect of some insecticides against *Chrysoperla carnea* (Stephens). *Pesticides Res. J.* 11: 97-98.

Sardana, H.R., Bambawale, O.M & Batra, P., 2008. Wider area validation and economic analysis of adaptable IPM technology in egg plant *Solanum melongena* L. in a farmers' participatory approach. *Pesticide Res. J.* 20: 204-209.

Sardana, H.R., Bambawale, O.M., Singh, D.K. & Kadu, L.N., 2006. Conservation of natural enemies through IPM in brinjal (*Solanum melongena* L.) fields. *Entomon* 31: 83-88.

Sathiamma, B., Jayapal, S.P. & Pillai, G.B., 1985. Record of *Ankylopteryx octopunctata candida* (Fabricius) (Neuropeta: Chrysopidae), as egg and larval predator on *Opisina arenosella* Wlk. the leaf eating caterpillar of the coconut palm. *Curr. Sci.* 54: 1128.

Sathiamma, B., Nair, K.R.C. & Soniya, V.P., 1998. Record of natural enemies of the lace bug *Stephanitis typica* (Distant) a pest on coconut palm. *Entomon* 23: 321-324.

Satpute, P.P., Taley, Y.M., Nimbalkar, S.A. & Kadu, N.R., 1986, Natural enemies of citrus black fly, *Aleurocanthus woglumi* Ashby in Nagpur. *Bull. Ent.* 27: 128-131.

Satpute, N., Katole, S., Nimbalkar, S. & Satpute, U., 2002. Attraction of seed treatment of imidacloprid and thiamethoam to the population of *Cheilomenes sexmaculata* (Fabr.) and *Chrysoperla carnea*(Stephesns) on cotton. *J.Biol.Control* 16: 81-83.

Satyanarayana, J., Singh, K.M., Singh, R.N. & Singh, S., 1998. Influence of intercropping and insecticide, II. Natural enemy complex in rice bean *Vigna umbellata*. *Indian J. Ent.* 60:69-73.

Saxena, R.C., 1981. Observations on some predators and parasites of *Thrips tabaci* Lin. *Bull. Ent.* 22: 97-100.

Selvakumaran, S., David, B.V. & Kumaresan, D., 1996. Observations on the natural enemies of the whitefly, *Kanakarajiella cardamomi* (Davi and Subramaniam), a pest on cardamom. *Indian J. env. Toxicol.* 6: 26-27.

Selvasundaram, R. & Muraleedharan, N., 1987. Natural enemies of certain leaf folding caterpillar pests of tea in Southern India. *J. Coffee Res.* 17 : 118-119.

Sen, S.N. & Verma, T.D., 1998. Population dynamics of endemic natural enemies of *Heteropsylla cubana* Crawford (Homoptera: Psyllidae) in Himachal Pradesh. *Indian J. Forestry* 21: 188-191.

Shanmugam, P.S., Balagurunathan, R & Sathiah, N., 2006. Biointensive integrated pest management in Bt cotton. *Intl J. Zool. Res.* 2: 116-122.

Shantibala, K., Somen, Debaraj, Y. & Singh, T. K., 1994. Development and predatory efficiency of larvae of *Micromus timidus* Hagen (Neuroptera: Hemerobiidae) on an oak aphid, *Cervaphid quercus* Takahashi. *Indian J. Hill Fmg,* 7: 212-214.

Sharanabasava, H., Kulkarni, K A., Mallapur, C. P. & Goud, K B., 2008. Population dynamics of sugarcane woolly aphid, *Ceratovacuna lanigera* Zehntner and its natural enemies. *Annals Pl. Prot. Sci.* 16: 384-388.

Sharanabasava, H. & Manjunatha, M. 1998. Reproductive biology and feeding potential of *Chrysoperla carnea* Stephens (Neuroptera: Chrysopidae) *on Tetranychus neocaledonicus* Andre (Acari: Tetranychidae). *Annals Pl. Prot. Sci.* 6: 115-118.

Sharanabasava, H., Manjunatha, M. & Lingappa, S. 1998. Predator prey interaction between *Chrysoperla carnea* (Stephens) (Neuroptera: Chrysopidae) and *Tetranychus neocaledonicus* Andre (Acari: Tetranychidae). *Karnataka J. Agric. Sci.* 11: 206-207.

Sharma, H.C., Arora, R., Pampapathy, G. 2007. Influence of transgenic cottons with *Bacillus thuringiensis* cry1Ac gene on the natural enemies of *Helicoverpa armigera*. *BioControl* **52**: 469-489.

Sharma, K., Khan, M.A. & Kumar, A., 2008. Effect of insecticides against *Chrysoperla carnea* eggs. *Annals Pl. Prot. Sci.*16: 223-224.

Sharma, S.K., Madan, Y.P. & Chaudhary, J.P., 1985. A record of hyperparasites on *Chrysopa* sp. and *Nimboa basipunctata* Withycombe in India. *Bull. Entomology* 26: 208-209.

Shivankar, V.J., Rao, C.N. & Singh, S., 2001. Studies on selectivity of botanicals against *Mallada boninensis* Okamoto. Pap. Abstr. *IPM in Horticultural Crops. New Molecules-Biopesticides and Environment.* p 213.

Shivankar, V.J. & Singh, S., 2001. *Citrus insect pests*. NRCC Nagpur, pp. 252.

Singh, D. & Singh, H., 1994. Predatory potentiality of immature stages of lacewing, *Chrysopa* sp. and syrphid *Episyrphus balteatus* (Degears) over mustard aphid *Lipaphis erysimi*. *Crop Res., Hisar* 7: 116-119.

Singh, J.P., Jaiswal, A.K., Monobrullah, M. & Bhattacharya, A., 2009. Response of some selected insecticides on neuropteran predator (Chrysopa lacciperda) of lac insect (Kerria lacca). *Indian J. Agric. Sci.* 79: 727-731.

Singh, J., Sohi, A.S., Dhaliwal, Z.S. & Mann, H.S., 1993. Comparative incidence of *Helicoverpa armigera* (Hub.) and other insect pests on okra and sunflower intercrops in cotton under Punjab conditions. *J. Insect Sci.* 6:137-138.

Singh, K., Sharma, U.S., Swaminathan, R. & Dashora, P., 2009a. Management of insect pests of *Vigna radiata* (L.) Wilczek. *Applied Ecol. Environ. Res.* 7: 99-109.

Singh, L.S., Devrajani, P., Debraj, Y. & Singh, T.K., 1994. Studies on the seasonal incidence of *Myzus persicae* (Sulzer, Homoptera: Aphiididae) on cabbage in relation to abitocic and biotic factors. *Proc. Nat. Acad. Sci. B Biol. Sci.* 64: 95-98.

Singh, P.P. & Varma, G.C., 1986. Comparative toxicities of some insecticides to *Chrysoperla carnea* (Chrysopidae: Neuroptera) and *Trichogramma brasiliensis* (Trichogrammatidae: Hymenoptera). Two arthropod natural enemies of cotton pests. *Agric. Ecos. Env.* 15: 23-30.

Singh, P.P. & Varma, G.C., 1989. Evaluation of the suitability of spent or dead *Corcyra cephalonica* (Pyralidae: Lepidoptera) moths as food for rearing *Chrysoperla carnea* (Chrysopidae: Neuroptera). *Biological Wastes* 28: 309-311.

Singh, S.K., 2008. Comparative feeding potential of aphidophagous pyralid, *Dipha aphidivora* (Meyrick) and brown lacewing, *Micromus igorotus* Banks on sugarcane woolly aphid, *Ceratovacuna lanigera* Zehntner. *Sugar Cane Intl.* **26**: 18-20.

Singh, S.K., Tripathi, G.M., Omkar & Mishra, G., 2009b. Life table analysis and intrinsic rate of natural increase of *Micromus igorotus* Bank (Neuroptera: Hemerobiidae), a predator of sugarcane woolly aphid, *Ceratovacuna lanigera* Zehntner (Hemiptera: Aphididae). *Sugar Cane International* 27: 26-29.

Singh, S.P., 1989. *Achievements of AICRP on Biolgical control of crop pests and weeds.* Technical Bulletin 2.Biological Control Centre, NCIPM, Faridabad. 20pp.

Singh, S.P., 1996. Biological Control in India. In *IPM systems in Agriculture.* Ed. Upadyay, R.K., Mukerji, K.G and Razak, R.L. Aditya Books Pvt Ltd, New Delhi. Pp 216-328.

Singh, S.P. & Narasimham, A. U., 1992. *Indian Chrysopidae.* Tech Bull. Biological Control Centre, NCIPM, Faridabad. 34 pp.

Singh, S.P., Bhummannavar, B.S., Jalai, S.K., Bakthavatsalam, N. & Pushpalatha, N.A., 1994a. *Production and Use of Chrysopid predators.* Technical Bull. Project Directorate of Biological Conrol, No 10: 14pp.

Singh, S.P., Bhummannavar, B.S., Jalai, S.K., Bakthavatsalam, N. & Pushpalatha, N.A., 1994b. *Chrysopids and Trichogrammatids: strain selection and Utilization.* Technical Bull. Project Directorate of Biological Conrol, No 9: 45 pp.

Singh, S.P., Rao, N.S. &Henneberry, T.J. 1993. *Leafhoppers and their natural enemies.* Technical Bull. Project Directorate of Biological Conrol, No 6: 65 pp.

Sontakke, B.K., Das, N & Swain, L. K., 2007. Bio efficacy of emamectin benzoate against boll worm complex in cotton *Annals Pl. Prot. Sci.* 15: 1-3.

Srinivasan, G. & Babu, P.C.S., 2000. Toxicity of certain insecticides to predator, green lacewing *Chrysoperla carnea* Stephens (Neuroptera: Chrysopidae). *J. Biol. Control* 14:5-7.

Srivastava, R.C., Quereshi, Q. G. & Mathur, N.M., 1996. A new field application technology for *Chrysoperla carnea. J. Biol. Control* 10: 53-59.

Surulivelu, T. & Kumaraswami, T., 1989. Effect of skip arow coverage of insecticides application on some sucking pests and their predators in cotton. *J. Biol. Control,* **3**: 77-79.

Suryawanshi, D.S., Bhede, B.V., More, D.G & Zanwar, P.R., 2007. Management of insecticide resistance in cotton bollworms in Hingoli and Amravati districts of Maharashtra (India). *Resistant Pest Mgmt Newsletter* 16: 16-18.

Tanwar, R.K., Bambawale, O.M., Jeyakumar, P., Dhandapani, A., Vikas Kanwar., Sharma, O.P. & Monga, D. 2007. Impact of IPM on natural enemies in irrigated cotton of North India. *Entomon* 32: 25-32.

Thakur, J.N. & Pawar, A.D. 1986. New records of *Syrphus confrator* Wiedmann and *Chrysopa* sp. as predators of apple wooly aphid *Eriosoma lanigerum* Hausmann from Kullu Valley (H.P) in India. *Proc. Nat. Acad. Sci.* 9:91-92.

Thakur, J.N., Singh, J.P., Verma, O.P. & Diwakar, M.C. 1995. Bioecological studies on gram pod borer *Heliothis* species under Jammu conditions. *J. Adv. Zool.* 16**:** 118-122.

Thakur, J.N., Pawar, A.D. & Rawat, U.S. 1988. Observations on the correlation between population density of apple wooly aphid and its natural enemies and their effectiveness in Kullu Valley (H.P). *Plant Prot. Bull.* 40: 13-15.

Thakur, J.N., Pawar, A.D. & Rawat, U.S., 1992. Apple wolly aphid, *Eriosoma lanigerum* Hausmann (Hemiptera: Aphididae) and post release impacts of its natural enemies in Kullu Valley. *Pl. Prot. Bull.* 44: 18-20.

Thilagavathy, G. & Ali, K.A. 1998. Effect of palmarosa oil on *Trichogramma chilonis* Ishii and *Chrysoperla carnea* (Stephens). In Reddy, P.P., Kumar,

K.K. and Verghese, A. (Eds). *Advances in IPM for Horticultural Crops. Proceedings of the First National Symposium on Pest Management in Horticultural Crops: Environmental Implications and Thrusts*, Bangalore India 15-17 October 1997. Pp. 213-214.

Thirumurthi, S., 1995. Insecticidal effects on predator populations in tobacco fields. *Insect Environment* 1: 10-11.

Thite, N.R. & Shivpuje, P.R., 1999. Biology and feeding potential and development of *Chrysoperla carnea* on *Aphis gossypii* (Glover). *J. Maharastra Agric. Univ.* 24:243-244.

Thontadarya, T.S. and Nangia, N., 1983. Natural enemies of tobacco caterpillar infesting soybean. *Curr. Res. UAS, Bangalore* 12: 105-106.

Tjeder, B., 1960. Neuroptera from Newfoundland, Miquelon, and Labrador. *Opuscula Entomologia* 25: 146-149.

Tripathi, G.M., 1995. Record of parasite and predator complex of sugarcane wooly aphid, *Ceratovacuna lanigera* Zehnt. in Nagaland. *Indian Sugar* 44: 839-841.

Vanderzant, E.S., 1969. An artificial diet for larvae and adults of *Chrysopa carnea* an insect predator of crop pests. *J. Econ. Ent.* 62: 256-257.

Varma, G.C. & Shenhmar, M., 1983. Some observations on the biology of *Chrysoperla carnea* (Stephens) (Chrysopidae: Neuroptera). *J. Res. PAU* 20: 222-223.

Venkatesan, S., Balasubramanian, G., Babu, P.C.S. and Sivaram, M.R., 1997. Use of nuclear polyhedrosis virus and green lacewing, *Chrysoperla carnea* Stephens for *Helicoverpa armigera* (Huebner) management on sunflower. *Pest Mgmt Econ. Zool.* 5: 63-66.

Venkatesan, T., Jalali, S.K., Murhty, K.S., Rabindra, R.J. & Lalitha, Y., 2009. Occurrence of insecticide resistance inf field populations of *Chrysoperla zastrowi arabica* (neuroptera; Chrysopidae) in India. *Indian J. Agric. Sci.* 79: 910-912.

Venkatesan, T., Poorani, J., Murthy, K S., Jalali, S K., Kumar, G A., Lalitha, Y & Rajeshwari, R 2008. Occurrence of *Chrysoperla zastrowi arabica* Henry et al. (Neuroptera: Chrysopidae), a cryptic song species of *Chrysoperla (carnea* group), in India. *J. Biol. Control.* 22: 143-147.

Venkatesan, T., Singh, S.P. & Jalali, S.K. 2000. Rearing of *Chrysoperla carnea* (Stephens) (Neuroptera: Chrysopidae) on semisynthetic diet and its predatory efficiency against cotton pests. *Entomon* 25: 81-89.

Venkatesan, T., Singh, S.P., Jalali, S.K. & Sadhana, P., 2002. Rearing of *Mallda astur* (Banks) (Neuroptera: Chrysopidae) on a semi synthetic diet. *Pest Mgmt Hortic. Ecosys.* 8: 121-126.

Vennila, S., 1998. Relationship between sucking pests *Amrasca biguttula biguttula, Aphis gossyii*, and their predators (*Cheilomenes sexmaculata, Chrysoperla carnea*) on cotton cultivars. *J.Ent. Res.* 22: 349-353.

Vennila, S., Biradar, V.K. & Panchbhai, P.R., 2007. Coccinellids and chrysopids as native predators of sucking pests in relation to rainfed cotton production system. *J. Biol. Control*. 21(Special): 65-71.

Vidya, M., Lingappa, S., Patil, R.K. & Ramegowda, G.K., 2007a. Host range, biology and feeding potential of brown lacewing, *Micromus igorotus* Banks. *J. Biol. Control*, 21(Special): 167-171.

Vidya, M., Lingappa, S., Patil, R.K. & Ramegowda, G. K., 2007b. A simplified technique for mass multiplication of *Micromus igorotus* Banks (Neuroptera: Hemerobiidae) on sugarcane woolly aphid, *Ceratovacuna lanigera* Zehntner. *J. Biol. Control*. 21(Special): 141-147.

Vidya, M., Lingappa, S., Patil, R.K & Ramegowda, G.K., 2008a. Biology and feeding potential of *Micromus australis* Hagen on sorghum aphid, *Melanaphis sacchari* (Zehntner). *J. Biol. Control*. 22: 467-470.

Vidya, M., Lingappa, S., Patil, R.K. & Ramegowda, G.K., 2008b. Toxicity of chemicals and entomopathogenic fungi to different stages of brown lacewing, *Micromus igorotus* banks (Neuroptera: Hemerobiidae). *Pesticide Res. J.* 20: 163-166.

Vidya, M., Lingappa, S., Patil, R.K. & Ramegowda, G.K., 2009. Intraguild predation studies between potential predators of sugarcane woolly aphid, *Ceratovacuna lanigera* Zehntner (Homoptera: Aphididae). *J. Ent. Res.* 33: 93-96.

Vidya, M., Patil, R.K., Ramegowda, G.K. & Lingappa, S., 2008c. Record of brown lacewing, *Micromus australis* Hagen (Neuroptera: Hemerobiidae) from cotton and sorghum ecosystem. *J. Biol. Control* 22: 189-190.

Vidya, M., Ramegowda, G.K., Patil, R.K. & Lingappa, S. 2008d. Studies on ovipositional preference of brown lacewing, *Micromus igorotus* Banks (Neuroptera: Hemerobiidae). *J. Ent. Res.* 32: 135-137.

Vidya, M., Lingappa, S., Patil, R.K. & Ramegowda, G.K.M., 2009. Seasonal influence on biology and feeding potential of the brown lacewing, *Micromus igorotus* Banks (Neuroptera: Hemerobiidae) on sugarcane woolly aphid, *Ceratovacuna lanigera* Zehntner. *J. Ent. Res.* **33**: 143-147.

Viji. C.P. & Gautam, R.D., 2005. Feeding potential and cost of production of green lacewing *Chrysoperla carnea* (Stephens) on alternate hosts. *Annals Pl. Prot. Sci.* 13: 271-274.

Wadhwa, S. & Gill, R.S. 2007. Effect of Bt-cotton on biodiversity of natural enemies. *J. Biol. Control* 21: 9-16.

Wells, M.M. & Henry, C.S., 1998. Song, reproductive isolation and speciation in cryptic species of insects : a case study using green lacewings. (In: *Endless Forms: Species & Speciation.* Howard, D. K. and Berloches, S. (Eds), pp. 217-233.

Yadav, D.N. & Patel, A.R., 1993. Effect of some botanical insecticides on oviposition of *Chrysopa scelestes* and their ovicidal action. *Botanical pesticides in Integrated pest management*. pp. 166-169.

Yadav, D.N., Patel, K.A. & Patel, R.C., 1983. Massive killing of *Chrysopa* sp. due to the ignorance of its beneficial role. *Indian J, Pl. Prot.* 11: 101.

Zade, S.M., Satpute, R.S., Bhagat, S.M. & Baheti, H.S., 2007. Predatory efficacy of *Mallada boninensis* (Okamoto) against *Aleurocanthus woglumi* Ashby on citrus. *Intl J. Agric. Sci.* 3: 164-166.

Chapter - 24

INDIAN BIODIVERSITY: PROFILE AND CONSERVATION STATUS

Vinod Khanna
Zoological Survey of India, Northern Regional Centre, 218, Kaulagarh Road, Dehra Dun - 248 195, Uttrakhand. India
E-mail: drvkhanna51@gmail.com

INTRODUCTION

Biodiversity, as the term stands today, was coined during the Earth Summit at Rio de Janeiro, Brazil in the early 90s. It is defined as "variety and variability between genes, species and ecosystems" or, to elaborate it further, it manifests itself at three levels, *viz*., **Species diversity** that refers to the numbers and kinds of living organisms, **Genetic diversity** that refers to the genetic variation within a population of species and **Ecosystem diversity that** is the variety of habitats, biological communities and ecological processes that occur in the biosphere.

India's immense biological diversity encompasses ecosystems, populations, species and their genetic make-up. This diversity can be attributed to the vast variety in its physiography and climatic situations resulting in a diversity of ecological habitats ranging from tropical, sub-tropical, temperate and alpine to desert. According to world biogeographic classification, India represents two of the major realms (the Palaearctic and Indo-Malayan) and three biomes (Tropical Humid Forests, Tropical Dry/Deciduous Forests and Warm Deserts/Semi-Deserts). Wildlife Institute of India has a modified classification – which divides the country into 10 biogeographic regions: Trans-Himalayan, Himalayan, Indian Desert, Semi-Arid, Western Ghats, Deccan Peninsula, Gangetic Plain, North East India, Islands and Coasts. Coasts, which too support the characteristic faunal and floral diversity that contributes to the richness of India's biodiversity. In fact, within only about 2% of world's total land surface, India is known to have over 7.45% of the species of animals that the world holds and this percentage accounts nearly for 91,484 species so far known, of which insects alone include 61,220 species.

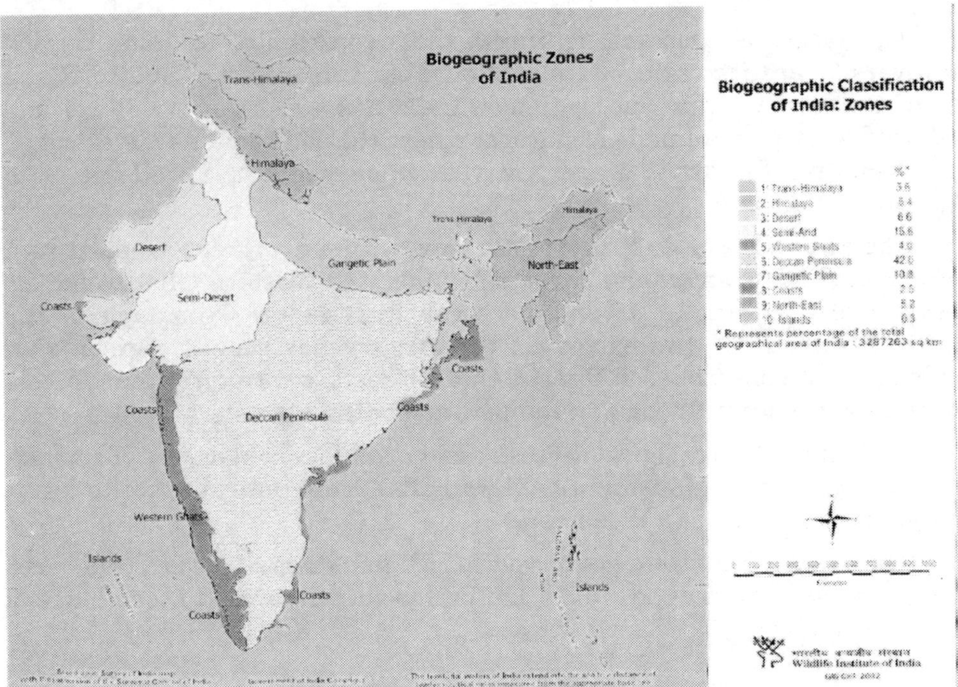

India is one of the twelve-mega biodiversity countries of the world and one of the four in Asia since it contains as much as 7-8% per cent of the world's species. The twelve countries that have been identified as mega biodiversity countries are: India, Brazil, Colombia, Ecuador, Peru, Mexico, Madagascar, Zaire, Australia, China, Indonesia and Malaysia. These countries being located around Tropic of Cancer support a richer assemblage of biodiversity than any other part of the world.

A. ECOSYSTEM DIVERSITY

1. Biogeography and Biomes

India's richness of biological diversity is contributed by many factors, in addition to simply being close to tropics. Although nearly half of India lies outside tropics in the middle latitudes, it is customary to speak of India as a tropical country (Mani, 1974), since the region is shielded of by the Himalayas in the north from the rest of Asia and has the same general type of tropical monsoon climate throughout the land. The moisture laden winds coming from south west hit the Himalayas in the North shed its water as copious rainfall in the north, and as snow in the upper reaches in north. The trans-Himalayan regions in Cold Desert of Ladakh, Pin Valley in Himachal Pradesh and a small segment in Uttarakhand receive precipitation in the form of snow brought by the winds transgressing over the high Himalayas. The area of Cold Desert does not receive regular rainfall.

The other important features that contribute to India's rich biodiversity are its variety in elevation from sub tropical Zone to Naval Zone, and local climate and wetlands. India has a rich variety of wetland habitats, may be manmade or natural where the soil remains waterlogged or submerged for whole or part of year upon which the wetland biota depends.

The panorama of Indian forests ranges from evergreen tropical rain forests in the Andaman and Nicobar Islands, the Western Ghats, and the northeastern states, to dry alpine scrub high in the Himalaya to the north. Between the two extremes, the country has semi-evergreen rain forests, deciduous monsoon forests, thorn forests, subtropical pine forests in the lower montane zone and temperate montane forests.

India also has ecological diversity from rainless cold desert of Ladakh to rainiest place in Mausingram (Garo Hills), from hottest place to arctic and alpine conditions in Himalaya.

The Marine Environment supports rich fish diversity and Coral reefs. A number of Islands opposite Sri Lanka too sustain a good assemblage of the marine biota.

2. India's Zoogeography and Geological History

The whole of the Indian sub-continent is rich not only in biological or ecological diversity but because it lies at the confluence of African, European and Indo-Malayan realms, the biota, therefore, includes, African, European, Eurasian and Mediterranean elements, which together with Indian and endemic elements contributes to the outstanding peculiarities and richness of the characteristic Indian biodiversity. The very idea for the above concept also came from the theory of **Continental Drift** that the continents of south and north America, Africa, Peninsular India, Australia and Antarctica were once united in one land mass (Gondwanaland) are now widely separated by southern Ocean and bear striking similarity of geological history and distribution of ancient and modern organisms.

(i) **The Assam Gate**: In early tertiary, the breakaway Gondwanaland in its northward drift first hit the Asian landmass at what is presently northeast India, served as the biogeographic gateway, "the Assam Gate", for dispersal and migration of much of the fauna and flora. The Northeast Zone is richest of biological resources, and has affinities with Indo-Chinese and Indo-Malayan regions in the east and southeast. From west came the Palaearctic and Ethiopian elements. Relatively young Himalayan mountain ranges opened up new southwards route of migration and acted as a two-way link between West Africa to South Asia. In peninsula there may be some cross over points between southern - Western Ghats and Eastern Hills.

On geomorphologic grounds the India is divided into Peninsular and extra peninsular divisions-correlated with climatic, historical and biogeographical conditions.

(ii) The Indian Peninsula or Deccan Peninsula is highest in the south and west and slopes eastwards. The Eastern and Western Ghats meet south of Mysore from which lofty plateau of Nilgiri is separated. Further to the south and separated from Nilgiri by Palghat Gap-the Cardamon Hills constitute the divide between the Eastern Ghats and Western Ghats.

(iii) Eastern Ghats: The Eastern Ghats (EGs) extend in NE and SW strike in the Indian Peninsula covering an area of 75,000 km2, with an average width of 200 km in the north and 100 km in the south. They extend over a length of 1750 km between river Mahanadi and Vaigai along east coast. The coastal area in the east limits its eastern part. EGs mainly spread through the states of Orissa Andhra Pradesh and Tamil Nadu. The EGs do not form a complete range because of the great rivers Mahanadi, Godavari and Krishna cuts across them. Most of the terrain in Northern part lies about 400m with a few peaks exceeding 1100m. The highest point being the Mahendragiri (1501m) in Ganjim. The EGs run in WSW direction meeting WG in Nilgiris. The region falls under tropical monsoon climate receiving rainfall from both SW monsoon and NE retreating monsoon. The area has (i) Evergreen forests, (ii) Tropical semi-evergreen forests, (iii) Tropical moist deciduous forests, (iv) Southern Tropical dry deciduous forests, (v) Northern mixed dry deciduous forests, (vi) Dry savannah forests, (vii) Tropical dry evergreen forests, and (viii) Tropical dry evergreen Scrub.

(iv) The Extra Peninsular Divide: In the north there is a lofty Satpura range running west to east and acts as a biogeographical barrier. This **Satpura divide** separates extra peninsular area in the north from peninsular area in the south.

(v) Aravali Range extends from South-West to North east through the state of Rajasthan and ends in Delhi. It slopes northwards from Vindhya Range gradually to Indo-Gangetic plains and is one of the oldest mountain systems in the world retaining some relief.

(vi) Indo-Gangetic Plains also known as the **Northern Plains** and **The North Indian River Plain** is a large and fertile plain encompassing most of northern and eastern India's land and named after the Indus and the Ganges, the twin river systems draining the north .The Indo-Gangetic plain is bound on the north by the abruptly rising Himalayas, which feed its numerous rivers and are the source of the fertile alluvium deposited across the region by the two river systems. The southern edge of the plain is marked by the Vindhya and Satpura Range, and the Chota Nagpur Plateau. On the west rises the Iranian Plateau.

(vii) Semi Arid Zones: Two zones of Semi-arid climate are recognized in India, one in North contiguous with the Desert of Thar, extending into Rajasthan, the Punjab, parts of Uttar Pradesh and Gujarat (Katchchh, Saurashtra) and other in the parts of Deccan in Tamilnadu, in south (Bharucha and Meher-Homji, 1965). Analysis of the flora of these regions reveals the 17 types of floral elements representing semi-arid zones. The southern zone has higher percentage of Indian elements, and to some extent of Indo-Malayan, on the other hand the northern zone is richer in North African– Indian Desert, Mediterranean as well as Oriental elements.

(viii) Ranges of Himalaya: Himalaya is the great range of mountains spread over west-northwest to east- southeast for a distance of about 2500 km covering Afghanistan, Pakistan, India Nepal, Bhutan and China. In India, it stretches from Indus Trench below Mt.Nanga Parbat (8125 m) in the west to Yarlungtsanpo-Brahmputra gorge below Namche Barwa Peak (7756m) in the east. In the south of the great Himalaya lies the Shiwalik Range and Lesser Himalaya comprising what is known as Cis-Himalaya and in the north lie the Zanskar, Ladakh and Karakoram comprising the Trans-Himalaya. The Shiwalik formation separates the Himalaya proper from Indo-Gangetic Plains and lies in front of the Himalaya from the bend of Brahmputra to the great defile of River Indus. The Lesser Himalayan Ranges lie between Great Himalaya and Shiwalik Ranges and comprise:

1. Nag Tibba
2. Dhaula Dhar
3. Pir Panjal
4. North Kashmir
5. Mahabharat
6. Mussoorie and
7. Ratanpir

Routes of migration of the fauna: These relatively young mountain ranges opened up new southwards routes for the migration of the species. At the eastern end through Assam gate came the Chinese and Malayan elements and from the west came the Palaearctic and Ethiopian elements. The Himalaya acted as a" two way link" between West Africa to South Asia. In the Peninsula there may be some cross over points between southern Western Ghats and Eastern Hills. The Satpura and Vindhya trends of the mountain ranges were formerly continuous with Assam Himalaya in the east and Western Ghats in the west and thus was assumed (Satpura hypothesis) to have served as a route of migration of specialized hill fauna and flora from east to west.

3. Altitudinal Diversity

Attitudinally the Himalaya has been classified into (i) Nival Zone– 4750m, (ii) Alpine Zone-4500m, (iii) Sub-alpine Zone-4000m, (iv) Montane Zone-300m, (v) Sub-montane Zone-2000m, (vi) Tropical Zone-2000m, and (vii) Subtropical Zone- 1000m. All these zone supporting a characteristic faunal and floral diversity-exclusively peculiar to the area.

4. Hotspots: India figured with two hotspots in an identification of 25 of the world's biologically richest and most threatened ecosystems. These 2 hotspots are the Western Ghats/Sri Lanka and the Indo-Burma region (covering the Eastern Himalayas).The hotspots are the areas with higher concentration of endemic species and which usually experience rapid rate of habitat modifications and loss.

(i) **Western Ghats:** Western Ghats or Sahyadri Hills stretch from River Tapti in the north to Kanya Kumari in the south, parallel to the west coast of peninsular India, through the state of Gujarat, Maharashtra, Goa, Karnataka, Tamil Nadu and Kerala.

- One of the major Tropical Evergreen areas.
- The second "Hot spot" for Biodiversity, the area is rich in endemics.
- All the major Peninsular Rivers have their origin in WGs and thus WGs forms a major watershed in Peninsular India.
- Unlike extra peninsular rivers the Peninsular Rivers depend exclusively on monsoons. Climate is chiefly monsoonal.

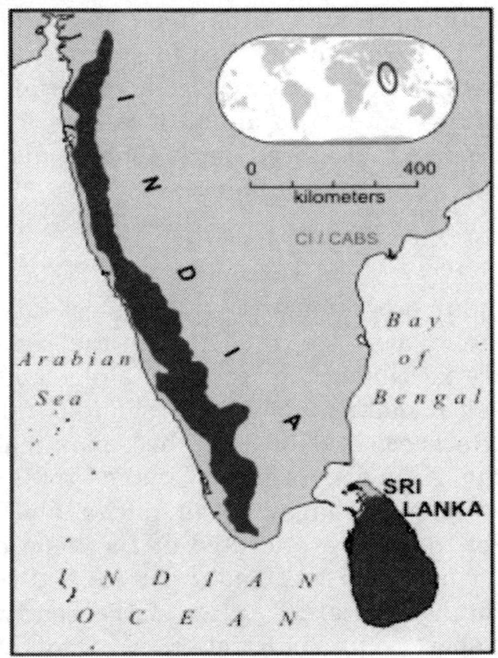

Geologically, since the Peninsular India once formed a part of Gondwanaland , some of its fauna and flora show relationship with now disjunct continents and islands of South America, Africa, Madagascar, Sri Lanka and even Australia. WGs fall under Palaeotropical (Indo-Malayan) biogeographical zone, divided into the province of Malabar Coasts/plains and WGs Mountains. Faced with tremendous population pressure the forests of Western Ghats and Srilanka have been dramatically impacted by demands for agriculture and Timber. The Region is home to rich endemic assemblage of Plants, reptiles and amphibians as well as elephants, tiger and endangered Lion tailed Macaque.

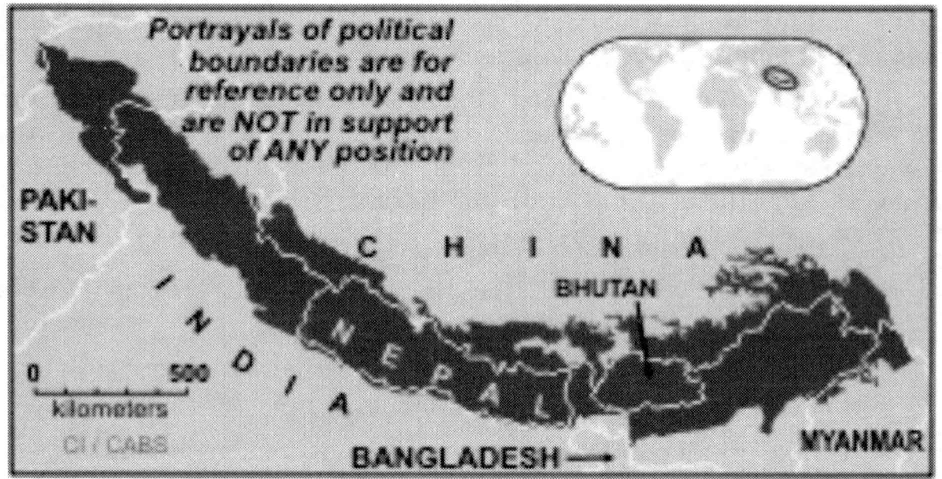

(ii) NE Himalayas: Himalayas is home to world's highest mountains, including Mt. Everest. Abrupt rise in rise of mountains results in diversity of ecosystems that range from alluvial grasslands and subtropical broad leaved forests to alpine meadows above the tree line. It's a home to a variety of large birds, mammals, including tiger, elephants, rhinos and wild water buffaloes.

5. Wetlands

These are complex ecosystems providing most effective life support system to mankind. They are crucial for the survival of natural biodiversity and are recognized as sources, sinks and transformers of biological and chemical matters, besides providing suitable habitats for endangered and rare species of birds, fishes, mammals and variety of Aquatic fauna and flora. In the form of Wetlands India have some of the largest rivers like Ganga, Ghagra, Brahmputra and Yamuna flowing across, with delta of these rivers located at the head of Bay of Bengal. India has a rich variety of wetland habitats with a total area of 58,286,000ha (excluding rivers) or 18.4% of the country, of which 70% comprises areas under paddy cultivation. A total of 1193 wetlands covering an area of 3,904,543 ha have been recorded by the Department of Science and Technology, of which 572 were natural .The countries wetlands are generally differentiated by regions into eight categories (Scott, 1989), (i) Reservoirs of the Deccan Plateau together with Lagoons and other wetlands of Southwest coasts, (ii) Saline expense of Rajasthan, Gujarat and Gulf of Katchchh, (iii) Freshwater lakes and reservoirs from Gujarat eastwards through Rajasthan and Madhya Pradesh, (iv) The delta wetlands and lagoons of India's east coast (Chilika Lake), (v) The fresh-water marshes of the Gangetic plains, (vi) The flood plains of Brahmputra, the marshes and swamps in the hills of NE India and Himalayan foot hills, (vii) The lakes and rivers of montane regions of Kashmir and Ladakh, and (viii) The Mangroves and other wetlands of the Island arcs of Andaman and Nicobar. These wetlands may be Freshwater, brackish water, estuaries, salt water, marine, salt lakes etc.

(i) **Brackish Water** is the admixture of salt and freshwater and it includes: (a) Sprawling Estuaries at the confluence of the river system with sea, (b) Large lagoons, (c) Brackish water lakes and back waters, (d) Intertidal mud, sand or salt flats, and (e) Intertidal salt marshes, salt pans and saline as well as brackish water impoundments and the vast areas of mangrove swamps containing tidal waters of fluctuating salinity.

Four Categories of Brackish water wetland habitats are defined as:

1. Brackish water Lagoon: Large, Shallow, inland, natural brackish or salt water lakes connected or separated from sea by dunes.
2. Brackish water lakes: Large, shallow, inland body of brackish water, usually natural, containing standing water, not connected by sea.
3. Brackish water impoundments: Man-made brackish water ponds including fish ponds and shrimp ponds.
4. Back waters: Currentles, usually coastal, water held back by dam etc., beside streams, rivers and estuaries, is known as back waters.

Very little is known about the brackish water floral and faunal diversity other than commercially important species. Chilika Lake and Pulicat Lake are relatively rich in faunal diversity.

(ii) Estuaries: The transitional ecotone which is the confluence of the river and the sea. Three main types of estuaries can be recognized namely:

1. Positive, where the evaporation from the surface is less than the volume of fresh water entering the estuary from the rivers and land drainage.
2. Negative, where the evaporation from surface exceeds the fresh water runoff entering the estuary

3. Neutral, where the fresh water exactly equals the evaporation and in such situation a static salinity regime is established.

Geomorphologically, the estuaries can be classified into (a) Drowned valleys, (b) Kayals, and (c) Deltas. Estuaries are generally rich in nutrients needed for plant and animal growth, especially nitrogen and phosphates.

(iii) Mangroves: covers a number of forest ecosystems (Alfred, Das and Sanyal, 2000), whose common feature is that the major constituent the mangrove vegetation tolerates both relatively high water salinity and flooding at regular intervals. The Indian Pacific region is known for its luxuriant mangroves. Some of these mangroves occur in the alluvial delta of Ganga, Godavari, Krishna and Cauvery and in the Islands of A. & N. They are the reservoirs of a large number of plants and animals species associated together over a long evolutionary time. The species composition of plants and animals vary from place to place.

(iv) Coral reefs: refers to a class of animals belonging to phylum Coelentrata, is a soft bodied radially symmetrical marine invertebrate which secretes a calcareous skeleton. The coral reefs are shallow water tropical marine ecosystems marked by a high biomass production and rich faunal and floral diversity. Major reef formations in India are restricted to Gulf of Manner, Palk bay, Gulf of Katchchh, A &N Isls., and Lakshadweep.

6. Marine Environment

The near offshore coastal waters of India are extremely rich fishing grounds. Coral reefs occur along only a few sections of mainland, principally the Gulf of Katchchh, off the southern mainland coast, and around a number of islands opposite Sri Lanka. Elsewhere, corals are found in Andaman, Nicobar and Lakshadweep island groups (UNEP/IUCN, 1988). Exploitation of corals, coral debris and coral sands is widespread on Gulf of Mannar, and Katchchh, while ornamental shells, chanks and pearls oysters are the basis of important reef industry in South India. Other notable areas are Seagrass beds, which are exploited as valuable habitat for commercially harvested prawns and mangrove strands.

7. Coastal Regions

The Indian mainland coast is divided into two parts – the West Coast fronted by the Arabian Sea and the East Coast is fronted by the Bay of Bengal. Other than these mainland coasts there are three Island Groups the Lakshadweep in the south Arabian Sea and Andaman and Nicobar in the eastern Bay of Bengal. The physical regime of the Indian coasts is characterized by different types of Coastal and shore ecosystems like (i)

Promotories (near Beypore in Kerala), (ii) Sand Pits (at Karnataka and Andhra Pradesh), (iii) Barrier Beaches (Along Kerala Coast), (iv) Embayments (Mirya Bay in Maharashtra), (v) Estuaries, and (vi) Offshore Islands. Besides, there are Mud flats, Rocky Coastland, Sandy stretches which are characterized by unique biotic and abiotic properties and processes.

8. The Islands

These are of two types (1) **Continental Islands,** which are detached segments of continents and inhabited by terrestrial mammals and amphibians, and (2) **Oceanic Islands** are originated in the ocean, comprising the volcanic rocks and corals, not inhabited by terrestrial mammals and amphibians. These islands may be (i) **Recent continental** islands or (ii) **Fringing archipelago**, or (iii) **Other Islands and Archipelago** according to their origin.

The Indian Islands include (a) the **Andaman and Nicobar Islands** in the Bay of Bengal consisting of a chain of more than 500 islands, islets and rocky outcrops running north to south, and (b) **Lakshadweep in the Arabian Sea**: This archipelago is the smallest territory in India which is irregularly scattered in the South Arabian Sea. The Lakshadweep, Maldives and Chagos archipelago form a continuous mountain ridge in the ocean. This ridge is believed to be continuation of Aravali mountain ranges of Gujarat and Rajasthan of the Indian mainland since late tertiary times. These islands are adorned with three ecosystems, viz., Forests, Mangroves and Coral reefs. The terrestrial part harbours mainly forests and mangroves. The mangroves harbour a rich diversity of terrestrial fauna.

9. The Deserts

Bounded by the Aravali Hills in east, the fertile Indus and the Nara valley in the west, the great salt marsh of the Katchchh in the south and Semi-arid districts of Haryana and Punjab in the north, the Thar Desert, one of the smallest in the world, exhibits a wide variety of habitat and ecological diversity. It is the only thickly populated desert with a population of over 80 persons per sq.km. and a heavy livestock population. Its uniqueness lies in having the crossover points for the Oriental, Palaearctic and Ethiopian elements of fauna and Flora. The Thar Desert is believed to be an extension of Sahara desert, through Arabian and Persian Deserts. In India it extends through Punjab, Haryana, and Rajasthan to Gujarat.

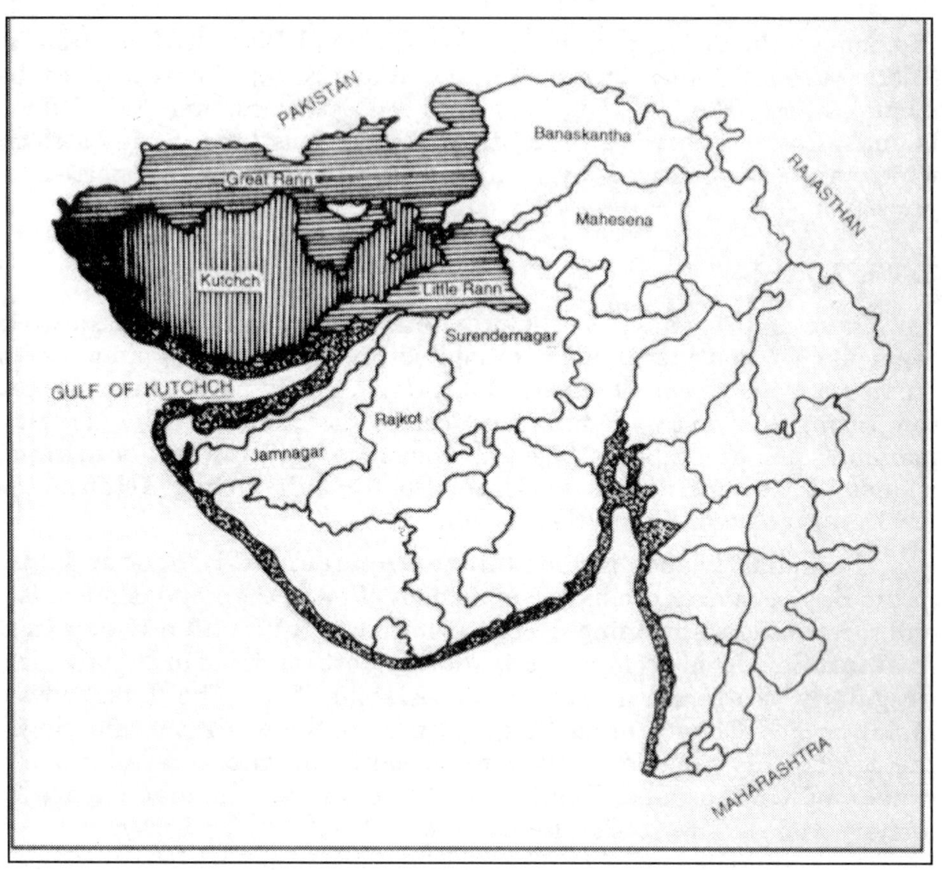

10. Rann of Katchchh

Or Kutch-Bhuj in Gujarat state forms a separate agro-climatic zone within the Thar Desert of India, characterized by low rainfall and sparse vegetation. It has a special topography because of its location near sea and low-lying areas by which the marine water enters into its vast expanse. Bordered with Sind (Pakistan) in north; in south with Surendranagar, Banaskantha and Mahsana in west by Arabian Sea, which also forms Gulf of Kutchch the ROK is divided into three zones: (i) Kutchch, (ii) Great Rann and (iii) Little Rann. Kutchch forms the shape of a Tortoise and, hence, it has its name derived. The area harbours several endemic and threatened species of plants and animals. Important rivers of ROK are Luni, Rakhari, Bhukhi, Banas, Nachhu and Demi.

11. Forests

Because of the great diversity of its natural ecosystems, India possesses a distinct identity. The panorama of Indian forests ranges from ever green tropical rain forests in Andaman and Nicobar Islands, the Western Ghats and the NE States, to dry alpine scrub high in the

Himalaya to the north. Between the two extremes the country has semi-ever green rain forests, deciduous monsoon forests, thorn forests, subtropical pine forests in lower montane zone and temperate montane forests (Lal, 1989). Champion, (1936) and Champion & Seth (1968) classified India into 16 major forest types and 221 minor types using structure, physiognomy and floristic as characters to define the types. The main areas of tropical forests are found in Andaman and Nicobar Islands the Western Ghats, which fringe the Arabian Sea coastlines of peninsular India and the greater Assam region in the northeast. There are substantial differences, both in fauna and flora between the three major forest regions (IUCN, 1986; Rodgers & Pawar, 1988).

12. Grasslands

Lands with plant community dominated by perennial grasses, with few or no Shrubs and trees. About 3.9% of the geographical area of India is grassland. The grassland diversity includes (i) Semi-arid and arid of Deccan Peninsula and Rajasthan, (ii) Water logged grassland of Terai belt, (iii) The rolling Shola-grasslands of hill tops of Western Ghats and 4. High altitude temperate-alpine grasslands of Himalaya. The grass lands in India, except in alpine regions are seral in nature and are maintained at various successional stages by grazing and burning. There is no climatic climax of grassland in India except the high Himalayan Meadows. These monsoon grasslands occurring in forest climate may be called tertiary communities. The grasslands primarily support a large number of herbi-vore species from minute insects to the largest land animal - the elephant. This in turn makes the grassland a happy homeland for carnivores.

13. Protected Area Network (PA Network)

The protection of wildlife has been a long tradition in Indian history. Wise use of natural resources was a pre-requisite for many hunter-gatherer societies, which dates back to at least 6000 BC. The adoption of a National Policy for Wildlife Conservation in 1970 and the enactment of the Wildlife (Protection) Act in 1972 lead to a significant growth in the protected areas network (Rodgers and Pawar, 1988 and Rodgers, Pawar and Mathur, 2002). To maintain rich biological diversity of the Indian Himalayan Region a Protected area network (PAN) has been established and biodiversity rich areas have been conserved as Sanctuaries, National Parks and Biosphere Reserves. The network was further strengthened by a number of national conservation projects, notably Project Tiger (initiated in April 1973) and the Crocodile Breeding and Management Project (Launched in April 1975). Presently, there are 661 Protected Areas in the country which cover around 4.8% geographical area: (155,978 km^2), which constitutes 99 National Parks with an area of

38,024.10 km^2 or 1.16% of the area, 515 Wildlife sanctuaries, with an area of 117913.42 km^2 or 3.59% of the geographical area of India, 43 Conservation Reserves and four community reserves.

14. Biosphere Reserves in India were created under the 'Man & Biosphere' (MAB) Programme by UNESCO in 1971 to conserve *in situ* all forms of life, along with its support system, in its totality, so that it could serve as a referral system for monitoring and evaluating changes in natural ecosystems. The first biosphere reserve of the world was established in 1979, since then the network of biosphere reserves has increased to 425 in 95 countries across the world (MAB, 2003). Presently, there are 14 existing biosphere reserves in India.

		Date of Notification	Area (Sq. Km.)	States
1.	Nilgiri	01-09-1986	5,520.00	Karnataka, Kerala & T.N
2.	Nanda Devi	18-01-1988	5,860.69	Uttaranchal
3.	Nokrek	01-01-1988	80.00	Meghalaya
4.	Great Nicobar	06-01-1989	885.00	A. & N.Isls
5.	Gulf of Mannar	18.02.1989	10,500.00	Tamil Nadu
6.	Manas	14-03-1989	2,837.00	Assam
7.	Sunderbans	29-03-1989	9,630.00	West Bengal
8.	Simlipal	22-06-1994	4,374.00	Orissa
9.	Dibru-Shikowa	28-07-1997	765.00	Assam
10.	Dehang-Debang	02-09-1998	5,111.50	A.P.
11.	Pachmarhi	03-03-1999	4,926.00	M.P.
12.	Khangchendzonga	07-02-2000	2,619.92	Sikkim
13.	Agasthyamalai	12-11-2001	1,701.00	Kerala
14.	Amarkantak	2005	3,835.51	M.P. and Chattisgarh

Of these 14, five namely, Nanda Devi, Uttarakhand, Manas, Assam and Dibru-Shikowa, Assam, Dehang-Debang, Arunachal Pradesh and Kangchendzonga, Sikkim are operational BR in the Himalayan region and Namdhapa, Arunachal Pradesh, Kaziranga, Assam, and Cold Desert, J & K are proposed Himalayan BRs.

15. World Heritage Sites

India's biodiversity richness is contributed additionally by the conservation of the species occurring in the areas declared as Heritage sites. UNESCO World Heritage Committee composed of 21 State Parties (countries) aims to catalogue, name and preserve sites of outstanding cultural or natural importance to the common heritage of humankind.

Since adopted by the General Conference of UNESCO on 16 November 1972, today 182 State Parties ratify the convention. The world body has listed 25 Heritage Sites in India, including the following five Protected Areas of great conservation significance to be a part of World Culture and Heritage: (i) Kaziranga National Park, Assam, (ii) Manas Wildlife Sanctuary, Assam, (iii) Keoladaeo National Park, Rajasthan, (iv) Sunderbans National Park, West Bengal, and (v) Nanda Devi National Park, Uttarakhand.

16. Sacred Groves

In India there is a very strong prevalence of the Sacred Groves. These Sacred groves are not only the sacred ecosystems functioning as a rich repository of nature's unique biodiversity, but also a product of the socio-ecological philosophy our forefathers have been cherishing since olden days. Sacred groves have well developed forest ecosystems and high degrees of species richness and a rich biodiversity in general. Sacred groves have well developed forest ecosystems and high degrees of species richness and a rich biodiversity in general. This, however, depends on the extent of preservation of the grove. Sacred groves do not just help conserve valuable biodiversity, soil and water, but are also critical in regulating weather and climate cycles so vital for life to blossom and flourish on the planet. Our forefathers held reverential attitude towards nature and its all life forms. Sacred groves have been flourishing in India from time immemorial. A number of plant and animal species described as rare, endangered, vulnerable or threatened are existing today thanks to the these sacred groves, which have also been pivotal in protecting and conserving endemic species. Protection of nature's biodiversity and especially the rare types of life forms through sacred groves is regarded as an obligation on the part of the society. They are storehouses of medicinal plants valuable to village communities as well as modern pharmacopoeia, and they contain wild relatives of crop species that can help to improve cultivated varieties. Sacred groves also provide for the water needs of nearby communities. Many sacred groves contain water resources such as ponds and streams, and the vegetative mass that covers the floor of a grove can absorb water during rainy seasons and release it during times of drought. Trees also improve soil stability, prevent topsoil erosion and provide irrigation for agriculture in drier climates. There has been no comprehensive survey of sacred groves in India, so their exact number and area are unknown. At least 13,720 sacred groves have been reported in various regions of the country, but experts estimate that the actual number is likely 100,000 to 150,000 (Pushpangadan *et al.,* 1998; Ramakrishnan, 1998, 2001; Sinha & Maikhuri, 1998).

B. SPECIES DIVERSITY

17. Faunal Diversity in India: While India has about 45000 species of the flora; it has a total of 91,484 animal species accounting for 7.45% of the faunal species in the world (Ramakrishna, 2010) and the flora accounts for 10.78% of the global total. The endemism of Indian biodiversity is high- about 33% of the country's recorded flora is endemic to the country and is concentrated mainly in the North-East, Western Ghats, North-West Himalayas and the Andaman and Nicobar islands. However, this rich biodiversity of India is under severe threat owing to habitat destruction, degradation, fragmentation and over-exploitation of resources.

Table 1. Faunal Diversity in India (Updated, January 2009) [Source: Animal Discoveries 2009: New Species and New Records, Zoological Survey of India, 5th June, 2010, pp. 1-64]

Taxonomic Groups	No. of Species		% in India
	World	India	
PROTISTA (Protozoa)	**31250**	**2577**	**8.24**
ANIMALIA			
Mesozoa	71	10	14.08
Porifera	4562	500	10.70

Cnidaria	9916	866	8.73
Ctenophora	100	12	12.00
Platyhelminthes	17500	1624	9.28
Rotifera	2500	330	13.20
Gastrotricha	3000	100	3.33
Kinorhyncha	100	10	10.00
Nematoda	30011	2866	9.55
Acanthocephala	800	229	28.62
Sipuncula	145	35	24.14
Mollusca	66535	5072	7.62
Echiura	127	43	33.86
Annelida	12700	841	6.61
Onychophora	100	1	1.00
Arthropoda	**998904**	**71323**	**7.14**
Crustacea	35536	2936	8.26
Insecta	867365	61220	7.05
Arachnida	73440	5829	7.90
Pycnogonida	600	17	2.83
Chilopoda	3000	100	3.33
Diplopoda	7500	162	2.16
Symphyla	120	4	3.33
Merostomata	4	2	50.00
Phoronida	11	3	27.27
Bryozoa(Ectoprocta)	4000	200	5.00
Entoprocta	60	10	16.66
Brachiopoda	300	3	1.00
Chaetognatha	111	30	27.02
Tardigrada	514	30	5.83
Echinodermata	6223	765	12.29
Hemichordata	120	12	10.00
Chordata	**48463**	**5045**	**10.41**
Protochordata	2106	119	5.65
Pisces	21723	2548	11.72
Amphibia	5162	289	5.60
Reptilia	5817	460	7.91
Aves	9026	1232	13.66
Mammalia	4629	397	8.58
Total (Animalia)	**1195534**	**88907**	**7.43**
Grand Total (Protista + Animalia)	**1226784**	**91484**	**7.45**

Insect Diversity: Insect as a group of animals is the largest of all in the world. Concluding from the above table it is interpreted that India has a very rich biodiversity, in terms of insects that numbers 61220 in species and comprise nearly 70% of the Indian faunal Diversity and nearly 86% of the Indian Arthropod diversity. They occur everywhere in any ecosystem. This high biodiversity among insects lies in their characteristics like size, presence of chitinous exoskeleton, high fecundity, proficiency to feed on variety of materials, capability of water retention, capacity for flight and dispersal, etc., that help them to survive in different ecosystems with success.

The Class Insecta in India are grouped into 28 orders, of which four are **Apterygota**, while remaining are **Pterygota**. Among apterygots, in India, the Collembola are known better amongst the apterygote orders. The other three groups, *viz.*, Thysanura, Diplura and Protura are known by 31 species, under 16 genera; 16 species under seven genera and 20 species under 8 genera, respectively (Ramakrishna & Alfred, 2007). The diversity in Coleoptera, Hymenoptera, Lepidóptera and Diptera is enormous. There are about 122 species of Lepidoptera that are protected in various schedules of Wildlife (Protection) Act, 1972. The order Trichopteras and Thysanoptera are represented though by a lesser number of species but the endemism in these groups is quite high. Odonata and Neuroptera too are quite differentiated and diversified groups of insects.

18. Endemic Species

India has many endemic vertebrate species (Khanna, 2004; Khanna and Kumar, 2003). Areas rich in endemism are northeast, the Western Ghats and the northwestern Himalayas. A small pocket of local endemism also occurs in the Eastern Ghats. The Gangetic plains are generally **poor in endemics.** Endemic Species are those whose distribution is restricted to certain limited area.

Table 2. Endemic Indian Fauna

Group	No. of species	%
Land Molluscs	878	
Freshwater Molluscs	89	
Insects	16,214	23.00
Amphibia	110	52.63
Reptilia	214	46.92
Aves	69	0.56
Mammalia	38	9.74

Source: MoEF (1999), Kumar & Khanna, 2003

C. GENETIC DIVERSITY

Amongst the tremendous diversity at ecosystem level and species level, the diversity at genetic level should be much more diverse and needs treatment at molecular level. The comprehensive studies on this front are lacking. The genetic diversity refers to any variation in the nucleotides, genes, chromosomes, or whole genomes of organisms (the *genome* is the entire complement of DNA (deoxyribonucleic acid) within the cells or organelles of the organism). Genetic diversity at its most elementary level is represented by differences in the sequences of four nucleotides (adenine, cytosine, guanine, and thymine) that form the DNA within the cells of the organism. The DNA is contained in the chromosomes present within the cell; some chromosomes are contained within specific organelles in the cell (for example, the chromosomes of mitochondria and chloroplast). Nucleotide variation is measured for discrete sections of the chromosomes, called *genes*. Thus, each gene compromises a hereditary section of DNA that occupies a specific place of the chromosome, and controls a particular characteristic of an organism. Different species can have different numbers of genes within the entire DNA or genome of the organism. However, a greater total number of genes might not correspond with a greater observable complexity in the anatomy and physiology of the organism (i.e. greater phenotypic complexity). The extent to which genetic variation between organisms is expressed in their phenotypes can be quite variable for different characteristics. Genetic variation between some features might be expressed as very subtle differences in their phenotype.

Table 3. Marine Biodiversity of India

	Name of the Group	No. of species
1.	Protista	750
2.	Animalia	
3.	Porifera	500
4.	Cnidaria	790
5.	Ctenophora	10
6.	Platyhelminthes	350
7.	Gastrotricha	88
8.	Kinorhyncha	99
9.	Annelida	440
10.	Mollusca	3370
11.	Bryozoa	170
12.	Entoprocta	8

13.	Phoronida	3
14.	Brachiopoda	3
15.	Arthropoda	
	Crustacea	2430
	Pycnogonida	16
	Merostomata	2
16.	Sipunculida	38
17.	Echiura	33
18.	Tardigrada	33
19.	Chaetognatha	5
20.	Echinodermata	30
21.	Hemichordata	12
22.	Chordata	
	Protochordata	116
	Pisces	1800
	Amphibia	3
	Aves	145
	Mammals	29
	Total	**12456**

Data for other phyla not available
(Source: ENVIS Newsletter, ZSI, 4 (1&2), 1997)

19. Acts and Convention to Conserve and Protect India's biodiversity

Additionally, there are several conventions already in action for conservation, management and monitoring of biodiversity, that contribute to the richness of the Indian biodiversity, like Convention on International Trade in Endangered Species of Fauna and Flora (CITES), Convention on Conservation of Migratory Species of Wild Animals (1979) (Bonn Convention or CMS, 1979), Convention on Wetlands of International Importance especially Waterfowl Habitat (Ramsar Convention, 1971), World Heritage Sites in India, Biosphere Reserves in India, Protected Area Network (PA Network), Wildlife (Protection) Act 1972. India has recently set up a National Biodiversity Authority of India (NBA) with its HQ at Chennai that will look into Biodiversity Conservation Strategies and Action Plans at National, State, Village and Panchayat levels, under the aegis of the Indian Biodiversity Act 2003.

(i) **Ramsar Convention (1971):** An inter-governmental treaty on wetlands for conservation and wise use of natural resources as also conservation of Waterfowl habitats (Ramsar, Iran, 1971). There are 25 wetlands in India that have been identified as Ramsar Sites.

1. Chilika Lake, Orissa
2. Harike Wetland, Punjab
3. Keoladeo National Park, Rajasthan
4. Loktak Lake, Manipur
5. Sambar Lake, Rajasthan
6. Wular Lake, Jammu and Kashmir
7. Kanji Lake Punjab
8. Ropar Lake, Punjab
9. East Kolkata wetlands, WB
10. Deepor Beel, Assam
11. Astamudi Lake, Kerala
12. Pong Dam Lake, H.P.
13. Kolleru Lake, Andhra Pradesh
14. Bhitarkanika Mangrovers, Orissa
15. Tso mori, J & K
16. Point Calimere WLS, TN
17. Sasthamkota Lake, Kerala
18. Bhoj Wetland, MP
19. Vembanad-Kol Wetland, Kerala
20. Chander Tal Wetland, H.P.
21. Hokera Wetland, J. & K.
22. Renuka Wetland, H.P.
23. Rudrasagar Lake, Tripura
24. Surinsar-Mansur Lakes, J. & K.
25. Upper Ganga River (Brijghat to Narora Stretch)

More Wetlands to be designated as Ramsar Sites: The Ministry of Environment and Forests, Government of India, has identified more new wetlands and started the process of designation as Ramsar Sites in consultation with the World Wide Fund for Nature-India (WWF-India). These sites represent different Aquatic habitats. The identified sites are: (1) Lali Sanctuary (Arunachal Pradesh), (2) Kabar Tal (Bihar), (3) Pulicat Lake (Tamil Nadu and Andhra Pradesh), and (4) Andaman and Nicobar Islands.

(ii) **Convention on Conservation of Migratory Species of Wild Animals (1979) (Bonn Convention or CMS, 1979):** Provides strict protection for a list of species and also provides a frame work for collaborative conservation agreements between the states through which each species on second list migrate. Mainly this is applied to birds, but also have bats and dolphins. The Bonn convention also calls for research and surveys.

(iii) **Convention on International Trade in Endangered Species of Fauna and Flora (CITES) is a** successful international conservation treaty in restricting international trade in endangered species through its three appendices of species, for each of which a different extent of trade is allowed.

1. Species in App-I are forbidden for international trade except with special permission.
2. App-II species have controlled international trade.
3. App.-III species lists species whose trade is forbidden by certain countries but are not listed in other two appendices
4. CITES members have to create National Management Authority, which co-ordinates with CITES Secretariat in Switzerland.

(iii) **Wildlife (Protection) Act 1972**: Provides for protection of-Wild animals, Animal articles, and - Plants. The WL (P) Act regulates the sale, barter etc of notified wild plants and animal species. It also provides control over keeping of wild animals in captivity. The 1991 amendment covers the possession of notified plant species. The Act exercise control under the Schedules I-VI. The Schedule I lists rare and endangered totally protected species, the Schedule II includes game species for which licenses can be issued under special circumstances, Schedule III and Schedule IV comprises species of small games and finally the Schedule V includes vermin, common crow, fruit bats, mice and rats. Bird trade was stopped in 1991 following an amendment to the Wildlife Protection Act, 1972. Amendment 2006 of the WL (P) Act deals with creation of a National Tiger Conservation Authority.

(v) **Convention on Wetlands of International Importance especially Waterfowl Habitat (Ramsar Convention, 1971):** Held at Ramsar, Iran, the treaty provides for international collaboration on wetland conservation, including Mangroves and Coral reefs. The contracting parties have four obligations.

1. Incorporate the consideration of wetlands conservation within their national land-use planning
2. Designate at least one wetland of international importance (Ramsar Sites) according to the specified criteria. Promote wetland conservation by creating nature reserves and
3. Train staff in wetland wardeneing, research and management and consult other countries especially for species or areas.

There are 116 countries participating and over a thousand Ramsar Sites. Ramsar provides small grants from a fund, international expertise and resources. Ramsar Convention Beaureo, rue Mauverney 28, CH-1196, Gland, Switzerland.

(vi) **Indian Fisheries Act, 1897:** (a) Prohibition on use of explosives and poison for fishing, (b) Regulation of fishing in private waters with consent of owners/right holders, (and c) Prohibition of all fishing in specified waters for two years.

(vii) **Indian Forest Act, 1927**

1. Setting and managing reserve forests, villages, restricted forests.
2. Protection of non-governmental forest lands.
3. Control of movement of forest produce and grazing and trespass.

(viii) **Prevention of Cruelty to Animal Act, 1960**

(ix) **Indian Forest (Conservation) Act, 1980**

(x) **Environment Protection Act, 1986**

1. General Measures to protect environment [3(1)]
2. Restriction of industrial and other processes/ activities in specified areas.

20. Red Data Book (RDB)

It was developed during 1960s and the species were placed under various **threatened categories** according to the severity of the threats faced by them and the estimated eminence of their extinction. World Conservation Monitoring Centre (WCMC) in collaboration with IUCN Species Survival commission network of the specialist groups compiles the IUCN Red list every two years since 1986. The Threatened species are those that are often impoverished of low fecundity, dependent on patchy or unpredictable resources, extremely variable in population density, persecuted or otherwise prone to extinction in human dominated landscapes.

IUCN Red List Categories

Extinct (EX) - A taxon is Extinct when there is no reasonable doubt that the last individual has died.

Extinct In The Wild (EW) - A taxon is Extinct in the wild when it is known only to survive in cultivation, in captivity or as a naturalized population (or populations) well outside the past range.

Critically Endangered (CR) - A taxon is Critically Endangered when it is facing an extremely high risk of extinction in the wild in the immediate future, as defined by any of the criteria.

Endangered (EN) - A taxon is endangered when it is not Critically Endangered but is facing a very high risk of extinction in the wild in the near future, as defined by any of the criteria.

Vulnerable (VU) - A taxon is Vulnerable when it is not Critically Endangered or Endangered but is facing a high risk of extinction in the wild in the medium-term future, as defined by any of the criteria.

Lower Risk (LR) - A taxon is Lower Risk when it has been evaluated, does not qualify for any of the threatened categories Critically Endangered, Endangered or Vulnerable or Data Deficient (LR/nt- near threatened, Lr/lc- least concerned, LR/cd-conservation dependent).

Near Threatened (NT) - A taxon is Near Threatened when it has been evaluated against the criteria but does not qualify for Critically Endangered, Endangered or Vulnerable now, but is close to qualifying for, or is likely to qualify for, a threatened category in the near future.

Least Concern (LC) - A taxon is Least Concern when it has been evaluated against the criteria and does not qualify for Critically Endangered, Endangered, Vulnerable or Near Threatened. Widespread and abundant taxa are included in this category.

Data Deficient (DD) - A taxon is Data Deficient when there is inadequate information to make a direct, or indirect, assessment of its risk of extinction based on its distribution and/or population status.

Not Evaluated (NE) - A taxon is Not Evaluated when it is has not yet been assessed against the criteria.

Endemics (E) - Species restricted to a particular geographical area or ecosystem.

21. Globally Threatened Indian Species

Kumar & Khanna (2006) in their overview of the threatened Indian fauna have listed 648 species of animals categorized as "Threatened "by IUCN (2002); it is approximately 8.91% of the world's total (7266 species) number of threatened faunal species.

Threatened Fauna from India by Taxonomic Groups [Source: Kumar & Khanna (2006) Globally Threatened Indian *Fauna. Zool. Surv. India,* pp. 1-104].

Group	No. of Species %
Mammalia	212
Aves	143
Reptilia	34
Amphibia	148
Pisces	78
Mollusca	05
Crustacea	12
Other Invertebrates	16
Total	648

Analysis of threat Categories: On analysis of the threat categories by groups (at global level); it is found that out of 648 species of threatened Indian fauna

Ext	Cr.En.	En.	Vu.	LRnt	NT	LR cd	DD
3	**44**	**109**	**195**	**63**	**91**	**9**	**134**

[Source: Kumar and Khanna, 2007].

22. Invasive Species in India

The impact of invasive species is second only to that of human population growth and associated activities as a cause of loss of biodiversity throughout the world. The invasions of non-native plants, animals and microbes are thought to be responsible for the decline of native species now listed as endangered or threatened The invasions potentially lead to an increase in species richness, as invasive species are added to the existing species pool. But the invasive species also leads to extinction, resulting in decrease of species richness. The negative interactions in these cases will primarily be competition, which may allow coexistence or lead to extinction more gradually than predation. The rate at which the invasions are occurring globally has been increasing exponentially over the course of human history. Khanna (2003a,b; 2004) has listed the invasive species to the Indian biodiversity. An "invasive species" is defined as a species, that is, (1) non-native (or alien) to the ecosystem, and (2) whose introduction causes or is likely to cause economic or environmental harm or harm to human health.

The Global Invasive Species Database developed by the IUCN/SSC Invasive Species Specialist Group (ISSG) as part of the global initiative on invasive species led by the Global Invasive Species Programme (GISP) provides global information on invasive alien species to agencies, resource managers, decision-makers, and interested individuals. The database focuses on invasive species that threaten biodiversity and covers all taxonomic groups from Micro-organisms to animals and plants. Species information is supplied by expert contributors from around the world and includes; species' biology, ecology, native and alien range, references, contacts, links and images.

The data in this document has been extracted from GISP database. There are 123 species of, both plants and animals; designated as Invasive Alien Species, in India, of which 56 are faunal elements and the remaining are herb, shrubs, weeds, grasses, trees or aquatic vegetation. Additionally, GISP has listed **"100 species as worst invaders"**

INVASIVE SPECIES IN INDIA CHECK-LISTED BY GLOBAL INVASIVE SPECIES PROGRAMME (GISP)

S.N	Name of the species	Common Name	Group
		FAUNA	
1.	*Mycale armata* Thiele	Orange Keyhole Sponge, Orange sponge	Porifera (Sponge)
2.	*Achatina fulica*	Giant African land snail	Mollusca
3.	*Euglandina rosea*	Cannibal Snail, Rosy Wolf Snail	Mollusca
4.	*Perna perna*	Brown Mussel, Mexilhao Mussel	Mollusca
5.	*Anoplolepis gracilipes*	Ashinaga-Ki-Ari, Crazy Ant, Gramang Ant, Long-Legged Ant, Maldive Ant, Yellow Crazy Ant	Insecta: Hymenoptera
6.	*Bemisia tabaci*	Cotton Whitefly, Mosca Blanca, Sweet Potato	Insecta: Homóptera
7.	*Monomorium pharaonis*	Pharaoh Ant	Insecta: Hymenoptera
8.	*Quadrastichus erythrinae*	Erythrina Gall Wasp	Insecta: Hymenoptera
9.	*Solenopsis geminata*	Fire Ant, Ginger Ant, Tropical Fire Ant	Insecta: Hymenoptera
10.	*Monomorium destructor*	Destructive trailing ant, Singapore ant	Insecta: Hymenoptera
11.	*Oryctes rhinoceros*	Coconut Rhinoceros Beetle, Palmen-Nashornkaefer	Insecta: Coleoptera
12.	*Xylosandrus compactus*	Ambrosia beetle, black twig borer, Dunkler-Holzbohrer	Insecta: Coleoptera
13.	*Culex quinque-fasciatus*	Southern House Mosquito	Insecta: Diptera
14.	*Paratrechina longicornis*	Crazy Ant, Hairy Ant, Long-Horned Ant, Slender Crazy Ant	Insecta: Hymenoptera

15.	*Trogoderma granarium*	Khapra Beetle	Insecta: Coleoptera
16.	*Xylosandrus mutilatus*	Camphor Shoot Beetle, Camphor Shot-Borer	Insecta: Coleoptera
17.	*Aristichthys nobilis*	Bighead carp	Fish
18.	*Carassius auratus*	Edible goldfish, Gold crucian carp, Goldfisch	Fish
19	*Cyprinus carpio*	Carp	Fish
20.	*Gambusia affinis*	Gambusia, Live-bearing tooth-carp, Mosquito fish	Fish
21.	*Gambusia holbrooki*	Eastern Gambusia, Gambusia, mosquito fish, plague minnow	Fish
22.	*Hypophthalmichthys molitrix*	Chinese Schemer, Silver Carp	Fish
23.	*Oncorhynchus mykiss*	Brown Trout, Coast Angel Trout, Coast Rainbow Trout, Coast Range Trout	Fish
24.	*Oreochromis mossambicus*	Common Tilapia, Java Tilapia	Fish
25.	*Salmo trutta*	Brook Trout, Brown Trout	Fish
26.	*Oreochromis* spp	Boulti, Freshwater Snapper, Mojara, Ngege, Pla Nil, St. Peters Fish, Tilapia	Fish
27.	*Tinca tinca*	*Tinca tinca*	Fish
28.	*Channa marulius*	Bull's-eye Snakehead, Cobra Snakehead, Giant Snakehead, Great Snakehead, Indian Snakehead, Kalumaha, Kubrah	Fish
29.	*Clarias batrachus*	Clarias Catfish, Climbing Perch, Freshwater Catfish, Thailand Catfish, Walking Catfish	Fish
30.	*Monopterus albus*	Rice eel, swamp eel, white rice field eel	Fish
31.	*Acridotheres fuscus*	Jungle Myna	Bird
32.	*Zosterops japonicus*	Japanese White-Eye, Mejiros	Bird
33	*Acridotheres tristis*	Common Myna,, Indian Myna, Indian Mynah, Martin Triste, Mynah	Bird
34.	*Cygnus olor*	Mute Swan	Bird
35.	*Passer domesticus*	English Sparrow, Europese Huismuis, House Sparrow, House Sparrow	Bird

36.	*Pycnonotus cafer*	red-vented bulbul	Bird
37.	*Axis axis*	Axis Deer, Chital, Indian Spotted Deer	Mammal
38.	*Rattus exulans*	Maori rat, Pacific rat, Polynesian rat	Mammal
39.	*Equus asinus*	African Wild Ass, Asino, Ass, Burro	Mammal
40.	*Herpestes javanicus*	Newla, Small Indian Mongoose	Mammal
41.	*Rattus rattus*	Black Rat, Blue Rat, Bush Rat, European House Rat	Mammal
42.	*Suncus murinus*	Asian House Shrew, Asian Musk Shrew, Brown Musk Shrew, Ceylon Highland Shrew, Common Indian Musk-Shrew, Grey Musk Shrew, House Shrew, Indian Grey Musk-Rat, Indian Musk Shrew	Mammal
		FLORA	
43.	*Acacia farnesiana*	Mimosa bush, needle bush, Small's acacia, sweet acacia	Shrub, Tree
44.	*Acacia mearnsii*	Australian acacia, black wattle, black wattle	Shrub, Tree
45.	*Acacia melanoxylon*	Australian Blackwood	Tree
46.	*Alternanthera philoxeroides*	Alligator weed, pigweed	Aquatic plant, Herb
47.	*Asparagus densiflorus*	Asparagus Fern, Sprengeri fern	Herb
48.	*Cabomba caroliniana*	Carolina water-shield, Fanwort, fish-Grass	Aquatic plant
49.	*Chromolaena odorata*	Agonic, bitter bush, chromolaena, hagonoy	Herb
50.	*Cirsium arvense*	Californian thistle, Canada thistle, Canadian thistle, creeping thistle, field thistle, perennial thistle	Herb
51.	*Cryphonectria parasitica*	Chestnut blight	Fungus
52.	*Cryptostegia grandiflora*	India rubber vine	Vine, climber
53.	*Eichhornia crassipes*	Water hyacinth	Aquatic plant
54.	*Elaeagnus umbellata*	Aki-gumi, Autumn Elaeagnus, Autumn-olive, silverberry	Shrub, Tree
55.	*Eugenia uniflora*	Barbados cherry, Brazilian cherry	Shrub, Tree

56.	*Eupatorium cannabinum*	Common hemp agrimony, water agrimony	Herb
57.	*Gymnocoronis spilanthoides*	Senegal tea plant, temple plant	Aquatic plant
58.	*Kappaphycus* spp.	Red Alga	Alga/Sea Weed
59.	*Lantana camara*	Ach man, angel lips, ayam, largeleaf plant-ana, prickly lantana, white sage, wild sage	Shrub
60.	*Leucaena leucocephala*	Wild mimosa, wild tamarind	Tree
61.	*Limnocharis flava*	Sawah-flower rush, sawah-lettuce, velvetleaf, yellow bur-head	Aquatic plant
62.	*Ludwigia peruviana*	Ludwigia, Peruvian primrose, Peruvian primrose-willow, Peruvian primrose bush, Water primroseian primrose, Water-primrose	Aquatic plant
63.	*Mikania micrantha*	American rope, Chinese creeper, mile-a-minute weed	Vine, climber
64.	*Parthenium hysterophorus*	Congress Grass, Santa Maria feverfew, White top weed	Herb
65.	*Pennisetum clandestinum*	Kikuyu Grass, Kikuyu pul, Kikuyu tana, Kikuyugras	Grass
66.	*Pittosporum undulatum*	Australian cheesewood, mock orange, Victorian box, Victorian laurel, wild coffee	Shrub, Tree
67.	*Prosopis juliflora*	Algarroba, ironwood , mesquite	Shrub
68.	*Prosopis spp.*	Vilayati babul, Vilayati and, Vilayati khejra, Vilayati kicker, Vilayati Kikar.	Shrub, Tree
69.	*Salvinia molesta*	African payal, giant salvinia, kariba	Aquatic plant, Herb
70.	*Solanum mauritianum*	Bug weed, flannel weed, wild tobacco, woolly nightshade	Shrub, Tree
71.	*Solanum viarum*	Juá, juá-bravo, tropical soda apple	Shrub
72.	*Spartina alterniflora*	Atlantic cordgrass, salt-water cordgrass, saltmarsh cordgrass, smooth cordgrass	Grass
73.	*Triadica sebifera*	Candleberry-Tree, chicken Tree, Chinese tallow, vegetable tallow, white waxberry	Tree
74.	*Vibrio cholerae*	Asiatic cholera, epidemic cholera	Micro-organism
75.	*Adenanthera pavonina*	bead Tree, coral bean Tree, red sandalwood Tree	Tree

76.	*Banana bunchy top virus*	BBTV, Bunchy top virus	Micro-organism
77.	*Cronartium ribicola*	White pine blister rust	Fungus
78.	*Phalaris arundinacea*	Reed Canary Grass, ribbon Grass, variegated Grass	Grass
79.	*Psidium guajava*	Amarood, guava	Shrub, Tree
80.	*Rubus moluccanus*	raspberry and blackberry	Shrub, Vine, Climber
81.	*Ulex europaeus*	Ajonc, Ginestra Spinosa	Shrub, Tree
82.	*Abelmoschus moschatus*	Gongul, Kamang, Kamwayang, Karereon, Wakiwaki	Herb, Shrub
83.	*Albizia julibrissin*	Mimosa, Powder-puff Tree, Silk Tree, Silky Acacia	Tree
84.	*Alliaria petiolata*	Garlic Mustard Plant, Garlic Root, Garlic-wort, Hedge Garlic, Jack-By-The-Hedge, Jack-In-The-Bush, Mustard Root, Poor Man's Mustard, Sauce-Alone	Herb
85.	*Arundo donax*	Arundo Grass, Bamboo Reed, Cana, Cane, River Cane, Spanish Cane, Spanish Reed	Grass
86.	*Azolla pinnata*	Ferny Azolla, Mosquito Fern, Water Velvet	Aquatic Plant
87.	*Caesalpinia decapetala*	Mysore Thorn, Shoofly, Wait-A-Bit	Shrub, Tree
88.	*Casuarina equisetifolia*	Australian Beefwood, Australian pine, Beach She-Oak, Beef Wood-Tree, Ironwood, Whistling-Pine	Tree
89.	*Caulerpa taxifolia*	Caulerpa, Killer Alga, Lukay-Lukay, Sea Weed	Alga/Sea Weed
90.	*Cinnamomum camphora*	Alcanfor, Alcanforero, Arvore Da Camphora, Campher, Camphor Laurel, Camphor Tree, Camphre, Camphrier, Canfora, Japanese Camphor, Kampferbaum, Kuso-No-Ki	Tree
91.	*Coccinia grandis*	Kundru, Scarlet-Fruited Gourd	Vine, Climber
92.	*Colubrina asiatica*	Asian Snake wood, Hoop With, Indian Snake wood, Lather leaf, Wild Coffee	Shrub
93.	*Dioscorea oppositifolia*	Chinese Yam, Cinnamon Vine	Herb, Vine, Climber
94.	*Elaeagnus angustifolia*	Russian olive	Shrub, Tree
95.	*Erodium cicutarium*	California Filaree, Cutleaf Filaree, Filaree, Heronsbill	

96.	*Hedychium flavescens*	Cream Ginger, Yellow Ginger	Herb
97.	*Hiptage benghalensis*	Madhalata, Madhavi, Madmalati	Shrub, Vine, Climber
98.	*Houttuynia cordata*	Chameleon-plant	Aquatic Plant, Shrub
99.	*Hydrilla verticillata*	Florida elodea, hydrilla, oxygen weed, water thyme, waterweed	Aquatic plant
100.	*Hygrophila polysperma*	East Indian hygrophila, hygro, Indian swamp weed, Miramar weed	Aquatic plant
101.	*Ipomoea Aquatica*	Aquatic morning glory, water bindweed, water spinach	Vine, Climber
102.	*Lespedeza cuneata*	Chinese Bush-Clover, Chinese Lespedeza, Himalayan Bush clover, Silky Bush-Clover	Herb, Shrub
103.	*Limnophila sessiliflora*	Ambulia, Asian marsh weed, limnophila.	Herb
104.	*Lygodium microphyllum*	Old World Climbing Fern, Small Leaf Climbing Fern, Small-Leaf Climbing Fern	Fern
105.	*Melia azedarach*	Chinaberry, Indian Lilac, Umbrella Tree, White Cedar	Shrub, Tree
106.	*Microstegium vimineum*	Asian Stilt Grass , Bamboo Grass, Chinese Packing Grass, Flexible Seagrass, Japanese Grass, Japanese Stilt Grass , Mary's Grass, Nepal Grass , Nepal Microstegium, Nepa-lese Brown top, Vietnamese Stilt Grass	Grass
107.	*Neyraudia reynaudiana*	Burma Reed, Cane Grass, False Reed, Silk Reed	Grass
108.	*Nymphoides peltata*	Asaza, Entire Marshwort, Floating Heart, Fringed Water Lily, Xing Cai, Yellow Floatingheart	Aquatic Plant
109	*Paederia foetida*	Chinese Fever Vine, Skunk Vine, Stinkvine	Vine, Climber
110.	*Panicum repens*	Creeping Panic, Millet Rampant, Panic Rampant , Torpedo Grass, Torpedo Grass, Victoria Grass	Grass
111.	*Pennisetum ciliare*	African Foxtail Grass, Buffalo Grass	Grass
112.	*Pennisetum polystachion*	Feathery Pennisetum, Thin Napier Grass, West Indian Pennisetum	Grass
113.	*Polygonum perfoliatum*	Asiatic Tear thumb, Devil's-Tail Tear thu-mb, Ishimikawa, Mile-A-Minute-Vine, Mile-A-Minute-Weed, Minute weed, Tear thumb	Weed

114.	*Rottboellia cochinchinensis*	Corn Grass, Dholu, Itch Grass, Jointed Grass, Kelly Grass , Kokoma Grass , Konda Panookoo,Paja Peluda, Prickle Grass, Raoul Grass, Rice Grass	Grass
115.	*Salsola tragus*	Russian Tumbleweed, Russian-Cactus, and Russian thistle	Shrub
116.	*Striga asiatica*	Asiatic Witch weed, Buri, Common Mealie Witchweed, Matabele Flower, Red Witchweed, Scarlet Lobelia, Striga, Witchweed	Herb
117.	*Syzygium cumini*	Black Plum, Damson Plum, Indian Blackberry, Jalao, Jaman, Jambu, Jambul, Jambulao, Jamelao, Jamelon-Guier, Jamelong	Tree
118.	*Trapa natans*	Bull Nut, European Water Chestnut, Water Chestnut, Water Nut	Aquatic Plant
119.	*Tussilago farfara*	Assfoot, British Tobacco, Bull's-Foot, Clayweed, Cleats	Herb
120.	*Verbascum thapsus*	Aaron's-Rod, Common Mullein, Flannel Mullein, Flannel Plant and Woolly Mullein	Herb
121.	*Xanthomonas axonopodis pv. citri*	Citrus Canker	Micro-organism
122.	*Zizania latifolia*	Manchurian Water-Rice, Manchurian Wild Rice, Manchurian Wild rice, Manchurian Zizania and Rice Grass	Grass
123.	*Ziziphus mauritiana*	Chinese Apple, Chinese Apple, Chinese Date, Coolie Plum, Crabapple, Indian Cherry, Indian Jujube, Indian Plum, Jujube	Shrub, Tree

23. Major Threats to Biodiversity

1. **Habitat Loss and Degradation**: Natural habitats are destroyed for settlement, Grazing grounds, agriculture, mining, industries, highway construction, drainage, dam building etc.
2. **Exploitation**: Commercial exploitation of biodiversity has invariably meant its overuse and eventual destruction, including hunting, collecting, fisheries and fisheries by catch, and the impacts of trade in species and species' parts, constitutes a major threat.
3. **Alien Invasive Species**; Native species are subjected to competition for food and space due to introduction of exotic species.
4. **Disturbance**, persecution and uprooting, including deliberate eradication of species, considered being pests.

5. **Incidental take**, particularly the drowning of Aquatic reptiles and mammals in fishing nets.
6. **Disease,** both exotic and endemic, exacerbated by the presence of large number of domestic livestock or introduced plant species.
7. **Limited distribution**, which may compound the effects of other factors.
8. **Pollution:** Mixing of undesirable elements and deleterious chemical compounds in environmental components
9. **Control of pests and predator**: Use of synthetic pesticides for plant protection in Agriculture and
10. **Other ecological factors** like distribution range, specialization, reproductive rate etc., and socio-economic and political factors overlooking conservation of resources.

In the majority of cases individual species are faced by several of these threats operating simultaneously, and it is often difficult or impossible to identify with confidence the primary cause of decline. However, the major category of threat, which affects 76% of species, is habitat loss and modification frequently due to cultivation and settlements.

24. Importance of Biodiversity in Human Welfare

Why Conserve Biodiversity- is an oft-asked question? Biological diversity affects us all. It has a direct consumptive value in food, agriculture, medicine, industry and it also has aesthetic and recreational value. Biodiversity maintains ecological balance and continues evolutionary process. The indirect ecosystem services provided through biodiversity are photosynthesis, pollination, transpiration, chemical cycling, nutrient cycling, soil maintenance, climate regulation, air, water system management, and waste treatment and pest control. The three preamble of Convention of Biodiversity (CBD) are (1) Conservation of Biodiversity, (2) Susta-inable use of Biodiversity and leaving enough for the future generations and (3) The fair and equitable sharing of profits arising out of the use of biodiversity.

Immense economic benefits accrue to man because of the prevailing biodiversity, therefore, continued improvement or impoverishment of Biodiversity, and hence the benefits will depend on new and enhanced resources from nature. Without biodiversity humans would perish but a variety of human actions, like cutting of forests, overgrazing, draining of wetlands, conservation of mixed forests into single species plantation, all these lead to extinction, loss of habitats that wild plants and animals depend upon. An ecosystem is an array of living things and the physical and chemical environment with which they interact like wetlands, forests, grasslands, streams and estuaries, desert and Himalaya. A healthy ecosystem provides the conditions and process that sustain human life, in addition to providing food, medicines, purification of air and water, accumulation of toxins, decomposition of wastes, mitigation of floods, moderation of storm surges, stabilization of landscapes and regulation of climates, detoxification of sediments of soil, maintenance of soil fertility, control of potential pest and disease-causing species, dispersal of seeds and aesthetics. By conserving ecosystem we conserve our biodiversity. This is the message, which is required to be spread across for awareness and for people's voluntary participation in this otherwise a very difficult mission for conservation and management of biodiversity. A respect for human life and human diversity is compatible with a respect for biological diversity. Each species has a right to exist and all species are inter-dependent. Humans must live within the same ecological limitations as the other species do. Resources should not be wasted. Peoples must take responsibility for their actions since they have the responsibility for the future generations. The nature has spiritual and aesthetic values that transcendent economic value. Biological diversity is vital to study the mechanism of origin of evolution of life. To conserve biodiversity human beings have to make adjustments with lifestyles collectively.

ACKNOWLEDGEMENTS

The author is thankful to Dr. B.K. Tyagi, Director in-Charge, Centre for Research in Medical Entomology (ICMR), Madurai (T.N.) and Dr. Vijay Veer, Joint Director, Defence R & D Establishment, Road, Gwalior (M.P.) for trusting the author to write and dedicate this article for the dedicatory volume in honour of Prof. T.N. Ananthakrishnan - the Father of Indian Entomology.

REFERENCES

Alfred, J.R.B., 1998. Faunal Diversity in India: An Overview, pp. 1-19. In Faunal Diversity in India. (Eds. Alfred *et al.*, 1998*) ENVIS Center, Zoological Survey of India.*

Alfred, J.R.B., Das, A.K. & Sanyal, A.K., 2000. Ecosystems of India, ENVIS- Zool. Surv, India, Kolkata, pp. 1-410.

Alfred, J.R.B. & Subba Rao, N.V., 1997. Biodiversity: Fauna-An Overview. *ENVIS Newsl. ZSI*, 4(1&2): 2-6.

Bharucha, F.R. & Meher-Homji, V.M., 1965. On the floral elements of semi-arid zones of India and their ecological significance. *New Phytologist* 64(2): 330-342.

Champion, H.G., 1936. A preliminary survey of the forest types of India and Burma. *Indian Forest Rec.* (New Ser.) 1:1-286.

Champion, H.G. & Seth, S.K., 1968. A revised survey of the forest types of India. Govt. of India Press, Delhi, pp.1-404.

FAO/UNEP, 1981. Tropical Forest Resources assessment Project. Technical Report No. 3.

FAO, Rome, Groombridge. 1993. The 1994 IUCN Red List of threatened animals. IUCN, Gland, Switzerland and Cambridge, UK, vi+ 286 pp.

Groombridge, B. & Jenkins, M.D., 2000. Global Biodiversity: Earth's living resources in the 21st century. U.K: WCMC, 246 pp.

Khanna. V., 2003. Invasive species- a threat to endemics and Biodiversity. *Ann. For.* 11(1): 145-153.

Khanna, V., 2003. Further notes on Global Invasive Species. *Ann. For.* 11(2): 311-318.

Khanna, V., 2004. Invasive Alien Species. Challenges to Biodiversity and Conservation. *National Wkshp. On Invasive Alien Species, Botany Deptt. Banaras Hindu University*, August, 18th-20th, 2004.

Khanna, V., 2004. Biodiversity Profile of India. National Symposium State of the art on Conservation of Biodiversity with particular reference to Himalaya, Kashmir, Srinagar, pp 1-23.

Khanna, V., 2004. Status of Biodiversity In India: Conservation, Management and Human Welfare, *Proceedings National Seminar on Zoology and Human Welfare*, SPM Mukherji Degree College, Allahabad, 22-24th Nov, 2004, pp. 308-317.

Khanna, V. & Kumar, A., 2000. Mountains: Western Himalaya, Ecosystems of India, ENVIS- *Zool. Surv. India*, Kolkata, pp.35-50.

Kumar, A. and Khanna, V. 2003. Threatened Indian Fauna. *ENVIS Newsl. Zool. Surv. India* 10 (1&2): 2-12.

Kumar & Khanna, V., 2006. Globally Threatened Indian Fauna: Status, issues and Prospects. *Zool. Surv. India*: pp. 1-104.

Mani, M.S., 1974. Ecology and Biogeography of India. Dr. W. Junk b.v., Publishers, The Hague, 1-772.

MOEF, 2001. Annual Report 2000-2001. New Delhi

MOEF, 2009. Protected areas in Country for Wildlife, Press Release dated 12th Nov, 2009, Ministry of Environment and Forests, New Delhi.

Myers, N., 1988. Threatened Biota: Hot Spots in tropical forests. *Environmentalist* 8: 187-208.

Pushpangadan, P., Rajendraprasad, M. & Krishnan, P.N., 1998. Sacred groves of Kerala synthesis on the state of the art of knowledge. In: Ramakrishnan, P.S., Saxena, K.G. and Chandrashekhara, U.M. (Eds) Conserving the Sacred for Biodiversity *Management*. Oxford and IBH Publishing Co. Pvt. Ltd., New Delhi, pp. 193-209.

Ramakrishnan, P.S., 1998. Conserving the sacred for biodiversity: the conceptual framework. In: Ramakrishnan, P.S., Saxena, K.G. and Chandrashekhara,

U.M. (eds.), *Conserving the sacred for biodiversity management*, Oxford and IBH Publishing Co. Pvt. Ltd., New Delhi, India,pp.3-15.

Ramakrishna & Alfred, J.R.B., 2007. Faunal Resources in India: 1-427 (Published by the Director, Zoological Survey of India, Kolkata)

Ramakrishna, 2009. Animal Discoveries 2009: New Species and New Records, Zoological Survey of India, 5th June, 2010, pp. 1-64.

Rodgers, W.A. & Panwar, H.S., 1988. Planning a Wildlife Protected Area Network in India (Vol. I) Dehra Dun, Wildlife Institute of India. 341 pp.

Rodgers, W.A., Panwar, H.S. & Mathura, V.B. 2002. Wildlife Protected Area Network In India: A Review (Executive Summary). Wildlife Institute of India. Pp.1-43.

Scot, D.A., 1989. A Directory of Asian Wetlands. IUCN, Gland, Switzerland and Cambridge, UK. Sutherland, W.J. 2004. The Conservation Handbook: Research, Management and Policy, Ixv: pp. 1-279.

Sinha, B. & Maikhuri, R.K., 1998. Conservation through socio-cultural-religious practices in Garhwal Himalaya: A case study of Haryali sacred site. A P H Publication Corporation, New Delhi, pp. 289-299.

Chapter - 25

THRIPS DIVERSITY AND THEIR FUNCTIONAL ATTRIBUTES

R. Varatharajan[1] and V.S. Chochong[2]

[1]Department of Life Sciences, Manipur University, Imphal – 79500, India
[2]Department of Education in Science and Mathematics, NCERT,
New Delhi – 110 016, India
E-mail: *rvrajanramya@gmail.com; rvrajanramya@rediffmail.com*

ABSTRACT

Thrips are minute insects of the order Thysanoptera and are endowed with certain special features like fringes on the wings, bladder feet and asymmetrical mouthparts with vestigial right mandible etc. Their occurrence has been known since the Triassic period of the Mesozoic era. About 6000 species have been recorded in the list of world fauna. They exhibit morphological, ecological and behavioural diversities. For instance, selection of different micro-habitats like flower, gall, litter, foliage etc. for colonization; diversity in feeding fungal spores, pollen, nectar, cell sap and soft bodied insects; manifestation of inter and intra species variations in terms of body organs; displaying the physiological propensity leading to parthenogenesis as well as sexual mode of reproduction; laying oviparous, viviparous and ovoviviparous types of progenies, dissemination of pathogen and inflicting damage to the crops on one hand and pollinating and protecting the plants on the other; showing stratified levels of distribution on the crop as well as at different altitude of a given habitat/ecosystem; having stenophagic and euryphagic types of host range etc. clearly highlight varied functional attributes of thrips.

INTRODUCTION

Thrips are one of the important groups of insects distinguishable from other insect orders by having long, fine fringes along the wing margin, protrusible bladder-like structure at the end of tarsus (Physapoda), in the possession of asymmetrical mouthparts with vestigial right mandible and having at least a pre-pupal stage between larva and pupa. Their body size range from 0.5 to 12mm in length. With the lacerating/ punching and sucking type of mouthparts, thrips feed on the contents of pollen grains, fungal spores and mycelia, besides sucking the cell sap. Therefore, in recent years, they have received better attention as invasive species and are recognized as potential pests of economically important plants, vectors of plant diseases, gall makers and a few acting as efficient predators (Ananthakrishnan, 1979; Bhatti, 1992; Mound, 2005).

Thrips exhibit remarkable diversity in terms of feeding habit and habitat. They inhabit grasses, flowers, galls, leaf litter, dry twigs and leaves. On the basis of the micro niche that they occupy, they can be categorized into grass thrips, flower thrips, litter thrips, gall thrips and so on (Fig. 1). Their occurrence in the respective niche describes definite functions like gall induction by gall thrips, pollen transfer by flower dwellers, predation by predatory thrips, spore feeding by litter inhabiting mycophagous forms etc. In addition, chemical communication leading to male – female aggregation for mating and discharge of diverse natural products through their anal droplets as defensive compounds, are some of the features recently documented in an array of thrips species (Ananthakrishnan, 1993). Such diverse functional attributes of various thrips species elevate the order Thysanoptera to a higher level. Although majority of them are terrestrial, certain species thrive on the leaf surface of the plants in the aquatic medium as instances like *Organothrips* (Bhatti, 1998). Thrips occur abundantly in tropical, subtropical and temperate regions and are distributed in all the zoogeographical regions of the world. The present article will throw light on the significance of this group from the view point of their functional approach.

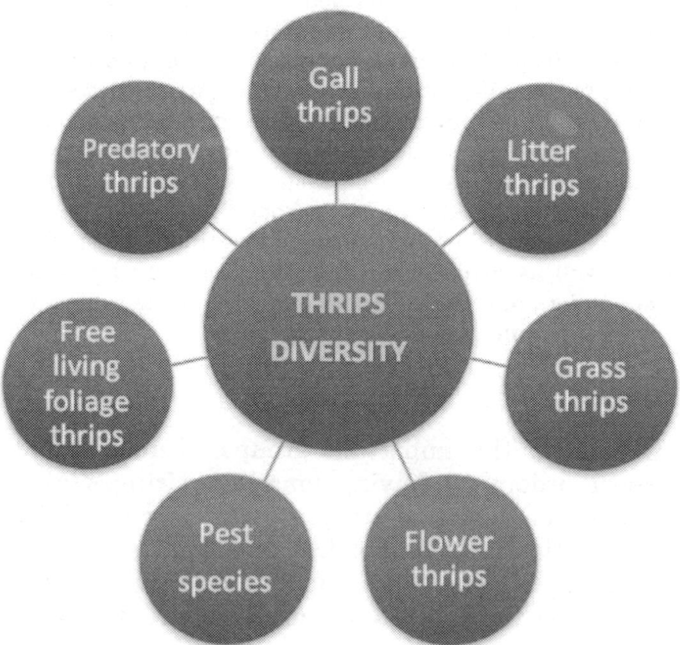

Fig. 1. Thrips diversity

Diversity of gall thrips and thrips galls

Thrips are well known cecidozoans, causing structural abnormalities on a number of gymnosperms, dicotyledons and even on selected monocots. Such induction of malformation varies from simple leaf roll to a complex pouch gall (Ananthakrishnan, 1992). Literature provides evidence of about 300 species of thrips capable of inducing galls on nearly 150 species belonging to 90 plant families. Cecidogenous thrips differ from the rest of gall insects in terms of producing appreciable density of their own members within the gall. The striking feature is that thrips are not only capable of modulating the gall growth, but their morphogenesis is also controlled by the causative agent, as a result they produce a variety of structures like marginal roll or complete roll, ceratonean gall, horn gall, button, pouch, rosette gall, blister, bud and tubular galls, etc. Such gall diversity depends on the specificity of the invading organism as well as the target plant host, besides population density and diversity of the gall maker (Raman, 1996; 2007a,b).

Vector potential

Thrips are efficient vectors of certain plant diseases caused by bacterial, fungal and viral pathogens. They disseminate either by biological or mechanical means. Tomato-spotted wilt viruses, bud necrosis virus of groundnut and banana, coffee and wheat leaf rust spores, are specific examples of thrips transmissible diseases (Ananthakrishnan, 1984, 1993). Further, species such as *Frankliniella schultzei* (Trybom), *Scirtothrips dorsalis* Hood, *Thrips palmi* Karny and *Thrips tabaci* Lindemann ravage a number of crops by transmitting the tospo virus (Ghanekar *et al.*, 1979). The vector status of thrips is attributed to their dynamic nature of infesting diverse plant hosts, swift movement, feeding and breeding propensity and possession of potential pathogen in/on their body (Ulman *et al.*, 2007). It is pertinent to state here that the international scientific communities meet quite often to discuss the issue of thrips - tospovirus interactions.

Thrips as pollinators

Although thrips are primarily recognized as pests, gall makers and vectors, yet at least a few flower dwellers assist in pollination. They visit the blossom to get pollen and enjoy nectar exudations oozing out of the secretory cells of florets. This process enables them to transfer pollen grains inadvertently from one flower to another. Moreover, the phenology of flowers and seasonal dispersal of the thysanopterans synchronizes in such a way, enabling the visiting herbivores to pollinate the flowers and in turn get the floral reward. The first report on the discovery of aggregation pheromone, R-lavandulyl acetate and neryl (S) – 2 methyl butate produced by the male western flower thrips, *Frankliniella occidentalis* (Pergande) throws further light on their functional attribute in attracting individuals for colonization (Hamilton *et al.*, 2005). A brief review on the pollination potential of flower thrips was given by Ananthakrishnan (1984), highlighting the mutual relation between thrips and the flower hosts, eventually leading to survival and propagation of both the pollinator and the flower species (Kirk, 1985).

Diversity of mycophagous thrips

Yet another significance of this group is the ability to feed on fungal spores, hyphae and their break down products present on the dead branches/twigs and fallen leaf litters. The association between thrips and fungus is so intricate that thrips belonging to over 50 genera feed on spores of diverse species of fungi belonging to Ascomycetes, Basidiomycetes, Coelomycetes, and Hypomycetes. Tropical and subtropical regions of the world harbor a large number of thrips species feeding on fungi and are categorized broadly under mycophagous group, which is divided into two subfamilies namely mycelia feeding phlaeothripines and spore feeding idolothripines on the basis of short thin maxillary stylet to the former and long, broad maxillary stylet for the latter group (Mound, 1974).

Thrips exhibit association with different types of plants. Although the terebrantians invariably inhabit inflorescence/flowers and leaves of the plant, the tubuliferans have adapted to both phytophagous and mycophagous habitat and quite a number of species are gall makers. The significance of the latter is that gall and litter dwelling thrips are specialized in their feeding habit and habitat, which in turn has reflected in their biology and body morphology resulting in speciation (Mound, 1974, 1976). The changing habitat coupled with environmental factors have contributed to the formation of diverse polymorphic forms especially among the tubuliferans showing the phenotypic plasticity of major, normal and minor individuals in females and oedymerous, normal and gynaecoid individuals in males (Ananthakrishnan, 2005). Such intra- and inter-species diversities of tubuliferans make the species recognition more complicated. However, the occurrence of such polymorphic forms reflects their functional attributes. For example, studies on the socio-biology of thrips with respect to their polymorphic forms revealed that the gigantic individuals of *Hoplothrips pedicularius* (Haliday) guard the oviposition sites. Further, these dominant males prevent repeated mating by other males; thereby defend the mated females (Crespi, 1992).

Feeding diversity

Although thrips, by and large feed on plants and fungal spores, a few species are obligatory predators feeding on pests like agromyzid larvae, aleyrodids, coccids, mites, psocids, and thrips, thereby acting as natural bio-control agents (Ananthakrishnan, 1993). Owing to their predatory habit, they are being marketed in Continental Europe and Israel as potential biocontrol agents of glass house pests (Larentzaki *et al.,* 2007). Among the terebrantians nearly 95% are plant feeders and the rest under the family Aeolothripidae are predators. Similarly 60% of the tubuliferans are fungivorous and the remaining are plant feeders. Phytophagous forms are generally associated with flowering plants and they feed on pollen, nectar and cell sap. Records on the incidence of thrips on bryophytes, pteridophytes and gymnosperms also exist (Mound, 2002; Mound & Masumoto, 2004). Moss, lichen and fern inhabiting species are very few as like *Lissothrips* & *Bournierothrips* on moss *Aptinothrips* on lichens, and *Monilothrips* & *Indusiothrips* on ferns. Several mycophagous species inhabit fungus infested dry or decaying vegetation, leaf litter, dead wood, bark etc., feeding on fungal spores or hyphae and the break-down products. Thus, it is evident that thrips have a wide range of feeding diversity as illustrated below.

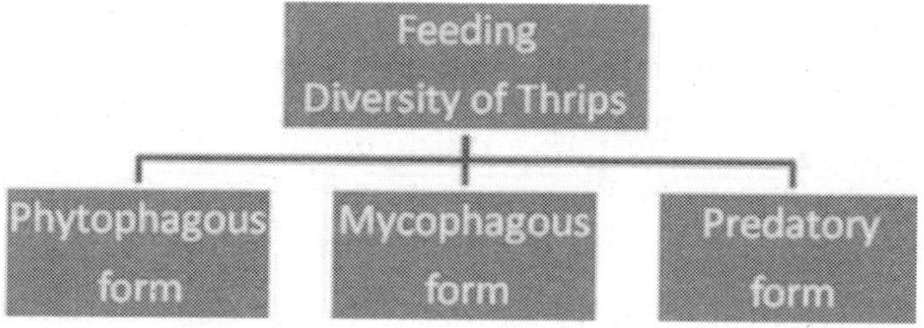

Generally the degree of thrips infestation depends on the plant age/stage, microclimatic conditions including protection level in a concealed niche, water stress and inter or intra specific competition besides the density of natural enemies (Lewis, 1973). Thrips feeding is expected to cause damage to plant parts. However, there appears to be a little difference in terms of the feeding pattern between the individuals belonging to the two suborders. For instance, feeding by terebrantian species is said to be superficial since they feed on the epidermis and upper layer of mesophyll cells with their short mouthcone which will not penetrate the leaf tissues. On the other hand, species with longer mouth-cone have a deep penetrative mode of feeding as like in tubuliferans, and their maxillary stylets have a tongue and groove system, which form a tube through which the liquid cell contents are sucked up.

Thrips diversity and host spectrum

Thrips are generally oligophagous inhabiting a narrow range of plant hosts belonging to closely related plant families. Nevertheless, monophagous as well as polyphagous species also exist among the phytophagous thrips. For instance, *Crotonothrips* is invariably found on *Memecylon, Gigantothrips* on *Ficus* sp., *Austrothrips cochinchinensis* Karny on

Calycopteris floribundus, Panchaetothrips indicus Bagnall on turmeric, *Leeuwenia ananthakrishnani* Varatharajan & Sen on oak, etc. In certain cases, the common names of thrips have been derived from their respective host such as Cardamom thrips (*Sciothrips cardamomi* [Ramakrishna]), Chilli thrips (*Scirtothrips dorsalis*), Turmeric thrips (*Panchaetothrips indicus*), Grapevine thrips (*Rhipiphorothrips cruentatus* Hood), Onion thrips (*Thrips tabaci*), Castor thrips (*Retithrips syriacus* [Mayet]) indicating their intricate relation with their concerned plant host for feeding and establishing the colony. Another level of host range exemplifies the orchid thrips (*Dichromothrips nakahari* Mound), composite thrips (*Microcephalothrips abdominalis* [Crawford]), legume thrips (*Megalurothrips distalis* [Karny]), grass thrips (*Anaphothrips sudanensis* Trybom) etc., wherein certain plant species of a particular family will form the hosts for thrips and hence categorized as oligophagous forms (Varatharajan, 1999). On the contrary, a few species have become extremely polyphagous having more than 30 species of plants belonging to diverse plant families as their hosts. Examples of this category include *Frankliniella schultzei, Thrips hawaiiensis* [Morgan], *T. flavus* Schrank, *T. tabaci*, *Haplothrips gowdeyi* [Franklin] and *H. tenuipennis* Bagnall etc. Although one can appreciate their adaptive spectrum to a wide range of hosts, they equally cause concern to entomologists in the context of plant protection (Varatharajan & James Keisa, 2000).

Thrips diversity in terms of distribution

Thrips distribution can be viewed from: (a) distribution of thrips on the same plant, (b) distribution of thrips at different habitats and (c) altitudinal distribution.

(a) Distribution on a plant: The distribution pattern of phytophagous thrips on the crop foliage is an essential factor to know the behaviour of thrips communities (Ananthakrishnan, 2010). For example, studies on soybean thrips in Urbana indicated the occurrence of three species of thrips, each one occupying a separate microhabitat, *viz*., terminal buds, blossoms and 3rd to 6th trifoliates below terminal bud (Irwin *et al.*, 1979). Similar sequence of thrips occurrence has also been observed on castor, wherein seven species could occupy different nodes with a characteristic pattern of vertical distribution on the same plant. Such variation at different strata of the same plant host specifies the ecological niche required for each species individually (Ananthakrishnan, 1993).

(b) Thrips distribution at different habitat: Abundance as well as distribution of thrips on a plant species varies from habitat to habitat. This aspect is invariably related to host density, phenology and physiological condition of the host, besides environmental conditions of the habitat (Lewis, 1973). For instance, the seasonal distribution of *Frankliniella tritici* Priesner and *Sericothrips varius* (Moulton) on soybean was found to vary with age of the plant, with the former showing infestation just after 30-40 days of planting while the latter infesting after 60 days of planting (Irwin *et al.,* 1979). In the case of flower inhabiting thrips, their distribution and abundance were found to be in tune with the flowering phenology of their respective flower host (Kudo, 1971; Varatharajan, 1984). In addition, thrips diversity in general tends to show a wide variation between plants occurring in natural habitat and plantation sectors. Studies relating to density and distribution of soil micro-arthropods have been attempted by Ananthakrishnan (1988, 1996) in natural and disturbed forest areas of Western Ghats, wherein the practice of monoculture in the plantation crops such as tea, eucalyptus and teak, significantly reduced the abundance of thrips. Similarly, studies on the density, diversity and distribution of thrips of Manipur have also revealed the richness of species diversity under the natural habitat in comparison to oak plantation sectors (Chochong, 2004).

(c) Altitudinal diversity of thrips: Thrips are weak fliers and therefore, they are invariably carried away by the wind. Flying individuals can be trapped by suction traps and sticky traps kept at different heights. The flight activities, population dynamics and altitudinal distribution in terms of trap-catch were studied by a number of scientists and the consolidated details were provided by Lewis (1973). The altitudinal distribution of about 110 thrips species was attempted by Chochong & Varatharajan (2006) on the basis of field collection within the provinces of the valley and hill terrains of Manipur and the collected specimens were classified into high, mid and low altitude inhabiting species as mentioned in Table 1. The low altitude (valley region) witnessed appreciably more number of thrips than the mid and high altitudes. Their occurrence at varying altitudes indicated their survival adaptation. The reason for maximum number of thrips species at the low elevation areas might be due to the availability of plant hosts and also conducive environmental conditions (Varatharajan, 1999).

Table 1. Altitudinal diversity of thrips in the hill and valley regions of Manipur

Thrips group	Number of thrips species collected		
	Low altitude (750 – 1250)	Mid altitude (1251 – 1750)	High altitude (1751 – 2250m MSL)
Terebrantians	42	33	11
Tubuliferans	66	47	14

Reproductive diversity

Some terebrantians reproduce both by sexual and parthenogenetic means, of which the latter method results in production of off springs either by arrhenotoky or thelytoky or at times by amphitoky, which may further be influenced by high temperature as in the case of *T. tabaci* leading to production of males through the process of arrhenotoky (Lewis, 1973) or induction by *wolbachia* (bacteria) on certain species of *Aeolothrips* resulting in female progeny by thelytoky (Warren, 1997). Although thrips in general are bisexual with varying proportion of males and females, reports on the occurrence of female dominated unisexual forms are also evident as in the case of *Franklinothrips vespiformis* (Crawford) (Mound and Marullo, 1998). In addition, gynandromorphs and intersexes are also recorded in *Oxythrips flavus* Morison and *Thrips fuscipennis* Haliday (Mound, 1970). Further, prevalence of ovarian polymorphism has also been noticed in mycophagous forms such as *Bactrothrips, Tiarothrips, Meiothrips* and *Elaphrothrips*, wherein they exhibit either oviparous or viviparous or ovoviviparous nature (Ananthakrishnan, 1987). Such varied strategies of reproduction enable them to maintain their density to an appreciable level in accordance with the climatic condition.

Diverse strategies of thrips management

As thrips occupy diverse niches like flower, leaf, plant gall and leaf litter, different methods need to be adopted to control them. The immature stages of thrips usually become victims of pathogens and natural enemies, while incidence of adult thrips is managed by chemical application and trapping mechanisms. Although a number of methods are adopted for pest control, role of climatic factors cannot be ruled out. Earlier studies have unambiguously indicated that low temperature below 12°C inhibits the growth and development of thrips whereas, their population build up enhance with increase in temperature. With regard to other factors, thrips density is regulated to some extent by humidity and rainfall (Varatharajan, 1984). On the basis of the above factors, it is possible that thrips population is naturally regulated by climatic factors, natural enemies and diseases, besides the availability of plant hosts.

A number of predators like anthocorid bugs, coccinellids, neuropteran and syrphid larvae, predatory mites and thrips exert control over thrips population especially under the field condition. Similarly a number of hymenopteran parasitoids belonging to the genera *Ceranisus, Tetrastichus* and *Thripobius* are parasitic on thrips pests. It is beyond the scope here to discuss about the natural enemy complex of thrips. Ananthakrishnan (1984) in his monumental volume on **Bioecology of thrips** has provided a detailed list of natural enemies. Pathogenic invasion by fungi appears to give encouraging effect under the lab bioassay. A dozen species of fungi has been reported to show appreciable efficacy against thrips, but the mycoinsecticides *Hirsutella thompsonii* and *Pochia lecanii* @ 10^8 conidia/ml exhibit good controls over thrips under field condition (Sithanantham *et al.*, 2007). In addition, periodical removal of weeds, leaf sheaths, old panicles etc as practiced by corporate estates cultivating cardamom, coffee, tea etc will keep thrips infestation under check (Muraleedharan, 1991; Varadarasan, 2003). Other measures such as trapping thrips using yellow sticky traps and applying neem products in the infested fields (Samiayyan & Chandrasekaran, 1995; James Keisa & Varatharajan, 1995) will significantly minimize thrips incidence. Field application of volatile organic compounds like jasmonic acid has been found effective in reducing thrips density to a significant level (Shimoda *et al.,* 1997). The recent discovery of aggregation pheromone in *F. occidentalis* has also given ample scope to the chemical entomologists to look for specific chemicals of thrips origin to combat thrips menace.

CONCLUSION

Thrips exemplify as a unique group of insects from the view point of multifaceted approach irrespective of their variation due to genetic or phenotypic, ecological or physiological, pest or benign, pollinivorous or sporophagous, predaceous or phytophagous. Their affinity with plants of diverse groups like gymnosperm and angiosperm, involvement in gall induction, pollen transfer dynamics, behavioural expressions, programmed reproductive strategies, ability to build up and maintain their population, pathogen transmissibility, and pestiferous nature etc appear to have given adequate support to treat thrips with respectable status. The recent discovery of basalmost thysanopterans namely *Triassothrips virginicus* Grimaldi & Fraser and *Kazachothrips triassicus* Shamakov of the Triassic period further adds feather to the phylogenetic status of thrips (Grimaldi *et al.,* 2004). Of late, the development of bar coding and nucleotide sequencing analysis of the mitochondrial cytochrome oxidase I (COI) gene, has given a fillip to identify adult as well as immature forms of thrips species to a maximum level of accuracy (Asokan *et al.,* 2007; Timm *et al.,* 2008). Although certain hidden secrets of thrips have been studied to some extent over the period of last century, still complexity exists to unravel the techniques of controlling the pests and understanding their intricacy with plants.

ACKNOWLEDGEMENT

This article is dedicated to Professor T.N. Ananthakrishnan on his 85th birthday. We are greatly indebted to Prof. TNA for introducing and encouraging us to work in this group of insects. The authors would like to thank the authorities of Manipur University and Department of Education in Science and Mathematics, NCERT, New Delhi for the support extended during the period of this work.

REFERENCES

Ananthakrishnan, T.N. 1979. Biosystematics of Thysanoptera. *Annu. Rev. Entomol.* 24:159-183.

Ananthakrishnan, T.N. 1984. *Bioecology of Thrips*. Indira Publishing House, Michigan, USA. pp. 233.

Ananthakrishnan, T.N., 1987. Implications of polymorphism in Thysanoptera. *Insect Sci. Appln.* 8: 435-439.

Ananthakrishnan, T.N., 1988. Dynamics of the litter ecosystem in natural and interfered forests. *Int. J. Ecol. Environ. Sc.* 14: 61 – 66.

Ananthakrishnan, T.N., 1992. *Dimensions of Insect – Plant Interactions*. Oxford & IBH. Publishing Co. Pvt. Ltd., New Delhi, pp.184.

Ananthakrishnan, T.N., 1993. Bionomics of Thrips. *Anun. Rev. Entomol.* **38**:71-92.

Ananthakrishnan, T.N., 1996. *Forest litter insect communities – Biology and Chemical Ecology*. Oxford & IBH Publishing Co. Pvt. Ltd., New Delhi, pp. 174.

Ananthakrishnan, T.N., 2005. Perspectives and dimensions of phenotypic plasticity in insects. In: *Insect Phenotypic Plasticity – Diversity of responses*, Eds: T.N. Ananthakrishnan and Douglas Whitman, Science Publishers, Inc., USA, pp. 1-24.

Ananthakrishnan, T.N., 2010. Biodiversity and ecosystem level consequences in relation to thrips – host interactions. In: *Insect Biodiversity: Functional dynamics and ecological perspectives,* Ed: T. N. Ananthakrishnan, Scientific publishers (India), pp. 179-198.

Asokan R., N.K. Krishna Kumar, Vikas Kumar & H.R. Ranganath, 2007. Molecular differences in the mitochondrial cytochrome oxidase I (mtCOI) gene and development of a species-specific marker for onion thrips, *Thrips tabaci* Lindeman, and melon thrips, *Thrips palmi* Karny (Thysanoptera: Thripidae), vectors of tospoviruses (Bunyaviridae). *Bull. Entomol. Res.* 97:461-470.

Bhatti, J.S., 1992. The Order Tubulifera (Insecta): Its characters and classification into families. *Zoology* (*J. pure & appl. zool.*) 3(2): 127-162.

Bhatti, J.S., 1998. Thrips in wetland habitats. *Oriental Insects* 32: 363-380.

Chochong, V.S., 2004. Biodiversity of Thysanoptera fauna of Manipur. *Ph. D Thesis,* Manipur University, pp.162.

Chochong, V.S. & R. Varatharajan, 2006. Diversity of thrips fauna in the hill and valley regions of Manipur. In: *Biodiversity and Insect Pest Management*, Eds: S. Ignacimuthu and S. Jayaraj, Narosa Publishing Co., New Delhi, pp. 133-136.

Crespi, B.J., 1992. Eusociality in Autralian gall-thrips, *Nature* 9:724-726.

Ghanekar, A.M., D.V.R. Reddy, N. Tizuka, P.W. Amin & R.W. Gibbons, 1979. Bud necrosis of groundnut in India caused by tomato spotted wilt virus. *Ann. Appl. Biol.* 93: 173-179.

Grimaldi, D., A.Shmakov & N. Fraser, 2004. Mesozoic thrips and early evolution of the order Thysanoptera (Insecta). *J. Palaeontol.* 78 (5): 941-952.

Hamilton, G. C. James., David, R. Hall & William D. J. Kirk. 2005. Identification of a Male-produced Aggregation Pheromone in the Western Flower thrips, *Frankliniella occidentalis*. *J. Chem. Ecol.* 31 (6): 1369 – 1379.

Irwin, M.E., K.V. Yeargan & N.L. Marston, 1979. Spatial and seasonal patterns of phytophagous thrips in soyabean fields with comments on sampling techniques. *Environ. Entomol.* 8: 131-140.

James Keisa, T. & R.Varatharajan, 1995. Efficacy of two neem products in the field control of *Scirtothrips dorsalis* Hood on *Capsicum annum*. *Indian J. Plant Prot.* 23: 166-168.

Kirk, W. D. J. 1985. Pollen feeding and the host specificity and fecundity of flower thrips (Thysanoptera). *Ecol. Entomol.* 10: 281 – 289.

Kudo, I. 1971. Observations on relative abundance, phenology and flower preference of Thysanoptera in Sapporo and the vincinity. *J. Fac. Hokkaido Univ. Ser., VI. Zool.* 17: 610- 627.

Larentzaki, E., Glen Powell Mike & J. W. Copland 2007. Effect of temperature on development, overwintering and establishment potential of *Franklinot-hrips vespiformis* in the UK. *Entomol. Exp. et Applicata* 124 (92): 143-151.

Lewis, T., 1973. Thrips. Their biology, ecology and economic importance. Academic Press, pp. 349.

Mound, L.A. 1970. Sex intergrades in Thysanoptera. *Entomol. Mon. Mag.* 105: 186-189.

Mound, L.A., 1974. Spore-feeding Thrips (Phlaeothripidae) from leaf litter and wood in Australia. *Aust. J. Zool.* 27: 1-106.

Mound, L.A., 1976. American leaf-litter Thysanoptera of the genera *Erkosothrips, Eurythrips* and *Terthrothrips* (Phlaeothripidae: Phlaeothripinae). *Bull. Br. Mus. (nat. Hist.). Ent. Jr.* 35: 27-64.

Mound, L. A., 2005. Thysanoptera: Diversity and Interaction. *Annu. Rev. Entomol.* 50: 247-269.

Mound, L.A. & R. Marullo, 1998. Biology and identification of Aeolothripidae (Thysanoptera) in Australia. *Invert. Tax.* 12: 929–950.

d, L.A. and Masumoto (2004) Plants associated with thrips. (Source: Mound, L.A. (2005): *Ann. Rev. Entomol.* **50**:247-269.

Muraleedharan, N., 1991. *Pest Management in Tea.* UPASI Tea Research Institute, Valparai, Coimbatore. pp.130.

Raman A. 1996. Nutritional diversity in gall-inducing insects and their evolutionary relationships with flowering plants. *Int. J. Ecol. Environ. Sc.* **22**: 150-160.

Raman A., 2007a. Insect-induced plant galls of India: unresolved questions. *Curr. Sci.* 92: 748-757.

Raman A., 2007b. Biogeographical implications in species richness, biological diversity, and evolution of gall-inducing insects of the Orient and the eastern Palearctic. *Oriental Insects* 41: 9-25.

Samiayyan, K. & B. Chandrasekaran, 1995. Efficacy of botanicals against rice thrips in nursery. *Madras Agri. J.* 82: 497-498.

Shimoda, T., J. Takabayashi, W.Ashihara & A.Takafuji, 1997. A response of predatory insect, *Scolothrips takahashii* towards herbivore induced plant volatiles under laboratory and field conditions. *J. Chem. Ecol.***23**: 2033-2048.

Sithanantham, S., R. Varatharajan, Chandish R Ballal & P.N.Ganga Visalakshy, 2007. Research status and scope for biological control of sucking pests in India: Case study of thrips. *J. Bio.Control.* 21 (Special Issue): 1-20.

Sreekanth, M., M.Sreeramulu, R., D.V.J. Prasada Rao, B. Sarath Babu & T. Ramesh Babu, 2002. Effect of sowing date on *Thrips palmi* Karny population and pea nut bud necrosis virus incidence in green gram (*Vigna radiata* L.Wilezek). *Indian J. Pl. Prot.* 30: 16-21.

Timm, A.E., M. Stiller & J.E. Frey, 2008. A molecular identification key for economically important thrips species (Thysanoptera: Thripidae) in southern Africa, *Afric. Entomol.* 16(1): 68-75.

Ullman. D., J. Moyer, R. Goldbach & G. Moritz, 2007. Eighth International Sympo-sium on Thysanoptera and Tospoviruses September 11–15, 2005 Asilomar, Pacific Grove, California. 49 pp. *J. Insect Sci.* 7: 28, available online: insectscience.org 7.28.

Varadarasan, S., 2003. Relevance of integrated pest management in cardamom cultivation. Spice India, Cochin, May 2003: 11-22.

Varatharajan, R., 1984. Bioecological studies on some Anthophilous thrips from Southern India. Ph.D. Thesis, Madras University.

Varatharajan, R., 1999. Biodiversity of Insects with special reference to Thrips (Thysanoptera) of Manipur. In: *"Biodiversity and conservation"*.Ed. B. Karbuli, NEHU, Shillong, pp.77-80.

Varatharajan, R. & T. James Keisa, 2000. Bioecology and Management of Thrips. In *"IPM system in Agriculture"* Eds. Rajeev K. Upadhyay, K.G. Muk-erji and O.P. Dubey. Aditya Books Pvt. Ltd. New Delhi, 219-234.

Warren, J.H., 1997. Biology of *Wolbachia. Annu. Rev. entomol.* 42:587-609.

Chapter - 26

BIODIVERSITY OF DERMESTID BEETLES (COLEOPTERA: DERMESTIDAE) FROM INDIA, WITH IDENTIFICATION KEY TO ADULTS OF INDIAN GENERA AND DESCRIPTION OF A NEW SUBSPECIES

Vijay Veer

Defense Research & Development Establishment, Jhansi Road, Gwalior-474002, India; E-mail: vijayveer50@yahoo.com

ABSTRACT

The biodiversity of Indian dermestids is now known by 83 species belonging to 14 genera in 3 sub families, Dermestinae, Megatominae and Trinodinae. Identification key to 14 Indian genera is given for the first time. These genera are *Dermestes*, *Thorictodes*, *Thorictus* (Dermestinae), *Anthrenus*, *Attagenus*, *Claviella*, *Katka-enus*, *Megatoma*, *Orphinus*, *Phardonoma*, *Thaumaglossa* and *Trogoderma* (Megatominae), and *Evorinea* and *Trinodes* (Trionodinae). *Anthrenus umbellatarum vijaii* new subspecies is described from Gwalior (Madhya Pradesh). It has 11-segmented antennae in both sexes instead of 10-segmented, a characteristic feature of *An. umbellatarum* and subgenus *Anthrenodes*. The updated information on distribution of each species is also given. A list of 24 species which are likely to occur in India is also given.

INTRODUCTION

Insects belonging to the family Dermestidae are commonly known as skin beetles. They are also referred as larder beetles, hide or leather beetles (*Dermestes*), woolly bears or furniture carpet beetles (*Anthrenus*), carpet beetles (*Attagenus*), and khapra beetles (*Trogoderma*). Most of them require animal proteins for their larval development but a few can develop on a mixed diet or plant material (e.g. cereals) alone. They are scavengers in xeric conditions and require little moisture for their development and occupying varied niches. Adults can range in size from 1–12 mm and have round oval shaped bodies covered in scales or setae, usually clubbed antennae fit into deep grooves. The hind femora also fit into recesses of the coxa. Larvae are scarabaeiform and with setae.

Species of *Anthrenus, Attagenus* and *Dermestes* are serious cosmopolitan pests of stored animal products such as woollen textiles, hides, fur and feather-lined garments, dried fish and beef, milk powder, cheese, casein, etc. Some of them are also serious pests in museums where they damage insect reference collections, animal exhibits, and human cultural artifacts. They also attack chocolate, copra, and cocoa beans. Most damage is done by the larval stage of the beetle, while adults feed on flowers for nectar and pollen grains. They attack products both during their manufacturing and storage, and cause loss in quality and quantity. Thus infestation of products can result in colossal financial loss in homes and business.

In nature they are found in bird nest, bee and wasp nests, rodent burrows, dead insects, fish or crab, spider webs, animal lairs, fur and feathers, animal carcasses, flowers, etc. Dermestid beetles after egg laying visit flowers and do pollination. Larvae during non availability of food undergo retrogressive metamorphosis and survive on their cast skins. Over the last 30 years or so after green revolution in India there are obvious changes in the agricultural practices and storage, and in the living standards; this change in habitat and insecticidal use are influencing the pest status of dermestid species. Some species like *Trogoderma grana-rium*, *Attagenus cyphonoides* and *Attagenus fasciatus* are more frequently infesting the domestic stores than commercial one.

Anthrenus flavipes, An. oceanicus, Attagenus fasciatus, Att. cyphonoides, and *Att. lobatus* are pests of stored woollen including wool-synthetic fibre blended textiles and garments, fur- and feather-lined aviation clothes, upholstered furniture, wool wall hangings, carpets, mattresses, brushes (made of hair), camel saddles, piano hammers, etc. Of these *An. flavipes* is very destructive pest of woollen textiles and caused serious losses in army ordnance depots, industrial and commercial stores in India and elsewhere. *Anthrenus coloratus* is a museum pest and attacking stuffed animals, insect collection, artifacts, etc. (Vijay Veer *et al.*, 1991). *Dermestes ater, D. maculatus,* and *Trogoderma halsteadi* are serious pests in silkworm cocoon stores in India (Vijay Veer & Rao, 1991, Vijay Veer *et al.,* 1996). *Dermestes ater* is also a serious pest in silkworm grainages (silkworm breeding places), dried fish and crabs. *Attagenus birmanicus* is a pest of muga silkworm cocoons in Northeast India. *D. ater* is also reported as intermediate host of tapeworms in poultry (Ueta & Avanccini, 1994).

Trogoderma granarium is a serious cosmopolitan pest of stored grain and cereal products. Species of *Evorinea* and *Thaumaglossa* are predatory on mites and eggs of mantids respectively, *Thorictodes heydeni* is found associated with stored cereal products in many parts of the world. *Thorictus* species are inquilines living with ants.

Dermestids are also important in forensic entomology as some species are known to occur in a predictable succession on decaying carcasses and offer investigators an estimation of the time since death, similar to the use of flies in forensic entomology. The arrival of *D. maculatus* to carrion occurs in a predictable succession. Adult of *D. maculatus* beetles generally arrive 5 to 11 days after death depending upon temperature. *Dermestes ater* was also reported on human corpse in Malaysia (Kumara *et al.,* 2009) and helping in estimation of postmortem interval of corpses. Larvae are also used in taxidermy and natural history museums to clean animal skeletons.

Major taxonomic works for identification of Indian species of this family, and directly or indirectly (the catalogue) sources of descriptions and covering most of the Indian locality records, are those by the following authors: Reitter (1881a, 1881b), Waterhouse (1883), Wasmann (1899, 1912), Arrow (1915), Pic (1910, 1916, 1923, 1937), Champion (1922), Lepesme (1939), Hinton (1945), Kalik (1952, 1954, 1987), John (1963, 1965), Mroczkowski (1963), Zhantiev (1976, 2006), Halstead & Green (1979), Vijay Veer *et al.* (1991, 1996, 2004), Vijay Veer & Rao (1994, 1995a, 1995b, 2000), Vijay Veer, Chauhan & Singh (2004), Háva & Kalik (1999, 2006, 2007), Háva (2000, 2001, 2002, 2003, 2006a, 2006b, 2009), Háva & Herrmann (2004), Háva & Kadej (2006, 2007), Kadej, Háva & Kalik (2007) and Herrmann & Háva (2007).

Mroczkowski (1968) listed 22 species belonging to 8 genera from India plus 8 others within these genera that are known to occur in India in his world catalogue of 882 species of Dermestidae. Later this total number of species from world was raised to 1200 in 45 genera by Háva (2003) in his world catalogue which includes 50 species plus 13 common species in 13 genera from India.

Despite of their economic importance, relatively less information is available on this group of insects in India. After the publication of the first checklist of 61 species in 12 genera of Indian dermestids species by Vijay Veer and Rao (1995b), 22 new species and 2 new genera were added to Indian fauna mainly by Háva and Kadej of Czech Republic and by Herrmann of Germany. Besides their distribution records and biological information were also updated. Thus the total number of species so far known from India is now raised to 83 belonging to 14 genera in 3 sub families including a new subspecies. An identification key to the Indian genera is also being provided for the first time. Full synonymies for all species are given by Háva (2003). Only those synonyms are mentioned here, which are frequently used in the Indian literature. Many species of dermestids are yet to discover from India and a list of expected 24 species is provided (Table 2). It is estimated that 150 to 200 species are occurring in India. The arrangement of the subfamilies and the genera is in accordance with their phylogenetic relationships as recently proposed by Kiselyova and McHugh (2006) and now genera *Thorictodes* and *Throrictus* are placed with subfamily Dermestinae and genera *Anthrenus* and *Attagenus* with subfamily Megatominae. Species in each genus are listed alphabetically.

Key to adults of Indian genera of Dermestidae

1. Compound eyes small or absent; median ocellus absent. Hind coxae not grooved for reception of femur. Beetles less than 1.30- 2.55 mm in length and usually found in ants' nests.......... 2

—. Compound eyes well developed; median ocellus mostly present (except in *Dermestes*). Hind legs more or less retractile and received into groove in coxae. Large or small beetles and not found in ants' nests. 3

2. Eyes very small with a few stemmata. Trichomes present on meso-, meta and occasionally on prosternum; lateral margins of the body flattened.......... *Thorictus* Germar

—. Eyes absent. Trichomes absent on meso-, meta- and prosternum; lateral margins of the body not flattened.. .... *Thorictodes* Reitter

3. Pronotum with longitudinal sublateral carinae. Dorsal surfaces covered with moderately long erect or suberect setae. Hind coxae abuting against inner margin of metepisternum and completely separated from metepimeron. Strongly convex beetles less than 2.5 mm in length.......... 4

—. Pronotum without sublateral carinae. Dorsal surfaces covered either with recumbent or suberect short setae or with scales. Hind coxae usually meeting metepimeron, but may extend only partly across posterior margin of metepisternum. More or less flattened or strongly convex beetles.......... 5

4. Elytra with alternate rows of erect, subrecumbent setae; prosternum not produced to form a pointed process. Female with a central pit having a cluster of long setae on abdominal sternite I .... *Evorinea* Beal

—. Elytra without such alternate rows of setae; prosternum produced into a strong processs which fits into a notch in the mesosternum. Female without a central pit having cluster of long setae on abdominal sternite 1.......... *Trinodes* Dejean

5. Head without a median ocellus. Procoxae large and contiguous at apices. Species 5.5 mm to 12 mm long .... *Dermestes* Linn.

—. Head with a median ocellus. Procoxae not contiguous at apices. Species less than 5.5 mm in length.. 6

6. First segment of hind tarsus much shorter than second segment. Prosternum not covering the gular area of the head, hence leaving the entire under surface of the head exposed. Lamina of hind coxae either toothed or broadened, laterad to insertion of femur.......... 7

—. First segment of hind tarsus as long as or longer than second segment. Prothorax forming a collor in which the prosternum covers the base of the labium when head is retracted. Lamina of hind coxa with margins subparallel or gradually narrowed laterally. 8

7. Antennal club with 3-segmented ....*Attagenus* Latreille

—. Antennal club with 4-segmented......................................*Katkaenus* Háva

8. Vestiture of flat conspicuously coloured scales. .... *Anthrenus* Geoffroy

—. Vestiture of setae, some of which may be slightly ensiform but never scale -like..9

9. Antennal fossa partially or completely closed behind...............................10

—. Antennal fossa broadly open behind..................... ....*Megatoma* Herbst

10. Pronotum with small, short diagonal, impunctate area on either side of basal lobe, Male antenna with greatly enlarged 11 segment; female antennal segment 11 about as long as length of segment 9-10 combined.. *Thaumaglossa* Redtenbacher

—. Pronotum without impunctate area on either side of basal lobe; antennal club composed of 2 or more segments.... 11

11. Antennal club composed of 2 segments... 12

—. Antennal club composed of 3 or more segments.. 13

12 Antennal club nearly circular in shape; penultimate segment equal to or shorter than terminal segment; antennal cavity completely margined behind by a narrow carina similar to those of *Trogoderma*, not visible from lateral side.. *Orphinus* Motsculsky

—. Antennal club elongated; penultimate segment longer than ultimate segment; antennal cavity deep, sharply defined similar to those of *Anthrenus*, visible from lateral side... *Claviella* Kalik

13 Front tibia with numerous, distinct arcute tooth on outer margin; antennal club 3-segemented; half of antennal cavity margined behind by a carina. ....*Phradonoma* Jacquelin du Val

—. Front tibia without distinct tooth; antennal club usually 6-8 segmented; antennal cavity usually completely margined behind by a narrow carina.. .... *Trogoderma* Dejean

INDIAN SPECIES OF DERMESTIDAE

Subfamily 1. DERMESTINAE

Genus 1. *DERMESTES* Linnaeus 1758

Subgenus 1. *Dermestes* Linnaeus, 1758

1. *Dermestes ater* DeGeer 1774

Dermestes ater DeGeer, 1774, *Mémoires pour servir à l'Histoire des Insectes. Tome IV.* Stockholm, 4: 223.

[syn. *Dermestes cadaverinus* Fabricius 1775, *Syst. Entomol....*, p. 55.]

Distribution. India: Mumbai (Maharashtra); Kalimpong (West Bengal); Dehradun (Uttarakhand); Boko (Assam); Dharwar and Hubli, Tumkur, Mysore (Karnataka); Anantpur (Andhra Pradesh); Chennai, Trichinapally (Tamil Nadu). **Cosmopolitan**.

Bioecology: D. ater is a common species in India and a serious pest of stored silkworm cocoons and in silkworm grainages (Vijay Veer *et al.*, 1996).

2. *Dermestes bicolor* Fabricius, 1781

Dermestes bicolor Fabricius, 1781, *Species Insectorum exhibentes eorum differentias specificas....*: 64.

Distribution: India: Punjab (in bird's nest). **Asia**: Afghanistan; Iran; Iraq; Japan; Kazakhstan; N Korea; Kyrgyzstan; Pakistan; Russia: Siberia; Syria; Turkey; Tadzhikistan; Turkmenistan; Uzbekistan.

3. *Dermestes lardarius* Linnaeus, 1758

Dermestes lardarius Linnaeus 1758, *Systema Naturae*, 10th edition : 354.

Distribution. India: widespread. **Cosmopolitan**.

4. *Dermestes leechi* Kalik 1952

Dermestes leechi Kalik 1952, *The Pan-Pacific Entomologist, San Francisco*, 28: 45.

Distribution. India: Bhadrawati, Dharwar (Karnataka); Gwalior, Jabalpur (Madhya Pradesh). **Asia**: Afghanistan; India; Iran; Nepal; Pakistan; Russia: SW Siberia; Tadzhikistan; Turkmenistan; Uzbekistan; **Europe**: England (intr.); Scotland (intr.); Spain; **Africa**: Egypt; Mali; Morocco; Sudan; Tunisia.

Bioecology: Adults of *D. leechi* were caught on fresh meat brought for blow fly culture. This species is quite common in Gwalior. Species was successfully cultured in the laboratory on dog's biscuits.

Subgenus 2. *Dermestinus* Zhantiev 1967

5. *Dermestes carnivorus* Fabricius, 1775

Dermestes carnivorus Fabricius 1775, Systema Entomologie, sistens Insectorum classes, ordines, genera, species...... Lipsiae, p. 55.

Distribution. India (Hinton 1945, Aitken 1975). **Nearly cosmopolitan**.

6. *Dermestes coronatus* Steven in Schönherr, 1808

Dermestes coronatus Steven, in Schonherr 1808, Synonymia Insectorum ..... Stockholm, p. 85.

Distribution. India: Yarkand and Sirikol (Jammu & Kashmir). **Asia**: Afghanistan; "Caucasus“; N India; Iran; Kyrgyzstan; Mongolia; Russia: SE European part, Stavropol; China: Xinjiang; **Europe:** Poland (intr.); Turkey.

7. *Dermestes frischii* Kugelann 1792

Dermestes frischii Kugelann 1792, *Neuestes Mag. Für die Liebhaber der Ent.*,1, 4: 478.

Distribution. India: Kolkata, Batanagar nr. Kolkata (West Bengal). **Nearly cosmopolitan.**

8. *Dermestes fulvicollis* Reitter 1881a

Dermestes fulvicollis Reitter 1881a, *Verhandlungen Naturforschenden Vereines Brünn* (1880), 19: 29.

Distribution. India: Kaulagarh, nr Dehradun (Uttarakhand).

9. *Dermestes maculatus* DeGeer 1774

Dermestes maculatus DeGeer 1774, *Mémoires pour servir à l'Histoire des Insectes. Tome* IV. Stockholm, 4: 223.

[syn. *Dermestes vulpinus* Fabricius, 1781, *Species Insectorum exhibentes eorum differentias specificas...* p. 64].

[syn. *Dermestes kurseongensis* Lepesme 1939, *Bulletin de la Société Entomologique de France,* 44: 192].

Distribution: India: Chakrata Hills, Dehradun, Haldwani (Uttarakhand), Agra, Izatnagar (Uttar Pradesh); Patiala (Punjab), Chennai (Tamil Nadu); South Mangalore (Karnataka); Kolkata (West Bengal); Kogyar (Jammu & Kashmir); Khasi Hills (Meghalaya); Naga Hills (Nagaland). **Asia**: Iran (Háva & Kalík, 1999), India; Sultanate of Oman; **South America**: Peru, Uruguay, Venezuela (Háva, 1999). **Cosmopolitan**.

Bioecology: In India *D. maculatus* was very common in the past but now its population is reduced. It is reported as pest of silkworm industry in Jammu & Kashmir and of dried fishes and beef in Manipur (Barwal & Devi, 1993).

10. *Dermestes tessellatocollis* Moschulsky 1860

Dermestes tessellatocollis Moschulsky 1860, *Coléoptères rapportés de la Sibérie orientale ....* St. Peterburg , p. 124.

10a. *Dermestes tessellatocollis pici* Mroczkowski 1963

Dermestes tessellatocollis pici Mroczkowski, 1963. *Entomologische Mitteilungen*, Hamburg 2: 377.

Distribution. India: Kulu, Shimla (Himachal Pradesh); Chakrata (Uttarakhand), Sikkim. **Asia**: India; N. China; Nepal.

11. *Dermestes undulatus* Brahm 1790

Dermestes undulatus Brahm, 1790, *Insektenkalender für Sammler und Oekonomen*, 1: 114.

Distribution. India: Shimla (Himachal Pradesh); Kashmir (Jammu & Kashmir). **Asia**: India; Nepal; China: Shaanxi. **Holarctic Region**.

Tribe 2. THORICTINI

Genus 2. *THORICTODES* Reitter 1875

12. *Thorictodes dartevellei* John 1961

Thorictodes dartevellei John 1961, *Rev. Zool. Bot.* Afr., 64: 41.

Distribution. India: intercepted on groundnut cake in Britain (Aitken, 1975). This record needs confirmation as cross infestation remains a possibility. **Africa**: Congo.

13. *Thorictodes erraticus* Champion, 1922

Thorictodes erraticus Champion 1922, *The Entomologist's Monthly Magazine*, 58: 73.

Distribution. India: "West Almora district of Kumaon" (Uttarakhand).

14. *Thorictodes heydeni* Reitter, 1875

Thorictodes heydeni Reitter 1875, *Coleopterologische Hefte*, 14: 46.

Distribution. India: Delhi; Kolkata (West Bengal). **Asia**: Turkey, Pakistan, Myanmar, Java; **Europe**: France, Spain, **South America**: Mexico; USA (Texas and Nebraska), Guadeloupe, S. Mariana Is., Saipan; **Africa**: North Africa; Senegal; Nigeria; Kenya; Tanzania; Uganda. **Cosmopolitan**.

Genus 3. *THORICTUS* Germar 1834

15. *Thorictus braminus* Wasmann 1912

Thorictus braminus Wasmann 1912, *Zeitschrift für Wissenschaftliche Zoologie*, 101: 108.

Distribution. India: Wallon (Ahmadnagar, Maharashtra).

Bioecology: Host ant: *Pheidole* sp., *Prenolepsis* sp. (Hetschko, 1926).

16. *Thorictus heimi* Wasmann 1899

Thoriclus heimi Wasmann 1899, *Deutsche Entomologische Zeitschrift*, 1899: 159.

16a. *Thorictus heimi heimi* Wasmann, 1899

Thoriclus heimi heimi Wasmann, 1899. *Deutsche Entomologische Zeitschrift*, 1899: 159.

Distribution. Wallon (Ahmadnagar).

Bioecology: Host ant: *Prenolepis* sp., *Pheidole* sp., *Triglyphothrix* sp. (Hetschko, 1926).

16b. *Thorictus heimi wroughtoni* Wasmann 1912

Thoriclus heimi wroughloni Wasmann 1912, *Zeitschrift für Wissenschaftliche Zoologie*, 101: 109.

Distribution. India: Walton (Ahmadnagar, Maharashtra). Hilly area nr Jabalpur (Madhya Pradesh) (Háva, 2009).

Bioecology: Host ant: *Pheidole* sp. (Hetschko, 1926).

17. *Thorictus indicus* John 1963

Thoriclus indicus John 1963, *Entomologische Abhandlungen und Berichte aus dem Staatlichen Museum fürTierkunde in Dresden* , 28 (5); 225. [= "*Thorictus indicus* Grouvelle" of authors.]

Distribution. India: Belgaon (Karnataka) and Malabar (Kerala).

Myanmar (formerly Burma); found on groundnut cake imported to the U.K. (Aitken (1975) a cotton seed cake loaded at Rangoon (specimen in NHM)]. Also recorded by Aitken (1975) on imports of rice bran and groundnut cake from India.

18. *Thorictus kejvali* Háva, 2009

Thorictus kejvali Háva, 2009, *Baltic Journal Coleopterology*, 9(1): 31-34.

Distribution. India: Orissa.

19. *Thorictus nilgiriensis* John 1963

Thorictus nilgiriensis John 1963, *Entomologische Abhandlungen und Berichte aus dem Staatlichen Museum fürTierkunde in Dresden*, 28 (5): 226.

Distribution. India: Nilgiri Hills (Tamil Nadu).

Subfamily 2. MEGATOMINAE

Tribe 3. ANTHRENINI

Genus 4. *ANTHRENUS* Geoffroy, 1762

Subgenus 1. *Anthrenodes* Chobaut 1898

20. *Anthrenus albonotatus* Pic, 1922

Anthrenus albonotatus Pic, 1922, *Mélanges Exotico-Entomologiques*, 36: 6

Distribution: India: South India. Vietnam

21. Anthrenus kejvali Háva, 2000

Anthrenus kejvali Háva, 2000. *Veröffentlichungen Naturkundemuseum* Erfurt 19: 167.

Distribution: India: Nilgiri Hills (Tamil Nadu).

22. *Anthrenus occultus* Herrmann & Háva, 2006

Anthrenus occultus Herrmann & Háva, 2006. *Biodiversity and Natural Heritage in the Himalaya II*: 465.

Distribution: India: Lohejang (Mandoli); Gangani (Uttarkashi Dist.) (Uttarakhand), Himachal Pradesh.

23. *Anthrenus semenovi semenovi* Zhantiev, 1976

Anthrenus semenovi semenovi Zhantiev, 1976, *Zhuki kozheedy fauny* SSSR, p.161.

Distribution: India: North India. **Asia**: Pamir Mts.

24. *Anthrenus maculifer* Reitter 1881b

Anthrenus maculifer Reitter 1881b, *Verhandlungen Naturforschenden Vereines* Brünn, (1880) 19: 58.

[syn. *Anthrenus globiger* Arrow 1915, *The Annals and Magazine of Natural History*, (8) 15: 447].

Distribution: India: Kolkata (West Bengal). **Asia**: China; Myanmar; Vietnam; Taiwan; Thailand; Japan; Borneo; Nepal; Lao.

25. *Anthrenus umbellatarum* Chobaut 1898

(cf. *Anthrenus umbellatarum* Chobaut 1898, *Revue d'Entomologie*, Caen 17: 74.)

25a. *Anthrenus umbellatarum vijaii* n. ssp. (Fig. 1-6, 7)

Description: Male: Body length (pronotum plus elytra) 1.6 -1.7 mm and width 1.1 mm. Dorsal integument of pronotum black and of elytra reddish brown with brownish black in anterior fourth or more. Ventral integument dark reddish brown or dark brown. Antennae and legs reddish brown, tarsi black. Dorsal surface with white, golden or golden brown to brown scales.

Head with white scales, eyes emarginated at inner mesal part. Antennae 11-segmented, club 3-segmented (Fig. l), length (width) of each segment of club in μm, IX 23 (46), X 42-46 (85), XI 70 (90-91), apical segment of club subequal to combined length of two posterior segment of club. Pronotum 1.3 times broader than long with predominant white scales and a few golden brown scales submedially at base and at middle of notum. Notum is anteriorly narrowed and anterior angles not visible from above. Antennal cavity occupying 0.4 part of lateral margin of pronotum, cavity 2 times longer than broad and margined. Prosternum channeled in 3/4 part.

Elytra 1.38 mm long, wider than base of pronotum at anterior end with three ill defined very narrow transverse bands of golden brown or brown scale each at subbasal, middle and subapical level; anterior band may extend to humeri, sometimes these bands are absent or represented by, discrete patches of golden brown setae. Elytra predominantly with white scales. Scales sparsely arranged leaving cuticle visible in narrow spaces, between them. Scales oval, 50 μm long and 23 μm broad.

Tibia of fore leg with 5-6 obvious spines on dorsal surface (Fig. 2) mid and hind tibia with also obvious spines. Metaepimeron with 6-7 scales in transverse width at narrowest point. Epipleura extending up to metepimeron.

Abdominal sternite with only white scales, first visible sternite with coxal lines at middle. Tergite of first periphilic segment (VIII segment) without sclerisation at middle (Fig. 6). Genitalia 360 μm long, paramere longer than adeagus and distally broad, 280 μm long. (Figs. 3-5). Aedeagus slowly narrowed towards distal end.

Female: Body length (pronotum plus elytra) 2.2 mm, width 1.5 mm. Female is similar to male except in size. Antennal club 3-segmented, length (width) of each segment of club in μm IX 60-62 (85), X 31 (75), XI 31 (62). Bursa copulatrix without structure.

Material examined: Holotype: Male, INDIA: Madhya Pradesh: Gwalior, 12.v.1990, ex flowers of *Capparis zeylanica* (Hindi common name, His), coll. Vijay Veer. Allotype: Female, with the same data as above. Paratypes: 2 M, 2 F. Holotype in author's collection and 2 paratypes will be deposited in collection of the Zoological Survey of India, Kolkata.

Etymology: This new subspecies is named after the author.

Distribution: India: Gwalior (Madhya Pradesh), Dholpur (Rajasthan).

Remarks: Dr D.G.H Halstead has examined the specimens from India and is of the opinion that they belong to a new form of *umbellatarum. Anthrenus umbellatarum vijaii* new subspecies has 11-segmented antennae in both sexes, whereas in *umbellatarum* the antennae are 10-segmented. It is readily distinguished from others by having obvious spines on the dorsal surface of tibia of fore leg as found in *Phradonoma* species. It is found out-of- doors in the wild in flowers of *Capparis zeylanica* in Gwalior and in the adjoining region

Variations: Population has adults with all dorsal scales grayish white in colour.

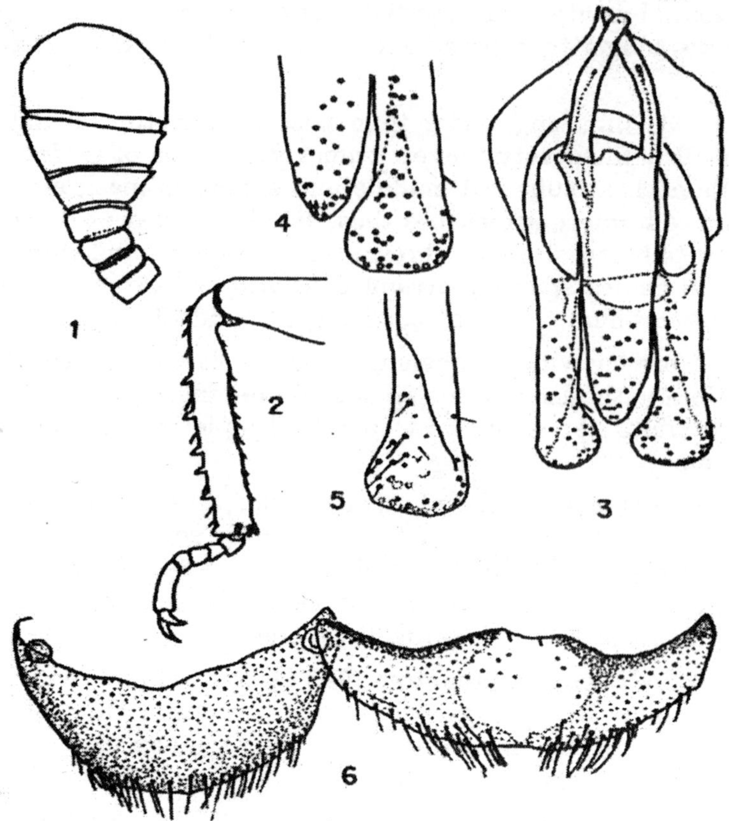

Figs. 1-6. *Anthrenus umbellatarum vijaii* n. ssp., male. 1, Antenna; 2, Fore leg tibia, tarsus; 3, Genitalia; 4, Paramere, adeagus; 5, Paramere distal end; 6, Periphilic Tergite and sternite.

Subgenus 2. *Anthrenops* Reitter 1881b

26. *Anthrenus coloratus* Reitter 1881b (Fig. 10-12)

Anthrenus coloratus Reitter 1881b, *Verhandlungen der Kaiserlich-Königlichen Zoologisch- Botanischen Gesellschaft in Wien*, 30: 91.

[syn. *Anthrenus rufescens* Pic 1923, *Mélanges Exotico-Entomologiques*, 40: 3. (synonymized by Hava, 2010].

Distribution: India: common and wide spread throughout India and a museum pest. **Asia**: Afghanistan; India; Israel; Kazakhstan; Kyrgyzstan; Saudi Arabia; Syria; Tadzhikistan; Turkmenistan; Namibia; **Europe**: Canary Is.; Turkey; **Africa**: Algeria; Egypt; Eritrea; Morocco; Sudan; Tunisia; **North Africa**: U.S.A.

Variations: Adults of *An coloratus* are identified by having elytra with three white transverse bands of scales, separated by darker bands of scales ranging from light yellow/organgish to dark brown. Colour variations of scales are as shown in Figs. 10-12.

27. ***Anthrenus subclaviger*** Reitter 1881a

Anthrenus subclaviger Reitter 1881a, *Verhandlungen Naturforschenden Vereines* Brünn, 19: 59.

Distribution: India: Kolkata, Darjeeling (West Bengal); N.W. Himalaya; Punjab; Jabalpur (Madhya Pradesh). **Africa**: Eritrea; **Asia**: Aden, Saudi Arabia, Yemen.

Subgenus 3. *Anthrenus* Geoffroy, 1762

28. ***Anthrenus flavipes flavipes*** LeConte 1854 (Fig. 8)

Anthrenus flavipes LeConte 1854, *Proceedings of the Academy of the Natural Sciences of Philadelphia*,7: 112.

[syn. *Anthrenus fasciatus* Reitter 1881, *Verhandlungen der Kaiserlich-Königlichen Zoologisch-Botanischen Gesellschaft in Wien*, 30: 75].

[syn. *Anthrenus vorax* Waterhouse 1883, *The Annals and Magazine of Natural History*, (5) 11: 61].

Distribution: India: common and widespread throughout the country. This species is a serious pests of stored woollen textiles in India and elsewhere. **Cosmopolitan**.

29. ***Anthrenus crustaceus*** Reitter, 1881a

Anthrenus crustaceus Reitter, 1881a, *Verhandlungen Naturforschenden Vereines Brünn,* 19: 58.

Distribution: India: Gujarat. **Asia**: Israel, Saudi Arabia, Syria, Yemen; **Europe**: Turkey; **Africa**: Egypt, Eritrea.

30. ***Anthrenus indicus*** Kadej, Háva & Kalik, 2007

Anthrenus indicus Kadej, Háva & Kalik, 2007. *Studies and reports of District Museum Prague-East, Taxonomical series,* 3 (1-2): 103.

Distribution: India: Himachal Pradesh.

31. ***Anthrenus maharastranus*** Háva 2002

Anthrenus maharastranus Háva 2002, *Acta Musei Moraviae, Sciëntiae biologicae*, Brno 87: 34-35.

Distribution: India: Mulshi nr Pune, Bhushi nr Lonavala, Wai nr Pune (Maharashtra).

32. ***Anthrenus oceanicus*** Fauvel 1903 (Fig. 9)

Anthrenus oceanicus Fauvel 1903, *Revue Entomologique,* Caen 22: 338.

[Syn. *Anthrenus fasciatus* Herbst 1797, *Natursystem aller bekannten in- und ausländischen Insekten* p. 337].

Distribution: India: Hyderabad (Andhra Pradesh); Coimbatore (Tamil Nadu). **Asia**: India; Indonesia; Madura; Malaysia; Sri Lanka; Mauritius; Hawaiian Is., **Europe**: Czech Republic; England; **Africa**: Egypt; New Caledonia, Nigeria.

Subgenus 4. *Florilinus* Mulsant & Rey, 1868

33. *Anthrenus hartmanni* Háva 2000

Anthrenus hartmanni Háva 2000, *Veröffentlichungen Naturkundemuseum Erfurt*, 19: 170.

Distribution: India: Almorha (Uttarakhand). **Asia**: India; Nepal.

34. *Anthrenus zhantievi* Háva & Kadej 2006

Anthrenus zhantievi Háva & Kadej 2006, *Mitteilungen des Internationalen Entomologischen Vereins* 31: 123.

Distribution: India: Madhya Pradesh.

Subgenus 5. *Nathrenus* Casey, 1900

35. *Anthrenus cardamom* Háva, 2001

Anthrenus cardamom Háva, 2001. *Entomologische Zeitschrift, Stuttgart*, 111: 58.

Distribution: India: Kallar Valley, Cardamom Hills (Tamil Nadu).

Subgenus 6. *Peacockia* Menier & Villemant, 1993

36. *Anthrenus taricus* Zhantiev, 2006

Anthrenus taricus Zhantiev, 2006. *Proceedings of the Russian Entomological Society*, St. Petersburg 77: 96.

Distribution: India: Rajasthan. **Asia**: Iran, Pakistan.

Subgenus 7. *Solskinus* Mroczkowski 1962

37. *Anthrenus sinensis* Arrow 1915

Anthrenus sinensis Arrow 1915. *The Annals and Magazine of Natural History*, (8) 15: 445.

Distribution: India: Chakrata (Uttarakhand); Shimla Hills (Himachal Pradesh); N.W. Himalaya; **Asia**: North China; N. India; Russia: Primorskiy Kray, Kitay.

Tribe 4. ATTAGENINI

Genus 5. *ATTAGENUS* Latreille 1802

Subgenus 1. *Aethriostoma* Motschulsky, 1858

38. *Attagenus bifasciatus* (Olivier 1790)

Dermestes bifasciatus Olivier 1790, Entomologie, ou histoire naturelle des Insectes,..., 2 (9): 13.

Distribution: India: Dehradun (Uttarakhand), Kashmir (J. & K.). **Asia**: Afghanistan, India; Iran; Iraq; Israel; Lebanon; Palestine; Syria; Turkmenistan; **Europe**: Turkey; **Africa**: Algeria; Egypt; Libya; Morocco; Tunisia.

39. *Attagenus birmanicus* Arrow 1915 (Fig.14a, b)

Attagenus birmanicus Arrow 1915, *The Annals and Magazine of Natural History*, (8) 15: 428.

Distribution: India: Boko (Assam); Mumbai (Maharashtra). **Asia**: Afghanistan; China; Myanmar; Russia (Trans-Baikal region); N & W China; Mongolia; E. Kazakhstan; Kyrgyzstan; Tajikistan; Turkmenistan; **Europe**: Sweden (intr.).

Remark: Adults of *Att. birmanicus* are with or without subbasal band on elytra. Both normal (unbanded) and banded adults were successfully cross bred in the laboratory as given in Table 1 below. No difference was observed in the male genitalia of both types of males. Thus presence of a subbasal band on elytra is a variable feature in *Att. birmanicus.*

Table 1. Cross breeding of normal adults with banded adults of *Attagenus birmanicus*

	Crossing between	Larvae found in each breeding tube		
i.	Normal ♂x Normal ♀	12	29	14
ii	Normal ♂ x Banded ♀	34	22	6
iii	Banded ♂ x Banded ♀	13	28	30
iv	Banded ♂ x Banded ♀	37	30	15

40. *Attagenus sparsutus* (Reitter 1881a)

Aethriostoma sparsutus Reitter 1881a, *Verhandlungen Naturforschenden Vereines Brünn,* 19: 36.

Distribution. India: Belgaon nr Mumbai (Maharashtra); Kurseong (West Bengal); Nilgiri Hills (Tamil Nadu). **Asia**: India, China Himalaya, Nepal, Tibet.

41. *Attagenus undulatus* (Motschulsky 1858)

Aethriostoma undulatus Motschulsky 1858, *Études Entomologiques*, Helsingfors, 7: 47.

Distribution: India: Kolkata, Pashok (Daljeeling distr.) (West Bengal); Puri (Orissa). Mulshi nr Pune, Bhushi nr Lonavala (Maharashtra). **Asia**: Buru Is.; Cambodia; S. China; Indonesia; Laos; Malayasia, Myanmar, Phillippines, Sri Lanka, Thailand, Vietnam; Africa: Comoros, Madagascar; Mauritius; Seychelles; **Austraila**: Papua New Guinea; S. Mariana Is.

Subgenus 2. *Attagenus* Latreille, 1802

42. *Attagenus bezdeki* Háva & Kadej, 2007

Attagenus bezdeki Háva & Kadej, 2007. *Genus*. 18(2): 216.

Distribution: India: Bhushi dam, Lonavala (Maharashtra).

43. *Attagenus chakouri* Pic, 1907

Attagenus chakouri Pic, 1907: *L'Échange, Revue Linnéenne* 23: 153

[syn. *Attagenus multinotatus* Pic 1916, *Mélanges Exotico-Entomologiques* 17: 15].

Distribution: India: locality unknown. **Asia**: India; Iraq; Saudi Arabia; **Africa**: Egypt.

44. *Attagenus cyphonoides* Reitter 1881a (Fig. 16)

Attagenus cyphonoides Reitter 1881a, *Verhandlungen Naturforschenden Vereines Brünn* 19 (1880): 34.

[syn. *Attagenus alfierii* Pic 1910, *L'Échange, Revue Linnéenne* 26: 17.]

Distribution: India: common and widespread throughout the country. **Asia**: Afghanistan; China; India; Iraq; Kazakhstan; Mongolia; Pakistan; Russia; Saudi Arabia; Tadzhikistan; Turkmenistan; Uzbekistan; Europe; **Africa**: Egypt; Morocco; Nigeria; Senegal; Sudan; Tunisia; **North America**: Canada; Mexico; USA. **Cosmopolitan**.

45. *Attagenus fasciatus* (Thünberg 1795) (Fig. 13)

Anthrenus fasciatus Thünberg 1795, *Dissertatio Entomologica, sistens Insecta Suecica*. Pars 9: 105.

[syn. *Attagenus gloriosae* (Fabricius 1798, p. 76)]

[syn. *Attagenus cinnamomeus* Roth 1851, *Archives Naturgesschichte*, 17: 122; cf. Halstead & Green 1979]

Distribution: India: Widespread and common throughout the country. Widespread throughout the tropical and subtropical regions of the world. **Africa**: Madagascar (Háva, 2003).

46. *Attagenus gobicola* Frivaldszky 1892

Attagenus augustatus gobicola Frivaldszky 1892, *Természetrajzi Füzetek*, Budapest 15: 122.

[syn. *Attagenus sikkimensis* Pic 1942, *L'Échange, Revue Linnéenne* Numéro spécial: 7. synonymized by Hava, 2010]

Distribution: India: Leh (Jammu & Kashmir); Sikkim. **Asia**: Afghanistan; N China; India: Kashmir, Sikkim; Kazakhstan; Kyrgyzstan; Mongolia; Tadzhikistan; Turkmenistan; **Europe:** Sweden (intr.).

47. Attagenus grisescens Pic 1937

Attagenus grisescens Pic 1937, *Mélanges Exotico-Entomologiques*, 69: 5.

Distribution: India: Mumbai (Maharashtra).

48. *Attagenus indicus* Kalik 1954 (Fig. 15)

Attagenus indicus Kalik 1954, *The Annals and Magazine of Natural History*, (12) 7: 368.

Distribution: India: Dehradun, Mussoorie (Uttarakhand), Dhampur, Bhadoi (Uttar Pradesh); Gwalior (Madhya Pradesh). **Asia**: India; Nepal.

49. *Attagenus lobatus* Rosenhauer 1856

Attagenus lobatus Rosenhauer 1856, *Die Thiere Andalusiens* .., p. 108.

Distribution: India: Dehradun (Uttarakhand), Dhampur, Kanpur (Uttar Pradesh); Ludhiana (Punjab); Gwalior (Madhya Pradesh); Kolkata (West Bengal). **Asia**: Afghanistan; China; India; Iran; Iraq; Kazakhstan; Kyrgyzstan; Mongolia; Saudi Arabia; Tadzhikistan; Turkmenistan; United Arab Emirates; Uzbekistan; **Europe**: Bulgaria; Greece; Italy; Spain; Turkey; **North Africa**; **America**: USA.

50. *Attagenus pellio* (Linnaeus 1758)

Demtestes pellio Linnaeus 1758, *Systema Naturae*, 10th edition, p. 355.

Distribution: Probably found in India, intercepted on rice bran from India at a British port (Aitken 1975). **Nearly cosmopolitan**.

51. *Attagenus unicolor* (Brahm 1790)

Dermestes unicolor Brahm 1790, *Insektenkalender für Sammler und Oekonomen*, 2: 144.

[syn. *Auagenus megatoma* (Fabricius 1798, p. 71).]

[syn. *Attagenus piceus;* - Olivier 1790, not Thünberg 1781.]

Also known in literature as *Attagenus megatoma* (Fabricius 1798).

Distribution: India: Punjab. **Cosmopolitan**.

Genus 6. *CLAVIELLA* Kalik 1987

52. *Claviella besucheti* Kalik 1987

Claviella besucheti Kalik 1987, *Revue de Suisse Zoology*, 94 (4): 656.

Distribution: India: Valara falls (Kerala).

Genus 7. *KATKAENUS* Háva, 2006c

53. *Katkaenus spectaculus* Háva 2006c

Katkaenus spectaculus Háva, 2006c. *Baltic Journal of Coleopterology*, 6: 30.

Distribution: India: Tamil Nadu (South India).

Genus 8 *MEGATOMA* Herbst, 1792

54. *Megatoma (Pseudohadrotoma) indica* Háva, 2000

Megatoma (Pseudohadrotoma) indica Háva, 2000. *Veröffentlichungen Naturkunde- museum* Erfurt, 19: 165.

Distribution: India: Jingchan 15 km West Leh (Ladakh).

Genus 9. *ORPHINUS* Motschulsky 1858

Subgenus 1. *Falsoorphinus* Pic, 1931

55. *Orphinus kejvali* Háva, 2006

Orphinus kejvali Hava, 2006. *Studies and Reports of District Museum Prague-east*, Taxonomical Series, 2: 40.

Distribution: India: South India.

Subgenus 2. *Orphinus* Motschulsky 1858

56. *Orphinus atrofasciatus* Pic 1916

Orphinus atrofasciatus Pic 1916, *Mélanges Exotico-Entomologiques* 17: 16.

Distribution: India: Madurai (Tamil Nadu).

57. *Orphinus biflexus* Reitter, 1881a

Orphinus biflexus Reitter, 1881a. *Verhandlungen Naturforschenden Vereines Brünn*, 19 (1880): 55.

Distribution: Belgaon (Karnataka).

58. *Orphinus fulvipes* (Guérin-Méneville 1838)

Globicomis fulvipes Guérin-Méneville 1838, *Revue Zoologique,* Paris 1: 138.

Distribution: India: Dehradun (Uttarakhand); Boko (Assam). **Nearly cosmopolitan**.

59. *Orphinus jucundus* Arrow 1915

Orphinus jucundus Arrow 1915, *The Annals and Magazine of Natural History*, (8) 15: 439.

Distribution: India: Mumbai (Maharashtra); Belgaon (Karnataka).

60. *Orphinus hartmanni* Háva, 2001.

Orphinus hartmanni Háva, 2001, *Veröffentlichungen Naturkundemuseum* Erfurt, 20: 213.

Distribution: India: North India. **Oriental region**: Bhutan; Nepal.

61. *Orphinus malabarensis* Pic 1916

Orphinus malabarensis Pic 1916, *Mélanges Exotico-Entomologiques,* 17: 16.

Distribution: Malabar Hills (Kerala).

62. *Orpmnus minor* Arrow 1915

Orphinus minor Arrow 1915, *The Annals and Magazine of Natural History*, (8) 15: 440.

Distribution: India: Mumbai (Maharashtra), Cardamom Hills (Tamil Nadu).

63. *Orphinus nilgiriensis* Arrow 1915

Orphinus nilgiriensis Arrow 1915, *The Annals and Magazine of Natural History,* (8) 15: 439.

Distribution: India: Nilgiri Hills (Tamil Nadu); Srinagar (Jammu & Kashmir).

64. *Orphinus notaticollis* Pic 1916

Orphinus notaticollis Pic 1916, *Mélanges Exotico-Entomologiques,* 17:16.

Distribution: India: North East India (= East India).

65. *Orphinus sikkimensis* Háva & Herrmann. 2004

Orphinus sikkimensis Háva & Herrmann. 2004. *Veröffentlichungen Naturkundemuseum Erfurt* 23: 201-202.

Distribution: India: Sikkim.

66. Orphinus tabitha Arrow, 1915

Orphinus tabitha Arrow, 1915 *The Annals and Magazine of Natural History*, (8) 15: 440.

Distribution: India: South India. **Oriental region**: India; Sri Lanka.

67. *Orphinus unifasciatus* Háva, 2006d

Orphinus unifasciatus Háva, 2006d. *Biodiversity and Natural Heritage in the Himalaya II*. p.463.

Distribution: India: Himachal Pradesh.

Genus 10. *PHRADONOMA* Jacquelin du Val 1859

68. *Phradonoma budha* Háva & Kadej, 2007

Attagenus buddha Harva & Kadej, 2007. *Genus* 18(2): 218.

Distribution: India: Bhushi dam, Lonavala (Maharashtra).

69. *Phradonoma nobile* (Reitter 1881b)

Trogoderma nobile Reitter 1881b, *Verhandlungen der Kaiserlich-Königlichen Zoologisch- Botanischen Gesellschaft in Wien* 30: 85.

Distribution: India: Dehradun (Uttarakhand), Saharanpur (Uttar Pradesh); Jabalpur (Madhya Pradesh). **Asia**: Afghanistan; "Caucasus"; India; Iran; Iraq; Israel; Jordania; Pakistan; Saudi Arabia; Syria; Tadzhikistan; Turkmenistan; United Arab Emirates; Uzbekistan; **Europe**: Cyprus; Greece; Portugal; Spain; **Africa**: Algeria; Egypt; Eritrea; Libya; Morocco; Namibia; Nigeria; South Africa; Sudan; Tanzania; Tunisia; Zimbabwe.

70. *Phradonoma piceum* Háva, 2002

Phradonoma piceum Háva, 2002. *Acta Musei Moraviae, Scientiae biologicae*, Brno 87: 33.

Distribution: India: Sanjay Gandhi park nr Mumbai (Maharashtra).

71. *Phradonoma tricolor* (Arrow 1915)

Trogoderma tricolor Arrow 1915, *The Annals and Magazine of Natural History*, (8) 15: 431.

Distribution: India: Kanpur. (Uttar Pradesh). **Asia**: India; Oman; Saudi Arabia; Yemen; **Europe**: Denmark (intr.); Netherlands (intr.);

Genus 11. *THAUMAGLOSSA* Redtenbacher 1867.

72. *Thaumaglossa hilleri* Reitter 1881a

Thaumaglossa hilleri Reitter 1881a, *Verhandlungen Naturforschenden Vereines* Brünn, 19: 42.

Distribution: India: Dehradun (Uttarakhand). **Asia**: China; Nepal; Japan; Philippines; Taiwan.

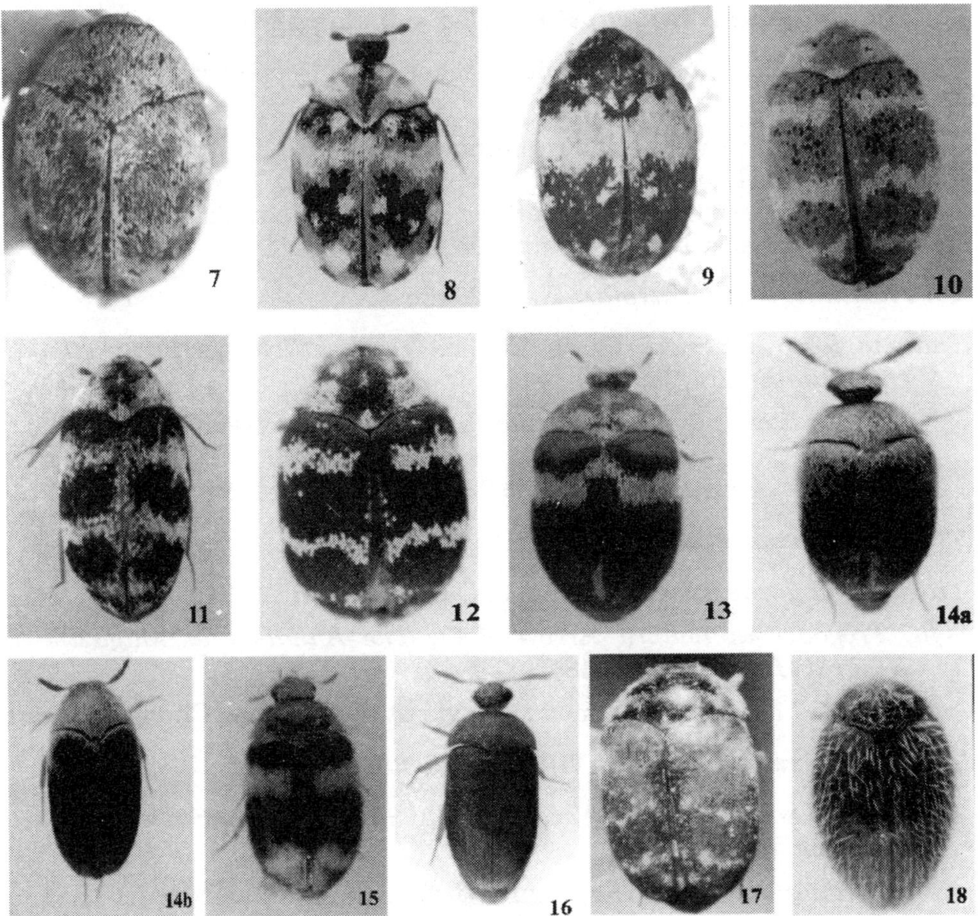

Figs. 7-18 Adults dorsal habitus. 7, *Anthrenus umbellatarum vijaii* n. ssp. female; 8, *Anthrenus flavipes*, female; 9, *Anthrenus oceanicus*, female, 10-12, *Anthrenus coloratus*, female with colour variation, a,b,c; 13, *Attagenus fasciatus*, female; 14a, *Attagenus birmanicus*, male with partial subbasal elytral band; 14b, *Att. birmanicus,* male with reduced setae on subbasal elytra 15, *Attagenus indicus*, male; 16, *Attagenus cyphonoides*, male; 17, *Trogoderma halsteadi*, female; 18, *Evorinea indica*, female.

73. *Thaumaglossa indiana indiana* Vijay Veer, 2004

Thaumaglossa indiana Vijay Veer, Chauhan & Singh 2004. *Oriental Insects*, 38: 389-394.

Distribution: India: Dehradun (Uttarakhand).

74. *Thaumaglossa pseudohilleri* Háva, 2006a

Thaumaglossa pseudohilleri Háva, 2006a. *Studies and Reports of District Museum Prague-east,* Taxonomical Series, 2: 42.

Distribution: India.

75. *Thaumaglossa rufocapillata* Redtenbather 1867

Thaumaglossa rufocapillata Redtenbacher 1867. *Coleoptera*. Zoological, 2: 44.

Distribution: India: locality unknown. **Asia**: China; India; Indonesia; Japan; Korea; Laos; Malaysia; Nepal; Taiwan; Thailand; Vietnam; **Europe**: Germany (intr.); Netherland (intr.); **Africa**: Cameroon; Congo; Kenya; Madagascar; Nigeria; Tanzania; Zimbabwe.

76. *Thaumaglossa wittmeri* Hava 2006b

Thaumaglossa wittmeri Hava, 2006b. *Veröffentlichungen Naturkundemuseum Erfurt*, 25: 233.

Distribution: India: Uttar Pradesh. **Oriental Region**: India; Nepal.

Genus 12. *TROGODERMA* Dejean 1821

77. *Trogoderma granarium* Everts 1898

Trogoderma granarium Everts 1898, Coleopt. Neerl., 1: 610.

[syn. *Trogoderma khapra* Arrow 1917, *The Annals and Magazine of Natural History*, (8) 19: 481].

Distribution: India: common throughout the country. **Cosmopolitan**.

78. *Trogoderma halsteadi* Vijay Veer & Rao 1994 (Fig. 17)

Trogoderma halsteadi Vijay Veer & Rao 1994, *Journal of Stored Products Research*, 30: 283-295.

Distribution: India: Dehradun (Uttarakhand); Bhadrawati (Karnataka). **Oriental Region**: Thailand.

79. *Trogoderma inclusum* LeConte 1854

Trogodemla inclusum LeConte 1854, *Proceedings of the Academy of the Natural Sciences of Philadelphia*, 7: 110.

[syn. *Trogoderma versicolor* of British authors, not Creutzer 1799].

Distribution: India: Dehradun (Uttarakhand); Jabalpur (Madhya Pradesh). Mediterranean subregion; **Europe**: England; **North America**: USA.

80. *Trogoderma unifasciatum* Pic 1942

Trogoderma unifasciatum Pic 1942, *L'Échange, Revue Linnéenne Numéro spécial*, p. 6.

Distribution: India: locality unknown.

Subfamily 3. TRINODINAE

Genus 13. *EVORINEA* Beal 1961

81. *Evorinea hirtella* (Walker 1858)

Trinodes hirtella Walker 1858, *The Annals and Magazine of Natural History,* (3) 2: 207.

Also known in literature as *Apsectus hirtellus* (Walker).

Distribution: India: Kottayam, Travancore (Kerala). **Asia**: India; Sri Lanka

82. *Evorinea indica* (Arrow 1915) (Fig. 18)

Apsectus indica Arrow 1915, *The Annals and Magazine of Natural History*, (8) 15: 450.

[syn. *Evorinea hisamatsui* Ohbayashi 1977, *Kontyû*, 45 (3): 357 (synonymized by Vijay Veer & Rao, 2000)].

Distribution: India: Mumbai (Maharashtra), Dehradun (Uttarakhand). Gwalior (Madhaya Pradesh). Japan, Taiwan.

Genus 14. *TRINODES* Dejean 1821

83. *Trinodes carinatus* Pic 1916

Trinodes carinatus Pic 1916, *Mélanges Exotico-Entomologiques*, 17: 17.

Distribution: India: locality unknown.

Table 2. The following species are expected to occur in India as they found in the neighbouring countries. Some common species may likely to introduce in the country through commerce.

Pakistan

1. *Anthrenus lindbergi* Mroczkowski, 1959
2. *Attagenus brunneus* Faldermann, 1835
3. *Mariouta stangei* Reitter, 1910
4. *Thaumaglossa indiana pakistana* Háva, 2006

Nepal

1. *Attagenus karnali* Háva, 2001
2. *Attagenus nepalensis* Háva, 2001
3. *Ctesias schawalleri* Háva, 2002
4. *Orphinus kresli* Háva, 2003
5. *Thaumaglossa weigeli* Háva, 2006
6. *Thaumaglossa yeti* Háva, 2003

Sri Lanka

1. *Anthrenus ceylonicus* Kadej & Háva, 2006
2. *Orphinus defectus* (Walker, 1858)
3. *Orphinus funestus* Arrow, 1915
4. *Orphinus guernei* Pic, 1916
5. *Orphinus horni* Pic, 1927
6. *Thaumaglossa tonkinea* Pic, 1916
7. *Trinodes cinereohirtus* Motschulsky, 1863

Invasive species likely to introduce through commerce

1. *Anthrenus museorum* (Linnaeus, 1761)
2. *Anthrenus pimpinellae pimpinellae* (Fabricius, 1775)
3. *Anthrenus verbasci* (Linnaeus, 1767)
4. *Dermestes haemorrhoidalis* Küster, 1852
5. *Reesa vespulae* (Milliron, 1939)
6. *Trogoderma variabile* Ballion, 1878
7. *Trogoderma versicolor* (Creutzer, 1799)

ACKNOWLEDGEMENT

The author is thankful to Dr. R. Vijayaraghavan, Director, Defence R.&D. Establishment, Gwalior for support and encouragement. I am indebted to Prof. J.S. Bhatti, Delhi University, Delhi, Dr. D.G.H. Halstead, United Kingdom and to Prof. R.S. Beal Jr., California for their kind help for identification, providing useful information and suggestions.

REFERENCES

Aitken, A.D., 1975. Insect Travellers 1. Coleoptera. *Ministry of Agriculture, Fisheries (U.K.), Technical Bulletin,* No. 31, xvi + 191 pp. London.

Arrow, G.J., 1915. Notes on the Coleopterous family Dermestidae, and descriptions of some new forms in the British Museum. *Annals and Magazine of Natural History* (Series 8) 15: 425-451. London.

Barwal, R.N & Devi, Jibanlata, 1993. Pests of stored food commodities and their management with particular reference to hide and skin beetle, Dermestes maculates (Deg.) in Manipur. *Journal of Insect Science* 6(2): 189-194.

Champion G. C., 1922. Some Indian Coleoptera (7). *The Entomologist's Monthly Magazine* 58: 68-76. London.

Halstead, D.G.H. & Green, M., 1979. *Attagenus woodroffei* sp.n. previously confused with *Attagenus fasciatus* (Thunberg) (Coleoptera, Dermestidae). *Notulae Entomologicae* 59: 97- 104. Helsinki.

Háva, J., 2001. New and little known species of the genus *Anthrenus* O.F. Müller, 1764 (Coleoptera: Dermestidae). *Entomologische Zeitschrift,* Stuttgart 111: 58-61.

Háva, J., 2002. Five new species of Dermestidae (Coleoptera), with notes on synonymy and distribution of the family. *Acta Musei Moraviae, Scientiae biologicae Brno* 87: 29-40.

Háva, J., 2003. .World Catalogue of the Dermestidae (Coleoptera). *Studie a Zprávy Oblastního Muzea Praha- východ,* Supplementum 1, 196 pp.

Háva, J., 2006a. Contribution to the knowledge of world Dermestidae (Coleoptera). *Studies and Reports of District Museum Prague-east, Taxonomical* Series 2: 35-50.

Háva, J., 2006b. New and interesting Dermestidae (Insecta: Coleoptera) from Nepal - IV. *Veröffentlichungen Naturkundemuseum Erfurt* 25: 321-234.

Háva, J., 2006c. Description of a new genus *Katkaenus* gen. nov. (Coleoptera: Dermestidae: Attageninae) from South India. *Baltic Journal of Coleopterology* 6 (1): 29-33.

Háva, J., 2006d. Descriptions of three new Dermestidae (Insecta: Coleoptera) from Pakistan and India. pp. 463-465. In: Hartmann, M. & Weipert, J. (eds.): Biodiversity and Natural Heritage in the Himalaya II. Verein der Freunde und Förderer des Naturkundemuseums Erfurt E.V., 524 pp., 983 b/w fig., XII coloured plates.

Háva, J., 2009. *Thorictus kejvali* sp.n., a new species from Eastern India (Coleoptera: Dermestidae: Thorictinae). *Baltic Journal of Coleopterology* 9: 31-34.

Háva, J., 2010. Dermestidae World (Coleoptera). World Wide Web electronic publication. http://dermestidae.wz.cz

Háva, J. & Kadej, M., 2006. Three new species of *Anthrenus* O.F. Müller, 1764 from Botswana, India and China (Coleoptera: Dermestidae). *Mitteilungen des Internationalen Entomologischen Vereins* 31: 119-130.

Háva, J. & Kadej, M., 2007. Contribution to the Dermestidae (Coleoptera) from Maharashtra State in India. *Genus*18: 215-220.

Háva, J. & Kalík, V., 2006. Redescription and new synonymy of *Attagenus multinotatus* (Coleoptera: Dermestidae: Attageninae) probably from India. *Entomological Problems* 36: 95-96.

Háva, J. & Herrmann A., 2004. A new *Orphinus* species (Coleoptera: Dermestidae) from India: Sikkim. *Veröffentlichungen Naturkundemuseum Erfurt* 23: 201-202.

Herrmann, A. & Háva, J., 2007. New faunistic records of Dermestidae (Coleoptera). *Acta Entomologica Musei Nationalis Pragae* 47: 123-125.

Hinton, H.E., 1945. *A Monograph of the Beetles Associated with Stored Products.* Vol. 1. viii + 443 pp. British Museum (Natural History), London.

John, H., 1963. Revision der Gattung Thorictus Germar (Thorictidae, Clavicornia Col.). Eine Bestandaufnahme, Kritik und Neubeschreibungen I Teil. *Entomologische Abhandlungen und Berichte aus dem Staatlichen Museum für Tierkunde in Dresden* 28: 177-255. (In German). [Revision of the genus *Thorictus* Germar (Thorictidae, Clavicornia, Col.). An inventory, revision, and new descriptions. Part I].

John, H., 1965. Revision der Gattung Thorictus Germar (Thorictidae, Clavicornia, Coleoptera). Eine Bestandaufnahme, Kritik und Neubeschreibungen. II. Teil. *Entomologische Abhandlungen und Berichte aus dem Staatlichen Museum für Tierkunde in Dresden* 31: 319-395. Dresden. (In German). [Revision of the genus *Thorictus* Germar (Thorictidae, Clavicornia, Col.). An inventory, revision, and new descriptions. Part II].

Kadej, M., Háva, J. & Kalík, V., 2007. A new species and a new synonym of *Anthrenus* Geoffroy, 1762 (Coleoptera: Dermestidae) from Palaearctic region, Studies and reports of District Museum Prague-East, Taxonomical series 3 (1-2): 101-108.

Kiselyova, T. & McHugh, J.V., 2006. A phylogenetic study of Dermestidae (Coleoptera) based on larval morphology. *Systematic Entomology* 31: 469-507. London.

Kalik, V., 1952. New and Interesting species of Dennestidae (Coleoptera} *Pan-Pacific Entomologist* 28: 43- 49. San Francisco.

Kalik, V., 1954. New and Interesting Dermestidae (Coleoptera). *Annals and Magazine of Natural History,* (Series 12), 7: 367-370. London.

Kalik, V., 1987. *Claviella besucheti* gen. n., sp.n. (Col., Dermestidae) aus Indien. *Revue Suisse de Zoologie,* 94 (4): 655-657. Geneve. (In German). [*Claviella besucheti* gen. n., sp.n. (Col., Dermestidae) from India].

Kumara, T.K., Abu Hassan, A., Che Salmah, M.R. & Bhupinder, S., 2009. The infestation of *Dermestes ater* (DeGeer) on a human corpse in Malaysia. *Tropical Biomedicine* 26(1): 73-79.

Mroczkowski, M., 1968.Distribution of the Dermestidae (Coleoptera) of the world with a catalogue of all known species. *Annales Zoologici* 26 (3): 15-191. Warszawa.

Lepesme, P., 1939. Note synonymique sur les Dermestes (Coleoptera) et description d'une espece et d'une varieté nouvelles. *Bulletin de la Société Entomologique de France* 44: 190-193. Paris. (In French). [Synonymic notes on *Dermestes* (Col.) and description of a new species and a new variety].

Pic, M., 1910. Descriptions ou diagnoses et notes diverses. (Suite). *L'Échange, Revue Linnéenne* 26: 17-19. Moulins. (In French). [Descriptions or diagnosis and various notes. continuations].

Pic, M., 1916. Diagnoses Spécifiques. *Mélanges Exotico-Entomologiques* 17: 8-20. Moulins. (In French).

Pic, M., 1923. Nouveautés diverses. *Mélanges Exotico-Entomologiques* 40: 3-32. Moulins. (In French).

Pic, M., 1937. Nouveautés diverses. *Mélanges Exotico-Entomologiques* 69: 1-36. Moulins. (In French).

Ranga Rao, G.V., Surender, A., Wightman, J.A. & Varma, B.K., 1987. *Attagenus fasciatus* (Thunberg) [Coleoptera: Dermestidae]—a new pest of stored groundnuts. *IAN* 2: 12-13.

Reitter, E., 1881a. Die aussereuropäischen Dermestiden meiner Sammlung. Mit 70 Diagnosen neuer Arten. *Verhandlungen Naturforschenden Vereines*

Brünn 19 (1880): 27-60. (In German). [The Dermestidae from outside Europe in my collection. With 70 diagnoses of new species].

Reitter, E., 1881b. Bestimmungs-Tabellen der europäischen Coleopteren. III Heft. I. Auflage. Enthaltend die Familien: Scaphidiidae, Lathridiidae und Dermestidae. *Verhandlungen der Kaiserlich-Königlichen Zoologisch-Botanischen Gesellschaft in Wien* 30: 41-94. (In German). [Keys for the determination of European Coleoptera. ill. Including the families: Scaphidiidae, Lathridiidae and Dermestidae].

Ueta, M.T. & Avanccini, R.M., 1994. Studies on the influence of age in the infection of chickens by *Railliatina laticanalis* and on the susceptibility to reinfection. *Veterinary Parasitology* 17: 157-162.

Vijay Veer, Chauhan, Neena & Singh, S., 2004. A new species of *Thaumaglossa* (Coleoptera: Dermestidae) from India. *Oriental Insects* 38: 389-394. Florida.

Vijay Veer, Negi, B.K. & Rao, K.M., 1996. Dermestid beetles and some other insect pests associated with stored silkworm cocoons in India including a world list of dermestid species found attacking this commodity. *Journal of Stored Products Research* 32(1): 69-89. Oxford.

Vijay Veer, Prasad, R. & Rao, K.M., 1991. Taxonomic and biological notes on *Atta-genus* and *Anthrenus* spp. (Coleoptera : Dermestidae) found damaging stored woollen fabrics in India. *Journal of Stored Products Research* 27 (3): 185-198. Oxford.

Vijay Veer & Rao, K.M., 1994. A new species of *Trogoderma* (Coleoptera: Dermestidae) found damaging stored silkworm cocoons in India. *Journal of Stored Products Research* 30 (4): 283-295. Oxford.

Vijay Veer & Rao, K.M., 1995a. Taxonomic and biological notes on three *Attagenus* spp. (Coleoptera: Dermestidae) not previously recorded as pests of stored woollen fabrics in India. *Journal of Stored Products Research.* 31(3): 211-219. Oxford

Vijay Veer & Rao, K.M., 1995b. Annotated checklist of Indian species of Dermestidae (Coleoptera). *Zoology* (*Journal of Pure and Applied Zoology*) 5(1): 1-12. New Delhi.

Vijay Veer & Rao, K.M., 2000. Descriptions of adult and larval morphology of *Evorinea indica* (Arrow) (Coleoptera: Dermestidae). *Oriental Insects* 34: 293-300. Florida.

Wasmann, E., 1899. Neue Termitophilen und Myrmecophilen aus Indien. *Deutsche Entomologische Zeitschrift,* pp. 145-169. (In German). [New termitophiles and myrmecophiles from India).

Wasmann, E., 1912. Neue Beitriige zur Kenntnis der Termitophilen und Myrmecophilen (No. 192). Zeit- schrifur wissenschaftliche Zoologie, 101: 70-115. (In German). [New contributions to the knowledge of termitophiles and myrmecophiles (No. 192)].

Waterhouse, C.O., 1883. Description of new species of Anthrenus from India (Coleoptera, Dermestidae). *Annals and Magazine of Natural History*, (Series 5) 11: 61. London.

Zhantiev, R.D., 1967. Opyt taksonomit~heskogo analiza roda *Dermestes* L. (Coleoptera, Dermestes). *Zoologicheskii Zhurnal,* 46: 1350-1356. Moskva. (In Russian). [On taxonomic analysis of *Dermestes* L. (Coleoptera, Dermestidae)].

Zhantiev, R.D., 2006. Novye vidy zhukov-kozheedov roda Anthrenus Müller (Coleoptera: Dermestidae) iz Azii. *Proceedings of the Russian Entomological Society*, St. Petersburg 77: 94-97 (In Russian, English abstract). [New species of the genus *Anthrenus* Müller (Coleoptera: Dermestidae) from Asia].